Business Strategies for Electrical Infrastructure Engineering:

Capital Project Implementation

Reginald Wilson
Redawil Engineering Company, USA

Hisham Younis
Wayne State University, USA

Managing Director:	Lindsay Johnston
Editorial Director:	Joel Gamon
Book Production Manager:	Jennifer Yoder
Publishing Systems Analyst:	Adrienne Freeland
Development Editor:	Austin DeMarco
Assistant Acquisitions Editor:	Kayla Wolfe
Typesetters:	Erin O'Dea
Cover Design:	Jason Mull

Published in the United States of America by
Business Science Reference (an imprint of IGI Global)
701 E. Chocolate Avenue
Hershey PA 17033
Tel: 717-533-8845
Fax: 717-533-8661
E-mail: cust@igi-global.com
Web site: http://www.igi-global.com

Library of Congress Cataloging-in-Publication Data

Wilson, Reginald, 1959-
  Business strategies for electrical infrastructure engineering: capital project implementation / by Reginald Wilson and Hisham Younis.
       pages cm
  Includes bibliographical references and index.
  Summary: "This book brings together research on informed-decision making within the strategic planning sphere of system integration, highlighting social responsibility and environmental issues"--Provided by publisher.
    ISBN 978-1-4666-2839-7 (hbk.) -- ISBN 978-1-4666-2840-3 (ebk.) -- ISBN 978-1-4666-2841-0 1. Electric power distribution--Planning. 2. Electric utilities--Finance. 3. Social responsibility of business. I. Younis, Hisham, 1965- II. Title.
  TK3101.W55 2013
  621.31068'4--dc23
                              2012032577

British Cataloguing in Publication Data
A Cataloguing in Publication record for this book is available from the British Library.

Table of Contents

Preface

The integration of renewable generation in the current U.S. electrical infrastructure is a complex and challenging endeavor. The process consumes resources across organizational boundaries, disciplines, and technical expertise within the spectrum of the industry. The characteristics of each renewable generator request are selectively analyzed and assessed to strategically integrate supply-side and demand-side options to meet the customer energy service requirements as well as environmental improvements in a least-cost approach. Currently, electrical system upgrades are required for the expansion and reliability of bulk transmission capital projects to accommodate the explosive growth of any renewable energy generation. The systematic analysis of the infrastructure upgrades are scrutinized on every level and scenario to alleviate the constraints, evaluate the alternatives, and implement cost avoidance strategies.

The Renewable Energy Resource is a supply that naturally replenishes over a human, not a geological, time frame and that is ultimately derived from solar, water, or wind power. The resource does not consist of petroleum, nuclear, natural gas, or coal. It is anticipated to originate from the Sun or thermal inertia of the Earth and minimizes the output of toxic material in the energy conversion process. Although these initiatives are forthcoming, various disadvantages to renewable energy development exist. Solar thermal generation, for example, necessitates large swathes of land, and typically impinges on the natural habitat of the indigenous wildlife. The surroundings are also affected when structures, electrical transmission lines, access roads, and transformers are constructed. Additionally, the solar or photovoltaic cells are manufactured by means of the same technologies as those used to produce silicon wafer chips for computer processors, a practice that may employ toxic chemicals. While the renewable power resource fails to discharge air pollution or utilize fossil fuels, it still has a significant consequence on the environment.

The wind turbine generator as a renewable energy resource, for example, has the perception of a raucous, visually intrusive, electromagnetic field inducing hazardous technology in some parts of the world. There are also concerns about the adjacent residential property value impacts, land use, and the affect on the local wildlife/habitat when the technology is employed (Blair,& Hand, 2008). Nevertheless, the participative approach in the wind turbine generator siting process was found to have a positive effect on the public attitude towards the project. The method leads to a decrease in public resistance when the local population was involved in the siting procedure, transparent planning processes, and high-information exchange levels (Damborg, 2003).

Business Strategies for Electrical Infrastructure Engineering: Capital Project Implementation exploits the conflicts and contradictions in various phases of the new technology integration development with the resolution delineated back to the functional requirements. It explores the major aspects involved within the strategic planning, data analysis, process improvement, environmental issues, and the technological development envelop of the electrical equipment installation schema. Specifically, it details cost-effective renewable energy system integration methods in electrical generation modeling, above/below grade construction processes, industrial management, transmission line reconductoring, and ordinal capital project prioritization as it is designed to optimize resources, the master schedule, and the overall venture strategy. The utilization of simulation modeling (Rockwell Arena®[1], Oracle Crystal Ball®[2], Decision Analysis by TreeAge®[3], and PowerWorld Simulator®[4]), the Theory of Inventive Problem Solving (TRIZ), and Quality Management mechanisms, are prevalent themes in the process improvement analysis of the renewable energy system integration efforts. Additionally, the decision processes are explored for each application for executive management as well as the project team member.

The foundation of the TRIZ methodology is to distribute all structure functions concurrently with the available maximum resources of an existing system (Altshuller, 1998). It is a negative opinion approach that originates with the contradictions/conflicts and terminates with its resolution. The methodology thoroughly examines the design parameters and systematically trace backwards to the functional requirements within the system (Taguchi, 1993). Conversely, the axiomatic (positive opinion) design approach observes the preferred output as a function of the original definition. It aims to maximize a perfect design impervious to pitfalls from the functional design requirements. In the robust design approach, energy transformation – or how a system can be optimized in order to minimize uncontrollable conditions – is the main concern. The methodology accentuates the exact output response and improves the practical functions. The TRIZ-directed evolution envisages what will be important in the future for customers within the electrical footprint. Moreover, the philosophy invokes higher-level thinking methodologies for problem-solving within the lower-level phases of the system integration process. If a problem exists in the event phase, it signifies that the original design process has serious imperfections. The Six Sigma methodology, utilized exclusively in the equipment installation process, concentrates on such tribulations at the event-level with pattern-level tools such as Failure Mode and Effects Analysis. Improvements are attained by drilling down (analyzed) to the details of a presented process to locate and eradicate the cause of the problem (improve). These are essential problem-solving methods specifically designed for the Industrial practitioner to implement within the scope of the system integration processes. It is important for top-management to recognize and support break-through quality improvement efforts/strategic planning and ask the forward-looking key questions (i.e., How may I assist with this endeavor? What is required for mobilization?), instead of the rearview mirror line of questions (i.e., we always performed at this level – Why bother?), to successfully transition into a renewable energy environment. Leaders must be equipped to eliminate the obstructions, knock down barriers, and drive out fear (promotes comprehensible thinking, innovation, and courageous actions). The importance of effective leadership is one of the critical elements embedded in the quality platform and strategic planning, and ought to be incorporated in every aspects of the decision-making development.

The conventional power plant involve several years of tedious planning, whereas wind turbine generators can be installed much more rapidly than it takes to plan, site, and construct the related transmission system improvements considered necessary to permit the wind turbines to operate safely as part of the power system and to carefully transport the power to customers without overloading transmission lines. Based on the current condition of the infrastructure, most of the proposed wind energy projects require transmission line upgrades. In some regional impact studies, the existing transmission system may have sufficient capability to handle the power from the planned renewable generating sources. Nevertheless, integrating large amounts of wind power (even small amounts in some areas) will require some form of infrastructure upgrades. Chapter One – *The Socioeconomic Environment in Wind Turbine System Integration* – explores the critical challenges, resources, and time requirements, via simulation, to implement a wind turbine generator within the current infrastructure.

The typical prerequisite of increasing transfer capability oftentimes involves many exotic alternatives within the bulk transmission system. The options are usually modeled by the Regional Independent Service Operator with the best case scenario today to include future system upgrade requirements. A classic example of this scenario derives from a demand to increase the transfer capability from 0MW to 100MW as the genesis of several wind turbine generator projects in the evaluation queue. Chapter Two – *Alternative Analysis: Equipment Installation Case Studies* – examines several alternatives available from a cost-benefit analysis perspective for renewable generator system integration.

Electrical utilities and process industries are challenged to find methods to reduce costs, from the raw material conversion process to the end-user. The continued cost overruns in electrical condition, capacity, and reliability project installation categories are an area where cost control methodologies are beneficial. Stricter regulations and environmental stewardship are the impetus for the continuous improvement efforts in the capital project arena to maintain consistent and reliable sources of power and minimize cost. Chapter Three – *Process Evaluation and Cost Analysis Mechanism for Equipment Installation Ventures* – explores how the lean quality principles and green initiatives are applied in capital project equipment installation of electric substations and line rehabilitation from the initial feasibility studies to the plant-in-service phase.

The key characteristic for the successful integration of new generation within the electrical grid is planning. The complexities involved with the inter-related components of the power system requires detailed analysis and models to quantify specific attributes in the electric footprint. The models are then evaluated and thoroughly assessed to determine how the system reacts to real-world conditions with various change. Hence, the models are an important asset in the planning process. The system analysis that formulates the basic upgrade of the generation and transmission entails integrated econometric linear-programming based models, which includes statistical analysis, optimization, simulation, and the calculus of variations. The integrated resource planning considers a full-range of feasibility supply-side and electrical demand-side options and evaluates each against a common set of objectives. The paradigm provides an opportunity for system planners to address complex issues in an inclusive, structured, and seemingly transparent method. The techniques utilized tend to be more diagnostic than prognostic in most cases where the health of the system is described but not necessarily *how to cure it*. Nevertheless, the primary objective of integrated resource planning from an electrical utility perspective is to develop a least-cost alternative (plan) that can meet the customer-energy service requirements and environmental improvements. Planning is an important tenet in the renewable energy generation implementation process and explored in Chapter Four – *General Modeling Practices for Renewable Energy Implementation*.

The quality control techniques and methodologies utilized in the electrical equipment installation arena are typically categorized as a supplementary service. The data analysis is generic and applicable to a wide range of conditions when implemented within a Quality Management System (Aft, 1998). It is analyzed to determine adverse affects on the quality of the service and identify preventive or corrective actions. In the electric utility industry, waste reduction and material integration cycle times are two of the key attributes for equipment installation, specifically within the renewable energy sector. When quality control techniques are put into practice and applied religiously in a quality plan, the components of the Lean Paradigm will automatically drive the capital project cost downward. The Lean Paradigm is a systematic approach to provide a path that specifies value and evaluates the best viable sequence while continually improving the process. It espouses a business process of doing more with less. Chapter Five – *Ordinal Capital Project Ranking and the Quality Component* – examines the quality component of the equipment installation process and the ordinal priority ranking of capital projects.

The below-grade process of a wind turbine generator implementation begins once the renewable energy integration challenges and criteria (meteorology, transmission, land, public relations, civic engineering development, permitting, and environmental studies) are satisfied. A comprehensive review of the geotechnical studies, surveys, and engineering heavy permits consume only a small portion of the pre-construction phase of the capital-intensive project. A more thorough analysis of the site work is expected to consume a greater share of time for the below-grade assessment. Most importantly, the prime contractors are secured to perform the work. All of the studies are expected to produce positive results after thorough evaluations of the construction site. The detailed progression and the related challenges are explored in Chapter Six – *Below Grade Process Analysis.*

Transmission lines are high-voltage conductors (60-kV or greater) that convey bulk power from remote generation sources to the electrical service area. Overhead high-voltage conductors are approximately an inch in diameter and are comprised of aluminum strands or a mixture of aluminum and steel strands. The lines are isolated electrically by the adjacent air and are not wrapped with insulation material (Fink & Beatty, 2001). The overhead transmission line is the standard technology to convey electrical power. Most damage to the overhead conductors is attributed to fault-current arcs. The conductors strands are significantly weaken from the enormous heat generated by the arc. The protection circuits are designed to prevent such occurrences; however, electrical utilities experience such burn downs (especially with covered lines) when instantaneous tripping schemes were not utilized or applied incorrectly (Barker & Short, 1996). Consequently, long outages may occur due to the high-impedance fault condition which is somewhat difficult to detect once the line plunges to the ground. Chapter Seven – *Transmission Line Reconductoring Process Analysis* – investigates the challenges faced with the reconductoring processes of the high-voltage transmission and distribution lines.

The various permits involved in the construction process are daunting. The parameters of the endeavor encompassed a variety of entities performing simultaneously to produce the essential output ~ the issuance of a permit. The object of the evaluation is to minimize the time required to issue the permit utilizing quality tools and methodologies. A tenet of this procedure is the Continuous Process Improvement technique employed to produce an efficient, streamlined application of the permit issuance progression. The strategic evaluation exploited several other quality techniques to include Lean (particularly the 5S philosophy to eliminate the non-value added activities), Earned Value Management, Statistical Process Control (for process tracking), and Six Sigma to improve the quality by reducing the variation in the process. These quality methods are explored within the permit issuance envelope in Chapter Eight – *Permit Issuance Process Analysis.*

The implementation of the Project Management Financial & Scheduling Optimization Tool in capital planning is required to control the individual project budget and corresponding schedule. The guidelines included checkpoint methodology, responsibility metrics, historical actual spend data, forecasting logic, and process management flow diagramming. Project managers not only must rely heavily upon the traditional on-time, on-budget, and on-quality performance measures, but also with the added on-strategy dimension central to managing project success and to alleviate leadership challenges. Many leadership tasks relate to developing a vision of the project outcome that is practical, yet capable of mobilizing and motivating team members to accomplish the project's goals and objectives. The leadership vision engages stakeholders who are not actively involved in the project; it also motivates them to sustain their support over the project's timeline. The solution-building negotiation approach to defining the scope of a project, and then clearly communicating this to the project team and other stakeholders defines a strategy for realizing the vision, and translating the strategies into operational plans and results. Chapter Nine – *Financial Analysis in Electrical Equipment Installation* – investigates the cost containment tactics utilized within the capital construction arena.

The execution phase is a key aspect of the project progress philosophy. The project manager ought to assemble the required resources and management plans soon after the "go" decision is made. By demanding that project team members link their own actions and decisions with the overall intended strategy of the project (an extension of the corporate strategy) can assist with *on-strategy* project execution. It extends a virtual leadership presence, which injects itself into every critical project event and decision. The associated administrative tasks involved in this phase (work orders, bid documents, status reporting, etc.) are extremely important in project management. Many of the key characteristics of the project (scope, cost, resources, procurement, quality, risk, communication) are identified and highlighted during this phase. Chapter Ten – *Project Management Substation Guidelines* – investigates the project management philosophy in the renewable energy system integration process.

The decision-maker has to exploit a disciplined project selection process for funding at the very minimum, especially with renewable energy generator grid integration. It may be sufficient to simply apply a threshold where projects with high Benefit-to-Cost Ratios are funded and ventures with lower ratios are discarded. In such cases, setting the appropriate threshold might be a reasonable objective for analysis in support of the Senior Management Team. The decision-maker(s) are evaluating data from numerous sources (economic, load growth, financial, performance, environmental) for the successful implementation of the power resource. For these reasons, it may be complicated to choose such a threshold. In such cases, unequivocally ranking schemes adds substantial significance. Judgment matrices and priority vectors are utilized in the typical least cost objective function. In decision-analytic practice, the ranking of select projects is often preceded by the formation of detailed estimates of the endeavor's worth. Such improved value estimates are also beneficial, but unless uncertainty is high, enhanced estimates are not nearly as imperative as the basic use of a regimented process. For the astute decision-maker, it is essential to distinguish between these two sources of value. The decision processes are explored in Chapter Eleven – *Electrical Contractor Comparison Evaluation*.

There are several power supply resource options available for the electric utilities. The renewable energy alternatives comprise of system trade-offs and environmental regulatory issues. The trend towards new power supply resources include wind, tidal, and solar arrays (among others) in an effort to primarily decrease the exploitation of fossil fuels. The implementation strategy of these relatively newer technologies is required to encompass the electrical region's economic situation, resource availability, and load growth. System planning and careful scrutiny of the resource alternatives are essential tenets for renewable energy penetration. Chapter Twelve – *Above Grade Analysis* – investigates the generation integration challenges from an operational perspective and an improved simulation model of the above grade analysis.

The book is designed to assist decision-makers with high-quality mechanisms for strategic processes of capital-intensive projects. It provides the essential tools to exploit problem areas and isolate specific portions of the overall equipment installation scheme to minimize cycle time as well as costs. Additionally, it develops the necessary framework for top-management to thoroughly analyze the power supply decision set based on the resource availability, dollars per megawatt-hour, economic conditions, and load growth as part of the optimal energy mix. It is an essential reference for the interested practitioner to study/review the cost containment strategies, project risk evaluation, and the wide-range of quality planning techniques utilized to integrate renewable generation and capital-intense endeavors in the current infrastructure. Moreover, the industry best practices are intertwined with selected real-world electrical transmission /generator problems and case studies for the advanced power-analysis student. The new generation of technologically-literate executives that migrated to upper-management positions is adept with analytical software and its related usage. The strong technical background possessed by these executives enable them to execute decisions based on data-driven strategies (Liberatore & Wenhong, 2010). Moreover, the executives tend to build teams of analytical people from industrial engineering, statistics, computer science, finance and operations research fields to leverage the analytic knowledge to assist with decisions based on data (Saxena, 2010). For the capital manager, engineer, or financial analyst faced with cost overruns, historical data and simulated modeling techniques are imperative tools to incorporate within the expenditure and time estimates to accurately formulate the annual budget.

The general concepts, strategies, and tools espoused in *Business Strategies for Electrical Infrastructure Engineering: Capital Project Implementation* are effortlessly adaptable across disciplines (i.e. financial, construction management, cost estimators, procurement personnel, engineering, regional project managers, corporate, business analyst, operation managers, system planners, and training directors) within the tentacles of the industry. Nevertheless, its main intent is to assist the renewable resource professional and capital-intense projects with detailed analysis to mitigate the generator integration and equipment upgrade challenges experienced in today's processes for a cleaner and greener tomorrow.

Reginald Wilson
Redawil Engineering Company, USA

REFERENCES

Aft, L. S. (1998). *Fundamentals of industrial quality control.* Boca Raton, FL: CRC Press.

Altshuller, G. (1998). *40 principles: TRIZ keys to technical innovation.* Worchester, MA: Technical Innovation Center.

Barker, P. P., & Short, T. A. (1996). *Findings of recent experiments involving natural and triggered lightning.* IEEE/Power Engineering Society Transmission and Distribution Conference, Los Angles, C.A.

Blair, N., & Hand, M. (2008). *Power system modeling of 20% wind-generated electricity by 2030.* National Renewable Energy Laboratory conference paper presented at the Power Engineering Society, Pittsburg, PA.

Damborg, S. (2003). *Public attitudes towards wind power.* Danish Wind Energy Association. Retrieved May 4, 2010, from http://www.windpower.org/en/articles

Fink, D. G., & Beaty, W. H. (2001). *Standard handbook for electrical engineers* (14th ed.). New York, NY: McGraw-Hill Inc.

Liberatore, M. J., & Wenhong, L. (2010). The analytics movement: Implications for operations research. *Interfaces, 40*(4), 313–324. doi:10.1287/inte.1100.0502

Saxena, R. (2010). As analytics subsumes O.R., will INFORMS subsume analytics? *OR/MS Today, 37*(1), 20-21, 24.

Taguchi, G. (1993). *Taguchi on robust technology development: Bringing quality engineering upstream.* New York, NY: ASME Press. doi:10.1115/1.800288

ENDNOTES

1. Arena Simulation Software is a registered trademark of Rockwell Automation Corporation.
2. Crystal Ball is a registered trademark of Oracle Hyperion Solutions Corporation.
3. Decision Analysis is a registered trademark of TreeAge, Inc.
4. PowerWorld Simulator is a registered trademark of PowerWorld Corporation.

Acknowledgment

The author is truly indebted to those individuals who influenced this book, including a list of associates too numerous to mention. My sincere thanks to my great friend and an insightful professional engineer, is granted to Mr. David A. Shafer, for invaluable support throughout this project. His pragmatic methodologies and electrical substation knowledge is overwhelming. The Commonwealth Associates, Inc., electrical consulting teams are true professionals.

A basic and distinctive debt is owed to the construction and industrial expertise of Lincoln Forbes, Ph.D. who provided the guiding light for this book. It was his deep passion for the construction industry that assisted with the incisive details for specific processes.

Special thanks are owed to the publishing team at IGI Global, whose priceless contributions and keen insights assisted with the completion of this book. Particularly the assistant development editor, Ms. Hannah Abelbeck, for the necessary guidance and support provided to see the project to its orderly end.

A genuine debt is owed to the reference staff at the regional technical libraries where time simply stood still during the research phase of this book. A special thanks to the reference librarian Mr. Sam Shipley for his assistance with detailed maps and historical data.

Finally, a heavy debt is owed to a true supporter and a pillar of strength throughout this entire work – my wife Monica. She is a Godsend.

Chapter 1
The Socioeconomic Environment in Wind Turbine Generator System Integration

ABSTRACT

The renewable energy generation initiatives contain a low-cost viable option for the electric consumer worldwide. The efficient methodologies of wind, solar, ocean current, and geothermal power generation are attractive alternatives to fossil fuels. Based on the current condition of the infrastructure, most of the proposed wind energy projects require transmission line upgrades. In some regional impact studies, the existing transmission system may have sufficient capability to handle the power from the planned renewable generating sources. Nevertheless, integrating large amounts of wind power (even small amounts in some areas) will require some form of infrastructure upgrades.

INTRODUCTION

A vast majority of end-users are keenly aware of the harmful influence caused by these carbon generators, yet are reluctant to embrace the change of renewable technologies. The resounding bawl of "Not-In-My-Backyard" (NIMBY) is heard in U.S. communities when major electrical capital projects are proposed. Nevertheless, the conveyance of these crucial megawatts from the generators to distant urban areas poses key challenges

with the current, constrained infrastructure in the U.S. The development of utility-scale or commercial wind energy systems in recent years created an incredible interest, mainly due to minimum renewable energy standard initiatives passed by a number of states for electricity providers. The move toward increased wind energy integration is stimulated by numerous aspects, to include state and federal policies (i.e., state renewable standards, federal production tax credit) as well as ambiguity over future limitations on carbon

DOI: 10.4018/978-1-4666-2839-7.ch001

emissions. The increase of installed wind energy systems in the U.S. and the considerable amount of applications by developers to interconnect planned wind projects to the electric grid is indicative of this trend. Nevertheless, this hasty enhancement in installed and proposed wind energy systems has presented significant challenge for utility planners. The conventional power plant require several years of tedious planning, whereas wind turbine generators can be installed much more rapidly than it takes to plan, site, and construct the related transmission system improvements considered necessary to permit the wind turbines to operate safely as part of the power system and to safely transport the power to customers without overloading transmission lines. Based on the current condition of the infrastructure, most of the proposed wind energy projects require transmission line upgrades. In some regional impact studies, the existing transmission system may have sufficient capability handle the power from the planned renewable generating sources. Nevertheless, integrating large amounts of wind power (even small amounts in some areas) will require some form of infrastructure upgrades. Currently, electricity is produced principally by large power plants by the conversion process of coal, natural gas, hydro or nuclear fission as the primary energy sources. These generation technologies are by and large inexpensive and reliable and utilized in power systems for decades. A significant drawback of the exploitation of fossil fuels and uranium nonetheless is their finiteness, creating power generation from these resources inherently untenable. A subsequent disadvantage that applies, especially to natural gas and uranium, is the unequal allocation of fuel provisions amid regions, creating energy dependencies between them and potential for exercising political influence. Another drawback is the emission of greenhouse gases, in particular CO_2, when burning coal, oil and natural gas for power generation. The nuclear fission conversion has the distinct disadvantage of nuclear waste and the develop-

ment of new installations is complicated in many countries. A large hydroelectric unit does not have the shortcomings of fossil fuel-powered generation since it uses a sustainable supply of rainfall for power generation. Nonetheless, its potential has previously been exploited, particularly in developed countries, and the construction of new large installations has considerable challenges of its own. New generation plants are most likely will be located in outlying areas from load centers, requiring bulk power transmission over great distances. Furthermore, the creation of hydro reservoirs necessitates flooding of vast areas, which has devastating effects on local environments. Unmistakably, there are restrictions to the extent that conventional generation technologies can be a branch included in a future, sustainable power supply. Another challenge is that the compulsory bulk transmission upgrades for energy systems have characteristically been developed and constructed on a case-by-case basis for a specific project. This technique is commonly appropriated for large, baseload generating facilities, yet it can prove problematical when numerous wind energy systems are constructed. The transmission system upgrades often requires longer planning, site, and constructing time than the typical lead time for a wind energy system, therefore the bulk transmission system can be wedged in a mode of continuous catch up ~ responding to the requirements of individual projects rather than the long-term needs of the system. The development of incremental improvements to interconnect every wind energy system may not be cost-effective in the long run, particularly given the time and cost involved in siting and constructing transmission lines. Other complicated challenges exist specific to diverse types of integrated resource in a renewable energy portfolio ~ specifically with distributed generators. The typical wind farm, for example, frequently requires expensive capital investments in transmission lines to connect the distant location of the farms to the grid and load centers. Fossil-fuel-based Distributed Generator resources commonly are required to meet federal or state air quality emission standards, which can be cost prohibitive and may thwart their operation when they are most desired. For example, during peak periods in the summer months, when Distributed Generator resources are most advantageous and energy prices are elevated, some fossil-fuel-based Distributed Generator resources may be prohibited from operating because of inferior air quality and high smog conditions that characteristically accompany hot humid weather.

Recent feasibility studies oftentimes neglect vital maintenance and outage issues (downstream and/or in parallel) associated with a newly connected generation source. The typical requirement of increasing the transfer capability of megawatts frequently involves many exotic alternatives within the bulk transmission system. The alternatives are typically modeled by the electric utility's regional Independent Service Operator with the best case system integration scenarios and little regard to the public impediments involved. Nevertheless, wind turbine generation has an extensive range of impacts on power system operation and design. The local impacts of wind power, (i.e. altered branch flows and power quality aspects), is capable of being managed. The system-wide characteristics of wind power integration become relevant at high wind turbine generator penetration levels. The impacts of modern wind turbines (i.e. variable-speed technologies) on short-term voltage stability and rotor-angle stability have been found to be minute, also for larger wind penetrations. The residual aspects of wind turbine generation that confront power system operation are related to its variable output and restricted predictability, and therefore to short-term and long-term frequency stability.

Planning for long-term system requirements (as opposed to explicit wind projects) circumvent some of prospective challenges. Recently several states, including California, Colorado, Nevada, and Texas, have attempted to concentrate on some of these challenges by recognizing the long-term infrastructure requirements for both wind energy

and bulk transmission in diverse areas or zones. Likewise, independent bulk transmission owners and utilities, in states such as Illinois, Iowa, Michigan, Minnesota, Wisconsin, and the Dakotas, are working with the Midwest Independent Transmission System Operator (MISO) on planning transmission improvements based on long-term wind turbine energy requirements, including the anticipated wind power to meet state renewable energy mandate.

The socioeconomic impact of the construction cycle has a negative and devastating effect on the local businesses and the community as a whole. The investigation evaluates the methodical pre-construction processes of the bulk transmission equipment and wind turbine generator system integration within the supplementary socioeconomic environment. Additionally, the chapter provides valuable insight into the human activities during the pre-construction cycle of large, capital intensive electrical wind turbine projects and the applicable social attitudes. Specifically, the chapter investigates multiple issues encountered in the community (including the associated turbine/rotor noise, visual intrusion and electromagnetic potential) when new wind turbine generators are implemented as well as the human attitudes/perception of the equipment during the construction phase. A secondary initiative the analysis purports how the construction cycle affects the local environment. Construction activities associated with the proposed action and alternatives may have the potential to increase sedimentation, affect water quality, and other fish/aquatic species. The major goals of the plan are outlined below:

- Evaluation of the pre-project general metrics to select the "best-of-the-worst" case construction cycles (minimize quality of life disruption) and its associated environmental impact.
- Public impediments and major issues involved in the pre-construction processes and correlations.

- Wind Turbine Generator System (WTG) Integration intense project delay/wait cost vs. the alternative cost success ranges.

A thorough investigation to compare and contrast the pre-construction sphere is anticipated to include the following areas:

- Pre-construction issues such as the potential of safety related hazards and community property values along the Right-of-Way (ROW).
- The negative effects on businesses based on the area's aesthetic appeal.
- Effect on the local quality of life.
- Results of trespass on nearby public/private land due to construction.
- The affect on tourism/recreation-related businesses based on the construction of the substation/transmission line aesthetics, visual qualities, and activities.

PREPARATORY WORK

The analysis of several U.S. Independent Service Operator (ISO) projects, and regional municipalities provided the impetus of further investigation in this area. Numerous components of the fundamental permitting and siting processes created project delays based on the misconceptions of the potential construction site. Typically, the alternative analysis were cost prohibitive and/or created additional problems when the project was implemented. Several of the pre-construction various studies (pre-aviation/aviation interference, visual impact, communication interference, cultural and historical resource, wetland, wildlife, agriculture, and endangered species studies) by various entities were not followed or implemented in major electrical capital intensive ventures. The majority (17%) of the effort involved in the wind turbine generator system integration are encompassed within engi-

neering issues. The general pre-planning metrics is shown in the pie-chart of Figure 1.

Comparative studies conducted of other U.S. regions reflect similar conditions. Numerous methods employed worldwide include diverse derivatives of the same theme. Nevertheless, the solutions include some form of the following techniques and strategies:

- Siting alleviation policies
  - Evaluate lifecycle effects of energy generation options,
  - Explore wildlife and habitat effects,
  - Characterizing risk,
  - Connect national leadership,
  - Concentrating on public concern.
- Humanizing awareness of social acceptance (Tegen & Lantz, 2009)
  - Stakeholder and public awareness (generate a catalog of surveys to load information gaps),
  - Preparation for expansion (assess the responsibility of state and local planning in supporting new development),
  - Distributional integrity (evaluate the current developer strategies for facilitating social acceptance/evaluate the allocation of benefits from wind energy ventures and how local ownership and/or community payments can reduce local opposition).
- Participative approach in the siting procedure (Damborg, 2003)
  - Has a positive effect on the public attitude towards the project (leads to decrease in public resistance),
  - Must have the local population involved in the siting procedure, transparent planning processes, and high informational level,
  - The current renewable energy initiatives are executed at the federal and state levels (Hankey, Liggett, & McNerney, 2000).

Figure 1. The pre-planning general metrics

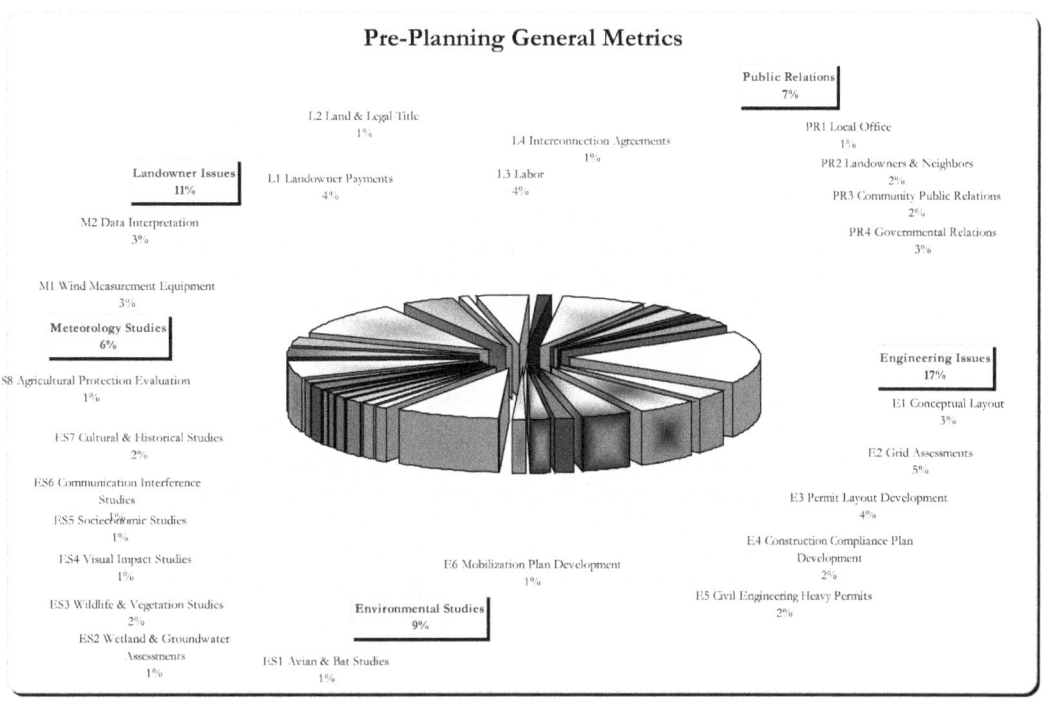

The legislation generally requires certain electric service providers to establish the promotion of renewable energy in the generation portfolio on a specified timeline (Figure 2)[1]. If the pre-planning process is pursued precisely, the decision-maker will acquire an influential economic instrument in the evaluation arsenal.

METHODS AND MATERIALS

The preparation of the substation layout for equipment installation is a vital element in the success of major capital projects. Equally important are the key integration processes involved in the equipment commissioning and plant-in-service phases of the endeavor. The proposed scheme is intended to provide front-end solutions to the continued development tribulations encountered in the field. The analysis aim is to avoid major safety and non-quality issues during critical system integration processes via simulated conditions and

viable alternatives for executive review. Many leadership tasks relate to developing a vision of the scheme's outcome that is practical, yet capable of mobilizing and motivating team members to accomplish the project's goals and objectives (Breyfogle III, et al., 2000; Schermerhorn, Jr., 1993). Teams of dissimilar groups/stakeholders are often the focus of strategic leadership to motivate and influence towards to common goal. The desires of the general public as well as the project's key stakeholders are espoused in the outcome (Levine, 2012). The integration of man, material, machinery, and methods require an intricate balancing act to maintain a specified budget, as well as timely component installation (Thomas, 1975). Consequently, the major objectives are to minimize the costs (labor, material, contingencies) and time (engineering, construction, contingency labor) for successful wind turbine generator system implementation. The strategy include checkpoint methodology, responsibility metrics, historical actual spend data, forecasting logic, and process

5

Figure 2. The renewable energy capacity

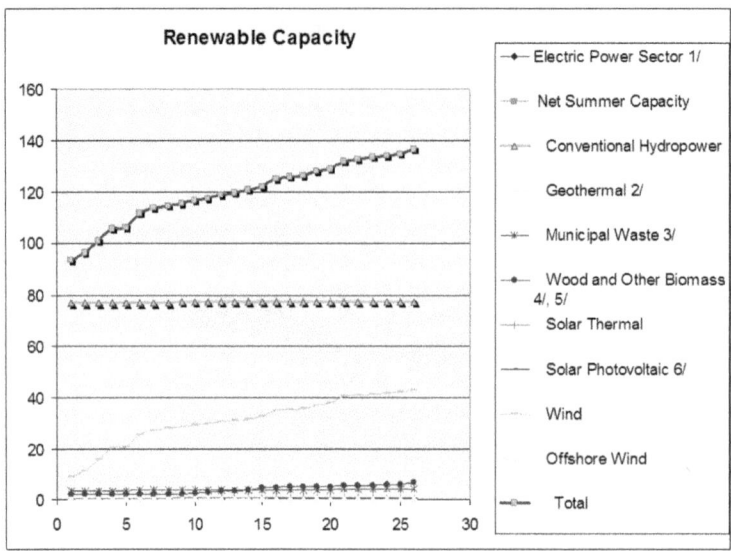

management flow diagramming (Besterfield, et al., 1999; Harry, & Schroeder, 2000). The intent of the investigation highlights a systematic approach to project management and aligned the objectives with the on-budget, on-quality, on-schedule, and on-strategy philosophy (Besterfield, et al., 1999). The process map, decision tree, and a Y-Chart are the main tools utilized to evaluate the major challenges involved in the progression.

The Initial Model Set-up

The Feasibility Studies and Pre-Assessments of wind farm system integration are deficient in standardization. Few companies partake in a formalized, detailed project evaluation and assessment studies (Feigenbaum, 1991). A thorough assessment of the WTG process sequence was performed to develop a logical work flow pattern in the model. The major aspects of the processes are not all inclusive for every wind farm project. However, the process is specific to the general system integration of parcel preparation based on the statistical analysis of 3600 maximum days allowed for the critical path for total project

completion. The high-level model and general process flow is depicted in Figure 3.

Generally, the major aspects of the pre-construction activities are shown in the diagram. The Meteorology Assessments are classified as one of the most important studies within the model and deemed as the keystone for any forward progression. The following processes (Grid Analysis, Land Assessments, Public Relations, Engineering Evaluation, and Environmental Studies) contain additional resources/activities encompassed in each step. Ideally, the actions are conducted in parallel as of a favorable meteorology assessment. Nonetheless, public opposition will impede any of the major processes and increase costs. The contingencies are executed in the model as a planning tool for such instances to aid amid the cost and manpower decision-making. The actual working model for the entire WTG process is shown in Figure 4.

The resources are shown to the left of the diagram at the initial stages of the process whereas the completion block is depicted on the right (pre-construction). The sequential steps of the main process include the individual resource and

Figure 3. The WTG high–level process map

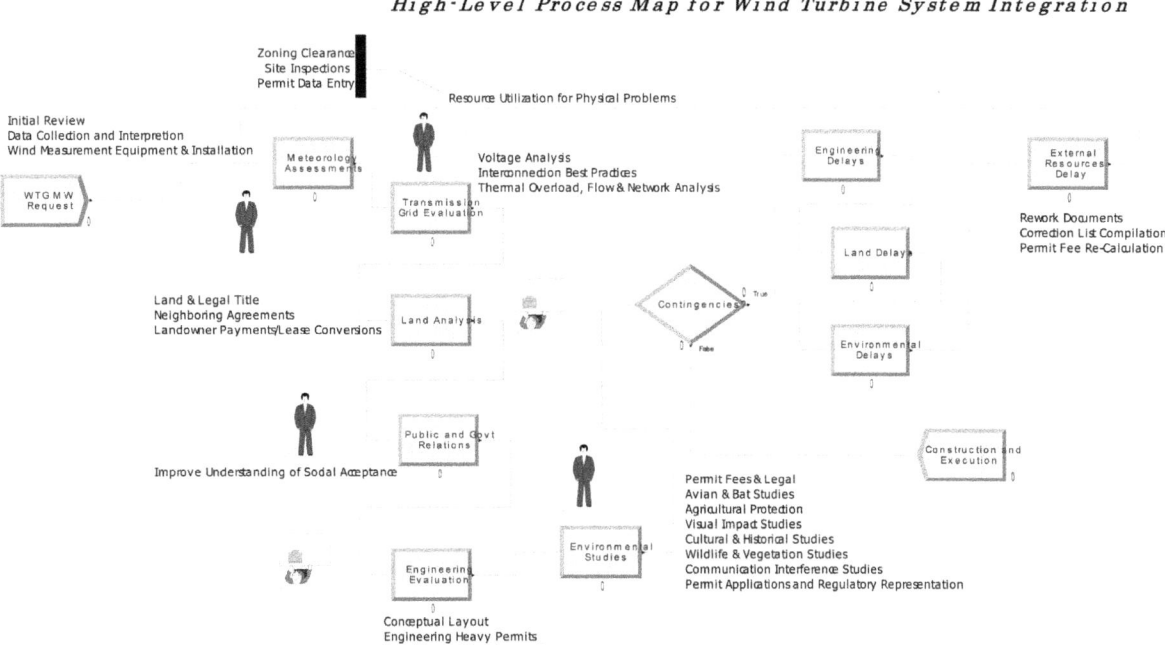

Figure 4. The WTG working model process map

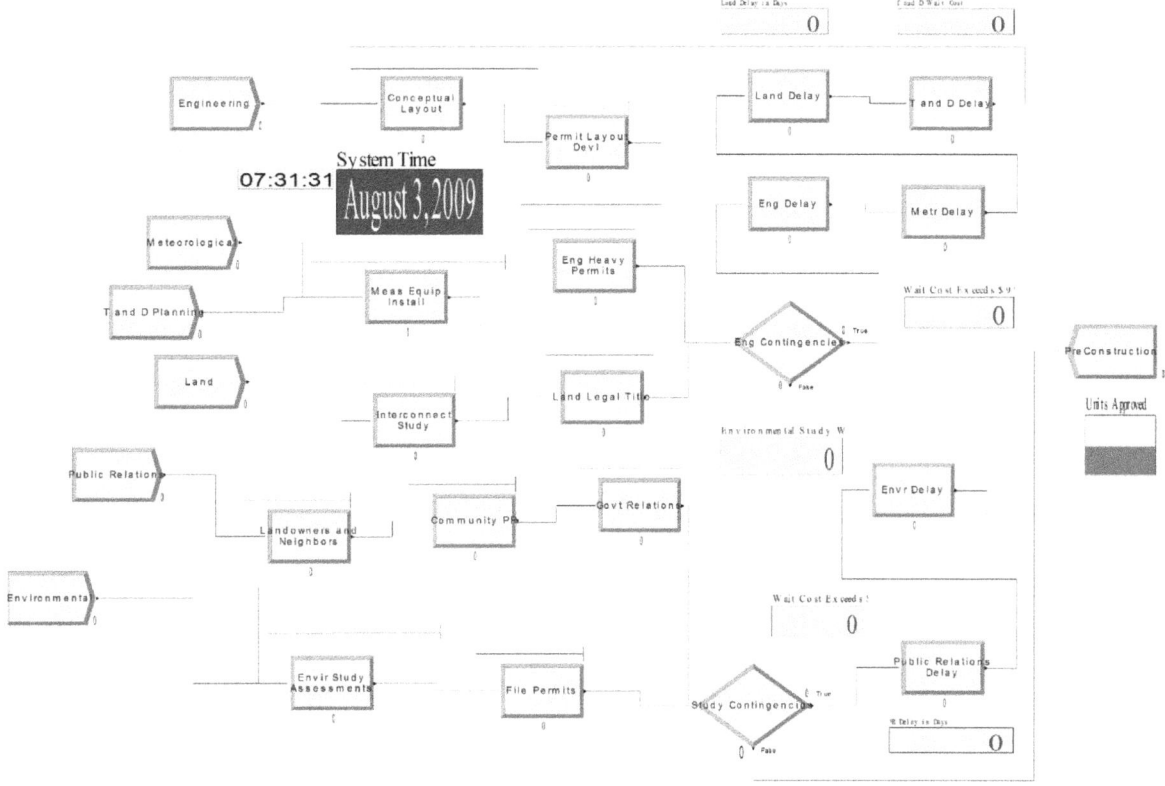

the associated work time to perform the task(s). Waiting queues were established at specific points to maintain continuity as well as to develop statistical data. The contingency delay area captures the material/resource via an exceed condition in the system.

The Entities

The general parameters of the six major entities are shown in Table 1.

The model is based on the process cost and its affect on the resources managing it (Blair, & Hand, 2008; Juran, 1999). The standard rates for the resource labor (engineering, meteorological, T&D, land, public relations, and environmental assessments) are captured in Table 1 as well to include the related (Value-Added, Non-Value Added, Hold, Wait, Other) costs, to produce one wind farm. Each entity was modeled to a prescribed schedule for the 3600 day work completion. The appropriate resource busy/hour and idle/hour allocation was also modeled from industrial statistics and utility good practice data for the general project implementation and shown in Table 2.

The Decision Parameter

The contingency delays are solely based upon the engineering wait cost (>=$999k) and the study wait cost (>= $499k). It consists of the time component of each resource in the scheme and the specific resource allocation rate from a wait perspective. The cost includes overhead expenditures of retaining the material in the system as well as the associated costs of continuing to hold any resources it owns. It is a significant stricture to determine the effectiveness of the preferred entities within the scheme. These parameters are shown in Table 3.

The contingency is realistically based on the resource availability and dedication to the project. It is depicted in the model as resource efficiency and schedule ramp rates for the WTG integration. The delay parameters include (but are not limited to) the following conditions:

- **Landowner Issues (p=0.314, $305k worst case ~ expenditure):** The decision to expedite the title negotiation process (p=0.063, $305k branch best case ~ cash retained). Stipulation for the Landowner Issues are resolved and proceed as scheduled (p=0.137, $170k ~ minimal cost).
- **Public Relations (p=0.275, $774k worst case ~ expenditure):** The decision to increase awareness of the WTG benefits in the community (p=0.055, $774k branch best case ~ cash retained). Stipulation for the community support is without predicaments (p=0.145, $196k ~ minimal cost).
- **Engineering Issues (p=0.160, $146k worst case ~ expenditure):** The decision to expedite the permit process (p=0.032,

Table 1. Entity parameters

Entity Type	Picture	Holding Cost/Hour	Initial VA Cost	Initial NVA Cost	Initial Waiting Cost	Initial Tran Cost	Initial Other Cost
Engr	**Man**	30	125	75	65	25	15
Meteorological	**Woman**	27	75	40	53	20	12
T & D	**Box**	32	95	53	60	21	10
Land	**Bike**	38	80	40	48	22	13
PR	**Truck**	28	65	53	47	23	14
Envir	**E-Mail**	35	85	43	38	17	11

Table 2. Entity utilization

Name	Type	Capacity	Schedule Name	Schedule Rule	Busy/Hour	Idle/Hour	Per Use
P Land	Fixed Capacity	1	1	Wait	0.62	0.35	0.02
P Engr	Fixed Capacity	1	1	Wait	0.83	0.12	0.05
P Metr	Fixed Capacity	1	1	Wait	0.73	0.22	0.05
P T & D	Fixed Capacity	1	1	Wait	0.52	0.45	0.03
P PR	Fixed Capacity	1	1	Wait	0.6	0.38	0.02
P Envir	Fixed Capacity	1	1	Wait	0.88	0.1	0.02

Table 3. Decision parameters

Name	Type		IF		Attribute Name		IS	Value
Eng Contingencies	2-way by Condition	50	Attribute	Variable 1	Entity. Wait Cost	Entity 1	>=	$999,000
Study Contingencies	2-way by Condition	50	Attribute	Variable 1	Entity. Wait Cost	Entity 1	>=	$499,000

$146k branch best case ~ cash retained). Stipulation for the Permitting Issues are resolved and proceed as scheduled (p=0.168, $81k ~ minimal cost).

- **Environmental Studies (p=0.220, $1.3M worst case ~ expenditure):** The decision to re-evaluate the specific unfavorable study (p=0.044, $1.3M branch best case ~ cash retained). The favorable environmental study results, in all aspects of the category, are imperative for successful WTG implementation (p=0.156, $184k ~ minimal cost).
- **Meteorology Studies (p=0.034, $1M worst case ~ expenditure):** The unfavorable results will create a developmental work stoppage at an exceedingly high cost. Conversely, favorable study results will escalate the WTG system integration progression (p=0.166, $851k ~ minimal cost).

Overall, the chance model produced an average cost of approximately $325k for the wind generator public impediments with equal branch probabilities. The major area of concern that will halt development is encompassed within the Meteorological Studies branch. It is the keystone of the entire scheme where unfavorable results produce project abandonment at the specific site. The cost-benefit model is depicted in Figure 5.

In order to determine the accurate quantity of resources to achieve the minimization goal (time and cost), a certain degree of sensitivity analysis is utilized. An increase of public relations personnel, meteorological expertise, engineering, and landowner/environmental assets is the main driver for the model comparison. Ideally, the WTG delays are diminished in sequential order (from the highest cost meteorology studies ~ average $530k to the lowest cost engineering issues ~ average $91k). Based on the complexity of the

Figure 5. The cost-benefit model for WTG public impediments

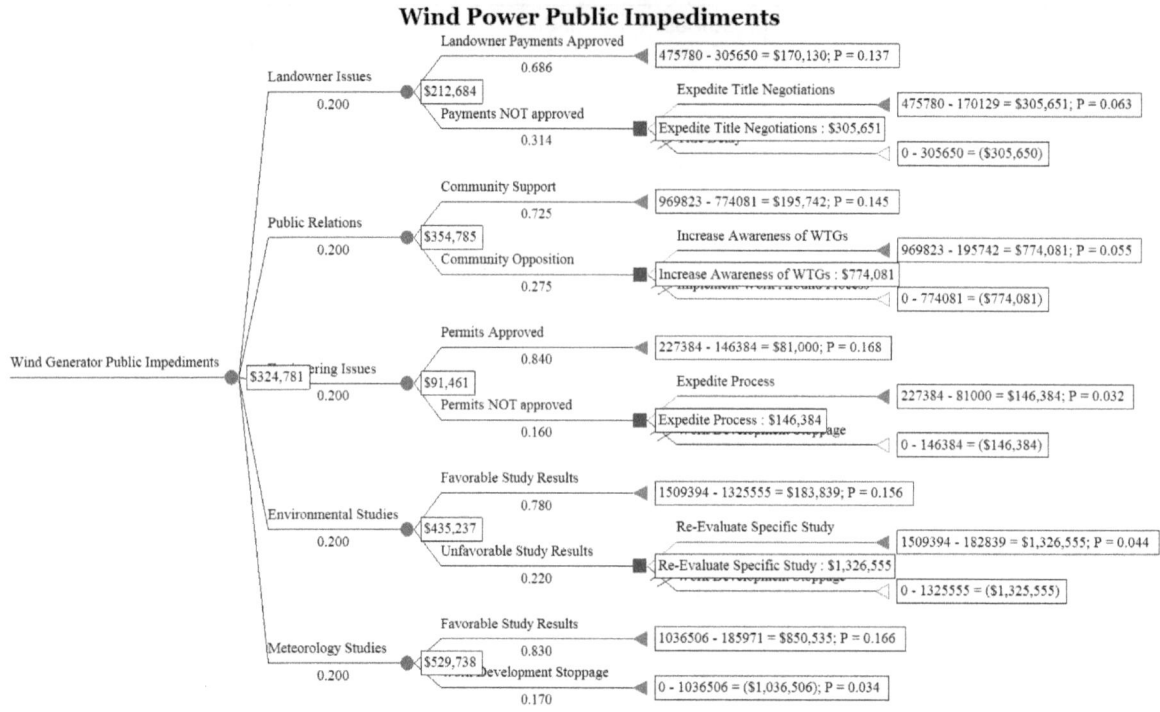

model structure, constraint mapping, and the process succession, the analysis generated predictable results.

Major Processes

The project parameters of the working model are based on the statistical analysis and heuristics data of historical wind turbine integration efforts. The specifics are shown in Table 4.

A mixture of high and medium priorities was utilized based on the work effort and system requirements. The resource allocation was divided by the single entity (non-value added) working within the process to multiple entities (value-added) performing the tasks. The most likely value for the Conceptual Layout delay (seize, delay, release), for example, is approximately 67 days (maximum 74) and involves several entities. The delay processes are all classified as "wait"

activities categorized as high priority progressions. A decision-maker may utilize these data to formulate minimum amount of time to complete the system integration work as a benchmark for contract incentives and budgetary inputs.

Key Performance Indicators

The total system cost for the model is approximately $25M (before resource leveling application) to complete the wind turbine system integration work effort. Further analysis of the potential vulnerabilities within the model and a modified resource schedule produced a more static and stable solution for the total system cost of approximately $11M (modified resource leveling application ~ an optimized model). Both versions of the model highlight the "wait cost per entity" perspective of the major targeted resources. The increased resources in these targeted areas eliminated sev-

Table 4. Entity allocation

Working Model								
Name	Action	Priority (urgency)	Resources (effort)	Units	Allocation (effort)	Minimum	Value	Maximum
Conceptual Layout	Seize Delay Release	High(1)	Metr & Engr	Days	Value Added	59	67	74
Permit Layout Devl	Seize Delay Release	High(1)	Engr & Land & Envir	Days	Value Added	112	125	138
Eng Heavy Permits	Seize Delay Release	Medium(2)	Engr	Days	Non-Value Added	78	89	100
Meas Equip Install	Seize Delay Release	High(1)	Metr (3)	Days	Value Added	205	240	262
Interconnect Study	Seize Delay Release	High(1)	TD & Engr	Days	Value Added	52	64	76
Land Legal Title	Seize Delay Release	Medium(2)	Land	Days	Non-Value Added	66	78	90
Landowners and Neighbors	Seize Delay Release	High(1)	PR & Engr & Land	Days	Value Added	135	150	165
Community PR	Seize Delay Release	High(1)	PR (3)	Days	Non-Value Added	95	120	145
Govt Relations	Seize Delay Release	Medium(2)	PR	Days	Non-Value Added	150	165	180
Envir Study Assessments	Seize Delay Release	High(1)	Envir & Engr & Land	Days	Value Added	185	210	230
File Permits	Seize Delay Release	Medium(2)	Envir	Days	Non-Value Added	62	93	112
Eng Delay	Delay	High(1)		Days	Wait	30	40	50
Metr Delay	Delay	High(1)		Days	Wait	20	30	40
T and D Delay	Delay	High(1)		Days	Wait	25	35	45
Land Delay	Delay	High(1)		Days	Wait	45	65	85
Envr Delay	Delay	High(1)		Days	Wait	45	60	75
Public Relations Delay	Delay	High(1)		Days	Wait	25	35	45

eral branches in the decision model as well as significantly diminishing the total system cost. Moreover, the average time consumed for each of the targeted entities declined proportionally as well. The increased awareness of the benefits of wind power, especially in the local community, has an extraordinary affect on the system integration progression. Participation and transparency of the process by the community is also a key positive aspect of good business practices.

THE ATTRIBUTE MATRIX DISCUSSION

The exploded pie chart in Figure 1 captures the general categories of the major attributes for wind turbine generator system integration. The foremost challenges are systematically prioritized in the chart. Because of their inferior controllability, renewable energy sources introduce additional uncertainty in the operation of power systems. This specific characteristic must be resolved by the power system operations, and is a technical challenge necessitating additional control actions from conventional generation units and of renewables themselves. While such control measures approach a certain associated cost, the system integration of renewables is furthermore an economic challenge as well. Nonetheless, these challenges are normally removed from the generation system since policy often devise regulation affirming that renewables are assigned as prioritized production. In essence, the renewable energy source comprises of first access to the system and the integration features are to be taken care of by the power system operator rather than the producer. In cases where the renewable energy is not prioritized, the integration cost has to be taken by the project developer. The variability and limited predictability of wind power have raised concerns about the impacts on the power system reliability and cost. The effects of a renewable energy wind farm on electrical operations can approximately be divided into local assessments and system-wide characteristics, when both the electrical aspects of wind turbine generator and the wind qualities are included. Additionally, the wind turbine generator integration challenges the planning and operation of the grid. Because of the variability and restricted predictability of wind power as a renewable energy resource, it increasingly challenges the real-time balancing of generation and loads in electrical power systems and necessitates supplementary secondary reserves. The methodology these reserves are provided and operated is no longer straightforward in market environments, and it is even more complicated if wind turbine generator is incorporated into these markets and subject to program accountability.

Sensitivity analysis of crucial resource components provides essential insight for the major attribute prioritization. The largest segment of the chart (engineering Issues, 17%) consumes the most resources as expected closely followed by Landowner Issues, 11% and Environmental Studies, 9%. Nevertheless, the importance of the Meteorology Studies segment, 6% is most beneficial to the entire system as incorrect data from the wind instrumentation/data interpretation may cause development stoppage. A chief contingency from the Environmental Studies segment, 9% and the Public Relations area, 7% may accelerate WTG system integration failure at a particular site. The overarching strategy is to focus a concerted effort of man, material, machines, and methods into an affirmative force of wind farm system integration and simultaneously mitigate the challenges. A logical approach to a complex problem is warranted, typically on a case-by-case basis. Process sequencing and resource-leveling techniques are required for successful implementation. As an example, the Engineering Issues Level-of-Effort (LOE) is minimal during the early stages of the endeavor and increases as the project progresses. The Permitting and Conceptual Layout areas are the most challenging efforts in the process. The Public Relations segment is beneficial throughout the integration to mitigate the opposition of visual intrusion, noise, and environmental concerns. The details involved in each of these major groupings (and the optimized results) are depicted in Table 5.

ENVIRONMENTAL STUDIES

The perceived negative effects of wind energy system integration include various government agencies, local landowners, and community organizations. Numerous summits and local meetings strive to educate the general public about the benefits of wind energy technology. Comments are wide and diverse as outlined below:

The wind turbines will move the clouds - Local resident (Flint Hills, KS. 2008)

Grow the trees before the construction of the substation.- Official at a Regional Planning Commission (Red Bank, N.J. 2007)

Table 5. Attribute matrix

Major Attributes:	Amount:	Percentage:	Positive Cost Factor (increased PR):	Negative Cost Factor (increased PR):	Positive Cost Factor (optimized):	Negative Cost Factor (optimized):	Negative Approval Probability (increased PR):	Positive Approval Probability (increased PR):	Negative Approval Probability (optimized):	Positive Approval Probability (optimized):
Landowner Issues	20	10.64%	$170,130		$108,329			13.70%		22.60%
L1 Landowner Payments	8	4.26%								
L2 Land & Legal Title	2	1.06%		($303,631)		($426,756)	6.30%		10.40%	
L3 Labor	8	4.26%								
L4 Interconnection Agreements	2	1.06%								
Public Relations	14	7.45%	$195,742		$0			14.50%		
PR1 Local Office	2	1.06%		($774,081)		$0	5.50%			
PR2 Landowners & Neighbors	4	2.13%								
PR21 newsletters	0	0.00%								
PR22 dinners	0	0.00%								
PR3 Community Public Relations	3	1.60%								
PR31 open houses	0	0.00%								
PR32 advertisements	0	0.00%								
PR33 charitable sponsorships	0	0.00%								
PR4 Governmental Relations	5	2.66%								
PR41 lobbyist	0	0.00%								
PR42 public meeting support	0	0.00%								
Engineering Issues	32	17.02%	$81,000		$219,003			16.80%		26.50%
E1 Conceptual Layout	6	3.19%								
E11 initial turbine siting	0	0.00%								
E12 constraint mapping	0	0.00%								
E13 transportation study	0	0.00%								
E2 Grid Assessments	10	5.32%								
E21 high voltage interconnection design	0	0.00%								
E22 system reliability interconnection study	0	0.00%								
E23 utility interconnection	0	0.00%								
E3 Permit Layout Development	8	4.26%		($146,384)		($631,932)	3.20%		7.50%	
E31 geotechnical study	0	0.00%								
E32 survey & staking	0	0.00%								
E33 electrical coordination	0	0.00%								
E4 Construction Compliance Plan Development	3	1.60%								
E5 Civil Engineering Heavy Permits	3	1.60%								
E51 road access	0	0.00%								
E52 spill control	0	0.00%								
E6 Mobilization Plan Development	2	1.06%								
Environmental Studies	17	9.04%	$183,839		$0			15.60%		
ES1 Avian & Bat Studies	2	1.06%								
ES2 Wetland & Groundwater Assessments	2	1.06%								
ES3 Wildlife & Vegetation Studies	3	1.60%								
ES4 Visual Impact Studies	2	1.06%								
ES5 Socioeconomic Studies	2	1.06%		($1,325,555)		$0	4.40%			
ES6 Communication Interference Studies	1	0.53%								
ES7 Cultural & Historical Studies	3	1.60%								
ES8 Agricultural Protection Evaluation	2	1.06%								
Meteorology Studies	11	5.85%	$185,971		$0			16.60%		27.40%
M1 Wind Measurement Equipment	6	3.19%								
M2 Data Interpretation	5	2.66%		($1,036,506)		($21,690)	3.40%		5.60%	
Total:	188	100.00%	$816,682	($3,588,177)	$327,332	($1,080,378)	22.80%	77.20%	23.50%	76.50%

One recent case involved the Pentagon's opposition to prevent the development of a wind farm in eastern Oregon (Meibom, P. et al., 2004). The 845-MW wind farm construction project (and other WTG's in the region) was denied its final Federal Aviation Administration permit based on the radar communication interference criteria (ES6 in the attribute matrix ~ Table 5). The proposed structures were not only deemed as a hazard to air navigation, but "would have an adverse physical or electromagnetic interference effect upon navigable airspace or air navigation facilities." The construction delay may cease the entire endeavor and jeopardize federal funding opportunities.

The Environmental Studies are a critical attribute for the successful completion of a wind farm. Public opposition continues to raise concern about the wildlife and visual impediments inherent in the WTG technology. Prolonged legal processes for the required U.S. Fish and Wildlife permits, specifically the avian/bat studies (ES1) further delays the system integration (Hammack, 2010). Concerns about the bird and bat eradication are addressed at the Federal, State Commission, and local levels. Wind turbine generators occasionally kill migrating passerines (warblers, vireos, and flycatchers). Bat mortality rates at WTG sites are typically three times higher, than the bird mortality in one particular Wisconsin study (Evans, et al., 2002). Three predominate migratory species were classified as the Hoary bat, Red bat, and the Silver-haired bat. These concerns affect the development of offshore wind turbines as well (Barrios, & Rodriguez, 2004; Breckenridge, 2010). Mitigation strategies for bird collisions at a specific wind farm site include improved structural and design features (minimize use of lattice towers in favor of tubular types) (Anaya-Lara, et al., 2009). The permits are applied well in advance to facilitate the construction schedule.

There are wind farm appreciators as well as detractors, specifically in the aesthetics arena (ES4). The 400 foot, 300 ton steel monopole structures that accommodate the wind turbine generators are considered geometrically-centered, serial uniformity beauty by the wind appreciators (Good, 2006). These large units are capable of producing 5 megawatts of power and utilize a three-bladed upward rotor system. The rotors are visually more acceptable than the earlier single blade design, although the high rotor speed reduced the gearbox ratio and increased aerodynamic noise/blade drag losses (Anaya-Lara, et al., 2009). The wind farm is judged as an "architectural essence" and embodies "mathematical rationality." Conversely, the NIMBY camp vehemently opposes this viewpoint and deems the wind farm as an ugly imposition of the natural structures of the region. It has an industrial-look that transmits an ideology of progress perceived as unnatural ~ hence unsightly (ES4). Strong opposition in this area may cancel the entire venture if social acceptance strategies are effectively utilized (Tegen, & Lantz, 2009). People generally support the green energy initiatives on a larger scale, yet on the local level when a wind farm project is proposed, the attitudes diminish tremendously (Damborg, 2003). The wind farm must contribute to the local economy while maintaining a sense of environmental stewardship are the top issues for key stakeholders in the system integration.

Landowner Issues

One of the disadvantages of wind energy generation involves excessive land use (Blair, & Hand, 2008; Farret, & Simoes, 2006). The average wind farm necessitates 17 acres to produce 1MW of electricity (approximately 750 to 1000 homes). Typically, the land area essential to sustain a 251MW wind farm is approximately 50,000 km², yet only 2% to 5% of that area is occupied by the turbine generators and ancillary materials. As the competition among wind developers for land in most states has increased, the land available to any single developer for a project might be dispersed over a wide area with other benefits holding land in between. Land availability can also be fragmented because of existing uses, natural features, or other considerations. The local zoning restrictions concentrating on wind facilities are highly inconsistent. Specified tower height limitations, the existing land in any given township could be close to zero or as high 9 percent. There are several types of expansion competing for land use, among which wind farm development is only one. Communities and landowners may have other ideas for the land recognized as potentially available for wind development under the analysis. Nevertheless, most farmers/ranchers are willing to harness the wind as a secondary source of revenue. Desert erosion is also a concern of wind farm system integration, as most are located on or just below ridgelines. The installations are commonly located in upland terrain to exploit higher wind speeds (Anaya-Lara, et al., 2009). Yet, the permitting and siting issues oftentimes are difficult as a trade-off between these high-wind speed sites and visual intrusion/environmental sensitivity become extraordinary. The placement of access roads for WTG maintenance subjects the landowner to crop field losses and decreased efficiency if positioned incorrectly. The property-owner must have a clear understanding of the issues and provide the necessary input during contract negotiations (Aakre & Haugen, 2009). Moreover, a considerable amount of land is impacted in excess of the requirement during the construction period, as the contract must include provisions to pay for the potential crop damage. The access roads, fences, and gates, must differentiate clear responsibility in all seasons between the utility and the farmer/rancher when a wind tower is on grazing land as well. The decision to implement a wind farm is a detailed and tedious process (usually 2-4 years), and can be cost prohibitive for the landowner. If, during the development and operation phases, the property-owner decides against the WTG, he/she must pay a termination fee of $25k/day for the crane alone. Other unique liability issues are outlined below:

- Vandalism
- **Ice Shedding:** As a human and/or animal hazard, falling ice from the tower blades may cause damage to structures or vehicles below. The ice, freezing rain, and fog can also cause the anemometers to fail, losing the ability to guide the direction of the blades to harness the wind effectively (Jacobson, 2010).
- **Blade drop/Throw:** Damage blades from imbalance and /or structural failure.
- **Air/Water pollution:** Experienced during the construction phase of the endeavor.
- **Shadow flicker:** The moving turbine blade creates a shadow effect and may cause health problems.
- **Stray Voltage:** Concern for humans and animals alike.

Ironically, people living in close proximity (550 yards) of a wind turbine generally possess a positive attitude towards the technology than others sited further away (Damborg, 2003). Notwithstanding, city dwellers tend to view the countryside as a romantic place and typically dread the sight of wind turbines dotting the landscape. In Europe (specifically Denmark and Germany), large-scale turbines dominate the landscape and create siting problems in densely populated areas. The offshore wind turbine alternative is most beneficial in these cases (Justus, 2005). Lease payments to the landowner are based on per tower, per megawatt, percentage of the gross revenue, or a combination of the three (L1). Furthermore, the property-owner is poised to receive bonus payments, base rent ($3k~$7k/MW per year), royalties, and renewal/extension payments (typically 20~25 year lease).

The issues are methodically mitigated if the WTG is appropriately sited, conceptualized, and integrated in the grid.

Engineering Issues

Currently, renewable energy initiatives such as wind power, geothermal, and solar supply less than 1% of power demand worldwide. If a substantially larger penetration was implemented, the inconsistency and periodic unavailability of renewable energy would present a formidable challenge for existing energy systems. The methodologies developed for the power system integration of wind power are in principle appropriate for a wide range of renewable energy initiatives, since most have an inadequate controllability of their respective energy sources. The current transmission infrastructure necessitates capacity expansion to advance the grid reliability for appropriate wind turbine generator system integration. The associated costs in the WTG endeavor include capital and financing expenditures (direct) as well as the operations and maintenance expenditures (indirect) that encompasses the integration expenses (Farret & Simoes, 2006). The investment costs occasionally surpass the conventional fossil-fuel plants. Normally, the grid integration expenditures are compensated by the network and allocated to the customers. Large wind turbine generator projects (particularly offshore wind farms) challenge the planning and operation aspects of the bulk transmission grid. The accessibility of wind energy is frequently unsurpassed in remote, open areas and distant from demand. The bulk transmission structures, utilized by the existing generation capacity, are often characteristically incapable to accommodate large-scale wind power or are minimally not available nearby. Grid connection challenges are not only technological, but also consist of economical concerns (cost for offshore wind energy connection), spatial planning aspects (extensive permission procedures), the low capacity factor of bulk transmission capacity for renewable wind resources and legal issues. The wind energy system integration may create probable conditions where transmission bottlenecks could occur, which may be solved by a number of solutions, including grid reinforcement or phase shifting transformers, wind energy curtailment, or even local storage. Wind power introduces a wide range of challenges for power system operation. Nevertheless, wind power integration into the electric system involves tedious planning and copious calculations of the operating reserve margins at the North American Electric Reliability Corporation (NERC) regional level. The preparation of generation units is performed based on load forecasts, the economics, and the technical characteristics of the available generators. The unit commitment and economic dispatch arena entails the computations of the optimal assortment of units for power generation within a specified period of time (hours to days). The essential parameters in unit commitment paradigm incorporate start-up and shut-down cost, minimum up- and downtimes and operating cost. The economic dispatch executes the accurate allotment of the total load among the committed units, and is optimized for each operating state while taking into consideration all cost-effective and technological aspects of the generation units. The products of the unit commitment and economic dispatch are the generation unit operation and timetables. From these, an estimation of the associated use of fossil fuels and emission of greenhouse gases can be calculated as well. The unit commitment and economic dispatch paradigm is challenged by wind energy because of its variability and restricted predictability, which is a higher priority than the existing disparities and uncertainties of the load. The generation impact studies are performed by regional independent service operators to evaluate the electrical system conditions during all stages of wind turbine generator integration. Throughout a low-load, high-wind phase, for example, the traditional generation units are usually taken out of operation in order to permit for the integration

of wind power. During such periods, a smaller number of conventional generation capacities are available to provide the primary and secondary reserves. If the circumstance persists, a generation unit outage may result in a larger excursion of frequency and /or the Area Control Error (ACE). A worst case scenario, examined for illustrative purposes, comprises of a generation unit outage during a high wind (11250 MW available), low load (10125 MW) condition. In order to avoid minimum load tribulations, traditional generation units are ramped down as far as technically achievable, and otherwise decreased to the operational lowest amount. Must-run nuclear and coal units at minimum load continue to supply a total of 4250 MW, and other must-run and distributed generation supplies an additional 5200 MW, resulting in a total of 9450 MW of generation in excess of the wind power. The situation results in a very large amount of wind power wasted in order to circumvent minimum load problems. Only a minimum requirement of wind power could be integrated in the current situation, which is very low. Therefore, even though it is not optimal for wind power integration, this particular case represents a rational worst case scenario for investigating frequency stability with wind power. Specific planning tools evaluate the impacts of wind energy in an electric system and scrutinize the effect on an hour-per-hour/seasonal basis (Meibom, et al., 2004). These planning tools are designed to assess the proposed wind power production forecast errors when making dispatch decisions for the power plants in the system. Aggregation of wind turbines, forecasting and modeling has been implemented as grid integration solutions to facilitate larger market penetrations of the technology (Justus, 2005).

The conceptual layout is another critical and time consuming area involved in the engineering section. Skilled engineers, technicians, and support staff must develop plans for constraint mapping, initial turbine siting, transmission/substation layout, identification of transportation routes (intersections, culverts, bridge improvements), and collection systems. The permit layout development engross the high voltage portion of the electrical design system, survey/staking, initial horizontal alignment, geotechnical study, area walk down (agency, landowner, environmental, engineering) as well as coordination with the planning departments. The pre-construction walk down occurs regularly and are intended to identify sensitive resources to avoid, limits of clearing, location of drainage features (culverts, ditches), and the layout for sedimentation and erosion control measures. Once these features are identified, specific instructions must be reviewed, and any modifications to the construction methods/locations must be agreed upon before production commencement. The site survey is performed to stake out the exact location of the wind turbines, access roads, electrical lines, and facility access entryways from the main highway(s), the O&M building laydown areas, electrical substation areas, and the electrical interconnection facility/switchyard. The geotechnical investigation engages a drill rig to bore to the required depths (typically 50 feet at each turbine location and lesser depths for the foundations of the ancillary structure). The goal of the investigation is to identify the subsurface soil/rock types and the associated strength properties by sampling and lab testing. Soil testing is also conducted to measure the electrical properties to ensure proper grounding system design. The geotechnical survey is conducted at each turbine location, at the substation and interconnection sites, along the access roads, and the O&M building site. Moreover, the permitting includes the department of transportation authorization for state road access/improvements, local permits for road enhancements/access into the local roads, and the spill control permits (assess the potential need for secondary containment around the transformers). Wetlands greater than 5 acres necessitate permitting and mitigation for construction of turbines and characterize additional engineering challenges and material costs for the installation of wind turbines. The essential pre-construction

planning also encompasses the negotiations with WTG equipment manufacturers, finalization of the long lead-time procurement processes for the WTG, finalization of the construction plans, development of the environmental construction compliance plan/mobilization plan, and securing the contractors. Numerous activities are conducted prior to the initiation of construction to including the aerial acquisition of contour interval topographic survey data. Many utilities employ high spatial resolution imagery as a source to determine fence lines (cadastral property lines) and the general footprint of the wind farm (Cowen & Jensen, 1999). During construction, warning signs, flaggers in high traffic areas, and security personnel at the turbine sites under development are employed for the safety of the public, construction workers, and for security at active work areas. Safety and warning signs are posted to inform the public of construction activities and recommending the general population to remain off the site. Off-road vehicle use is likely to remain unchanged during the construction period. The projects predominately on public land with wide open access are designed and planned to co-exist with current and future land uses.

Meteorological Issues

Wind energy as a renewable resource is one of the fastest growing technologies to date. However, because of the inconsistency and limited predictability of the wind speed, the turbine output cannot be controlled to the same point as traditional generation technologies. Conventional generation provide an essential function in maintaining the energy balance between generation and demand. Wind energy in particular challenges the electrical system balancing in two ways: 1) wind power introduces additional variations and uncertainty, and 2) if the wind is obtainable for longer periods of time, the existence of the resource actually diminishes the amount of traditional generation capacity planned and available for balancing purposes. The effects of wind energy on electrical system operation comprises of diverse time scales ranging from seconds to weeks. The shorter time-scale, varying from seconds to minutes (from an operations and planning perspective), the wind energy resource has a direct impact on system frequency, the fundamental constraint for the power balance between generation and load. Primary and secondary generation reserves are used to sustain this balance. The longer time-scale, varying from hours to weeks, the wind energy system provides a greater influence on the economic dispatch and unit commitment of the traditional generation units. Currently, the wind energy resource in particular, lessens the output level and/or operating hours of the conventional generation units. Typically, the fossil-fuel-based units, scheduled to maintain the electrical system reliability, are critical for the compensation of the wind turbine generator's inconsistency and narrow predictability. The improved magnitude of wind energy projects and the development of large offshore wind farms convey a numerous prospects and challenges. Opportunities comprise of larger power and energy outputs and enhanced technical capabilities. The increased outputs however concurrently present fundamental challenges because of the uncontrollability of the primary resource ~ the wind. The major challenge of utilizing the wind as a primary energy resource for power generation is its changeability as wind speeds oscillate on timescales varying from seconds to seasons. It equates to an output of wind turbine fluctuations as well, depending on the association amid wind speed and wind turbine output power. An additional challenge is that wind speeds depend on abundant meteorological factors that can only be forecast up to a restricted extent. As wind speed disparities can only be predicted with accuracies declining with the forecast lead time, it is not possible to precisely evaluate the wind power output for longer time ranges. As the wind energy system installation efforts increases, the

impacts of wind power's variability and restricted predictability become momentous as well from the perspective of a dependable operation of the electrical grid. The major constraints of wind farm integration from a meteorological perspective are outlined below:

- The variability of wind energy potential by geographic region and daily weather conditions.
- The intermittency of wind energy typically is asynchronous with the load demand.

The availability of wind resources may not correspond with the peak demand of electrical usage. Wind turbines are typically subjected to callous weather conditions. The winds are strongly influenced and modified by local terrain, bodies of water, weather patterns, vegetative cover, and other factors. In order to harvest the wind suitable for electrical production, wind turbine site location is crucial. The location must have a minimum annual average wind speed of 11 – 13 mph for serious consideration. The weather data is available from area airports, state wind maps, and meteorological stations. The wind resource may or may not be conducive for comprehensive energy expansion. A small increase in wind speed results in a large increase in power output from the WTG; therefore the windiest spot is ideal. The wind speed increases with altitude and is reduced by surface roughness elements such as trees, coarse undulating terrain, and buildings. Currently, the existing energy generation system has a baseload equipment that operate at the same level all the time and load-following equipment that is designed to vary its output to match the load. When wind turbines place electricity onto the grid, the load-following equipment will respond by backing down, as if the load had been decreased. This automatic system is capable of compensating for some percentage of wind energy added to the system.

Generally, wind power generation is modeled using actual wind speed measurements, numerical weather-prediction data, and the relationship between wind speed and power. A high-quality model has contributions of ten minute wind speeds measured with an accuracy of 0.1 m/s at several locations in a two year (minimum) interval. First, the wind speed time-series are exploited to establish periodic effects such as daily wind patterns. The use of time-series assures correlations of wind speed (variations over space and time) are automatically taken into account. Secondly, the wind speed data are conveyed from the measurement height to the wind turbine hub-height. Thirdly, the data are then transferred from the measurement sites to existing and foreseen locations of wind parks by linear interpolation, taking into account the spatial correlation between the sites. Finally, wind speed forecast errors are estimated using the same method and subsequently used to develop wind speed forecasts for each location. Wind speed and the forecast time series are then utilized for the generation of wind power time-series and wind power forecasts. Inaccurate data will halt the entire development process.

COMPARATIVE DECISION TREE MODEL RESULTS

The improved chance model (Figure 6) produced an average cost of *only* $176,793 (reduced from $325k) for the wind generator delays with equal branch probabilities. The elimination of the Public Relations and Environmental Studies branch is a direct result of the increased resource effort. The favorable results of the Meteorology Studies branch to continue the work effort produced minimal cost in the improved model. The Engineering Issues involve tedious operations for proper system integration. Thorough reviews are expected to achieve this goal. The increased community support is expected to accelerate the Landowner Issues for the Wind Generation implementation thereby decreasing the delay time and associated cost within the branch.

The Attribute Ordinal Ranking

The attribute matrix (Table 5) represents the major tasks involved and the proposed work effort to achieve the wind farm system implementation. It also includes the before and after (optimized) costs

Figure 6. The improved cost-benefit model for WTG public impediments

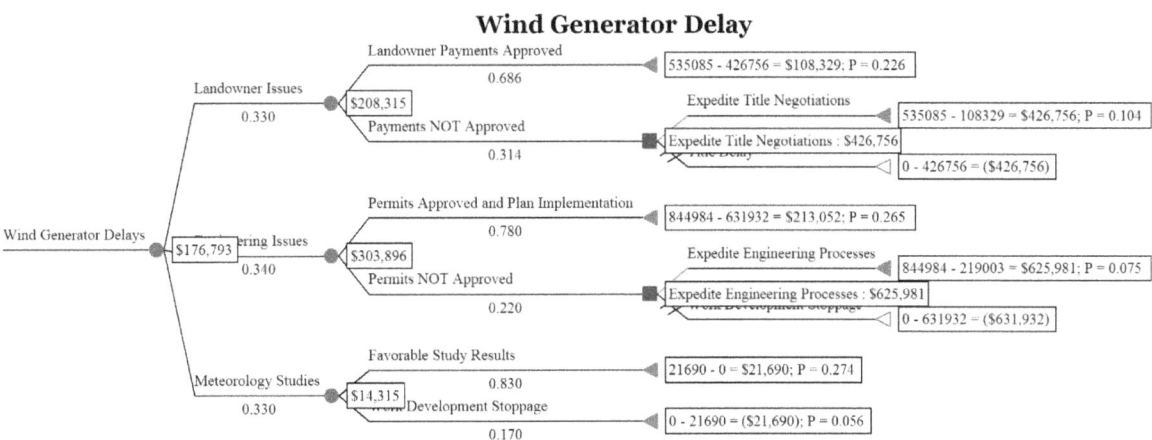

of the model for comparison purposes. Nevertheless, the level-of-effort and urgency categories are not readily apparent. The goal of the Y-Chart is to display the work effort, urgency, and need level in a logical and methodical priority sequence.

The optimized model reduced the decision tree (Figure 5) to only three branches (Figure 6). The increased resources and effort effectively eliminated the Public Relations and Environmental Issues branches of the tree. The three optimized remaining areas are depicted in the Y-Chart. The level-of-effort and sense of urgency calculations are exhibited as well. A cursory inspection of the residual data reveals an intense urgency within the Landowner and Engineering general categories with binary levels of effort (Figure 7).

The Engineering Issues segment requires a large level-of-effort whereas the Landowner Issues only necessitates a small level-of-effort. The outline of the Title, Agreements, Payments, and Associated Labor, is in process within the system, thereby equating to little or no additional effort. Nonetheless, the intense urgency in two out of the three categories warrants high priority status. Most importantly are the Grid Assessments (utility interconnection, system reliability, and conceptual collection system) and permit layout development (high voltage electrical coordination, survey & staking, environmental/agency/land-

owner walk down). Each activity requires a tremendous amount of effort and expertise for a successful WTG implementation. Conversely, the meteorology studies are in the final phases (consultation/advice) and require minimal effort.

Sample Calculations

The sample calculations and the Y-Chart Key are described below and in Box 1.

Landowner (20) + Labor (8) = 28 > 30 == Small LOE & 28 > 21 == Intense Urgency

Engineering (32) + Concept Layout (6) = 38 > 30 == Large LOE & 38 > 21 == Intense Urgency

Meteorology (11) + Wind Measurement (6) = 17 < 30 == Small LOE & 17 < 30 == Indifferent Urgency

Landowner Issues (20) + Category Sum (20) = 40 > 26 == Strong Total LOE & Severe Total Urgency

Meteorology (11) + Category Sum (11) = 22 < 35 == Weak Total LOE & Indifferent Total Urgency

Figure 7. The y-chart of WTG specific task areas

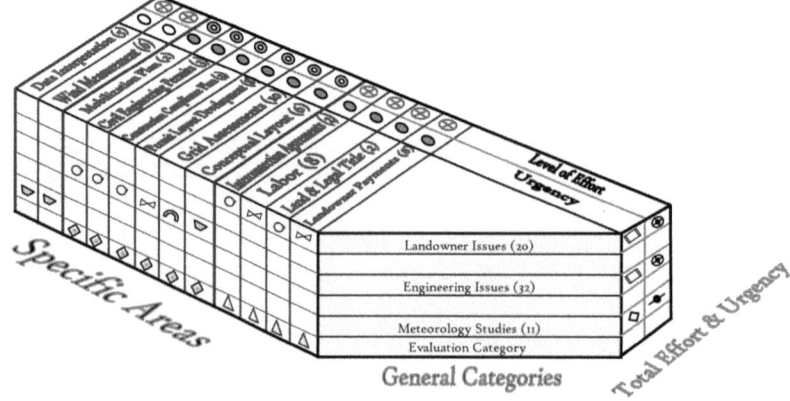

Box 1.

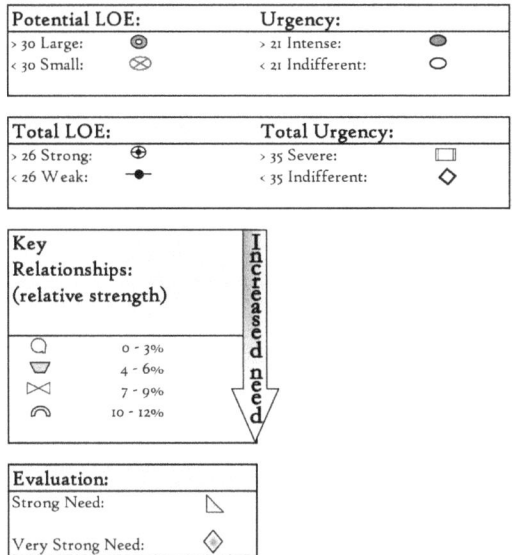

The Comparative Process Model Results

The Political, Economic, Social, and Technological (PEST to include Environment and Legal issues ~ PESTAL in European circles) analysis affect impacts system performance, and deemed as intangible costs typically excluded in many corporate environments. The thorough comprehension of values, assumptions, and limitations of analytical solutions is the foundation of user analyst within the organization (Liberatore, W. and Luo, W., 2010). The user analyst (business analyst, MBA concentration) is adept with the use of analytical models and capable of sound decision-making in the PEST evaluation in any global corporation. The model incorporated several aspects of these parameters and successfully reduced the potential hazard. Nevertheless, a compilation of the European results and trends driving new distributed generation/renewable energy sources are shown in Table 6 below (ENIRDGnet, 2003):

Several of the challenges in the model are mitigated via increased resources, specifically in the Meteorological and Public Relations campaigns. The model displayed the average wait time per entity compared the resource increase with an optimal replica (3 additional meteorologists, 3 public relations, 3 engineers, 3 landowner & neighbors personnel, and 3 environmentalists) for improved efficiency. Appreciable changes were detected in four major areas as depicted in Figure 9.

The Conceptual Layout segment wait time decreased from 182 days to 78 days and the Permit Layout delay area experienced a more profound decrease (2081 days to 456 days). This area alone consists of the geotechnical studies, surveys/staking, and electrical coordination from the grid assessment subdivision. The Engineering Heavy Permits (605 to 268 days) and Landowner & Neighbors area (981 to 565 days) experienced significant reductions as well. The wait time throughputs for the other major for the other

Further analysis of the prioritization process yields severe urgency and strong level-of-effort in the total evaluation categories of the Y-Chart. The Engineering Issues category requires additional effort and/or resources to mitigate the "very strong" need differentiation. The possible alternatives to implement parallel processes or sequencing techniques lien the system cost boundaries as well as the "wait time" tolerances for WTG implementation. It is obvious that the engineering evaluation must be performed in conjunction with the Environmental and Public Relations studies as a cost containment measure. However, the variables involved in a WTG project implementation front-end analysis inhibit highly-skilled engineers to perform specific tasks until key parties are in agreement. Nevertheless, the data exhibits a very strong sense of urgency (need) in the engineering category for completion. The ordinal ranking prioritization results are depicted in Figure 8.

Table 6. PESTEL survey results

Political	Economic
• A clear renewable energy policy • Various tax and pricing options • Promotion strategies for renewable energy based on rebates & feed-in tariffs • Funding R&D sector to find economical and performing solutions • Difficulty of siting and receiving permits for central generation • Lack of national coordination	• Money supply • Reduction of investment on new transmission and distribution structures • Low price of electricity • Energy trading agreements as a barrier penalize non-predictable energy sources • Investment costs to ensure safety and power quality • Low energy demand in some areas due to economic activity reduction
Social	**Technological**
• An increase in employment for wind turbine manufactures • Enhancement of electricity demand due to population increases. • Increase in income that allow people to invest in power plants • Public acceptance of renewable energy technologies and resistance to new distributed generation • Consciousness of sustainable development, demand side management, and interest in renewable energy technologies • Educational and vocational qualifications to promote the development of distributed generation and widen its utilization	• Research is expected to promote distributed generation and develop new business • Speed of technology transfer that is supported to promote the results of research • Information & Communication Technology/enhancing storage technology/mew materials & processes/improved wind energy forecasting software • Barriers include: The cost of the technology required and the long-term return of investment/Poor promotion of research results/ Technical restriction on the network/Existing network design procedures/Safety issues/Potential disturbance and need for auxiliary devices/Requirements for sophisticated metering and control protocols • Lack of infrastructure for distributed generation built in an area far away from an existing network (necessitates a link requirement)
Environmental	**Legal**
• Reduction of greenhouse gas emissions to increase the quality of life • Elimination of pollution sources • Land use zoning restrictions for wind turbine generation • Noise level reduction	• Restriction on trade and product (emission, control, import, license) standards • Lack of interconnection standards • Renewable energy obligation certificates as purchasing agreements • Utilize the law as a tool to promote distributed generation and renewable energy technologies

Figure 8. The ordinal ranking prioritization results

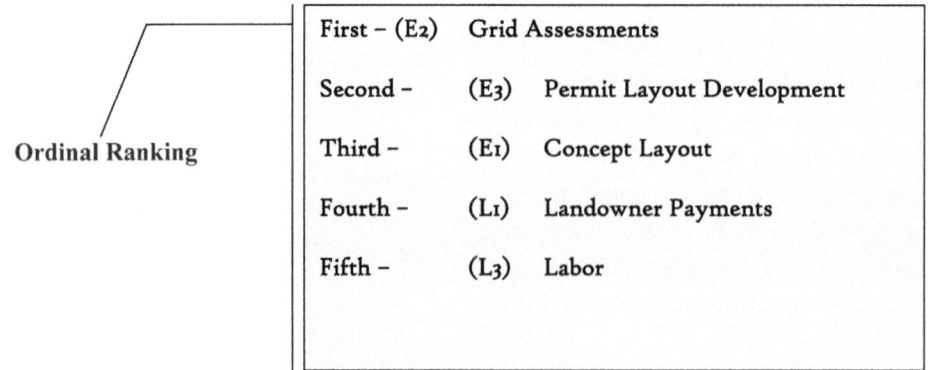

Figure 9. The average wait time comparison

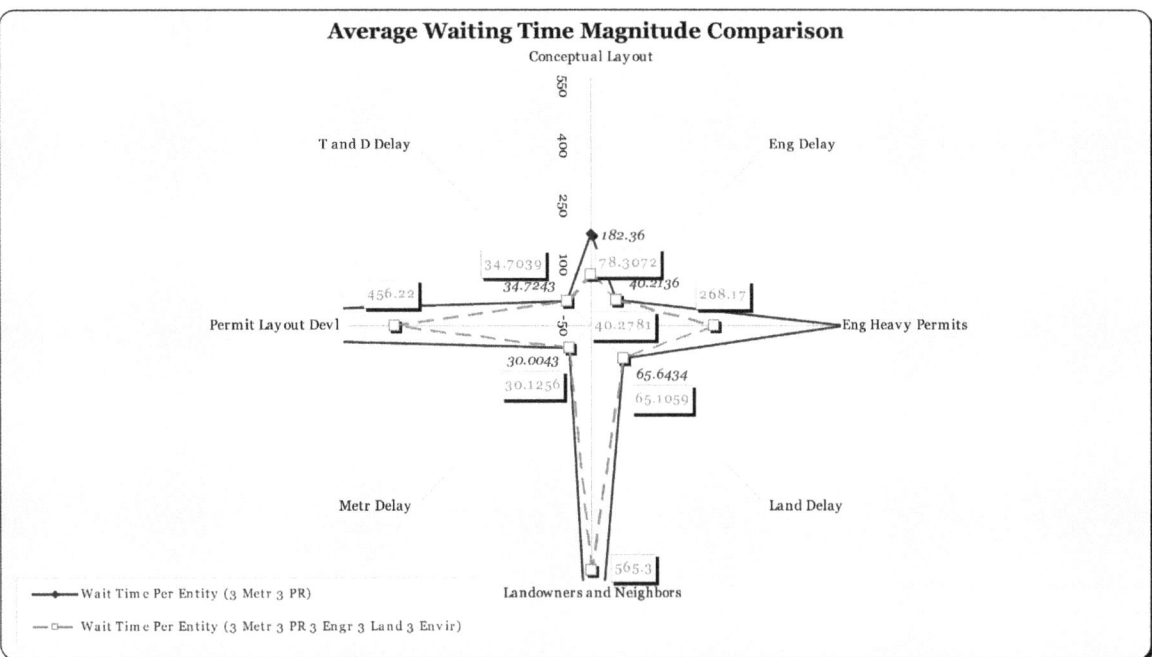

major entities were effectively eliminated as the additional resources dissipated the issues. However, the Engineering Issues area demarcates three of the four major segments in both wait time and total time (Figure 10) radar chart.

The complexity of the Conceptual Layout, which consists of the initial turbine siting, constraint mapping, and transportation studies, requires additional planning and skilled resources. The minimum average time for this segment improved profoundly as well. The Engineering Heavy Permits segment encompasses the road improvement areas at the state & local levels as well as the permitting in support of potential requirements for secondary containment around the substation transformers. The total time reduction of the optimized model maintained the resource efficiencies of the wind turbine generator system integration without compromising its integrity.

The bar charts (Figures 11 and 12) depicts the comparison of the average (expected) improvement percentage verses the minimum average improvement (optimistic) in the integration model. The importance of the Landowners & Neighbors areas is crucial in the success of the wind farm system integration process.

The minimum average in the Permit Layout Development area displays the most improved performance in the optimized model. Although the segment is encompassed in the Engineering Issues area, the vital area includes successful geotechnical study results, survey & staking, and electrical coordination. Another profound improvement in the permitting is outlined in the Engineering Heavy permit category. Nevertheless, the Landowners & Neighbors issues must be resolved in the early stages of the process for successful WTG implementation. The delay catego-

Figure 10. The minimum average time comparison

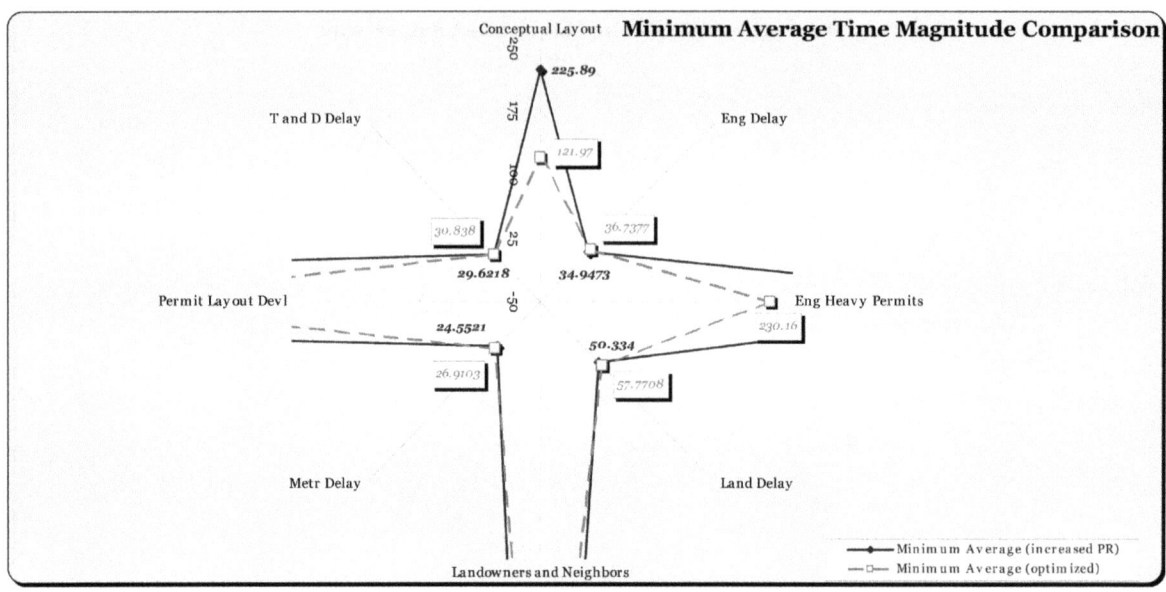

Figure 11. The optimized total time improvement percentage comparison

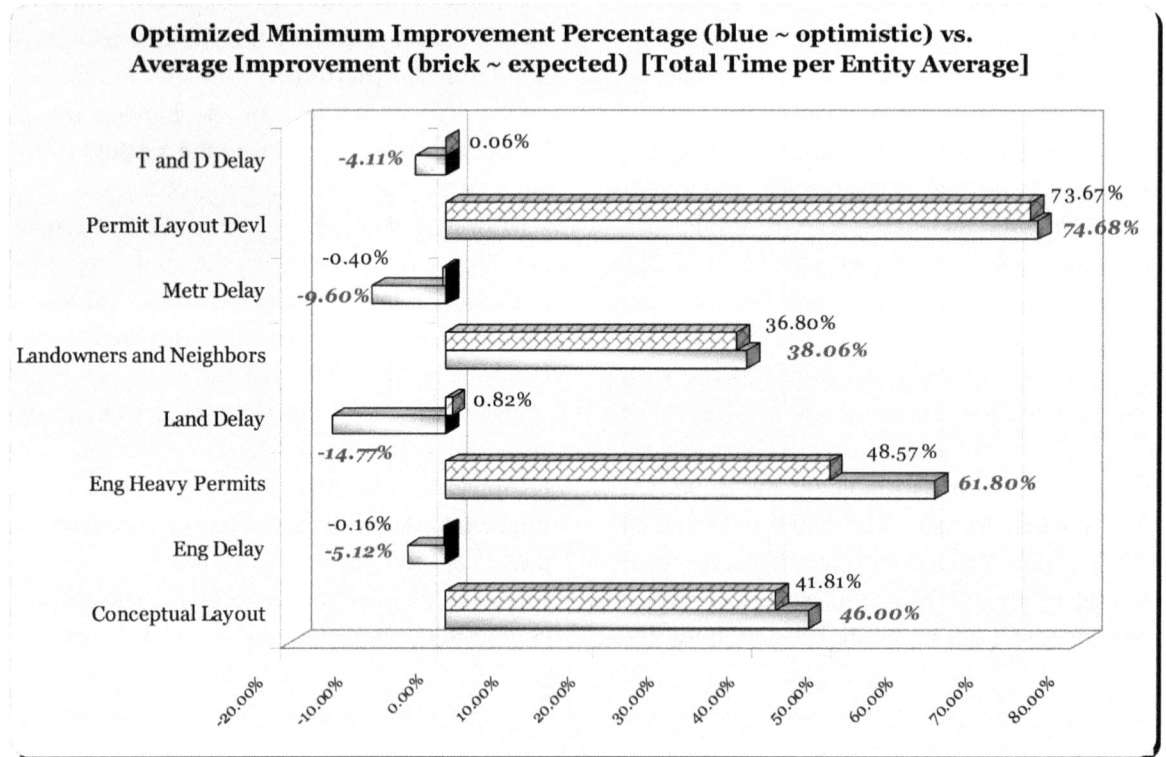

Figure 12. The optimized wait time improvement percentage comparison

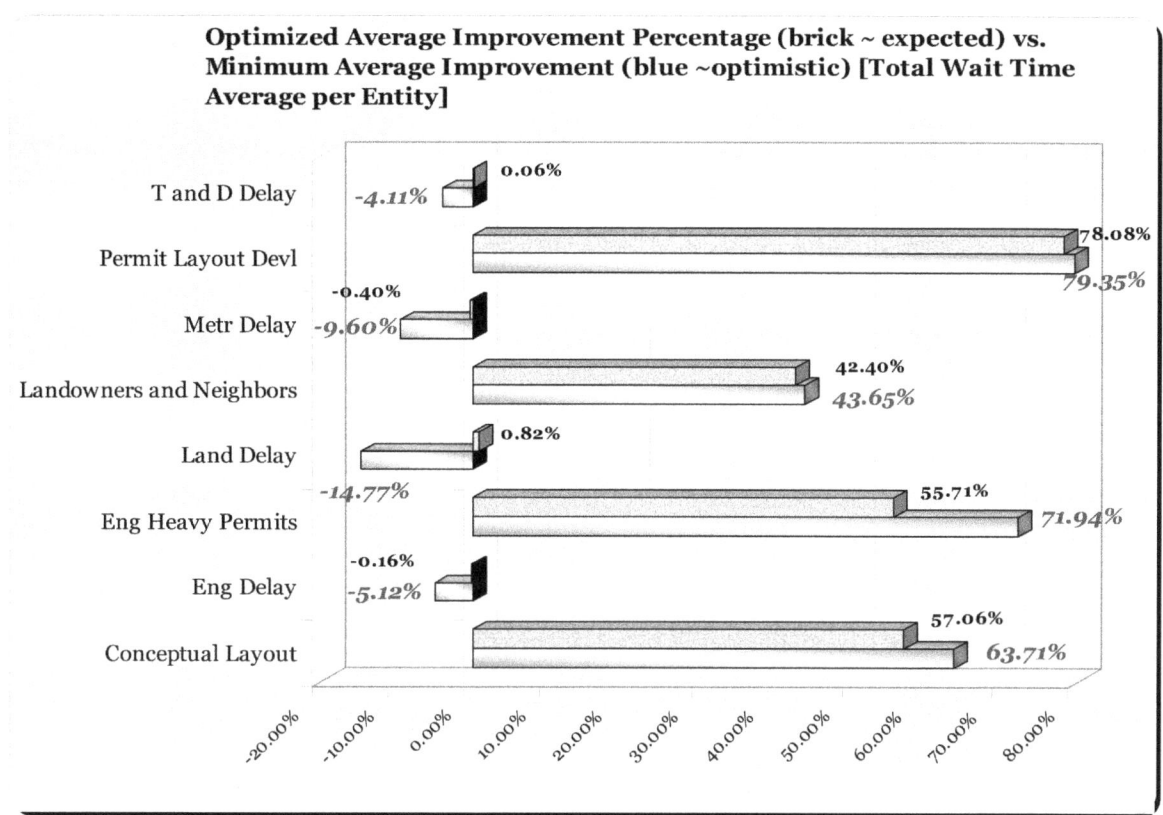

ries are negligible in the expected categories. The data is exploited for cost and scheduling rationale in the wind turbine system integration development.

The optimized model produced a decrease in the resource waiting time. Consequently, the minimum wait time average for the key categories (conceptual layout, heavy permits, landowners & neighbors, and permit layout development) displays an austere enhancement. A careful, yet meticulous trade-off balance is required by informed decision-makers to continue the optimization process, especially in the skilled labor categories.

Once the resources are accurately implemented in these critical areas, the WTG system integration is expected to prevail without incident. Similarly, the community awareness issues are theoretically eliminated (as demonstrated in the model) with the increased lobbying/education/public relations resources of the green energy initiatives. The communications and public affairs department within the pre-construction envelope are essential resources in the organization. The technique transcends the boundaries of WTG implementation and can be applied across technical disciplines with large, capital-intensive endeavors.

REFERENCES

Aakre, D., & Haugen, R. (2009). *Wind turbine lease considerations for landowners.* North Dakota State University Extension Service Report supported by the U.S. Department of Energy Wind Powering America Program. Award Number DE-FG36-07GO47010.

Anaya-Lara, O. (2009). *Wind energy generation: Modeling and control.* Chichester, UK: John Wiley & Sons, Ltd.

Barrios, L., & Rodriguez, A. (2004). Behavioural and environmental correlates mortality at on-shore wind turbines. *Journal of Applied Ecology, 41,* 72–81. doi:10.1111/j.1365-2664.2004.00876.x

Besterfield, D. H. (1999). *Total quality management* (2nd ed.). Upper Saddle River, NJ: Prentice Hall.

Blair, N., & Hand, M. (2008). *Power system modeling of 20% wind-generated electricity by 2030.* National Renewable Energy Laboratory conference paper presented at the Power Engineering Society, Pittsburg, PA.

Breckenridge, T. (February 13, 2010). Lake Erie wind turbine backers ready to study wildlife impact, court manufacturer. *The Plain Dealer.* Retrieved 19 May, 2010, from http://www.plain-dealer.com

Breyfogle, F. W. III (2000). *Managing Six Sigma.* New York, NY: John Wiley & Sons.

Cowen, D., & Jensen, J. (1999). *Remote sensing of urban/suburban infrastructure and socio-economic attributes* (pp. 611–622). Columbia, SC: American Society for Photogrammetry and Remote Sensing.

Damborg, S. (2003). *Public attitudes towards wind power.* Danish Wind Energy Association. Retrieved 4 May, 2010, from http://www.wind-power.org/en/articles

ENIRDGnet. (May 31, 2003). *The driving European forces and trends: Concepts and Opportunities of DG.* Retrieved 4 May, 2010, from http://www.wp1.leader@dgnet.org

Evans, W., Howe, R. W., & Wolf, A. (2002). *Effects of wind turbines on birds and bats in Northeastern Wisconsin.* University of Wisconsin-Green Bay submitted to the Wisconsin Public Service Commission and Madison Gas & Electric Company, Green Bay, WI.

Farret, F. A., & Simoes, M. G. (2006). *Integration of alternative sources of energy.* Chichester, UK: John Wiley & Sons, Ltd.

Feigenbaum, A. V. (1991). *Total quality control,* 3rd ed. New York, NY: McGraw-Hill Green Energy News. (April 19, 2010). *Pentagon attempts to stop wind farm development.* Retrieved 28 April, 2010, from http://www.renewable-energy-news.info

Good, J. (2006). The aesthetics of wind energy. *Human Ecology Review, 13*(1), 76–80.

Hammack, L. (April 12, 2010). Construction of state's first wind farm to resume this month in Highland. *The Roanoke Times.* Retrieved 10 May, 2010, from http://www.roanoke.com

Hankey, R. S., Liggett, W. D., & McNerney, R. A. (2000). *The changing structure of the electric power industry 2000: An update. Energy Information Administration, Office of Coal, Nuclear, Electric and Alternative Fuels.* U.S. Department of Energy.

Harry, M., & Schroeder, R. (2000). *Six Sigma, the breakthrough management strategy.* New York, NY: Currency/Doubleday.

Jacobsen, J. (April 2010). *Wind power company gets to the root of an icy issue.* Making the Case for Quality: The Knowledge Center. Retrieved 1 April, 2011, from http://www.asq.org/knowledge-center/index.html

Juran, J. M. (1999). *Juran's quality handbook* (5th ed.). New York, NY: McGraw-Hill.

Justus, D. (2005). *Case study 5: Wind power integration into electricity systems,* International Energy Technology Collaboration and Climate Change Mitigation. Organization for Economic Co-operation and Development International Energy Agency. Retrieved 2 May, 2010, from http://www.oecd.org/env/cc/

Levine, E. S. (March-April 2012). Challenges for public sector analytics. *Analytics Magazine,* pp. 27-29.

Liberatore, M., & Luo, W. (2012). The analytics movement: Implications for operations research. *Interfaces, 40*(4), 313–324. doi:10.1287/inte.1100.0502

Meibom, P., Ravn, H., Soder, L., & Weber, C. (2004). *Market integration of wind power.* RISØ National Laboratory conference paper presented at the 8th International Conference on Probabilistic Methods Applied to Power Systems, Ames, Iowa.

Schermerhorn, J. R. Jr. (1993). *Management for productivity* (4th ed.). New York, NY: John Wiley & Sons.

Tegen, S., & Lantz, E. (2009). *Social acceptance of wind power in the United States: Evaluating stakeholder perspective.* National Renewable Energy Laboratory poster session at the Wind Powering America Conference, Chicago, IL.

Thomas, K. (1975). *The handbook of industrial and organizational psychology* (Dunnette, M., Ed.). Chicago, IL: Rand McNally.

ENDNOTES

[1] Source: Energy Information Administration (EIA) 2006 renewable energy sector data (Washington, DC).

Chapter 2
Alternative Analysis:
Equipment Installation Case Studies

ABSTRACT

The point of reference of an exceptional design in a system is the lack of a crisis. Electricity is expected to flow effortlessly with the flick of a light switch in a home or industrial setting. The equipment behind the switch – the generators providing the electric source – is meaningless to the end-user. It is designed well and only noticeable in a crisis (blackout as a worst case scenario). However, when the renewable energy technology is incorporated in the present infrastructure, it ought not to create additional problems in the electrical footprints. The cost-effectiveness of the renewable generation must provide substantial eco-friendly benefits to justify the capital expenditures. Only when the renewable energy system is designed well and seamlessly integrated in the grid to minimize the fossil-fuel dependency, the potential crisis will be averted.

ALTERNATIVE ANALYSIS OVERVIEW

The implementation of major equipment within an existing electrical system necessitates a thorough comprehension of environmental policies, integrated planning, and organizational interaction in the sphere of the endeavor. In an alternative energy system integration methodology, there are various optional processes and detailed analysis encountered throughout the implementation. The selection of a wind turbine generator, for example, requires several layers of evaluation to assess the boundaries of the technology within the existing electrical system. Moreover, the decision-makers must understand the effects of capital project implementation and utilize system thinking throughout the process. The principle of the system boundary examines the

DOI: 10.4018/978-1-4666-2839-7.ch002

interaction and issues of importance regardless of the organizational demarcation (Senge, 2006). This boundary principle espouses organizational interaction beyond the limiting constraints and requirements of the venture where one solution creates another problem. An example of this phenomenon is the effects of a wind farm installation that created a loop flow problem during winter peak loading conditions when there is little or no wind. Supplementary problems with the wind farm installation involve the negative impacts on avian and the bat populace. The actual bird/bat mortality estimates in one particular region was 1.29 birds/turbine/year and 4.26 bats/turbine/year, where the national estimate is 2.19 birds based on the meta-analysis (Erickson, et al., 2001). The bat mortality is associated with migration period. The carcasses are consistently found below the wind turbines germane to the seasonality of the migratory habits of specific species. The alternative analysis processes is beyond the financial, performance, energy security, and environmental criteria established in the industry, and must include the impact on biodiversity, the ecosystem as well as the livelihood of the local population. The service business must provide an enhanced platform to attract complimentary external companies to further its cause (Chesbrough, 2011).

In a learning organization where advance technology is proposed, key personnel must comprehend the invisible fabrics of interrelated activities. The body of knowledge espoused in system thinking is designed to assist the learning organization with keener observations of the overall endeavor. The renewable energy implementation requires overall system observations from planning of the technology to the plant-in-service phase and beyond. The undammed Mekong River in Asia is scheduled to receive a series of eight hydropower units and several dams within the next ten years (Grumbine & Xu, 2011). The projects are proceeding largely without detailed analysis

of the biodiversity and ecosystem in an area distinguished by high paucity and low development. Consequently, the increased costs created by the proposed dams – fish reduction, loss of nutrients for the floodplain, crop growing, deluge of the river bank plots – is equivalent to \$0.5B/year. The human source of revenue would affect well over two million people in losses. The investors are focused solely on profits with little regard to feasibility studies and recommendations. The alternative analysis for the future power system must include key stakeholders in a transparent, decision-maker process. Long-term objectives are jeopardized when managers focus solely on short-term financial gains (Gojanovic, 2012). Systems thinking compel these key entities to focus on the chief issues important to the renewable energy implementation. A company engaged in the system integration efforts must include these associated external costs in addition to the traditional accounting methodologies for capital project assessments (Atkinson, 2000). Generally, these costs focus on the external economic activities imposed by an entity as a by-product on third-parties. The incorporation of a modified balanced scorecard to include the dimensions of sustainability – economic prosperity, environmental quality, and social justice – is a basic obligation of the socially conscious company operating in the 21st century. The comprehensive management mechanism is designed specifically for organizational sustainability (Figge, et al., 2002). The tool involves several layers and domains to include performance measurement, cost management, environmental quality, and strategic management, to provide an integrated approach to model and evaluate important areas.

A particular case involved a 500-kV transmission line proposal and considered three routing alternatives[1]. The proposed high-voltage line also included a new substation to increase the electrical capacity in the region. The company its desired goals. Theses specific goals must not erode over time where gaps are allowed to creep in between the current situation. Consequently, the goals are lowered (or abandoned) with the two sets of pressures, and creates a *shifting the burden* situation. However, by exploiting the gap between the vision and current reality – called *creative tension* – the incisive decision-maker can channel it as a source of energy instead of a sea of hopelessness. The only options are to resolve the tension (pull the reality toward the vision) or release it (pull the vision toward the reality). When a tidal wave turbine generator is proposed for implementation, for example, the traditional fossil-fuel culture is forced to strategically think on the renewable energy system level. The control, maintenance, and adaptability of the new technology in the current infrastructure will have a steep learning curve in specific regions of the electrical footprint. Nevertheless, in order to maintain the vision, the creative tension cannot be associated with discouragement, hopelessness, or other negative emotions. Astute decision-makers will escape the emotional tensions to reduce the gap and bring the reality closer to the vision.

Experienced decision-makers typically rely on intuitive analogies, the ability to recognize patterns, and parallels to other seemingly disparate situations (Agor, 1984). Systems thinking integrate reason (linear logic) and intuition with an emphasis on the cause-and effect paradigm. Decision-makers with enhanced intuition have the ability to determine how obvious solutions to complex systems will produce additional impairment than high-quality, and temporary modifications generate lasting problems. These same leaders can sense the eroding goals in an organization and provide the necessary reinforcement mechanisms to maintain the standards. Conversely, if the team is focusing on simply deliberate signs of performance and disguise deeper tribulations, it may further exacerbate the conditions. The incorrect indicators will generate alternatives to produce enhanced results within the organization.

THE UTILIZATION OF MENTAL MODELS

The first method for estimating the intelligence of a ruler is to look at the men he has around him (Niccolo Machavelli)

The utility planner is deeply involved in power supply issues and future electric demand inside the region. The utilization of scenario analysis as a practical planning tool is essential for renewable energy implementation within the current infrastructure. The methodology employs extensive use of alternative evaluations is well as generalizations about the characteristics of the electrical footprints. Technically savvy executives, strategist, and application analyst can produce fact-based decisions with the availability of good data for performance management (Liberatore & Luo, 2010). The development of these scenarios is designed to assist decision-makers with a metal image of the proposed reality based on future assumption (Wack, 1985). However, most of the identified assumptions are usually utopian by nature. If the decision-makers believe their views are facts instead of sets of assumptions, the openness to challenge these views are nonexistent. The *leaps of abstract* concept utilize substitution for simplicity in an object for details, which limits learning. It generally occurs when one moves from direct observation (empirical data) to generalization without testing. This process tends to obstruct learning because it is axiomatic. A previous assumption is treated as fact. Decision-makers must force themselves to questions and test the generalizations; explicitly separate it from the data which produced it. These same decision-makers and systems planners are encouraged to incorporate practicing reflection methodologies in the strategy sessions and be conscientious of the leap of abstract concept in the systems thinking paradigm. The fear of the unknown can develop if the status quo is not challenged (Cokins, 2012). The *no decision* is in fact a decision often omitted by key executives. Well-established mental models will prevent changes derived from systems thinking.

Consider the integrated resource plan scenarios of a typical electric utility company in one particular case, where the expected loads are forecasted with planned/existing power supplies (generators). The economic inputs and fundamental growth assumptions are developed based on employment, predicted population, income, housing stock, and inflation grounded in historical relationships. The social indicators are imperative to establish trends and formulate conclusions for the electrical footprint (Budget of the U.S. Government, FY2013). The simplistic model relies heavily on the past to predict future growth in the electric servicer area. The assumptions are generally focused within four major scenarios:

- **Severity Model:** Ten year sluggish economy, high unemployment, businesses exit the area, slow income growth, diminished electric purchases.
- **Conservative Model:** Implementation of a green lifestyle, more home-based life, popular renewable energy adoption (rooftop gardens, solar panels, wind turbines), downsizing to one vehicle.
- **Pre-Optimistic Model:** Influx of new business focusing on green technology, stringent emission standards adopted, residents are inclined towards a more energy conscious lifestyle, strong economy, and increase demand for electricity.
- **Optimistic Model:** Strong business, population, and personal income growth, smart grid technology, affordable electric vehicles, peak manufacturing orders, implementation of a carbon tax.

Testing the assumptions will lead to incisive and conducive inquiries into the reasons behind the actions. For example, full manufacturing with peak orders in the optimistic model equates

to a skilled workforce and supplemental area universities to supply the managerial labor pool. Moreover, an efficient plant requires fewer workers and technological advances (automation, robotics) to meet the demand. The skilled workers receive continuous training at local colleges to operate and maintain the high-tech equipment to manufacture a viable product. Yet, with fewer workers in this sector alone in conjunction with a conservative, efficient plant operations (diminished electrical consumption, *smart devices*); the optimistic model assumptions in this area alone are seriously flawed and limit the learning process. Conversely, in the severity model, the sluggish economy will shut down manufacturing and shift the labor pool overseas as ancillary businesses struggle to survive. Consequently, the lack of energy consumption in the industrial sector equates to a decrease in electrical loads and a surplus in capacity where capital resources are not being utilized efficiently.

Consider a second scenario in the conservative model, where the general population readily adopts renewable energy initiatives. The home-based lifestyle promotes an energy conscious family that is inclined to install technologically advanced generators to locally produce electricity. In fact, an approximately thirty-five hundred square foot positive impact house designed to pull water and cool air from the environment can sustain five people and send electricity *back to the grid*. The house employs both active (solar power, vertical axis wind turbines, solar thermal water heating, grey water recycling and bio-filtration, ground source/sink loop for cooling water/air, and automated building management) and passive (wind tower, green roof garden, compressed Earth wall, low-energy casement windows) renewable energy systems to maintain its power requirements. Based on an average household production of approximately 125 kilowatt-hours per day, the home can easily support itself (the typical electric consumption for its size is 40~60 kilowatt-hours per day) without any external sources. The utilization of an atmospheric water generator to run refrigerant through metal coils (or harvest humidity from the air) attracts condensation that is channeled to a purifying holding tank. Two installed generators in the house can produce enough freshwater for drinking, laundry, and showering where it is recycled for flushing the toilets and growing foodstuff. The water demand is estimated at ten gallons per day per person (three gallons for faucets/potable and seven gallons for dishwasher and /or four gallons for toilets and six gallons for plant irrigation) or fifty gallons for a family of five (the atmospheric water generator can generate five gallons per hour). Moreover, an additional 35-40 kilowatt-hour is produced by the airflow from the installed wind turbine generator via the house ductwork forced from the roof. Highly-concentrated (eighteen hundred times more focused) installed photovoltaic solar panels can produce additional electricity for the house, as needed for electric vehicles or sold back to the utility. If one housing unit can sustain itself and produce surplus energy to the grid, an entire residential area can follow suit. The utilization of distributed generation and micro-turbines designed for industrial use is another option of continual energy consumption in the district. The vision expounds on the self-sustaining community willing and able to support itself without any external energy sources. Once this renewable energy threshold is encroached, the populace is more inclined to maintain their way of life and abandon the electric utility altogether (as espoused in the pre-optimistic and optimistic models) instead of reaching back to the traditional methods of energy utilization.

The details expand the opportunity to explore the simplistic generalization to solidify the leaps of abstraction concept where this assumption is deemed as factual. The contradictory scenarios of predictable growth establish an unrealistic representation of the future electric consumption. The assumption set must include realistic details likely to become true instead of *fairy tales*. Metal models are powerful tools designed to affect ones perception of the details and their related interpretations. The ability to test the assumptions develops an

accurate picture of the scenario. Deeply embedded negative mental models in the organization's traditional are counter-productive in the areas where advanced technology is the norm. A robust generation expansion model utilizes empirical data in the scenarios with a set of realistic constraints designed exclusively for decision-makers at the strategic level. Details are systematically evaluated and weighed to obtain a viable solution for renewable generation implementation. The model assesses power system expansion efforts under uncertain load forecasts. Its objective is primarily composed of three conditions:

- The expected expenditures over the probable scenarios
- The cost variance
- The penalty function for feasibility deviations
    - **Surplus Capacity:** Inefficient use of capital resources
    - **Unmet Demand:** Shortage of electric supply (adverse societal effect)

The model evaluates the regional load duration curve (peak and base in hours) to develop the optimal resource plan with consideration of the impact of uncertainty in the forecast of electricity demand (Malcolm & Zenios, 1994). For example, the system considers five demand scenarios over two operating modes and assumes equally in a chance model of 20% for each in Table 1.

Table 1. Power supply resource option

Option:	($/MW)	($/MWh)
A	High Capital Cost	Low operating expenditures
B	Low Capital Cost	High operating expenditures
C	Medium Capital Cost	Medium operating expenditures
D	Low Capital Cost	Low operating expenditures
E	Zero Capital Cost	Very high operating expenditures

The approach to the solution will provide an optimization process that encompasses the reference resource plan to minimize the total expected cost (stochastic programming). Planners determine the constraints of the cost increase to reduce the variances and excess installed capacity to a preferred level (target) to define the tradeoff region for robust optimization. The resulting scenarios will reveal the improved performance of the expected cost for comparative purposes. The model depicts an accurate representation of the required generation expansion from a financial perspective. Moreover, the econometric inputs will provide a detailed analysis of the forecasted demand to assist with the elimination of the leaps of abstract concept and facilitate continued learning in the organization. Ultimately, scenario thinking is designed to arrive at a deeper understanding of the world where the organization operates to formulate enhanced and more informed planning decisions (Olgivy, 2002).

LIMITING CONSTRAINTS OF THE RENEWABLE TECHNOLOGY: CASE STUDY

How well a new, advanced technology is integrated in the current electrical infrastructure is based on its limiting constraints and its ability to seamlessly merge into the operating environment with negative impact. The decision-makers are tasked with the evaluation of advanced technologies while simultaneously upgrading the existing electrical system. Characteristics and distinct attributes of renewable energy technology vary widely across the selected industry (wind, solar, geothermal, tidal, etc.). The performance of the proposed sustaining units must convert mechanical energy to electricity without disruption to the interconnected grid. Moreover, the cost-effective technology is expected to enhance efficiency throughout the system. The induction generator case evaluates the general applications and limiting constraints

of selected renewable technologies to facilitate the alternative approaches in the decision-making process. The integration issues involve grid stability, the increased requirement for balance and reserve power, congestion, and temporary overproduction of generation. The technical concerns are based on the proposed technology to facilitate low-voltage ride through, reactive power correction, and frequency support.

Consider the major induction generator configurations currently utilized in wind turbine technologies. There are two general categories: 1) fixed-speed turbines and, 2) variable-speed turbines. The characteristics of these generators are classified – based on the limiting constraints, requirements, and to some extent, the cost – within these general categories. The basic questions posed by the decision-makers certainly include the paradigm of *what, why, where, when, and how.* Yet, the incisive organization has already performed the requisite due diligence on the technology that included the environmental studies, transmission assessments, engineering issues, public relations, and landowner issues. Nevertheless, the alternative analysis of the selected technology (beyond the cost-benefit evaluations) must be conducted in the strategy sessions. The detailed questions such as (*Why did we select the doubly-fed induction generator over the brushless type?*) will produce alternative paths to better solutions. Astute decision-makers often find themselves holding the assumptions and re-evaluating the options.

The fixed-speed turbine utilizes a squirrel-case induction generator. The simplicity of the cage winding as well as its ruggedness prompted the ease to manufacture and reliability over other machines. Consequently, the squirrel-cage winding is cost-effective when compared with other technologies. However, the fixed-speed turbine tends to impede voltage recovery after a fault is experienced in the system. Moreover, the technology requires reactive power compensation in the form of fixed shunt capacitors. As a voltage regulation technique, it is regard as rather coarse to the system operation as these capacitors are switched in discrete steps with some time delay. Consequently, the compensation is not as smooth or speedy enough in most cases.

The fundamental induction generator operates above the synchronous speed over its range of power output. Theses units provide the greatest amount of electrical energy in the form of an A.C. supply at 50~60Hz. The excitation is provided by the reactive power from the point-of-interconnection of the electric grid. However, self-excitation can be unstable and results in system over-voltages (Morren, et al., 2006). The unit cannot sustain fault current because of the lack of an independent excitation system. The requirements for the point-of-interconnection with the induction generators are to employ a converter configuration (AC/DC devices) capable of absorbing or supplying reactive power (kVAR). These converters are rated by the limitation on the fault contribution (kA). Some classes of converters (Insulated Gate Bipolar Transistors, Gate Turn Off devices) are expensive and must be protected against over-currents/over-voltages. The management strategy based on these over-voltages/over-speeds is to provide constant stator voltage control instead of reactive power. The advantage of this technique permits the generator and converter to operate at the designed rated voltage. However, this method increases the converter rater proportionally. The asynchronous generators utilize a solid-state inverter device. All of these general schemes for the induction generator must be considered in the component portion of the alternative analysis process. The popular variable-speed doubly-fed induction generator is commonly used in wind turbine technology. The configuration is illustrated in panel A of Figure 1.

The stator winding is directly connected to the network in this configuration. Moreover, the rotor winding is connected to the network via a four quadrant power converter comprised of two back-to-back PWM-VSI. The silicon controlled rectifier is limited in this configuration (Muller, et al., 2002). The torque is regulated by the rotor-side

converter and supplies part of the reactive power to maintain the magnetization of the machine. Additionally, the supply-side converter regulates the D.C. link (Rodriguez-Amenedo, et al., 2002). The reactive power (which cannot perform work) is calculated by:

$$P = I * V \sin\theta$$

This reactive power is derived as a function of the load angle δ when the armature resistance is negligible where the output power per phase is given by:

$$Q = VE/X_d \cos\delta - V^2/X_d \cos^2\delta - V^2/X_q \sin^2\delta$$

The C_{grid} converter is used to regulate the voltage of the D.C. bus capacitor. A disadvantage of the variable speed configuration is based on the additional cost and complexity required to interface the generator to the grid. The variable-speed doubly-fed full controller induction generator is illustrated in panel B of Figure 1. The grid interface method can utilize a cycloconverter to transfer the power to the network (Brown, et al., 1992). The cycloconverter apply rectified voltage to the load and operates in four electrical quadrants to accomplish this task. The converter can supply leading, lagging, or unity power factor loads as its input is constantly lagging. Hence, the synchronous machine matches the converter as it draws power factor current from it. Nevertheless, the unit output voltage has inherited complex harmonics and may create losses and torque pulsations. Another proven technique to integrate the technology to the grid employs a matrix converter (Zhang & Watthanasarn, 1998). It consists of nine switches that connects the input phases directly to the output phases and usually controlled by PWM to produce variable voltages at variable frequencies. When the turbine is configured for offshore applications, the utilization of submarine cables facilitate the connection to a land gateway (Hofmann & Okafor, 2001). While these methods vary in dimensions

Figure 1. Induction generator configurations

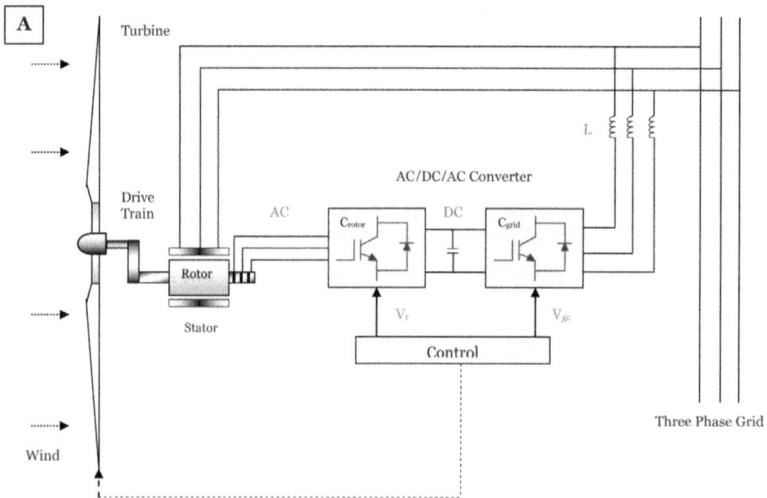

Wind Turbine and Doubly-Fed Induction Generator System

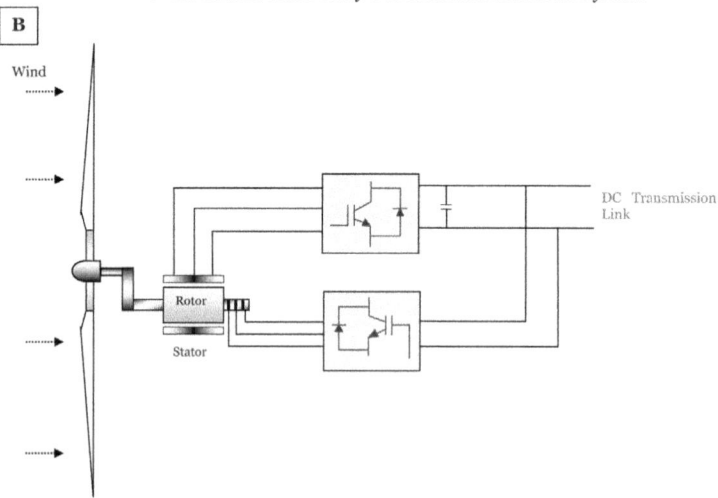

Variable Speed Doubly-Fed Full Controller Induction Generator System

and requirements, each carry related advantages and disadvantages. The major drawbacks for the technology in general are:

- High harmonic distortion in the transmission line
- Poor line power factor
- More complex and less mature technology

The doubly-fed induction generator utilizes slip rings, which can reduce the life time of the machine. Yet, the slip power of the rotors is con-

trolled by the converters *only*, thereby reducing the inverter cost (rating is typically 25% of total system power). However, these same slip rings require periodic maintenance, especially at offshore sites. An alternative to this method is to employ brushless technology to reduce maintenance and the associated costs.

The direct-drive gearless generator is designed to reduce maintenance and failures in gearboxes (Polinder, et al., 2006). The technology utilizes a multi-pole, synchronous generator and a power electric converter with a rating equal to the unit's nominal output. The new technology comprises of a compact permanent magnet with a robust design and requires no slip rings or excitation power. These characteristics equates to higher efficiencies at low loads and maximum power production with low to moderate wind speed. When coupled with lighter rotor blades, the noise levels are only 105 decibels (dB). Wind turbines create a steady, low-decibel sound that can be heard within a short distance of the units. The noise level from each wind turbine generator increases as the wind speed at the site increases. The background noise typically increases proportionally under these conditions and can mask the noise of the turbine generator. Consequently, a wind farm could comply with local base level audible noise standards yet exceed it when the wind speed increases. Most international jurisdictions set a base noise level for low wind speeds that includes the wind farm background noise. This standard typically does not exceed 5 dbA above the rated noise level as the wind speed increases.

The power losses based on given output enable the efficiency (η) of a synchronous machine too be obtained from the follow relationship (Ramshaw & Van Heeswijk, 1990):

$$pu, \eta = 1-(P_{cu}+P_f+P_s+P_{nl}+P_{wf}/ mP_o+P_{cu}+P_f+P_s+P_{nl}+P_{wf})$$

where:

m= Number of armature winding phases

P_o= Power output per phase ($I*V \cos\theta$)

P_{cu}= Armature winding ohmic losses (mI^2R_a)

P_f= Field winding ohmic loss ($I^2_f R_f$)

P_s= Stray load loss

P_{nl}= No load core losses due to hysteresis and eddy currents

P_{wf}= Windage and friction loss

The separations of these losses are illustrated in Figure 2.

These are the generator losses calculated and compounded with other losses before the point-of-interconnection. How well the proposed electrical equipment is seamlessly integrated in the current infrastructure with minima negative impact must be the goal of the renewable energy system

Figure 2.

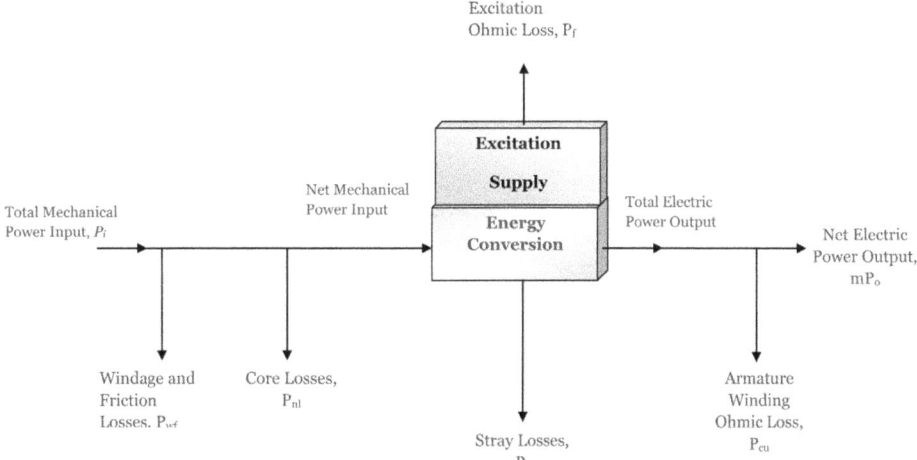

implementation. How organizations achieve this goal based on the realities of the current processes formulates the basis of the gap analysis (creative tension) in the systems thinking paradigm. Table 2 highlights some of the major characteristics of the technologies.

The options for the generator types are unique and specific for the installation. A wind farm designed for the high plains region of the U.S. may not perform satisfactorily in the Pacific Northwest. The limitations and capabilities of the turbine generators require a detailed, top-down review from all stakeholders within the system organization. The technicalities of the equipment integration will involve several iterations and phases before a final technology is approved and selected. The weights of the maintenance versus annual expense, investment optimization, environmental impact analysis, and overall production costing scenarios must be evaluated to develop the precise alternatives. The suspension of established mental models and connecting in a process of collective inquiry based on direct experience of the system is the essential component to incorporate the technology successfully. Moreover, the incisive decision-maker must engage in dialog involving many points of view based on the reality of the equipment installation. In essence, the

Table 2. Induction generator characteristics

Generator Type:	Advantages:	Disadvantages:	General Requirements:	Converters:	Cost:	Comments:
Fixed-Speed Turbine	■ Electrically efficient ■ Utilized active stall components ■ Rugged, brushless, reliable, and robust	■ Governed by passive-stall regulated control ■ Fixed angle blades tend to impede voltage recovery after a fault ■ Aerodynamically less efficient ■ Stiff power train dynamics as the electrical generator is locked to the grid o Only a small variation of rotor shaft is allowed ■ Mechanical Stress – can cause voltage and rotor speed instability (either fault initiated or tripping of an adjacent synchronous generator). ■ Turbulence and tower shadow that induces load fluctuation (decrease life of the turbine generator) ■ Fails to extract maximum power from the wind ■ Not for low speeds (too large reactive power demand) ■ Noisy	■ Up to 1.5MW ■ Requires reactive power compensation (fixed capacitors) ■ Determined by the intrinsic relationship between active power, terminal voltage, reactive power, and rotor speed of the squirrel-cage induction generator	■ Must be oversized (30~50%) with respect to rated power for magnetization requirements of the device ■ Inverter ~ also called DC-AC converter or static inverter ■ Can be operated as a VAR/harmonic compensator when spare capacity is available	Economical	■ Multi-stage gearbox ■ Asynchronous (not synchronized to the frequency of the power operating it) squirrel-cage induction generator (SCIG) directly-coupled to the grid. ■ *Danish* concept ■ Limiting methods: o Passive/Active stall regulation o Pitch regulation o Furling regulation

continued on following page

utilized a transparent process and solicited input from federal, state, and local agencies as well as landowners, concerned citizens, tribes, and interest groups. The tenets of the project encompassed the potential environmental impacts to land use, socio-economics, fish and wildlife affect, wetlands, water resources, visual quality, vegetation soils, cultural resources, as well as health and safety issues within the proposed corridor. The input was designed to assist with the refinements of the alternatives and improve the scope of the capital project. These options produced fundamentally conducive outputs by moving one proposed route slightly east and along property lines, the utilization of different transmission tower types to accommodate both existing and proposed lines, and the addition of a fiber optic cable on an existing segment of the line to meet the communication requirements. Moreover, the elimination of a proposed western route alternative for the transmission line required the purchase of additional right-of-way with the removal of several homes. Consequently, these houses were spared because of the incisive planning within the systems thinking paradigm. The organization observed the entire system for capital project upgrades and implementation.

The no action alternative must encompass the potential impacts of nit providing the upgrade service or new installation for a proposed capital project. Identification of optional transmission routes and substation sites (as discussed in the previous case) is required where condemnation may be necessary for easements. The minimum agency contact for environmental impacts applicable to capital-intense projects during the alternative analysis process and system evaluation is listed below:

- National Conservation Resource Service (wetlands, important farmland, prime rangeland, and forest land)
- State Historic Preservation Officer (cultural resources)

- U.S. Fish and Wildlife Service (threatened and/or endangered species, their critical habitat and wetlands)
- Army Corps of Engineers (floodplains and wetlands)
- State Wildlife Agencies (threatened and/or endangered species and wetland concerns)

There are other areas of potential impacts germane to the venture include socioeconomics, land use, vegetation, and coastal barrier areas. The utilization of the boundary principle enacted within the realm of the venture, includes the external entities and concerns of the citizenry that is beyond the limiting constraints of the proposed technology implementation. The importance of systems thinking in the strategic phases of the project development is designed to avoid costly errors downstream *before* the resources are on site.

Most organizations are faced with a plethora of information with little or no structures to properly manage its key elements. The components generally reflect patterns of vulnerabilities (too much variability) within certain business units of an organization. The utilization of system mechanisms is designed to assist personnel with the visualization of patterns to frame strategic change options (i.e. reduce variability, re-modification of measures, re-creating interaction patters) to derive ideas from complexity theory (Axelrod, 2000). When a new technology is proposed, the alternatives are thoroughly vetted and evaluated within the organizational structure. The proposed technology must successfully integrate in the existing infrastructure without disruptions to the system and external entities. How this technology incorporation is accomplished is based on the organization's ability to learn, interact, and improve its position in the ecosystem. The innate ability of systems thinking is the drive for humans to learn (Hall, 2007). The model to manage and lead change in an organization is difficult. Nevertheless, a business that is willing to adopt new technologies and work together as a whole can achieve

Table 2. Continued

Generator Type:	Advantages:	Disadvantages:	General Requirements:	Converters:	Cost:	Comments:
Variable Speed Turbine	■ Rotor may operate at any speed ■ Variable rotor resistance used for power output & pitch control ■ Aerodynamically efficient ■ Can operate constantly near its optimal tip speed ratio ■ Robustness/ Stable response against disturbance ■ Ability to control rotor speed ■ Increase annual energy production (some cases ~ 10%) ■ Reduced mechanical stress (absorb the turbulence and wind shear) ■ Reduced output power variation based on the decoupling of the mechanical & electrical system ■ Improved power quality by the reduction of power pulsations (decrease voltage variation from rated value at point-of-interconnect and increase penetration in the electrical footprint) ■ Reduced pitch control complexity ■ Fault response is determined by the system protection settings ■ Reduced acoustical noise	■ Periodic maintenance of slip rings (especially at off-shore sites) ■ Less efficient electrically	■ 1.5MW and up	■ Complex power converter to interface generator to grid ■ Electric power connected to turbine's rotor winding ■ Rating: approximately 30% of rated power of generator	■ Additional cost of converters ■ Reduced inverter cost (DFIG) ■ Increased maintenance cost ■ Costly system (DFIG)	■ Configured with variable rotor resistance ■ Configured as doubly-fed induction generator (DFIG) ■ Multi-stage gearbox ■ Wound-Rotor Induction Generator ■ Brushless DFIG is an alternative to slip rings (where the slip rings reduce the lifetime of the machine)
Direct-Drive Gearless Generator	■ Aerodynamically efficient ■ Designed to reduce maintenance and failures in gearbox ■ Less mechanical stress ■ Less noisy	■ Less efficient electrically ■ Heavy and large generator ■ Complex	■ Requires a large generator	■ Electric power with a rating equal to the generator's nominal output	Expensive	■ Multi-pole synchronous generator

learning organization involved with the renewable energy technology integration in the current infrastructure is expected to design and direct profound cooperative learning processes. The management and strategic techniques encompasses the sensing (deep inquiry and the ability to see mental models beyond filters), pre-sensing (move the purposing and visioning phase), and realizing (translate the vision into working models) area of the systems thinking paradigm (Senge, 2005).

Consider the marine turbine generator configuration in Figure 3. The permanent magnet synchronous generator (PMSG) has a preference to operate as a wind turbine or as a tidal generator. Nevertheless, the unit has a higher efficiency based on the gearless construction and permanent magnet configuration (Grauers, 1996). The elimination of the gearbox also allows for low speed operation (15~20 rpm). The addition of the variable speed controller increases the system

availability for the unit configuration. Moreover, the weight reduction of the technology as well as the reduced maintenance aspects produces an attractive renewable energy alternative.

The technology requires non-linear and robust control for effective operation. The generator-side converter typically consists of passive elements (i.e. diode rectifier) or active elements (i.e. Insulated-Gate Bipolar Transistors, or Gate Turn Off devices) where it is best controlled. The insulated-gate bipolar transistor (IGBT) is a semiconductor device that incorporates a gate control and a bipolar current flow mechanism. Yet, the IGBT, the device that combines high current, high voltage, and high input impedance, is expensive and must be protected against over-currents and over-voltages. Another major drawback of the technology is the high cost of the permanent magnet materials and fixed excitation. The tidal resource characteristic creates uncertainties in the turbulence and swell aspects of the source wave. Con-

Figure 3. Permanent magnet synchronous generator: marine configuration

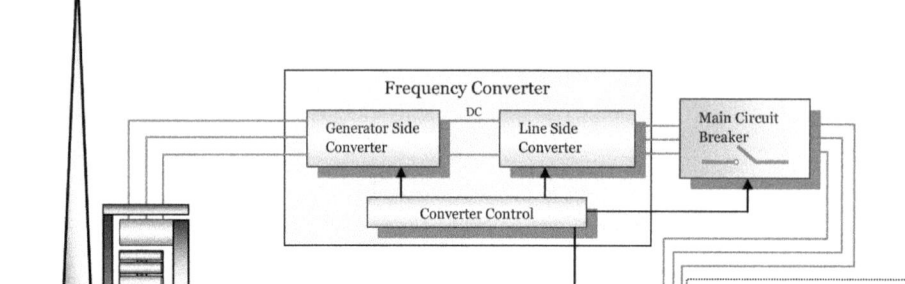

sequently, this effect produces poor system performance and low reliability with the linearization approach (Chinchilla, et al., 2006). However, a sliding mode control will assist with the system uncertainty to maintain the constraints with high-frequency control switching (Matas, et al., 2008). The configurations and distinct characteristics of the renewable energy technology engender a plethora of options for system implementation.

The incisive learning organization will include the key stakeholders of the venture as well as essential *positional leaders* from each business unit involved to assess the implications of the proposed generator integration. The overall strategic plan is subsequently transformed at this level for regional implementation by the project engineering, construction management, maintenance, regional operations, asset management, planning, environmental, and financial analysts to facilitate the proposed system integration. The decision-makers are empowered to make requisite changes and improvement to the overall strategy to decrease the gap between the vision and reality. The most plausible equipment and procedures are produced from various formats of these strategy sessions. Consequently, the acceptable mechanism is selected for the appropriate environment with the expected performance to supplement the current generators in the electrical footprint.

ALTW-WPS CASE STUDY

The General Synopsis

The typical requirement of increasing transfer capability oftentimes involves many interesting alternatives within the bulk transmission system. The options are usually modeled by the Independent Service Operator with the best case scenario today without little regards to future system requirements. A classic example this scenario derives from a demand to increase the transfer capability from 0MW to 100MW as the genesis of the ALTW-WPS Project.

The Alternative Analysis

Four major alternatives were presented to achieve the objective with a cost ranging from $1.25M to $10M. Two of the most intriguing cost options were separated by only $100k. The actual work effort and processes for each alternative were not evaluated as presented. The problem was approached as an absolute cost solution-based initiative. The estimates are extremely inaccurate generalizations loosely derived from ad hoc historical construction data and senior project manager's "best guess." The standardization of each alternative work process is the impetus of the study to eradicate heuristic methods in the industry. Given this environment, a cost-benefit analysis was conducted with the alternatives statistically dispersed within a 95% confidence level to resolve the voltage limits. The public sector in particular, is considered to be a forerunner in cost-benefit analytics (Levine, 2012). The model also included a decision node in each branch to re-evaluate the system loads if the alternative solution falls outside of the prescribed tolerances. The assumption of the paradigm is to decide on an alternative plan (at 5% outside limits) if the original endeavor fails. This decision branch formulates the basis of a failure correction process (an important attribute) of the entire model. A related cost was attached to each terminal node to specify the additional work effort involved as part of the failure component. The ranges are depicted in Table 3 and pane A of Figure 4.

Since this is the "best of the worse" case scenario in the failure branches, the $90k alternative for the cost-benefit analysis is the logical path. A close examination reveals the reconstruction of the Rice-Saratoga 69-kV line in seven years to meet the 44 MVA minimum requirement with an

Table 3. Cost failure ranges

	Worse Case	Best Case
Cost-Benefit Analysis	($296,849)	($89,864)
Minimum Cost Analysis	$2,044,293	$1,588,586

Figure 4. Cost failure range analysis comparisons

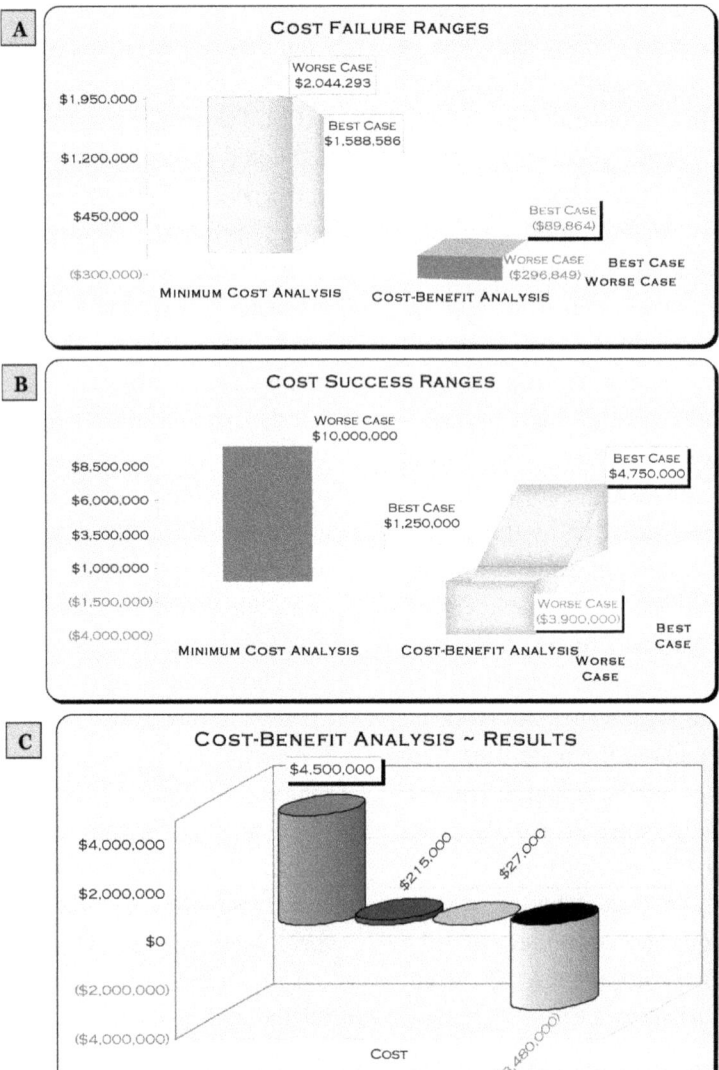

additional load increase expected (p=0.025). The Cost-Benefit Financial Model is the analysis method of choice for the utility industry. If the project falls outside of its rated tolerances during that period (p=0.05), the $90k payout is a small consolation cost based on the overall benefit ($4.5M) of the program. The analysis of the probable success factors (p=0.95) in the Cost-Benefit Model as compared with the Minimum Cost Analysis is depicted in Table 4 and panel B Figure 4.

The branch to achieve a minimum summer rating of 35 MVA by reconductoring multiple spans of Rice-Saratoga 69-kV line is the viable alternative in the Cost-Benefit Model ($4.75M). When the alternative is compared with the other prescribed choices, it clearly offers more "bang for the buck" ($4.5M) in today's dollars. The results are displayed in Table 5 and panel C Figure 4.

Table 4. Cost success ranges

	Worse Case	**Best Case**
Cost-Benefit Analysis	($3,900,000)	$4,750,000
Minimum Cost Analysis	$10,000,000	$1,250,000

Table 5. Alternative results

Rank	Cost	**Alternative Description**
A	$4,500,000	Re-conductor multiple spans of the 69-kV Rice-Saratoga line
B	$215,000	Addition of a second 345/161-kV transformer at Pleasant Valley
C	$27,000	Rebuild the Genoa-La Cross tap 161-kV line
D	($3,480,000)	Construct an additional 161-kV line from Pleasant Valley to Byron

ALTW-WPS Case Summary

Both analysis (Cost-Benefit and Low Cost) produced the same results ~ *Reconductoring of Multiple Spans of the Rice-Saratoga 69-kV Line (recommended)*. This option is the most suitable cost-benefit solution given the alternatives (Figure 5). A detailed ordinal/cardinal analysis is required to determine the exact cost estimates for a more accurate assessment of the work effort. Nevertheless, the best alternative available "today" to increase

the transfer capability from 0MW to 100MW for the ALTW-WPS Project is to Reconductor multiple spans of the Rice-Saratoga 69-kV line. The temporary solution will provide the desired capacity increase for seven years until other projects are designed and modeled in the system.

TITTABAWASSEE REACTOR INSTALLATION CASE STUDY

The General Synopsis

One of the main challenges for the planning engineer and/or project manager lies within the depths of the endeavor feasibility study alternative selection. Several methods are employed to determine the most viable option for the project in question. The Tittabawasse reactor installation project utilized the alternative evaluation method under uncertainty. The optimal solution objective function equated to a threshold of $42,250. As a minimization problem of man, material, methods, and the environment, (and several in-depth interviews with planning engineers & industry project managers), a probability distribution was established as a benchmark. The alternatives were developed as depicted in Table 6.

Figure 5. Decision tree analysis results: ALTW-WPS case study

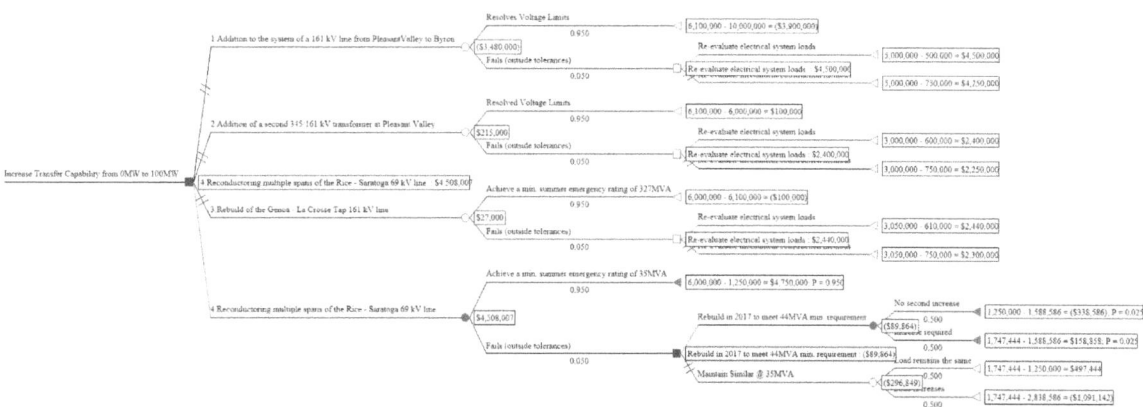

Table 6. Installation alternatives

Alternative:	Method:	Consequences:
A	Do Nothing:	Transformer Failure:
B	Wash transformer down during high loads	Eventual transformer failure
C	Install Reactor	Equipment extended life

The current method of "washing down the transformer" during high loads elevated the project to the summer critical list. The requirement was only recognized after several flow gate evaluation studies and recommendations by the regional independent service operator. Detailed analysis revealed an increase in manpower and material (subsequent costs) when the wash down was utilized. The method provided "temporary" relief for the high loads at extraordinary costs: The chance nodes in the decision tree (panel A of Figure 6) equated these costs to approximately \$68k/per occasion when the 60/40% man/material paradigm was utilized. Moreover, the cost prohibitive method exceeded the minimal objective threshold by \$47k. Further analysis revealed a cost reduction of \$25,805 in manpower as well as the \$62,305 in the material category. A catastrophic failure analysis was revealed in the "Do Nothing" scenario, and produced an equipment replacement cost in excess of \$261k. The decision tree produced the low-cost option (\$36k one time installation cost) via alternative C (Install Reactor). Although this is a temporary measure, its main objectives extend the life of the equipment, decreases, resources, and reduce costs. The potential reserve of \$384,066[2] is realized in the front-end decision analysis of the Tittabawassee Reactor Installation alternative methodologies.

The alternative analysis is classified as a "conditional" project, where it is typically ranked near the bottom of the utility priority scale. Nonetheless, the impact study elevated the endeavor to a higher level based on the overload status of the transformer. The situation is usually deemed as *normal* in the realm of the current infrastructure business practices. In fact, a specific overload condition subsisted on a 115-kV line in New Mexico where a pre-existing condition for a reactor installation exacerbated the overload in 2006. Several costly alternatives were considered (> \$25M) to mitigate the problem. Another case of overloaded conditions occurred on the Alberta Electric System in the late summer of 2009. The suspected overload condition caused equipment failure of two 138-kV lines and isolated the utility from the interconnection. Consequently, the frequency of the islanded area decreased to 59.729 Hz and tripped a neighboring interconnected transformer. The three hour outage affected nearly 5000 customers in the region. Equipment failure was the cause of yet another outage on October 30, 2009 that affected 11,000 customers in the Louisiana region. A bushing failure on the high side of the main transformer on the 115-kV tie was the major cause of the four-and-a-half hour event[3]. The probability cost distribution is shown in pane B of Figure 6. Seventeen percent of annual project proposals for an independent service operator were in the overload category (see Ordinal Project Ranking Chapter).

The Background

The purposes of the line reactors are to control power transients, provide line protection, and isolate loads at a smaller size and considerably reduced cost than its counterpart, the isolation transformer. The key to the device is its ability to provide isolation of the drive from the power system. The reactors are used to reduce voltage notching caused by the rapid change in current draw, and provide short-circuit protection, reduce spikes, sags, and harmonics. The increase in the reactor impedance will reduce the current in the line. The higher the frequency of the line, the current is lower. The DC resistance of the reactor

Figure 6. Decision tree analysis results and probability cost distribution: Tittabawassee case study

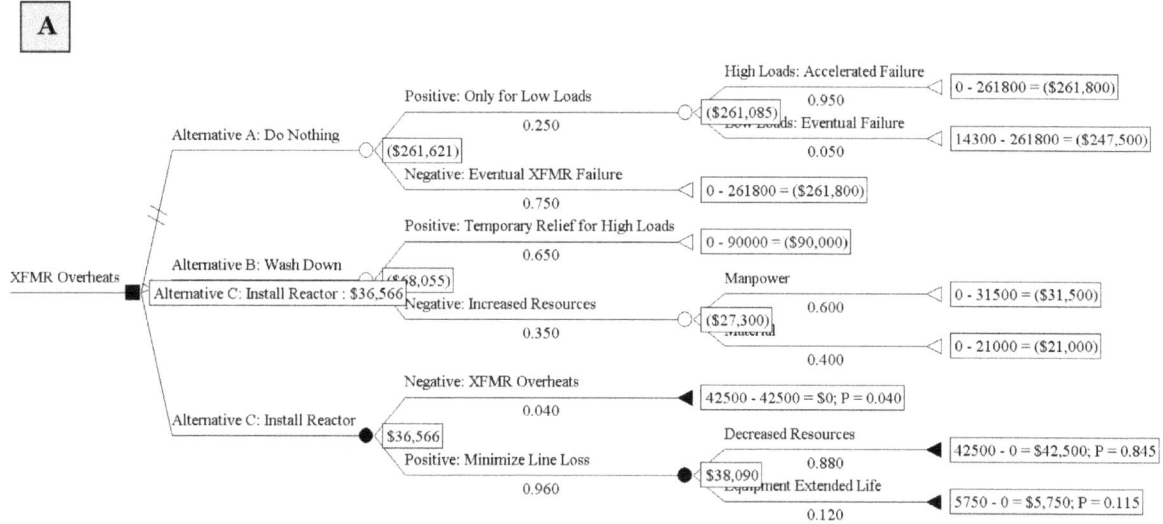

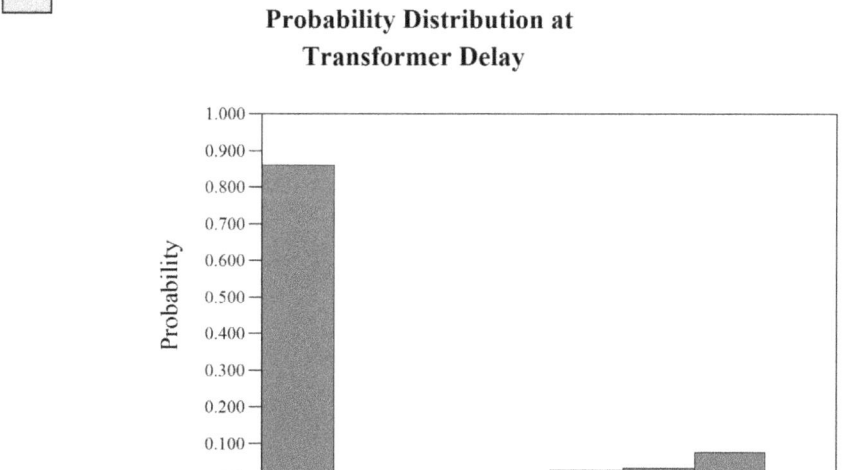

is very low by design so that the corresponding power losses are low.

An overloaded transformer is a unit that operates at or above the desired temperature. A unit that requires frequent corrective of on-condition maintenance (due to a variety of problems) is classified as a "bad actor" of lemon. The transformer

that has improper/inadequate design features and requires frequent maintenance is placed in the "poor design" category. The general solutions in these cases are a modification/re-design instead of routine maintenance. Moreover, a harsh environment is one of the tenets as a driver of maintenance for transformers. An environment where additional, atypical routine maintenance is required (i.e. a transformer near a coal plant that requires routine bushing cleaning, or one near a road that is heavily salted in the winter) must be considered when a maintenance program is implemented. In retrospect, the Tittabawassee transformer was subjected to several of the above conditions.

Tittabawassee Reactor Case Summary

Condition projects are allowed to "run-to-failure" mode if it does not jeopardize other critical circuits in the system. Obviously, the equipment had reached a *point of no return* for complete repair and was elevated to the replacement realm in the organization. If a reliability-centered maintenance program was implemented and utilized to its maximum potential, the transformer overheating problem would essentially be non-existent. The comparison of the regular inspection data with the predictive maintenance information was an obvious missing component of the evaluation scheme. Other adverse conditions (improper application, bad actor, poor design, slow/worn mechanisms) may be directly related to the demise of the transformer. If these options were explored during the early stages of the maintenance cycle, the minimum cost-benefit alternative analysis would not be implemented.

RAMSEY 230/115-KV TRANSFORMER UPGRADE CASE STUDY

The General Synopsis

Feasibility studies oftentimes neglect vital maintenance and outage issues (downstream and /or in parallel) associated with a major grid problem. A loop flow problem existed in the Ramsey (Forks Lake) project that was created by the addition of a new wind farm connection on the Langdon-Hensel 115-kV line. The problem was further exasperated during winter peak load conditions when little or no wind generation was produced. In the general pre-planning metric of the capital projects, the predicament (overload category) was discovered (new expansion category) *after* the connection of the wind farm. The equipment overload condition is the major segment of the metric. The Independent Service Operator studies unsuccessfully captured and modeled the current condition when the new expansion connection was implemented. This is a classic example of how one proposed solution created additional problems in the existing system. Detailed simulation studies, current equipment assessments, and experimental analysis were required in the electrical footprint to determine the electrical practicability of the wind farm during all conditions and seasons of the new technology. Moreover, the thermal overload study, flow-based analysis, network evaluations, and voltage investigations should have been assessed from different perspective(s) to avert the rework efforts of the integration. Some organizations are subjected to the "group think" syndrome where the entire "group" feels that they are on the correct path no matter the conditions. Generally, an external entity (or expert) will dislodge the group think mentality and set the order of sequences in place to successfully integrate the wind farm. The positive aspects of the group think condition engages the organization in unique approaches to learning, innovating, and inventing that builds on the diversity of the individuals with the common strategy as a roadmap. Nevertheless, the experts must parley the strategies to the learning organization to increase the capacity of the business unit and continue the implementation process for subsequent ventures. The costly re-work effort is essentially performing the analysis in reverse. This effort is evaluated in the decision trees to facilitate the loading relief effort and future equipment installation with the current wind farm.

The Alternative Analysis

The three major cost-effective, solution-based alternatives are compared (depicted in Figure 7) and outlined below:

Cost-Benefit Analysis Solution (examination of the "Best of the Worst" case scenarios) (Utility-based methodology):

The solution is based on the alternatives to eliminate the constraints.

1. The "Do Nothing" perceived low-cost solution ($0) allows the equipment to deteriorate at a faster pace with detrimental (loss) of $5.5M as a replacement cost.
2. Replace the 84 MVA 230/115-kV transformer to maintain capacity (p=0.80 based on industry standards) at a modest benefit (payoff) of $200k. *Recommended*
3. Stabilize the loop flow with the installation of an additional 230/115-kV transformer in parallel with the existing unit at a financial loss of $210k.
4. Minimize outage time of Forks Lake-Prairie (Grand Forks) 230-kV line to reduce loop flow. Strict coordination and procedures is required for maintenance work at a financial loss of $912K.

The Minimum Cost Analysis Solution (examination of the "Least Cost Option" scenarios) (Highest-to-Lowest methodology):

1. The "Do Nothing" yields a high-cost of $4.3M for the loop flow problem as the equipment deteriorates.
2. The replacement of the existing unit will cost and estimated $3.9M to maintain the capacity as the second highest alternative.

3. Installation of an additional 230/115-kV transformer in parallel with the existing unit as the third highest alternative of $3.43M.
4. Minimize outage time of Forks Lake-Prairie (Grand Forks) 230-kV line to reduce loop flow is *recommended* immediately as the least-cost option $2.77M. However, this is recognized as a short-term solution based on the alternative methodology and an absolute cost perspective.

The Cost-Effective Analysis Solution (Initial Cost vs. Overall Effectiveness payoff):

1. The "Do Nothing" alternative is not cost effective.
2. The replacement of the existing unit is not cost effective.
3. Installation of an additional 230/115-kV transformer in parallel with the existing unit is *recommended in this alternative scenario based on the given criteria* (very high cost vs. effectiveness payoff).
4. Minimize outage time of Forks Lake-Prairie (Grand Forks) 230-kV line to reduce loop flow is not cost effective.

Ramsey Transformer Upgrade Summary

The largest payback based on the major alternatives within the investigation elucidation brackets is derived from the Cost-Benefit (utility-based methodology) solution ~ Replacement of the existing 84 MVA 230/115-kV transformer (to 112 MVA unit) at an estimated cost of $3.9M. It is designed to maintain capacity and alleviate the loop flow problem at a modest overall benefit (payoff) of $200K.

Figure 7. Decision tree analysis results: Ramsey transformer upgrade case study

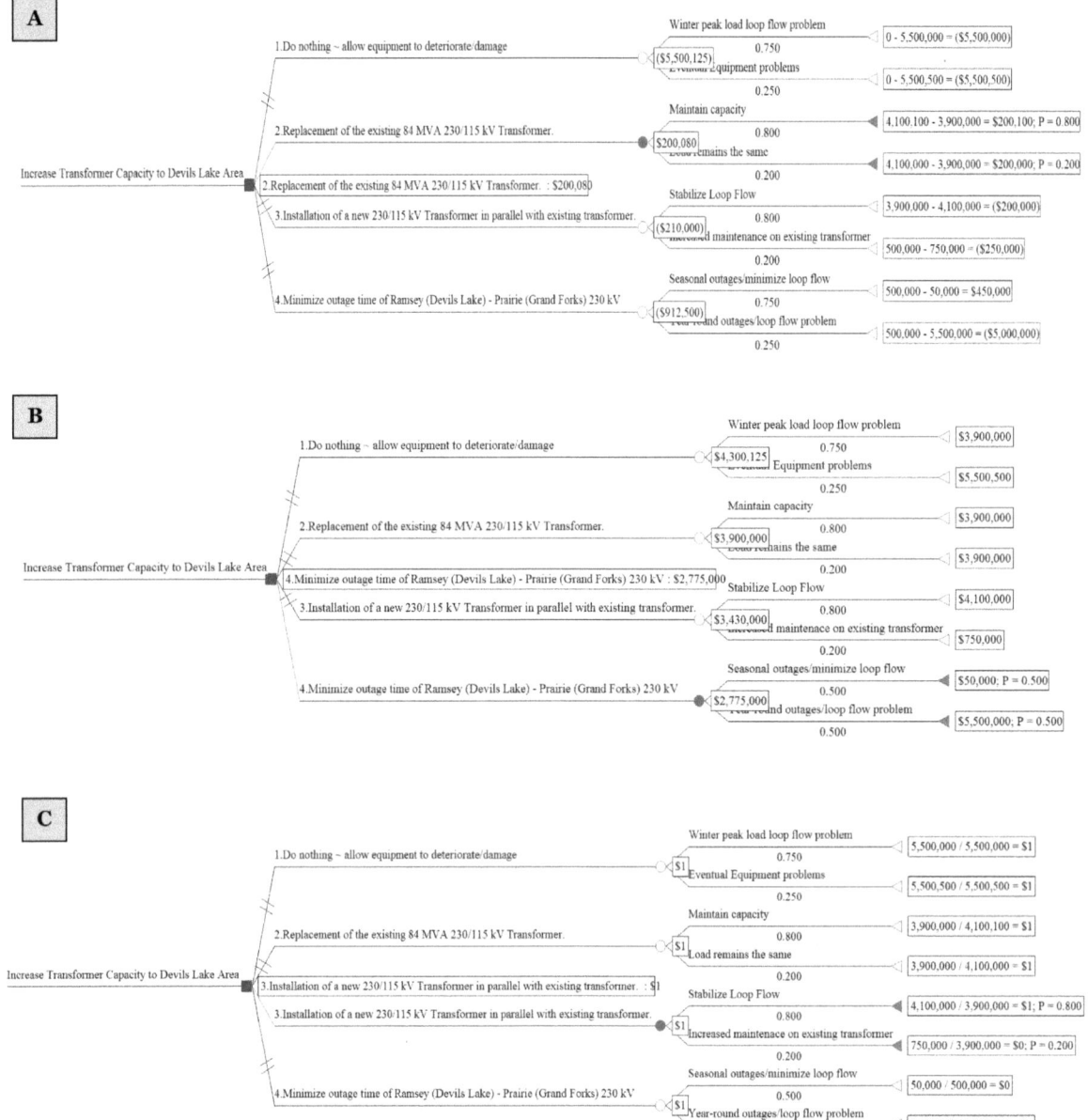

THE CIRCUIT BREAKER ALTERNATIVE ANALYSIS CASE STUDY

The economic analysis of a substation component is equally important as the alternative evaluation for an entire wind farm implementation. In order to upgrade the current infrastructure to accommodate generation expansion, a thorough assessment of the installed electrical equipment is warranted. For example, the economic life of a circuit breaker will assist the decision-maker in the *make versus buy* arena as the cost-effective alternative to replace or repair the unit based on future electric demand.

Analytical solutions span functional boundaries and focus on the entire business process. Consequently, many connected problems and decisions are relevant to a broad group (financial, supply-chain, field operations, electrical, mechanical, and analytical) of users. Conversely, the traditional operations research methodology (identifying, formulating, solving) is only applicable to a few decision-makers. The acquisition of new analytical skills and credentialing are viable topics for the operations research profession (Nestler & Leong, 2009). Consider a circuit breaker that has an internal cost of $63,500 with no salvage value after installation. The manufacture warranty will pay for all of the first year maintenance and repairs. The cost for the second year maintenance and repairs is $975 and will increase on a $975 arithmetic gradient in subsequent years. The useful life of the unit is determined based on the duty cycle and frequency of repairs/demand maintenance. Moreover, the twenty-five year project lifespan of the unit (as well as the interest rate) is utilized to determine its economic life in the system and annual value throughout the period. The data in panel A of Figure 8 illustrate the annual cost of capital recovery, the uniform cost of annual maintenance, and the total equivalent uniform annual cost of the unit. Where the maintenance cost encroaches on the present value of the circuit breaker is the strategic decision point within the repair or replace realm.

This particular point is the lowest dollar values ($13,490.53) on the total equivalent annual cost tend line. Consequently, the device has a projected useful life of fourteen years. The decision-makers evaluate the current system components and related historical data *before* the addition of newer technology is integrated. From a practical perspective, the penetrating questions must be asked to include:

- Can this particular circuit breaker (transformer, capacitor bank, etc) accommodate new generation in year X?
- Will the new technology necessitate replacement of all key elements in a substation upon approval and construction execution?
- What are the cost avoidance alternatives?

Obviously, the maintenance history plays an integral role in the decision-making process. If the wind turbine generator is implemented by year eight and no parallel electric flows, increased mechanical duty cycle, or other negation action is encountered to hasten the circuit breaker's replacement, then the life of the unit continues to its projected end (and maybe beyond). However, if the new technology exacerbates the present system and negatively affect installed equipment, the utility will replace a practically new circuit breaker well before its useful life is complete.

There are times when financial decisions are made to replace an old circuit breaker based on its current useful life. The decision point occurs when the equivalent uniform annual cost of the old unit is compared to the proposed circuit breaker in a specified time horizon as depicted in Table 7.

In this case, it is best to retain the old circuit breaker as the preferred alternative in the five year horizon scenario. The model incorporates the annual benefit of greater usefulness where the value for the present unit is zero and the new circuit breaker value is $850. The financial decision is based on the cash outlay of each unit and cost avoidance parameters. Decision-makers may elect to evaluate the limiting constraints, historical maintenance data, and the financial benefits of the equipment replacement characteristics when new generation technology is introduced in the current infrastructure.

Consider the case where demand maintenance is prevalent in the region. The replacement analysis

Figure 8. Economic life and overhaul cost graphs: circuit breaker case study

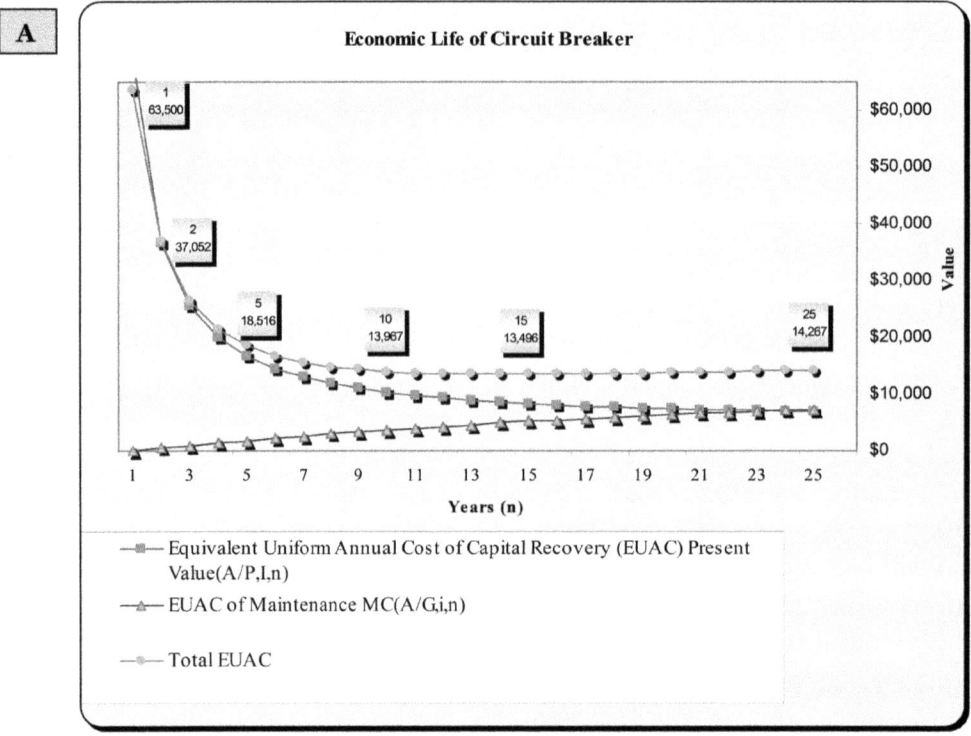

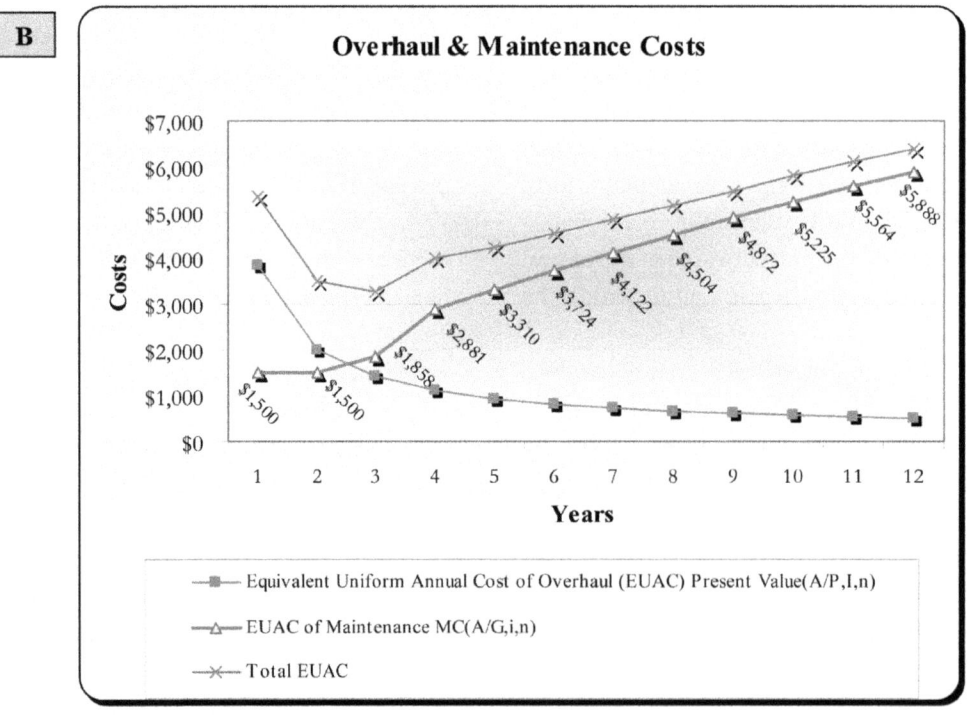

Table 7. Remaining life of old circuit breaker equal the useful life of a new circuit breaker

	Old Circuit Breaker	New Circuit Breaker
Present Value	$21,875.00	$63,500.00
Future Salvage Value	$0.00	$18,500.00
Annual Maintenance	$975.00	$975.00
Interest	10.00%	10.00%
Years	5.00	5.00
Annual Benefit of Greater Usefulness	$0.00	$850.00
EUAC	$6,745.57	$13,845.89

of the substation electrical equipment is again performed to determine whether to retain the current unit or replace it when immediate maintenance is imminent. An economic analysis is used to assess the old circuit breaker for a suitable replacement. A $3500 overhaul must be performed *now* if the unit is to be retained in service. The estimated maintenance expenses are $1500 for the next two years after which it is expected to increase on a $1000 arithmetic gradient. The present and future salvage value for the unit is zero, with twelve years remaining on the unit's projected useful life, at a ten percent interest rate. The alternatives to evaluate are:

- Replace the old circuit breaker *now*
- Retain the unit in its current state

The present value of the circuit breaker in this case is $3850 and the wind turbine technology is expected to integrate in the system within the next eighteen months. The data is illustrated in panel B of Figure 8. For the minimum equivalent uniform annual cost data, the remaining economic life of the current circuit breaker is three years (minimum total EUAC). Therefore, a new unit must be installed *now* as a suitable replacement in this case.

Circuit Breaker Alternative Analysis Summary

Current maintenance issues are important elements in the strategy sessions of the decision-making processes. Oftentimes, these concerns are inconsequential until the issues are elevated to a *fix-it-now or demand maintenance level* where the project is recognized on the priority list. The current assets in the electrical infrastructure re assumed to be in moderate-to-fair condition when expansion projects are proposed (i.e. renewable energy projects). The health of the system of the proposed generation integration must be thoroughly evaluated for anomalies. The vital area is equally important as the new technology implementation itself. This is why the periodic checks must be performed as scheduled to collect pertinent data for the system assets. It is also why the maintenance organization, project engineering, planning, operations, corporate as well as the vested stakeholders is privy to every facet of the proposed technology integration – in a large forum to group breakout sessions. Once the new technology is installed, a problematic circuit breaker with a minimum useful life will create excessive re-work and additional operational problems in the electrical region unless it is addressed in the appropriate evaluations and alternative analysis.

CONCLUSION

The point of reference of an exceptional design in a system is the lack of a crisis. Electricity is expected to flow effortlessly with the flick of a light switch in a home or industrial setting. The equipment behind the switch – the generators providing the electric source – is meaningless to the end-user. It is designed well and only noticeable in a crisis (blackout as a worst case scenario). *How* the electricity is produced and *what* source of energy the generator is using is equally important to the twenty-first century end-user as the design to implement renewable technologies.

The innocuous light switch is now a symbol for the energy efficient home as well as the power that provides to it. The incisive decision-maker (leader) can benefit from a profound equipment integration design if implemented correctly in the organizational structure. However, when the renewable energy technology is incorporated in the present infrastructure, it ought not to create additional problems in the electrical footprints. Effective leaders work to facilitate a different order of things and advocate change in an organization, especially in an area where the aging infrastructure decries system upgrades and newer technologies. The incisive leader focuses on the new and emerging technologies, the dimensions of sustainability, and environmental impact of the proposed generators during the equipment implementation while evaluating the alternatives. The importance of assumption testing by positional leaders is a basic requirement to mitigate potential problems during scenario analysis of the strategic phases. It is here where ideas are tested in a simulated and/or experimental practice to assess the validity of the scenarios and the capabilities/limitations of the proposed technology. This approach is applicable across disciplines (energy, heavy construction, transportation, and manufacturing) where supposition testing is indispensable. Nevertheless, the decision-makers must not be too consumed with the simulation itself (programming, plausibility of input scenarios, mechanism span and control, etc) and lose sight of the problems the proposed technology is designed to solve. The integrated resource plan, for example, requires detailed econometric inputs for viable scenario analysis. Yet these conditions ought to encompass realistic situational analysis to accomplish an accurate forecast for load growth and potential power supply availability in the region. Mitigation of the creative tension (minimizing the gap between the vision and the reality) is crucial during theses evaluations. The astute executive leader is influential and has the unique ability to convert local approaches to broader organizational strategies. Included in these

policies and procedures are distinct environmental, biodiversity, and performance criteria to assist decision-makers with organizational sustainability initiatives of the electrical system integration efforts. Conversely, organizations must not solely rely on the utilization of "experts" to solve difficult problems, as self-reliance is essential. The subsequent negative effect of this action reveals a company that cannot expand its capacity with the renewable energy efforts. By always relying on experts, the organization will become no smarter when faced with difficult situations. The performance of the proposed technology is expected to pass through a series of tests and standards, as well as adhere to the electrical limitations of the grid. The cost-effectiveness of the renewable generation must provide substantial eco-friendly benefits to justify the capital expenditures. Only when the renewable energy system is designed well and seamlessly integrated in the grid to minimize the fossil-fuel dependency, the potential crisis will be averted.

REFERENCES

Agor, W. (1984). *Intuitive management: Integrating left and right brain management skills.* Englewood Cliffs, NJ: Prentice-Hall.

Atkinson, G. (2000). Measuring corporate sustainability. *Journal of Environmental Planning, 43*(2), 235–252.

Axelrod, R., & Cohen, M. (2000). *Harnessing complexity: Organizational implications of a scientific frontier.* New York, NY: Basic Books Publishers.

Brown, G. M., Szabados, B., Hoolbloom, G. J., & Poloujadoff, M. E. (1992). High-power cyclo-converter drive for doubly-fed induction motors. *IEEE Transactions on Power Electronics, 39*(3), 230–240.

Budget of the U.S. Government. (FY 2013). *Performance and management section*. Analytical Perspectives Volume. Retrieved 29 May, 2012, from http://www.whitehouse.gov/omb/budget/analytical-perspectives

Chesbrough, H. (2011). *Open services innovation: Rethinking your business to grow and compete in a new era*. San Francisco, CA: Jossey-Bass. doi:10.1007/978-88-470-1980-5

Chinchilla, M., Arnaltes, S., & Burgos, J. C. (2006). Control of permanent magnet generators applied to variable-speed wind energy systems connected to the Grid. *IEEE Transactions on Energy Converters*, *21*(1), 130–135. doi:10.1109/TEC.2005.853735

Cokins, G. (May-June 2012). Why do large companies fail? *Analytics Magazine*, pp. 16-19.

Erickson, W. P., Johnson, G. D., Strickland, D. P., Young, K. J. Jr, Sernka, K. J., & Good, R. E. (2001). *Collisions with wind turbines: A summary of existing studies and comparisons to other sources for avian collision mortality in the United States*. Washington, D.C.: Western Ecosystems Technology, Inc. National Wind Coordinating Committee Research Document.

Figge, F., Hahn, T., Schaltegger, S., & Wagner, M. (2002). The sustainability balanced scorecard – Linking sustainability management to business strategy. *Business Strategy and the Environment*, *11*(5), 269–284. doi:10.1002/bse.339

Gojanovic, T. (April 2012). Theory of evolution, *Quality Progress*, pp. 12-13.

Grauers, A. (1996). Efficiency of a three wind energy generator system. *IEEE Transactions on Energy Conversion*, *11*(3), 650–657. doi:10.1109/60.537038

Grumbine, R. E., & Xu, J. (2011). Mekong hydropower development. *Science*, *332*, 178–179. doi:10.1126/science.1200990

Hall, E. (2007). *Beyond culture*. New York, NY: Anchor Publisher.

Hofmann, W., & Okafor, F. (2001). Optimal control of doubly-fed full controlled induction generator with high frequency. *Annual Conference of the IEEE Industrial Electronics Society*, Vol. 2, (pp. 1213-1218).

Levine, E. S. (March-April 2012). Challenges for public sector analytics. *Analytics Magazine*, pp. 27-29.

Liberatore, W., & Luo, W. (2012). The analytics movement: Implications for operations research. *Interfaces*, *40*(4), 313–324. doi:10.1287/inte.1100.0502

Malcolm, S. A., & Zenios, S. A. (1994). Robust optimization for power system capacity expansion under uncertainty. *Journal of Operations Research*, *45*(9), 1040–1049.

Matas, J., Castilla, M., Gueirrero, J. M., Garcia de Vicuna, L., & Miret, J. (2008). Feedback linearization of direct-drive synchronous wind turbines via a sliding mode approach. *IEEE Transactions on Power Electronics*, *23*(3), 1093–1103. doi:10.1109/TPEL.2008.921192

Morren, J., de Haan, S. W. H., Kling, W. L., & Ferreira, J. A. (2006). Wind turbines emulating inertia and supporting primary frequency control. *IEEE Transactions on Power Systems*, *21*(1), 433–434. doi:10.1109/TPWRS.2005.861956

Muller, S., Deicke, M., & De Doncker, R. W. (2002, May-June). Doubly-fed induction generator systems for wind turbine applications. *IEEE Industry Applications Magazine*, *8*(3), 26–33. doi:10.1109/2943.999610

Nestler, S., & Leong, J. (January-February 2009). O.R. credentialing. *OR/MS Today Magazine*, pp. 32-36.

Ogilvy, J. (2002). *Creating better futures: Scenario planning as a tool for a better tomorrow*. New York, NY: Oxford University Press.

Polinder, H., Vanderpijl, F. F. A., De Vilder, G. J., & Tavner, P. J. (2006). Comparison of direct-drive and gear generator concepts for wind turbines. *IEEE Transactions on Energy Conversion, 21*(3), 725–733. doi:10.1109/TEC.2006.875476

Ramshaw, R., & van Heeswijk, R. G. (1990). *Energy conversion: Electric motors and generators*. Philadelphia, PA: Saunders College Publishing.

Rodriguez-Amenedo, J. L., Arnalte, S., & Burgos, J. C. (2002). Automatic generation control of a wind farm with variable speed turbines. *IEEE Transactions on Power Electronics, 17*(2), 279–284.

Senge, P. M. (2005). *Presence: An explanation of profound change in people, organizations, and society*. New York, NY: Doubleday/Currency.

Senge, P. M. (2006). *The fifth discipline: The art and practice of the learning organization*. New York, NY: Doubleday Publication. doi:10.1002/pfi.4170300510

Wack, P. (1985). Scenarios: Uncharted waters ahead. *Harvard Business Review*, 72.

Zhang, L., & Watthanasarn, N. C. (1998). A matrix converter excited doubly-fed induction machine as a wind power generator. *7th International Conference on Power Electronics and Variable Speed Drives,* Vol. 456, (pp. 532-537).

ENDNOTES

1. Source: Fact Sheet, *Big Eddy-Knight Transmission Project Update*. Bonneville Power Administration DOE/BP-4131, December 2009
2. Based on Reduced Costs: High Loads=$219,550 Low Loads=$8550 XFMR Fail=$20,106 Temp Relief=$47,750 Manpower=$23,805 Material=$62,305
3. 2009 NERC Disturbance Report

Chapter 3
Process Evaluation and Cost Analysis Mechanism for Equipment Installation Ventures

ABSTRACT

This chapter explores how the lean quality principles and green initiatives are applied in capital project equipment installation of electric substations and line rehabilitation from the initial feasibility studies to the plant-in-service phase. Equipment installation process predicaments induce cost overruns and unnecessary extended electrical outages as well as extraneous strain on the bulk transmission system. Demand side management issues are best mitigated through astute planning of the power supply resources, load curtailment initiatives, and transmission flow contingency schemes.

OVERVIEW

Electrical utilities and process industries are challenged to find methods to reduce costs from the raw material conversion process to the end-user. The continued cost overruns in electrical condition, capacity, and reliability project installation categories are an area where cost control methodologies are beneficial. Stricter regulations and environmental stewardship are the impetus for the continuous improvement efforts in the capital project arena to maintain consistent and reliable sources of power and minimize cost. This chapter explores how the lean quality principles and green initiatives are applied in capital project equipment installation of electric substations and line rehabilitation from the initial feasibility studies to the plant-in-service phase. Equipment installation process predicaments induce cost overruns and unnecessary extended electrical outages as

DOI: 10.4018/978-1-4666-2839-7.ch003

well as extraneous strain on the bulk transmission system. The overall expenditures dedicated to capital-intensive projects are generally randomly prioritized as other rehabilitation and maintenance ventures take precedence. In fact, revenue and expense data compiled from one hundred and eighty investor-owned electric utilities (Table 1) exemplify the declining operating income relative to the rising expenses.

Further analysis of the data reveals some interesting relationships. The operating revenue equates to the operating expenses less net operating income. Although the utility operating revenue increased during the period, the net operating income decreased precipitously. The entities are allocating excess expenditures in various demand categories to maintain the infrastructure while retaining less (from 17% to 11%) of a profit margin. The relationship is illustrated in Figure 1.

In terms of relative expenses and nominal dollars, the expenditures allocated for the transmission category increased by one percent whereas the maintenance category experienced a two percent decrease in the expense budget. The boost in the transmission expenditures demonstrates an escalating requirement to reinforce the present-day infrastructure. Supplementary scrutiny of the data revealed an increase in production costs and relatively flat fuel expenditures for the period. These data is illustrated in Figure 2.

Additionally, the customer peak load is increasingly higher with each passing year. The various electronic devices households utilize continue to strain the electric grid. The interconnection of electric vehicles and other fuel efficient technologies necessitates a rapid upgrade in the aging electrical infrastructure. Nevertheless, demand side management issues are best mitigated through astute planning of the power supply resources, load curtailment initiatives, and transmission flow contingency schemes. The summer peak load, for example, will reach its zenith during the hottest portions of the day when demand for air conditioning, large appliances, and hot water tanks are performing at their maximum levels. Specific low-cost programs offered by the local utilities allow for electrical interruptions in times of high demand to reserve power for the grid and avoid shedding load. Since most of the electrical generators are operational during this summer critical period, as well as expensive standby gas turbine peaker units, the potential for an unscheduled outage is extremely high. Moreover, the cost of interconnected emergency power as a replacement of lost generation is highly competitive for such a reserve. The hydroelectric pumped storage facility is a viable alternative resource for short-term regulation as well as the wind turbine generators (depending on the intensity of the wind) and, to a lesser degree, the photovoltaic solar resource.

Table 1. Selected revenue and expense statistics for major U.S. investor-owned utilities: 1995 to 2007[1]

Item	1995	2000	2003	2004	2005	2006	2007
Utility Operating Revenue	199,967	235,336	226,227	240,318	267,534	277,142	282,875
Utility Operating Expenses	165,321	210,324	197,459	207,161	238,590	247,170	252,216
Production	68,983	107,352	96,181	104,287	121,058	128,016	128,914
Cost of Fuel	29,122	32,555	26,476	28,678	36,161	38,158	42,178
Transmission	1,425	2,699	3,585	4,519	5,687	6,185	6,095
Maintenance	11,767	12,185	11,141	11,774	12,058	12,879	13,675
Other Expense	54,024	55,533	60,076	57,903	63,626	61,932	61,354
Net Utility Operating Income	34,646	25,012	28,768	33,158	28,944	29,972	30,659

(In millions of nominal dollars (199,967 represents $199,967,000,000))

Figure 1. Operating expenses relative to net operating income

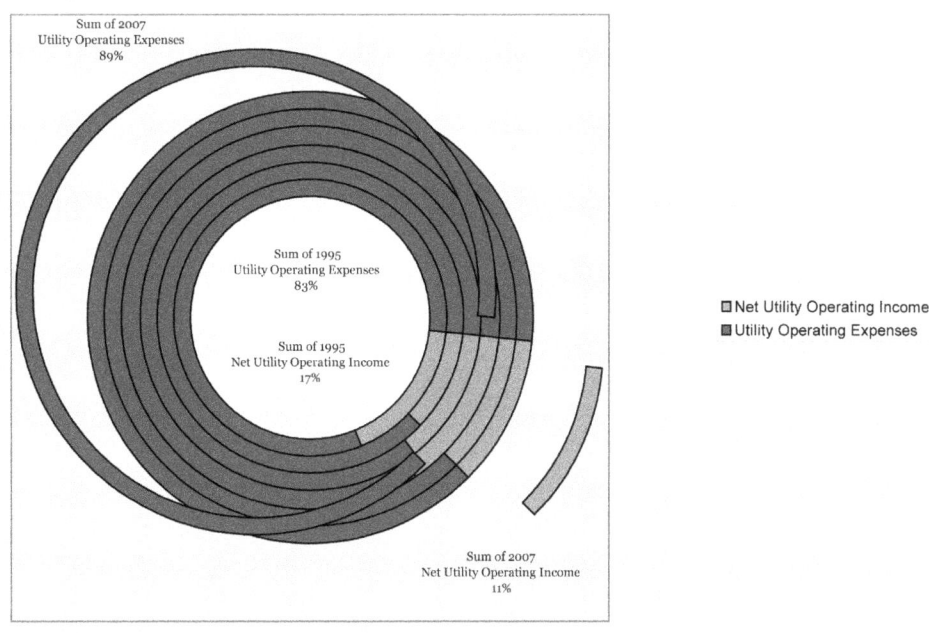

Figure 2. Operating expenses breakdown

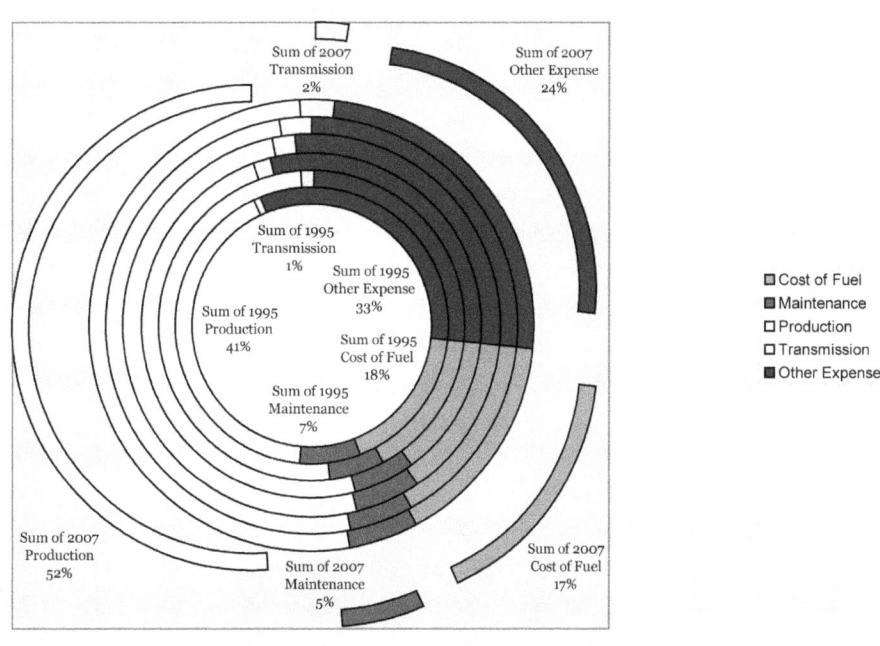

Nonetheless, there is adequate generation to meet the current requirements of the system, albeit it is tenuous and problematic on its best days. The constraints are characteristically instituted within the transmission and distribution system of the highly-interconnected electrical grid.

The system reliability function is a major tenet of the North American Reliability Corporation (NERC). The organization implemented compliance monitoring and enforcement programs as a positive initiative to publicly disclose recalcitrant issues in the utility industry. The entity provides a systematic response to a scheme disturbance, complaint, or possible violation of a reliability standard identified by any other means. Unscheduled outages and system disturbances are thus recorded and acted upon either during the event or after-the-fact. Examples of the typical grid constrained problems are outlined below:[2]

Weather Related

- On June 25, 2009 at 15:16 in the Texas region where high electricity demand was expected to break the all-time peak record based on a heat advisor in the area, a public appeal was issued. The reliability coordinator issued a "Power Watch – conservation needed" notice as a courtesy. The power system did not proceed to an emergency condition where a major generation inadequacy was probable. Additional public appeals in the same region in subsequent months (July 8, 2009 at 13:30 and August 5, 2009 at 15:00) experienced similar conditions with restricted generation reserves available because of several unplanned outages.
- The opposite of the heat-related issues are public appeals and warnings on behalf of cold weather. On December 7, 2009, the California Independent Service operator issued a grid warning notice at 18:05 as the forecasted load was running 800 MW to 1000 MW over and the loss of 300 MW due to an outage of a 500-kV inter-tie line. Operational reserves were utilized to recover the system within seventy-five minutes.

Acts of God Related

- On September 19, 2009, at 15:55, the Comisio'n Federal de Electricidad in the Western Electric Coordinating Council territory experienced an earthquake (5.1 magnitude on the Richter scale) tripping all 549 MW of transmission line generation and 14 MW of load at a substation. The loss of generation created a system frequency excursion which was returned to normal within nine minutes. The spinning reserves and cold start units were immediately utilized and pressed into service. Additionally, 200 MW was requested from neighboring utilities to supplement the reserves. A total of 110 MW of firm load was shed. The restoration activities were completed within forty-two minutes.

Equipment Related

- On July 4, 2009 at 12:52, the Alberta Electric system operator's generating station "A" tripped based on unknown causes. Consequently, 364 MW was immediately dropped from the system which decreased the frequency to 59.926 Hz (a violation of the frequency excursion policy greater than -0.05 Hz). The electric system recovered within fifteen minutes utilizing internal and external reserves as the interconnection was not affected.

The unscheduled loss of generation based on unknown reasons is extremely difficult to process from an operational standpoint. Replacement power or load shedding schemes is quickly enacted to make up for the loss. Nevertheless, the root cause of the problem (weather, animal contact, human error, equipment malfunctions) is typically revealed well after the event. Ideally, the system recovery time is within ten minutes after the disturbance to maintain the Area Control Error. However, this situation is usually not attainable shortly after the unscheduled outage as the spinning reserves, cost-prohibitive emergency interchange power, and the high-cost/high-emission quick start units are rapidly replaced with additional baseload generation ramped up to the required level. Once the system is eventually stabilized, the reliability and robustness of the electrical grid migrates back to the economic generation dispatch philosophy. The intent of the front-end analysis is to examine the equipment installation process and provide decision support mechanisms to evaluate parallel activities integration that affects the electrical operations of the grid. It is designed to alleviate potential cascading operational quandaries *before it happens* by utilizing thorough investigative process methodologies for electrical equipment implementation. Summer critical capital projects are seamlessly placed into operation in the system before the need arise. The consistent factor is correlated to the system peak load conditions and its ability to cope with system changes during unscheduled events. The electrical capability, non-coincident peak, and capacity margin data is illustrated in Table 2 below.

The summer capacity margin (cushion) for 1980, for example, is 131, 179 kW (558,237 – 427,058), and has hovered around this figure throughout the period. The summer peak load increased by 7,688 kW in 2008 from the previous year and well over 362,857 kW since 1980. Ideally, there must be a sufficient capacity margin to withstand the loss of the largest generator on the system. The capability factor at the time of the summer/winter peak loads are proportional to the non-coincident peak loads as the electrical infrastructure continues to reach its constraints and expand. The data is illustrated in Figure 3.

Conversely, the winter non-coincident peak load reached its maximum in 2008 (656,989 kW)

Table 2. Selected capability, peak load, and capacity margin: 1980 to 2008[3]

Year	Summer Peak Load Capability (1000 kW)	Winter Peak Load Capability (1000 kW)	Summer (NCP)	Winter (NCP)	Summer Amount Capacity Margin	% of Capability (S)	Winter Amount Capacity Margin	% of Capability (W)
	Capability at the time of ---		Non-Coincident Peak Load (1000 kW)		Capacity Margin			
					Summer		Winter	
1980	558,237	572,195	427,058	384,567	131,179	23.5	187,628	32.8
1985	621,597	636,475	460,503	423,660	161,094	25.9	212,815	33.4
1990	685,091	696,757	546,331	484,231	138,760	20.3	212,524	30.5
1995	714,222	727,679	620,249	544,684	93,973	13.2	182,995	25.1
2000	808,054	767,505	678,413	588,426	129,641	16.0	179,079	23.3
2005	882,125	878,110	758,876	626,365	123,249	14.0	251,745	28.7
2006	906,155	899,551	789,475	640,981	116,680	12.9	258,570	28.7
2007	915,292	913,650	782,227	637,905	133,065	14.5	275,745	30.2
2008	929,338	927,781	789,915	656,989	139,423	15.0	270,792	29.2

(558,237 represent 558,237,000 kW)

Figure 3. Summer peak load and capabilities

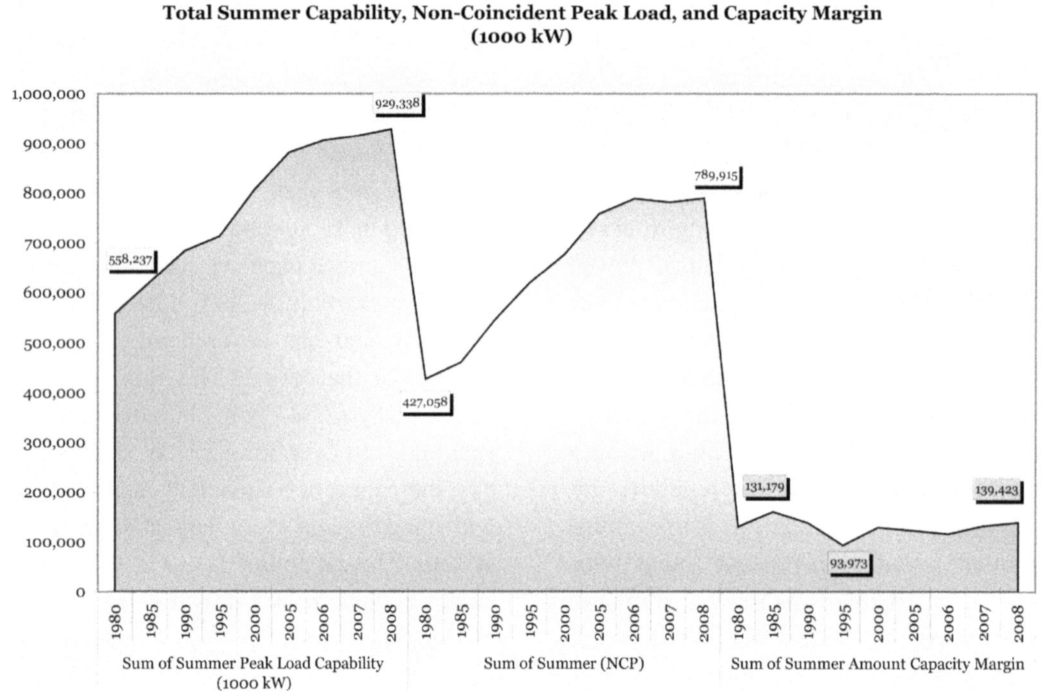

Total Summer Capability, Non-Coincident Peak Load, and Capacity Margin
(1000 kW)

with a much larger capacity margin (Figure 4). Nevertheless, the stress on the electrical grid is evident and requires strategic reinforcement with efficient technologies and innovative equipment implementation techniques.

Decision theory, alternative evaluation, and process optimization disciplines are required as a front-end analysis mechanism to ensure constructive integrity in the electrical infrastructure. The chapter investigates the equipment system integration problem specifically encountered by electrical utilities, process industries and the end-user (society as a whole). The major focus of the venture examines the optimization methodologies germane to the power generation industry. The project evaluation and analysis study encompasses detailed project cost estimates, capital project assessments, and process flow investigation for investor-owned utilities, municipalities, independent power producers, and selected government entities. The value-added study is de-

signed as a decision support mechanism for senior management in front-end analysis of high expenditure projects (>$2M) that reinforces the utility grid. The venture purports financial accountability, green stewardship, and detailed assessments while maintaining a safe and high quality work process to protect the current infrastructure. The project is expected to specifically focus on principal summer critical ventures, process analysis, and project alternative assessments to supply a comprehensive analysis of the assets for front-end decision support. The evaluation is intended to exploit vulnerabilities and identify potential benefits of the plan.

PROBLEM DEFINITION

A detailed understanding of the fundamental business requirements of an organization is required to adequately assess a capital system integration

Figure 4. Winter peak load and capabilities

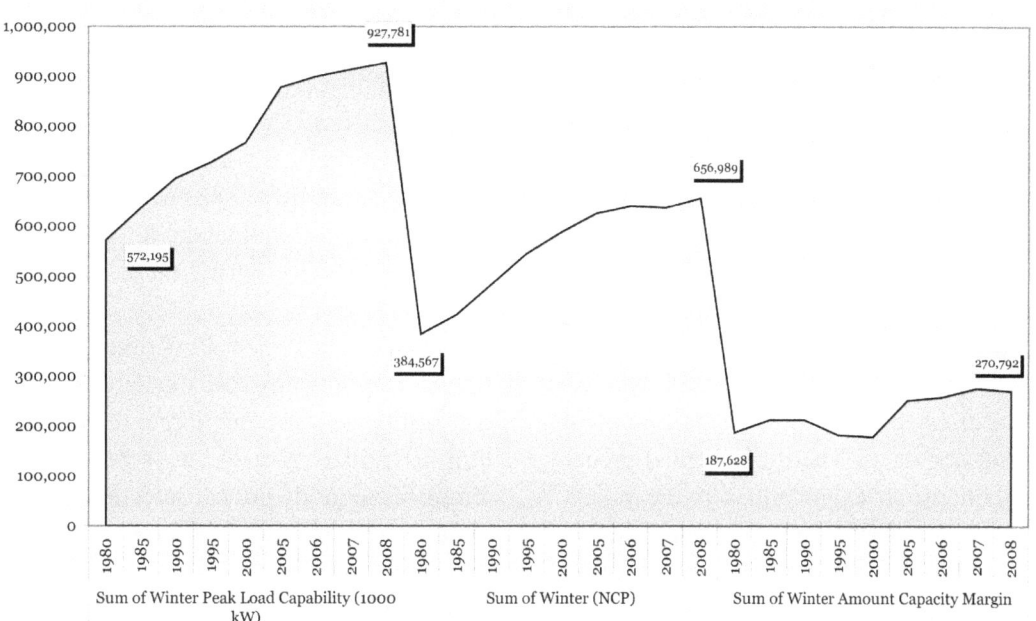

Total Winter Capability, Non-Coincident Peak Load, and Capacity Margin
(1000 kW)

problem. The thorough evaluation of cost over-runs within the project management organization based on the finite capital budget within the Asset Management Department is consumed with "out-of-scope" projects well behind schedule. Several resources are displaced in various regions across the three state footprints with no systematic procedure in place to control the cost over-runs. The project manager's ad hoc procedures lacked consistency within the organization. The implementation of the Process Evaluation & Cost Analysis Mechanism for Equipment Installation Ventures in capital planning to control the individual project budget and corresponding schedule is necessitated. The guidelines include checkpoint methodology, responsibility metrics, historical actual spend data, forecasting logic, and process management flow diagramming (Besterfield, et al., 1999). Project managers not only must rely heavily upon the traditional on-time, on-budget, and on-quality performance measures, but also

with the added on-strategy dimension central to managing plan's success and to alleviate leadership challenges. Many leadership tasks relate to developing a vision of the scheme's outcome that is practical, yet capable of mobilizing and motivating team members to accomplish the project's goals and objectives (Breyfogle III, et al., 2000). The management vision engages stakeholders who are not actively involved in the project; it also inspires them to maintain their support over the project's timeline. The solution-building negotiation approach to defining the scope of a project, and then clearly communicating this to the project team and other stakeholders defines a strategy for realizing the vision, and translating the strategies into operational plans and results. The core problem is the insufficiency of instability of strategy to properly develop and express project vision connected through measurement to tangible business outcomes. The organization links planned outcomes to their corporate strategy using a measurement framework, referred to as performance management, and is common within corporations (Scholtes, Joiner, & Streibel, 1996). The implementation of these formal stringent guidelines provides clear project vision as well as comprehensible goals and objectives. It also actively involves the executives and sponsors in the venture. The alignment of the project manager's resources via systematic tracking, contingency fund management, construction overheads, mitigation strategies, and scheme status communication generates profound results in the overall operating budget. Additionally, the project manager ensures that the accounting and planning departments are provided with accurate equipment data, costs, and plant-in-service dates. The implemented procedures will impose strict guidelines for the project manager to follow that includes a final technical report (significant results, discussion of objectives and deliverables, examples of progress), a financial report (final invoices, cash flow), and resource performance evaluations (technical difficulties, feedback reports, activity verification, lessons-learned) (Thomas, 1975). Moreover, the financial and scheduling optimization instrument allows the project manager to utilize essential monitoring and control techniques during equipment installation. The intent of the investigation will provide a systematic approach to project management and aligned the company's objectives with the on-budget, on-quality, on-schedule, and on-strategy philosophy (Harry & Schroeder, 2000).

The current capital project selection & cost estimation process varies immensely across the electric utility sector. Some companies employ a regional discipline methodology to select annual capital projects on one end of the spectrum whereas other investor-owned utilities utilize an indiscriminate "fire drill" approach on the other end of the realm. The electric utilities typically absorb the feasibility costs in the project's budget with the selection based on the Reliability Projects (expenses incurred to improve/reinforce the reliability of the infrastructure assets), Capacity

Projects (new load and system reinforcement associated with ventures required to improve, relieve, or correct an existing or projected voltage/thermal condition) and Condition Projects (costs associated with replacement of equipment because of the inability to obtain parts or obsolesce). The work is scheduled around an arbitrarily due date of June 1st and considered a "summer critical project," since most substation and transmission/distribution equipment is prohibited from removal of service during these peak electrical billing periods.

The Current Project Initiation Process is deficient in uniformity and standardization across the utility sector. The larger investor-owned utilities utilize spreadsheet software program templates to guide project selection and cost estimation. The methodology rarely reconcile with the plant-in-service (final costs) criteria upon project completion. The actual expenditures of the ventures are not properly tracked, evaluated, and/or analyzed. In fact, a contingency fund is established at the outset of the project in the range of ten to thirty-five percent. The material & labor costs are "generally" calculated in these estimates with construction/supervision overheads as well as "Allowable Funds Used during Construction (AFUDC)." These calculations are woefully inaccurate (order of magnitude of 75%~90%) for the majority of transmission & distribution projects. The non-value added labor costs, transportation, storage, engineering expenditure review, were typically neglected in the final project financial tally. Moreover, the investor-owned utility companies direct various unallocated costs in a variety of financial "buckets" which may or may not be associated with the project.

The Feasibility Studies and Pre-Assessments of capital projects are deficient in standardization. Less than five percent of the major investor-owned utilities participate in a structured benchmarking study in the industry. Fewer companies partake in a formalized, detailed project evaluation and assessment studies (Feigenbaum, 1991). The bulk of the project cost/schedule estimation tasks are derived from the company's internal engineering planning departments and/or contracted Architectural & Engineering (A&E) companies. The scope of the pre-assessment work provided by these shops is depicted below:

1. Field Surveys
2. Turnkey proposal preparation ("General cost estimates," layouts, Sub-contractor list, Request for Proposals, etc)
3. Design compliance (civil-structural Review, physical, electrical, Schematic wiring, etc)
4. Construction management services (Field audits)

The companies that elect to participate in the benchmarking surveys seize the added advantage to evaluate/select projects based on the historical data (i.e. outage data to determine the worst performing circuits). An electrical utility will benefit from all of these services if the focus provided a unitized theme for project selection, cost estimates, and schedule evaluation. The fragmented services and evaluation often leads to the decision-maker's inadequate judgment of project selection based on erroneous costs estimates and measurements.

METHODOLOGY

The proposed project is intended to provide front-end solutions to the continued installation problems encountered in the field. The analysis aim is to avoid major safety and non-quality issues during critical system integration processes via simulated conditions and viable alternatives for company executive review (Schermerhorn Jr., 1993). The potential for electrical blackouts based on the assigned resource inattentiveness to detail is crucial for the investor-owned utilities/process industries as well as society as a whole. Intensive project evaluation via this method is essential for the substation capital equipment installation, transmission line upgrades, and environmental ventures. The objective of the plan is to thoroughly re-evaluate the alternatives, resources, and available technologies specifically during the summer critical construction period. The existing problems specific to the electrical utility business that create delays typically involve the conditions outlined in Table 3.

These conditions are not inclusive; several derivatives of these conditions create bottlenecks throughout the entire electrical grid.

Table 3. Typical conditions

Problem:	Current Alleviation:	Potential Consequences:
Transformer Overheats	Crew "wash-down" during summer critical peak periods	Eventual equipment failure: Power outage
Incorrect phasing during installation	Field review & design clarifications by company or contract Engineering	Regional outage ~ one transmission line carried two schools and commercial sites.
Crew Encounter Endangered Species	Work stoppage ~ re-evaluate parcel, Right-of-Way, and environmental issues	Load exasperation, equipment fatigue, and eventual blackouts
Incorrect Parts and/or Material Preparation	None	Load exasperation, equipment fatigue, and eventual blackouts

The Project Approach

The unique differential advantage the study offers is based on a five point approach (with executive recommendations) as outlined below:

1. Project Alternative Evaluation (Decision-Support Software)
2. Portfolio Decision Analysis (Project Ranking Software)
3. Process Evaluation (Labor- Assessment Software)
4. Detailed Cost Analysis (Financial Forecasting Software)
5. Schedule Assessments (Probability Software)
6. Recommendations

The entire project evaluation process proposed is grounded in the "resource tracking" and "activity-cost" methodologies. The paradigm encompasses intense project analysis of the activity of individual cost elements, resource utilization, and project gap-fit evaluation. It leverages the cutting edge technologies of industrial engineering to include quality/financial control, risk assessments, and cycle-time evaluation (Lamprecht, J.L, 2005). The project evaluates the elements under uncertainty as part of the feasibility study and provides executive decision support. It is a valuable tool for "front-end" analysis *before* the project's funding, and aids the company's prioritization process of all (condition, reliability,

capacity segmentation) capital ventures. The study project's variables are given considerable weight regardless of cost estimates (Juran, J.M. 1999). Once the projects are vetted at this stage, the detailed process evaluation is conducted. The key performance indicators are essential to the intended project evaluation strategy. This stage provides accurate assessments of construction/ material installation to include contingency delay costs and related consequences/penalty factors. A scenario analysis (what-if) is provided that will examine the project "crashing" (increased resources) cost/schedule, resource leveling, and material delay to eliminate the guesswork.

DISCUSSION

High-Level Process

An example of the methodology is discussed in this section. The original transformer estimates were investigated in a post-mortem study of an installation process (Figure 5). The results were then compared with the tangible invoices/disbursements and the actual final asset in-service costs.

A preliminary financial outlook illustrates a $299k total system labor cost reduction ($737,347.85 actual cost - $438,567 simulated results data) in *only* the transformer evaluated process with nominal resources. A detailed exploration of the optimized processes and resource-

Figure 5. High-level transformer installation process

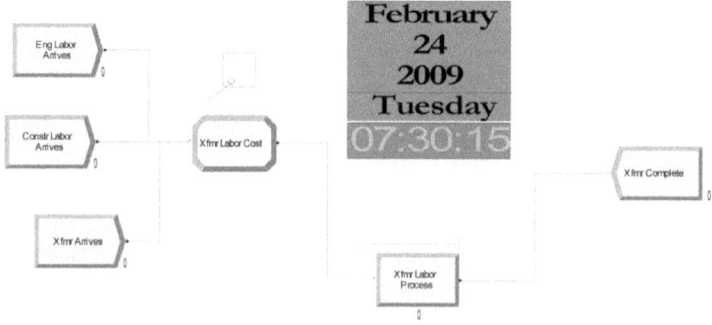

leveling will further reduce the cost. The most intriguing portion of the study investigated the various cost scenarios ("what-if analysis") and its affect on specific outputs. It is a simulated form of stimulus-response testing exploited in the financial realm. The control functions utilized the divergent degrees of the engineering & construction labor aspects and probed the responses encompassed in the idle costs (ideally minimum), process accumulated costs, and the labor/material wait costs. Additional resources, commonly referred to as "crashing the project," will certainly complete the production in a timely manner at the expense of increased costs. However, the challenge is to ascertain the optimized point(s) of the available resources within the budget constraints. A venerable statistician once said about variance analysis:

First you plot the dots...then you plot the dots... and plot the dots. (Ellis Ott, 1975)

The analysis, in retrospect, provided a definitive methodology and insight into the substation transformer cost overruns. The trade-off between the schedule/cost minimization and the revenue maximization criteria is a reality in major capital improvement project installation processes. The general installation parameters are depicted below:

- The material (transformer), engineering & construction labor arrives as scheduled. The maximum value is 8064 hours.
- The labor costs are assigned in two categories:
  - Labor ($747,348)
  - Material ($2,370,500)
- The transformer labor process is a value added activity with an expected completion time of 8064 hours:
  - Minimum (6500 hrs)
  - Maximum (12000 hrs)

The installation process (and cost) varies by the investor-owned utility, municipality, and region. The range and scope of the installation project fluctuates on a periodic basis as well. Re-cycled and surplus materials from other work sites are rarely implemented in the summer critical projects. In fact, several electrical utilities are uninformed and oblivious of the stock levels within their jurisdiction. Recent substation audits discovered circuit breakers, capacitor banks, and wood poles, fully-equipped and prepared for installation in a required critical region. The inadequate inventory and inspection methods usually categorize the material as unallocated with questionable ownership issues. The typical material and associated resource requirements for this particular transformer installation are listed in Figure 6.

Figure 6. Typical resources

Asset Description:	Original Material Estimates:
Transformer	$2,370,499.35
Wood Pole	$13,120.00
Fee Land	$300.00
Lighting of Yard	$1,832.00
Yard Stoning	$76,882.52
Misc Steel Structures	$442,011.66
Foundations	$287,899.00
138 kV GCBs	$84,504.82
345 kV GCBs	$550,258.01
Reactor	$12,340.00
138 kV Disconnect	$28,157.00
345 kV Disconnect	$265,968.00
138 kV Potential	$19,446.00
345 kV CCVTs	$62,880.00
Wave Trap	$25,418.00
138 kV Lightning	$4,110.00
258 kV Lightning	$43,242.30
345 kV Lightning	$12,849.00
Switchboards	$381,317.00
Station Wiring & AC/DC Aux Power	$231,429.96
Insulators, Fittings,	$53,169.00
Install Bus Supports	$401,232.00
Steel Tower	$24,177.28
	$374,222.51
Material Totals:	**$5,767,265.41**

The heuristics cost estimates produced short-comings in several categories, specifically in the resource allocation and non-value added areas. A cursory evaluation of *only* the final transformer installation expenditures revealed a 31.11% increase ($3,107,847.20) from the original estimate. An accounting practice of allocating the project overhead dollars percentage-wise across the largest asset first, then subsequently in a descending order was also evaluated for efficiency as well. The cash flow analysis for selected material and resource disbursement is discussed in the following paragraphs. The gas circuit breakers financial outlays are illustrated data in Figure 7.

The installation of the gas circuit breakers (GCB) incorporated moderate-to-excessive cash disbursements during the 4th and 8th cycles. More of the work effort and material cost was encompassed in the higher voltage unit (based on quantity) than the lower voltage unit by comparison. The series of peaks and valleys in the disbursement is attributed the cash flow decisions and tepid departmental communication between the asset management group and the accounting department. The TRIZ[4] principle #12 for Equipotentiality simply states:

- In a potential field, limit position changes (i.e. change operating conditions to eliminate the need to raise or lower objects in a gravity field).
- This is achieved through resource leveling by smoothing peaks and valleys, curtailing the effect of conflict in demand for the equivalent resources, as well as scheduling activities during slack periods.

The feat is accomplished well before the summer critical period begins with essential planning and fierce stakeholder involvement in the front-end analysis process. The total labor cost equated to 11% of the total work effort for the project. Some of the charges were liberally dispersed across the asset ($45.77 in the second disbursement in the 138 kV GCB) as others were followed the regulations in the Federal Electric Reliability Commission codes. There was absolutely very little or no activity during the early stages of the project. However, as the capital venture progressed, the cash disbursements were more frequent. There are resource utilization, communication quandaries, and material issues experienced in the field during the equipment installation process. The

Figure 7. Gas circuit breaker expenditures

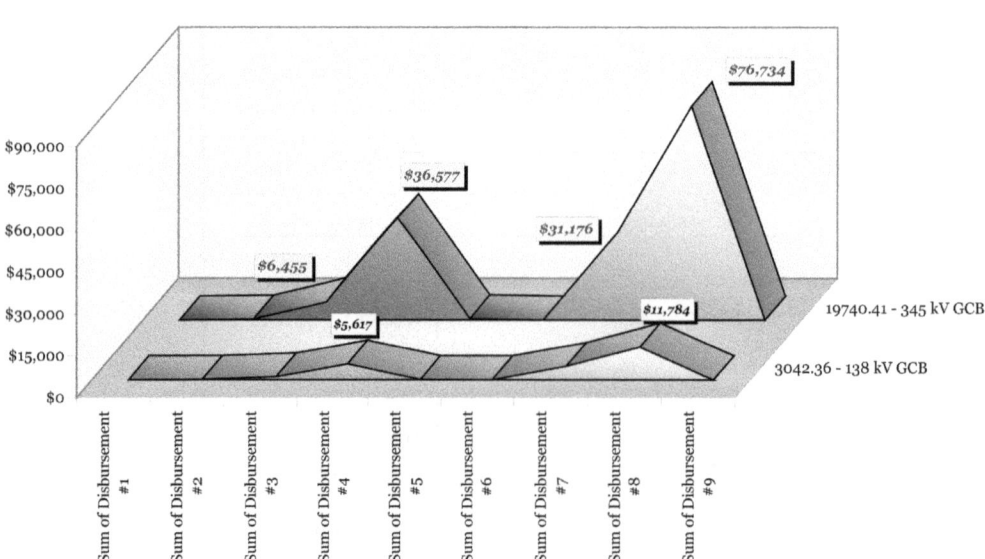

irregular cash flow is a testament to the speculative problems encountered in the field. The seemingly large final payment ($19,740 for the 345 kV GCB) demonstrates a hurried completion to a project inadequately planned. The Steel Tower and associated Structure Support expenditures are illustrated in Figure 8.

The installation of the Steel Tower and associated Structure Support materials exemplifies a reasonably timely and symmetrical dispersion of the data. The work efforts coincide with each other as the field crews followed a pseudo plan. The data reflects very little to no disbursements was dispatched until a concerted effort near the middle stages of the project. However, there were no disbursements (save a small transformer cost of $2,058) during period six and period nine (with the exception of the Steel Tower disbursement $4,036) of the study. As a summer critical project, astute planning is essential for efficient operation of the power grid. The project experienced a temporary work stoppage (with minor exertion on the tower) during this time according to the data. Enhanced preparation would facilitate ongoing parallel activities throughout the project's lifecycle. The resource-leveling methodology as

well as the Lean Quality techniques will create an evenly-dispersed cash flow in capital intensive ventures. The wait/idle scenario of the equipment and resources are consequents of poor departmental communication throughout the organization. The foundation and disconnect switches were categorized in Figure 9.

The disconnect switch category clearly depicts erratic spending in the fourth, fifth, seventh, eighth, and final disbursements. Specifically, the 345-kV disconnects switch cash dispersion ranged from the low of $143 (disbursement #2) to a high of $37,089 (disbursement #8). The data displays the final "rush-to-the-finish-line" approach as more funding was utilized near the end of the project. The payment lag time in accounting system attributed to some degree to the situation. Nevertheless, the peaks and valleys depicted in the diagram are clear indicators of unplanned activities with managerial support unfortunately. The timing of critical material integration with the available resources on station and prepared to for installation are essential components of the Lean Quality paradigm. The lighting arresters cash flow followed the same cash flow pattern, albeit in varying magnitudes of invoice payments. The data

Figure 8. Steel tower and miscellaneous steel structure and support expenditures

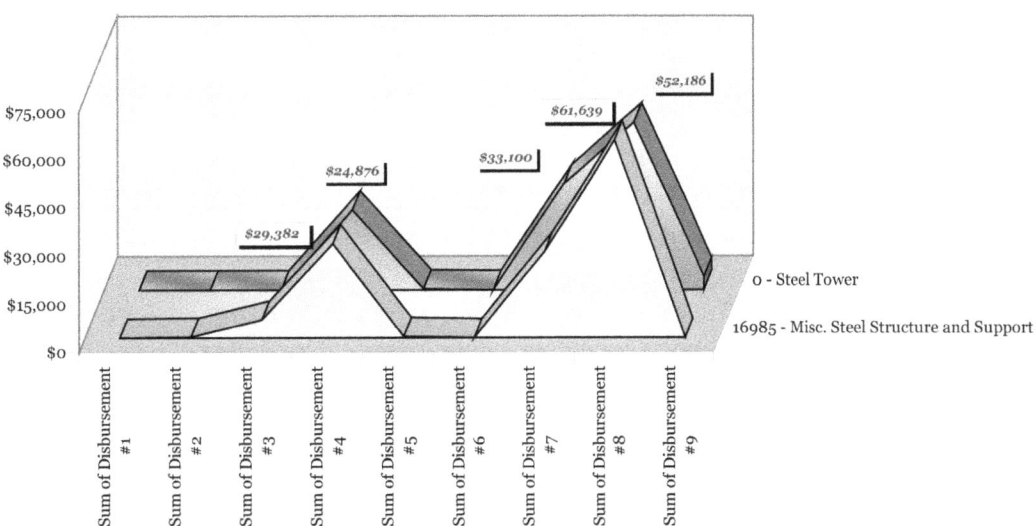

Figure 9. Foundation and disconnect switch expenditures

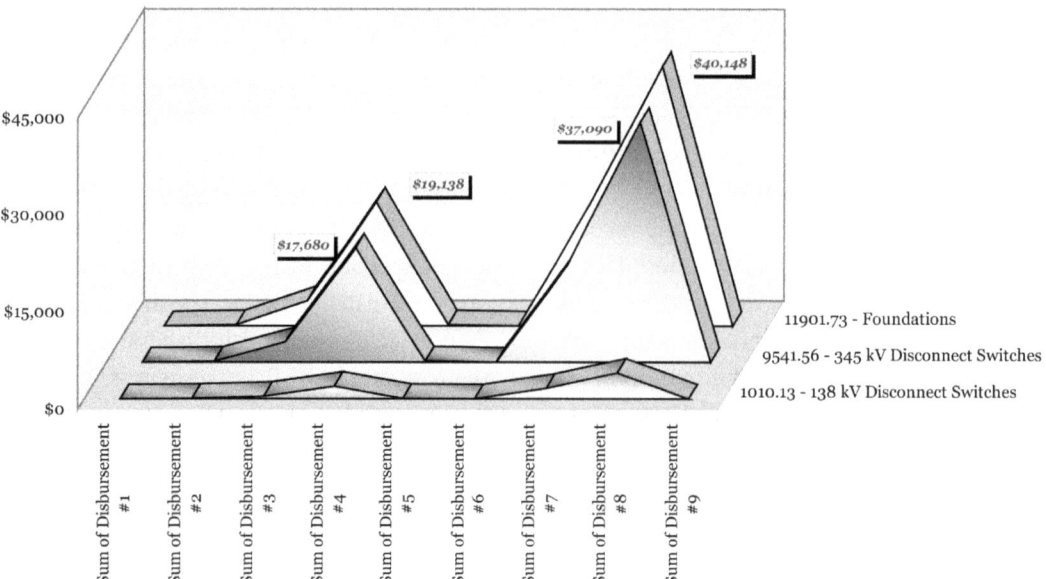

for the lightning arrester material and labor costs were categorized along with a potential transformer in Figure 10.

On average, the lightning arresters exhausted a very small amount (from as low as $2.22 to as high as $6,030) of the capital project cash outlay. However, the final cumulative cost of the category exceeded the original estimates by over 30%. The entire labor cost for the 258-kV lightning arrester *alone* was in excess of ten thousand dollars to install the forty-three thousand dollar part. There have been several instances where charges were applied erroneously to a project and consequently driven up the cost. These simple mistakes create unnecessary work stoppages in an organization where communication is deficient. A defined financial disbursement system working with the activity-based costing methodology is a phase-forward technique designed to track spending and cost overruns. When the system is merged with a front-end process analysis mechanism, the planning assessments for capital-intensive project are significantly simplified for the decision-maker. The complexity of electrical equipment installa-

tion must utilize Lean Quality techniques as well in order to diminish the wait/idle time of the resources by employing the just-in-time paradigm for material integration. Moreover, a viable cash-flow plan with established maximum limits and median spend ranges are necessary for each line item to control costs – especially for high visibility, system reliability capital projects. Computational modules are well equipped to perform the tasks. The reduction of contingencies has a direct correlation with decreased project expenditures related to the resources, and must be included within the cost containment strategies. The transformer was the major delivered equipment and work effort for the electrical installation at this particular project site. These cash flow data, along with the Switchboards, Insulators, Fittings, and Bus are illustrated in Figure 11.

The transformer cost category is one of the most important indicators of the project progression within the capital venture envelop. Similarly, the station wiring, switchboards, insulators, fittings and bus are significant ancillary activities performed within the substation. The transform-

Figure 10. Lightning arresters and potential transformer expenditures

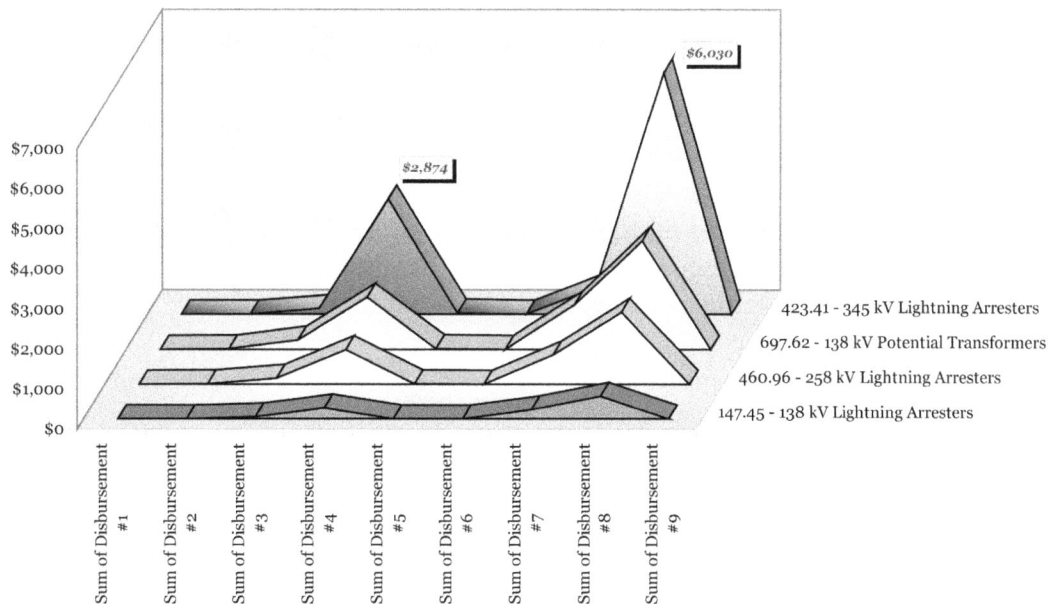

Figure 11. Transformer, switchboards, insulators, fittings, and bus expenditures

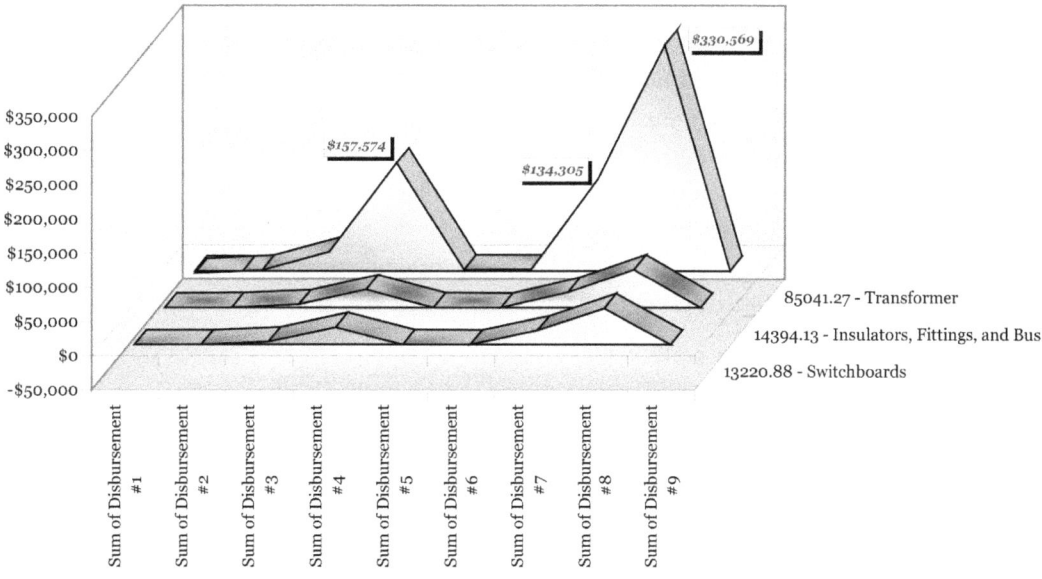

er expended all of the funding resources for each disbursement periods (save #9). The largest one ($330,569) was disbursed near the final phases of the equipment installation. The usual peak and valley data dispersion is depicted in the figure. When an established budget is implemented with appropriate managerial oversight, the costs are expected to stabilize in accordance with the equipment installation precedence and resource availability for implementation. Further analysis produced excessive wait and idle costs, as well as vulnerabilities within the entire system. However, by exploiting the green techniques in conjunction with lean technologies, the total transformer installation cost were reduced dramatically (especially when the project experienced "crashing" in the simulated model).These changes are illustrated in Figure 12.

Strictly from a cost perspective, the improvement in the installation process is noticed immediately with the use of a specialized display. Further examination of the idle cost is depicted in Figure 13.

The *box and whisker plot* (the specialized display in Figure 13) graphically provides additional detail of the information simultaneously with multiple sets of data in the same illustration (Lamprecht, J.L, 2005). The format is efficient and easy to read. The box and whisker plot is used for data comparison from usually different categories to facilitate decision-making. The box plot is typically developed from five statistics as outlined below:

1. **Minimum Value:** The smallest value in the data set
2. **Second Quartile:** The value below which the lower 25% of the data are contained
3. **Median Value:** The value in a range of numbers
4. **Third Quartile:** The value above which the upper 25% of the data are contained
5. **Maximum Value:** The Largest value of the data set

The box plot (as it is commonly referred) is a histogram-like method to summarize and graphically display the data. The outliers (whiskers) are data points out of line with the rest of the data set. The whisker values extend outside of the box, based on the extremities of the data.

The additional manpower, as expected, increased the total system cost. However, the transformer installation required *fewer* resources in a *shorter* period of time. Consequently, the lean techniques produced a summer critical project that was on schedule and well under budget. Figure 14 depicts the various cost scenarios subjected to the time paradigm:

LIFECYCLE ANALYSIS AND THE WAIT/IDLE SCENARIOS

The organization involved with equipment installation of the substation materials is obliged to consider the inherit lifecycle processes and associated manufacturing emissions of the product as

Figure 12. Transformer installation responses

Properties:	Controls:										
Scenario	Eng:	Constr:	Total System Cost:	Xfmr Labor Process Total Accum Cost:	Wait Costs:	Value Added Cost:	Idle Cost:	Xfmr Queue Wait Cost:	Busy Cost:	Xfmr Total Time (hrs):	Xfmr Labor Process Total Accum Time
Increase Manpower Cost	2	4	$874,246.55	$486,596.25	$62,746.53	$486,596.25	$324,903.77	$20,915.51	$373,716.63	---	4,982.89
Base Case	1	1	$438,567.28	$381,362.86	$57,204.43	$381,362.86	$0.00	$28,602.21	$286,022.14	---	3,813.63
Normalized Cost	1.125	2.75	$599,847.28	$381,362.86	$57,204.43	$381,362.86	$161,280.00	$28,602.21	$286,022.14	---	3,813.63
Xfmr Crashing Cost	4	8	$1,573,574.55	$547,371.61	$0.00	$547,371.61	$1,026,202.94	$0.00	$423,261.77	4,412.90	5,643.49

Figure 13. Transformer installation idle cost scenarios

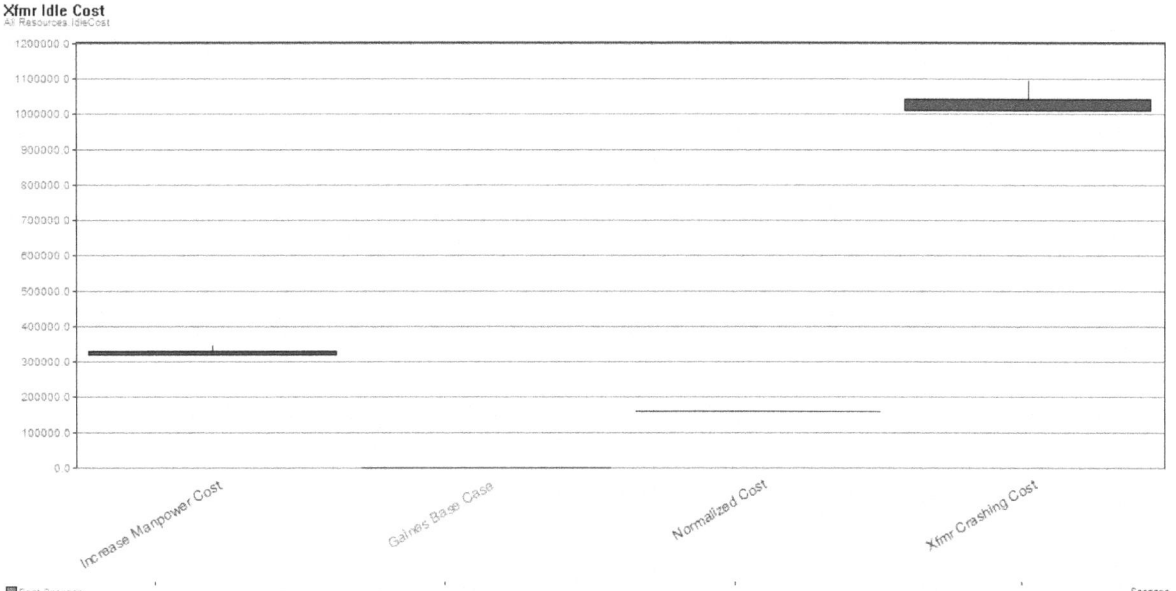

Figure 14. Transformer cost vs. time scenarios

Xfmr Idle Cost	Maximum Outliers	Minimum Outliers	Second Rank Number	Third Rank Number	Median
Increase Manpower Cost	$345,658.67	$322,560.00	$320,650.62	$329,156.92	$324,903.77
Gaines Base Case	$0.00	$0.00	$0.00	$0.00	$0.00
Normalized Cost	$161,280.00	$161,280.00	$161,280.00	$161,280.00	$161,280.00
Xfmr Crashing Cost	$1,093,037.61	$1,008,000.00	$1,009,180.59	$1,043,225.29	$1,026,202.94

stewards of the environment. The TRIZ principle #25 for Self-Service merely affirms (Mann, D. and Domb, E. 1997):

Use waste (or lost) resources, energy, or substances by:

1. *Re-cycle packaging material*
2. *Re-hire retired workers for positions where experience is required*
3. *Use scrapped or dummy parts for experiments*

The green processes for electrical equipment installation certainly employ numerous tenets of this principle. There is several waste boundaries (recycling) embedded within the material before it is place into service. The typical practice of most utilities are to retire a unit "in place" at the end of its useful service life. However, the method is not an effectively green technique from an environmental standpoint for most substation materials. The general waste boundaries of the lifecycle analysis are outlined below:

- Raw materials, including production waste
- Manufacturing of components, including electric consumption
- Painting of components
- Transportation

- Disposal (recyclable/land filled materials)

For the green company, the manufacture, utilization, and end-of-life environmental impact of a component are encompassed within the material installation decision-policy. An example of the material emissions germane to the electrical utility business evaluates the lifecycle of the switchgear. The general processes are illustrated in Figure 15.

The greatest environmental impact is experienced during the manufacture of the component at 130 kWh per unit. However, based on the uncertainty and inconsistency of the electrical grid (discussed in the previous disturbance section), the highest emissions are from the *generation of electrical energy to replace the losses of a substation transformer*. This profound activity occurs in the "end-user" phase of the lifecycle analysis. There is also minimal impact to the environment during this stage of the equipment's life unless a spare part is required or the substation experiences an electrical arc; yet rework activities during the initial installation will certainly increase emissions. Nevertheless, the example switchgear produces emissions to the air (SO_2, NO_x, CO_2, Methane), water (sulphate, solids), and the soil (mineral waste, ashes). Transportation is involved in every aspect of the lifecycle – from manufacturing of the product to the dismantling phase – and produces undesirable emissions as well. The switchgear dismantling consists of heavy machinery (additional emissions) on station performing specific tasks to assist with the product recycling. The increased fuel consumption utilized to haul the materials to and from the site contributes to the CO_2 emissions and adds another component for organizational consternation. The used wood scraps of broken pallets, for example, are burned in a waste-to-energy boiler of an independent power producer as well as recycling of the packaging materials derived from the newly installed electrical equipment (TRIZ principle #25 ~ Self-service).

Another example of the material emissions involves the most expensive component in the layout – the transformer. The transformer installation in a substation is designed for an operational life between thirty (distribution) and forty (transmission) years with an additional ten year period if the unit is adequately monitored and maintained (ideally, it is a maintenance-free device with no moving parts). The major materials that comprise the transformer are aluminum and steel. The material and weight per unit for a typical three phase distribution transformer are illustrated in Table 4.

The approximate thirty-two hundred pounds of material for this one functional unit is required for efficient operation. The transformer weight and complimentary packaging material are important logistical and installation attributes. The specialized equipment and associated resources used to handle, store, and transport the unit based on the weight are considerable issues in a cost containment strategy. The major electrical pieces prepared for installation in the substation are usually equipped with a special alignment/handling apparatus and tools to assist with the one time operation. Unfortunately, several of those additional parts are left unused and typically discarded instead of recycled or implemented in a training program. Although the transformer oil is recycled at a 100% rate, the CO_2 contributed to 98% of the emitted gases to the atmosphere during the raw material processes and assembly production phases. Moreover, approximately 10kWh of hydroelectric power and over sixteen thousand pounds of coal was expended to produce one unit. A summary of the natural resources to manufacture the transformer, in terms of energy and material utilized are outlined in Table 5.

The emissions expended to produce a "clean energy" component for a renewable energy substation is a major concern for the truly green orga-

Figure 15. The lifecycle of the switchgear

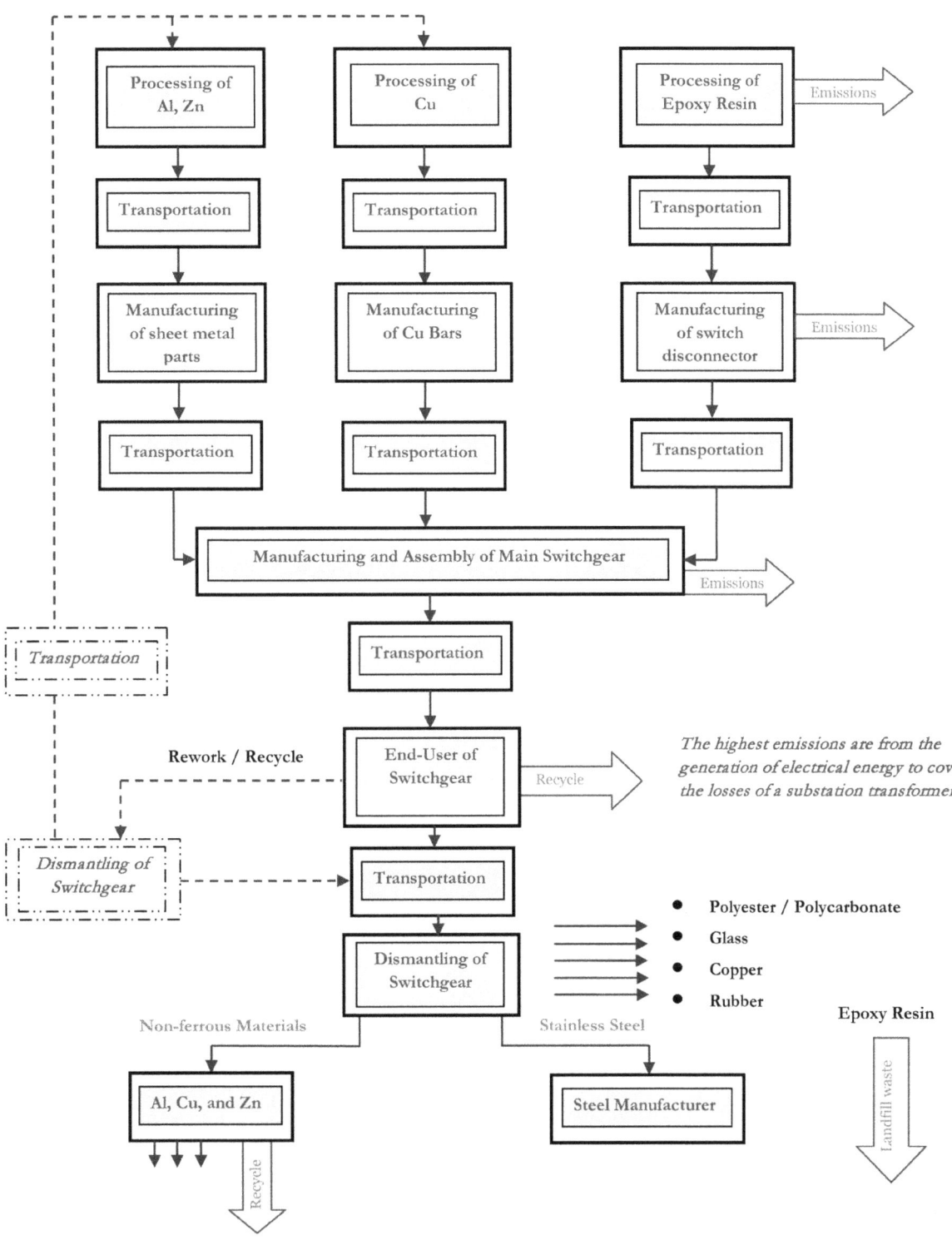

Table 4. Transformer material and associated weight

Material	(lbs)
Core Steel	1,175.10
Transformer Oil	749.57
Steel (tank)	714.30
Aluminum Wire	250.25
Aluminum Sheet	190.26
Insulation Material	132.28
Porcelain	24.25
Other Material	19.84
Total:	3,255.85

Table 5. Natural resources utilized to manufacture a typical transformer

Resource	(lbs)/kWh
Coal	16,627.00
Oil	3,254.30
Fe	1,561.30
Gas	1,149.60
Cu	758.17
Al	586.10
Hydropower	10.45 kWh
Si	42.26

nization. Ideally, the materials are produced, transported, and installed employing green technology throughout the supply-chain. Presently, this optimal path has not been achieved and off-set credits (i.e. tree planting in Africa) were instituted as a replacement method to placate the environmental policies.

In the simulated model, the utilization of common equipment on-site, as well as the as the "just-in-time" methodologies expedited the project and brought it to an orderly end. The proper equipment staging via the supply-chain and logistical material scheduling assisted in the project's success. Although the base case yields the best scenario and the least cost, it is not

"green" effective. Several methods are employed to mitigate the risks, specifically in transmission line rehabilitation to include:

- Heavy Equipment Vehicles remain within the ROW
- No vehicle/construction equipment storage
- No fuel storage
- No refueling and/or servicing of vehicles/ construction equipment
- No line component storage
- No construction debris disposal
- Minimize borrow/filling
- Cross on existing roads ~ two tracks if possible
- Use of the Timber mat road

The detailed material logistics for the transformer to arrive at the substation is another element of the Lean Quality methodologies. Since most of the electrical equipment is manufactured overseas, a discussion of the major components of the transportation network within the supply-chain is warranted. The substation equipment is first transported to the construction site via the inter-coastal waterways. The selected waterborne commerce is illustrated in Figure 16.

The transformer as freight on a cargo ship, for example, is grouped in the "All Manufactured Equipment Machinery and Products" category characterized by the 50% increase in tonnage over the period. The typical substation equipment requires special handling and warehousing during the transportation stage. Consequently, the green logistics paradigm (cost, time/speed, reliability, and warehousing) is jeopardized utilizing the ship-rail-truck transport methods. Ships and railways traditionally inherited a reputation of poor satisfaction based on the on-time delivery aspects as well as equipment damage. For example, a long awaited transformer for the Northeast corridor (which carried twenty-five percent of the load) was delayed for a second time when the freight returned to port for reasons unknown. An expe-

Figure 16. Selected waterborne commerce by type of commodity: 1997 to 2007[5] (in millions of short tons (324.5 represents 324,500,000)) ~ one short ton equals 2,000 pounds

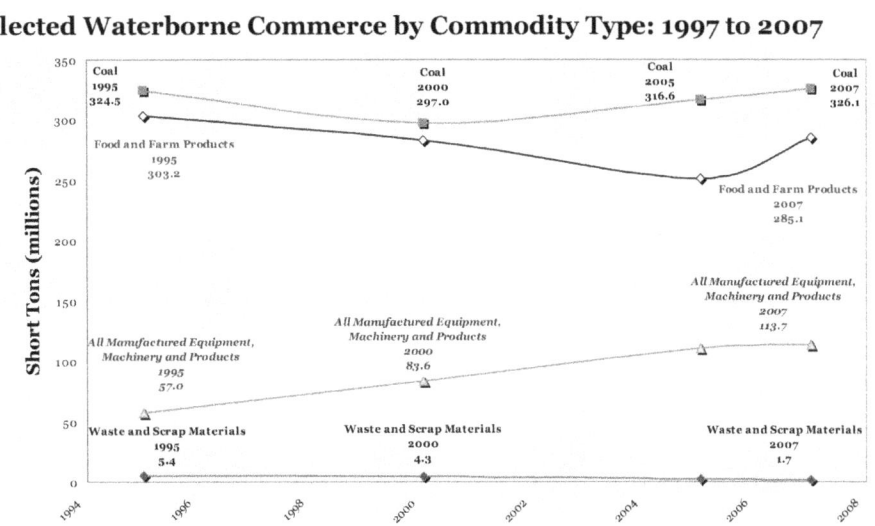

rienced retiree from the electric utility was dispatched to South America during the Christmas holiday season to expedite the shipment of the critical transformer. The efficient use of a lost resource (retiree) demonstrated yet another facet of the self-service statement #25 of the TRIZ principles. Nevertheless, the costly supply-chain logistics for this type of overseas manufactured equipment is a standard in the utility industry. The increased emissions in the form of carbon monoxide, especially by the highway transport vehicles, are supposedly offset over a period of time with the installation of the renewable energy resource. Moreover, a positive indicator in the data illustrates a sixty-nine percent decline in scrap material over the same period. Once the transformer freight is off-loaded from the ship, it is typically transported by rail. The selected rail freight based on carloads and tons is illustrated in Figure 17.

Considerably more carloads of coal, (23.35% increase), food (13.25% increase), fabricated metal products (46.55% increase), machinery (11.36% increase), and scrap materials (39.78% increase) were transported in 2008 than any

other year in the study period. Further inspection of the data revealed a 34.12% tonnage increase in coal transport, 22.85% tonnage increase in food and kindred products, as well as 42.85% tonnage increase in the waste/scrap material categories. However, the fabricated metal products and machinery categories remained the same in the tonnage value (one million) throughout the period. Some of the substation equipment is conveyed via this method as economies of time aspects for improvements in service reliability, cost, and flexibility. Ideally, the selected mode of transportation will provide on time freight delivery with the least threat of equipment breakage and pollution. A particular high-voltage transformer fell off of a semi-truck during a material transfer process in the Midwest region, resulting in severe damage to the unit. Equipment handling by skilled labor crews is the essential intangible elements in the process control envelope. Specific handling and installation procedures are developed by system engineers solely for this purpose. By reducing the time flows, the speed of the distribution system within the supply-chain framework are increased, and consequently its efficiency. The

Figure 17. Selected railroads, class I line-haul-revenue freight originated by commodity group: 1990 to 2008[6] (5,912 represent 5,912,000,000 and 579 represents 579,000,000 tons)

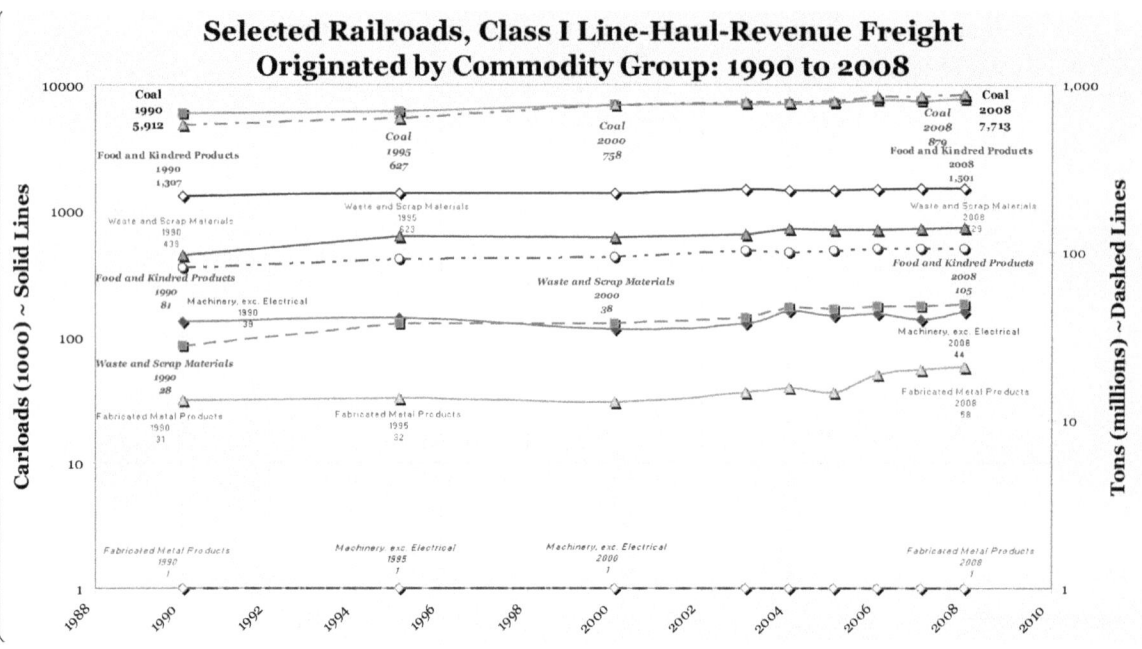

rail transportation mode is an additional element embedded within an efficient just-in-time strategy.

Truck transportation is the primary mode to convey heavy equipment to remote substation construction sites. Most of the emissions in the equipment installation process emanates from this source of conveyance. In fact, highway vehicles rated the highest in the carbon monoxide (Figure 18) and nitrogen oxide (Figure 19) air pollution emission categories for the evaluated period.

The carbon monoxide for all highway vehicles produced 41,610,000 tons of pollution. By contrast, the electrical utilities only produced 689,000 tons of the pollutant (excessive from an environmental perspective). A potential pitfall and environmental tradeoff with the just-in-time methodology are the increase freight in transit issues. The reduction of warehouses has a direct correlation with the inventories increase in the transportation industry. The in-transit freight contributes to congestion and additional pollution. The ex-

ternal pollution costs are absorbed by the environment and society as a whole if left unchecked.

The nitrogen oxide for all highway vehicles produced 5,563,000 tons of pollution. By contrast, the electrical utilities produced 3,331,000 tons of the pollutant. The largest offender of sulfur dioxide pollutants originates in the electric utility industry (8,973,000 tons). By contrast, the highway vehicles only produced 91,000 tons of the pollutant (yet, an excessive number from an ecological perspective). These data is illustrated in Figure 20.

In an ideal situation, green vehicles are required to facilitate the reduction of emissions in order to implement the renewable energy resource. Nevertheless, the trucking industry primary purpose for the utility business is to move precariously heavy and oddly-shaped (turbine blades, etc.) freight loads to the construction site. The annual revenue to provide this service increased precipitously during the period. The data is illustrated in Figure 21.

Figure 18. Selected air pollutant emissions by carbon monoxide pollutant and source: 1970-2002[7] (In thousands of tons (5,304 represents 5,304,000 tons))

Carbon Monoxide Pollutant Emissions: 2007

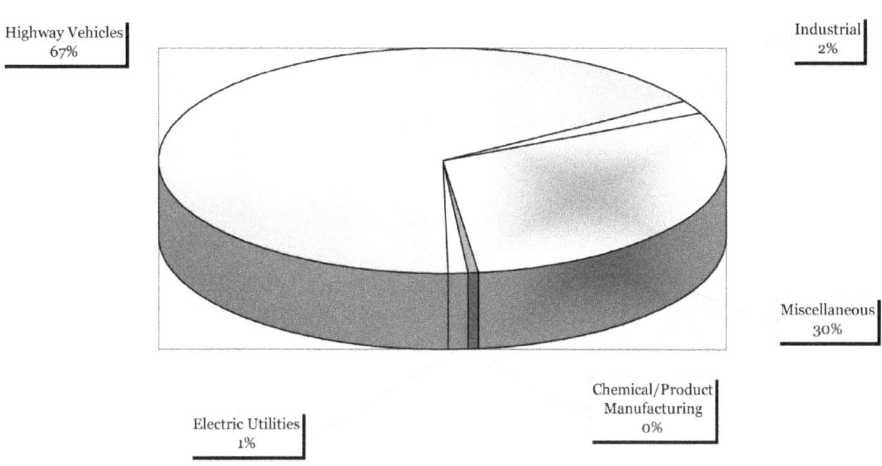

Figure 19. Selected air pollutant emissions by nitrogen oxide pollutant and source: 1970-2002

Nitrogen Oxide Pollutant Emissions: 2007

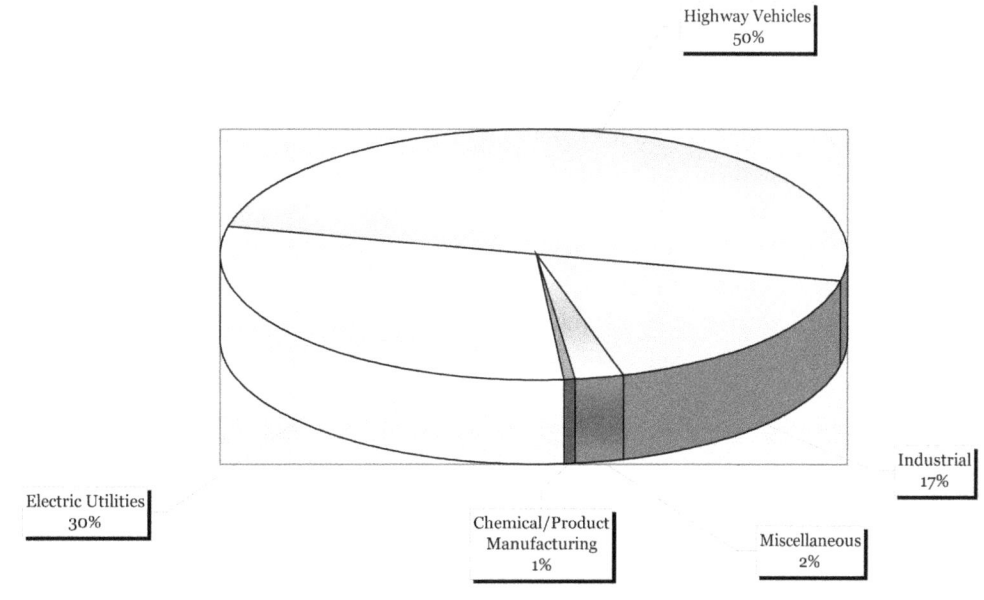

Figure 20. Selected air pollutant emissions by sulfur dioxide pollutant and source: 1970-2002

Sulfur Dioxide Pollutant Emissions: 2007

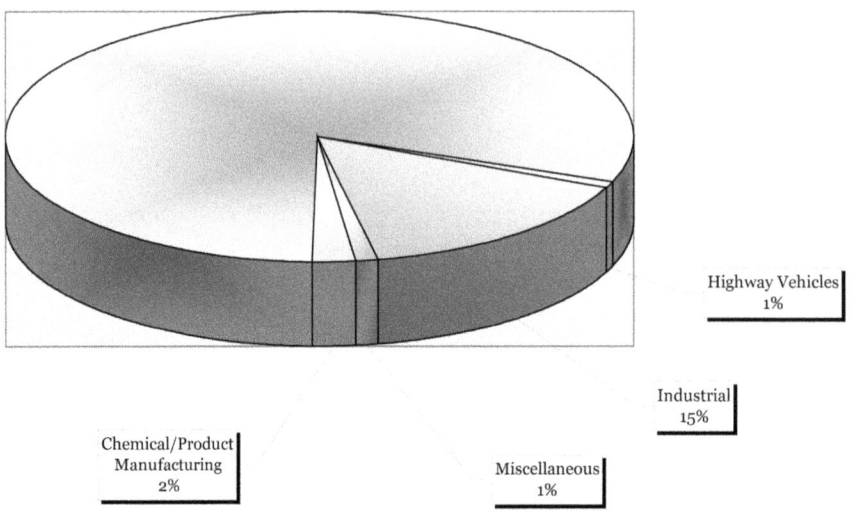

The increased truck traffic and related detours creates more congestion and additional fuel consumption (emissions) to transport the requisite machinery and equipment to the work site. Moreover, the machinery utilized to excavate, compact, backfill, and repave the road leading to the construction site consumes fossil fuels and emit harmful pollutants as well. The reduction in emissions associated within the construction horizon are achieved based on shorter job durations, strategically located resources, fewer pieces of construction machinery, and limited/no disruption to traffic flow. The green logistics methodologies are designed to reduced transportation cost and align the organization with compatible environmental issues.

When the transformer installation wait and busy costs are examined and compared, a trade-off appears between the base case (idle) cost and the crashing (wait) costs. Depending on the utilities tolerance for the time (material delay, resource reallocation, etc), the implementation of a summer critical transformer project ahead of schedule far outweigh the idle costs. This specific condition is depicted in Figure 22 and Figure 23.

An example of a summer critical construction project wait cost scenario applicable to green technology is described in the following paragraphs. A front-end analysis was required to assess the biological impact of the Tippy Dam to Hodenpyl 138-kV electrical transmission line rebuild project in Michigan. Various agencies (U.S. Fish and Wildlife, The Michigan Department of Natural resources Wildlife Division, Eastern Michigan University, and the Federal Energy Regulatory Commission) were involved with the project's evaluation of the local wildlife particularly the endangered Indiana bat (*Myotis Sodalis*) and the

Figure 21. Selected truck transportation summary: 2000 to 2007[8] (in millions of dollars (5,812 represent 5,812,000,000))

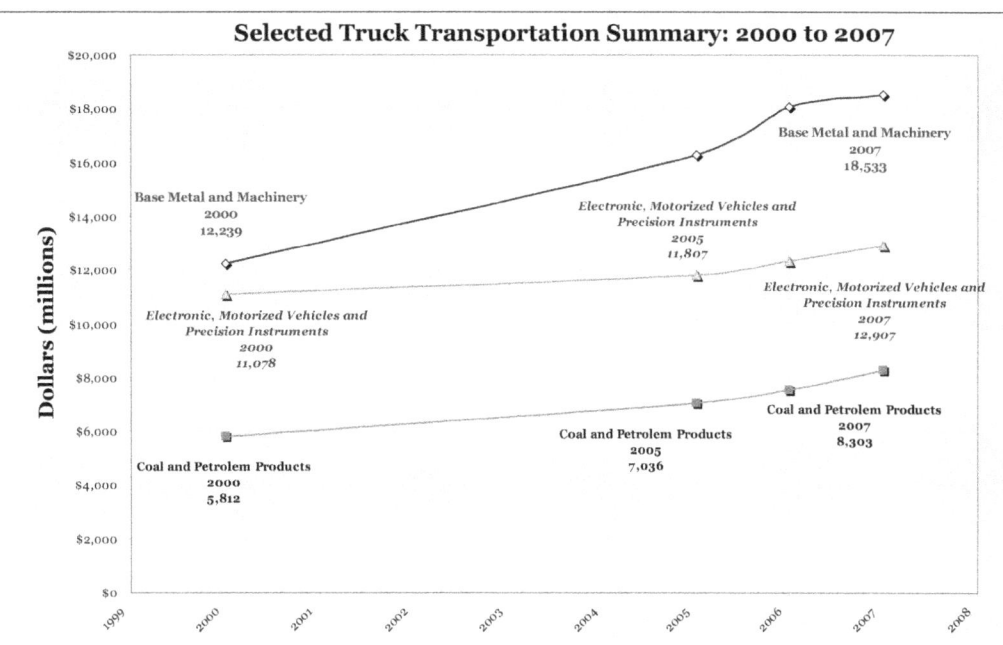

Figure 22. Transformer idle cost vs. wait cost scenarios

Xfmr Wait & Busy Cost	Wait Costs:	Idle Cost:	Busy Cost:
Increase Manpower Cost	$62,746.53	$324,903.77	$373,716.63
Gaines Base Case	$57,204.43	$0.00	$286,022.14
Normalized Cost	$57,204.43	$161,280.00	$286,022.14
Xfmr Crashing Cost	$0.00	$1,026,202.94	$423,261.77

threatened Bald Eagle (*Haliaeetus Leucocephalus*). The Endangered Species Act was initially passed by congress in 1973 and reauthorized in 1988. It is designed to set policies in conjunction with the efforts to rescue specific species that are in peril of extinction because of human action. Furthermore, the policies in the act aim to conserve the species and their associated ecosystems. The endangered animals and plants are listed by the U.S. Fish and Wildlife Service, where recovery plans are developed to ensure the species receives no additional harm by actions of the U.S. Govern-

ment or its citizens. *Endangered species* are one in danger of becoming extinct throughout all or a significant part of its natural range. *Threatened species* are one likely to become endangered in the foreseeable future. A selected count of the threatened and endangered wildlife and plant species are shown in Table 6.

After World War II, the Bald Eagle was vulnerable to the rampant spraying of the powerful pesticide DDT (dichlorodiphenytrichloroethane), which the birds picked up from contaminated fish. The DDT was used by farmers, foresters, and

Figure 23. Transformer installation wait vs. idle cost scenarios

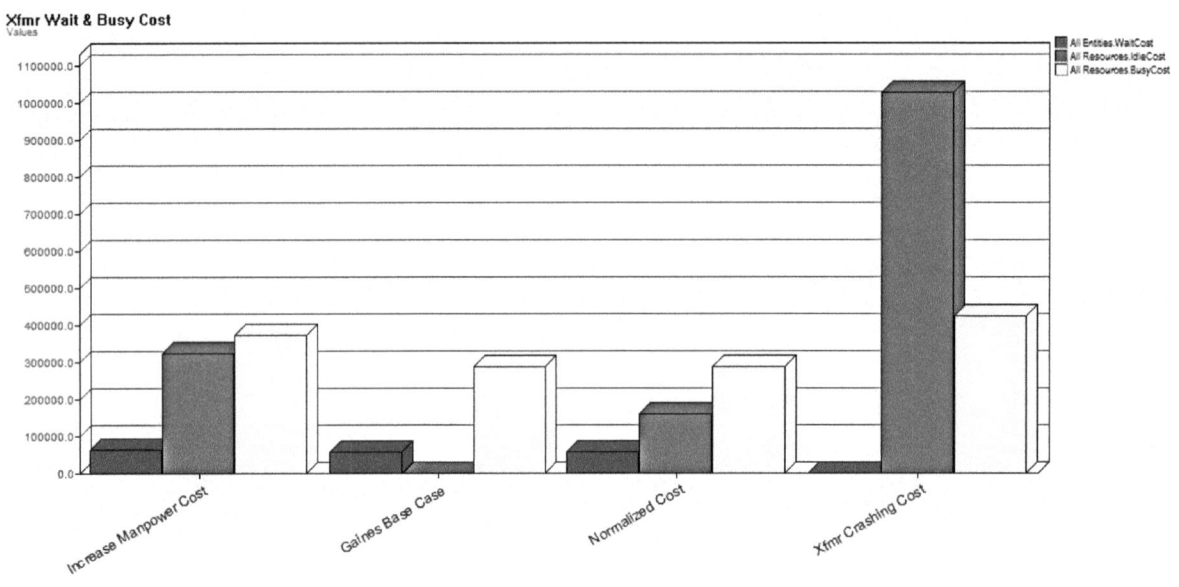

Table 6. Selected threatened and endangered wildlife and plant species: 2009[9]

Item	Mammals	Birds	Reptiles	Amphibians	Fishes	Snails	Clams	Crustaceans	Insects	Plants
Endangered Species (U.S.)	69	75	13	14	74	24	62	19	47	600
Threatened Species (U.S.)	13	15	24	11	65	11	8	3	10	146

others to kill weeds, insects, rodents, and other pests that harmed crops. A metabolite of DDT – DDE – causes the eggshells of the birds to thin, since the chemical does not dissolve in water or breakdown in the environment. Consequently, the thin eggshells are inadvertently broken by the parent eagle during incubation. The DDT and other pesticides, combined with over-hunting in the early twentieth century led to the decline of the eagle population by the late 1950s. Once the chemical was banned in the U.S. in 1962, the Bald Eagle population began to rise. The U.S. Fish and Wildlife Service officially down-listed the bird from "endangered" to "threatened" in 1995. Nevertheless, habitat destruction and illegal hunting

are still major threats to the Bald Eagle. One aspect of the project involved the removal of old towers near Tippy Dam and replace the units with new support structures and wires. The correspondence with these entities and subsequent communication with the general contractor facilitated an accord in the construction window. The section of right-of-way was approximately 4.5 miles west of a wetland identified in the biological assessment as containing high-quality Indiana bat habitat. Moreover, a potential roost tree, proposed for removal prior to the construction data and a considerable distance from the wetland, was deemed to adversely affect the Indiana bat. Wetlands are the most diverse and fertile areas in the United

States. The wetlands improve the quality of life of unique plants and animal species as well as the water passing through it. The water is filtered through dense vegetation as inorganic ions and toxics are removed via settling. The wetlands act as storage areas for floodwaters to protect urban regions and reduce erosion. The proposition to implement temporal tree removal restrictions and conduct activities away from the high-quality habitat was a positive effort to reduce the adverse affect of the Indiana bat's environment. Also, it was determined that the project had no effect on the local bald eagle population because no nests were in the immediate vicinity of the Tippy Dam. The biological assessment of the wetlands and the delay of the project start assisted with the success of the line-rebuilding endeavor. The evaluation provided key personnel to utilize extreme tact, technical abilities, project management skills, and business acumen in the research phase of this project.

Endangered butterflies have served as key species in challenging the construction of major dams, the spraying of dangerous pesticides, and the clearing of land for transmission line right-of-ways. The increase in greenhouse gases are mainly derived from human activities of burning fossil fuels (coal, oil, natural gas) and clearing of land. Scientist at the University of Melbourne in Australia definitively linked the timing of a natural event to human-induced global warming (Wexler, M. (2010)). It came in the form of the common brown butterfly (*Heteronympha Merope*) and its emergence from their chrysalises 1.6 days earlier per decade over the past seventy years. Researchers studied data dating to 1940 and correlated the temperature increase (elevated 0.25 degree F/decade) to the butterfly's emergent date, now 10.4 days earlier than the beginning of the evaluation period. The shift in the emergent date was demonstrated in temperature-controlled chambers by researchers and can improve the

forecasting ability of the climate change impact on biodiversity. The Karner Blue (*Lycaeides Melissa Samuelis*) butterfly issue is another example of the construction wait cost scenario and incisive planning. The butterfly, about the size of a postage stamp, was listed as an endangered species by the U.S. Fish and Wildlife Service in 1992, and found in only seven states (a 42% decrease since the early twentieth century). Essential to the Karner Blue (named after a town in New York where it was first described) habitat is the presence of a prairie or oak-pine savanna and prairie – first to farmland and later to forest and residential development – most of the butterfly's habitat has been lost. Wild Lupine play a vital role in the Karner Blue's lifecycle as the plant act as a host for the female eggs. These plants are found in open, sunny areas with sandy soil, essential elements of the native habitat. Subsequently, licenses issued by the Federal Energy Regulatory Commission (FERC) for hydroelectric projects include provisions for managing and protecting the Karner Blue habitat on project land – including the rights-of-way of transmission lines – which must be maintained to prevent woody undergrowth. Ongoing efforts include vegetation monitoring of sites identified as suitable habitat, population monitoring of the butterfly, and special management projects to increase the Wild Blue Lupine at project sites. Although the management effort by the U.S. Forest Service involves the utilization of controlled wildfires to assist with the recreation of the native prairie habitat, the option is not practical beneath electrical transmission lines. Consequently, the utility focuses on strategic use of herbicides, hand removal of competing vegetation, and seeding of Wild Blue Lupine as a basis for the Karner Blue management plan on project area development.

RESULTS

The box plot diagram in Figure 24 is used to summarize the data set. The best case scenario for the minimization problem is the base case (ideal with zero idle cost) followed closely by the normalized case (minimal cost across the responses).

The increased manpower control parameter slightly doubles the resources from the base case. However, the accumulated total process time *only* increases by 23.43% (well below the standard time of 8064 hours). The median total system cost reflects the expected idle cost increases and associated wait costs (Figure 25). The transformer wait cost in this control parameter experienced a 26.87% decrease when the lean supply-chain processes were implemented. Although the crashing cost category increases the engineering and construction efforts, the elevated expenditure (72.12%) may or may not be justified by the decision-makers depending on the summer critical project priority. Nevertheless, the median cost is far lower than the actual expenditures (by 49.36%) when the system is implemented with the appropriate planning evaluation mechanism.

Example Calculation

The example calculation of the manpower increase cost scenario is depicted below (Schermerhorn, Jr., J.R. 1993):

Median= $(n-1)/2$ (second quartile)

= $(4+1)/2$

=2.5^{th} number (the second value plus 50% of the difference between the second and third values or: ($1,268,540-479,953)/2=$394,295.50

Hence:

479,953.08 +394,295.50=$874,246.58

The first quartile (q1) is the $(n-1)/4$ rank:

= $(4+1)/4$

=1.25^{th} position

Hence:

322,560+0.25(479,953-322,560)

1st quartile=$361,908

The third quartile (q3) is the $(3n+3)/4$ rank:

= $((3*4) +3)/4$

=3.75

Or the 3rd ranked number plus 75% of the difference between the 3rd and 4th numbers. Hence:

$1,268,540+ (0.75) (640,035)

3rd quartile=$1,748,566

Fifty percent of the values are below/above the median ($874,246) and one quarter of the values are below or equal to the first quartile ($361,908). Moreover, one quarter of the values are above or equal to the third quartile ($1,748,566). This equates to 50% of the values must be within the inter-quartile range=q3-q1 or: ($1,748,566-$361,908=$1,386,658). The data is summarized in Figure 26.

Project funding opportunities must include accurate and timely information for decision support. A detailed scenario analysis provides an additional mechanism to evaluate high-end capital projects. The data in Figure 26 displays the median, 1st quartile, 3rd quartile, outliers, and inter-quartile ranges associated with each scenario. An examination of the labor cost is in order to assess the alternative *best case* scenarios. The 3rd quartile is used as a funding threshold level

Figure 24. Transformer installation system total cost by scenario

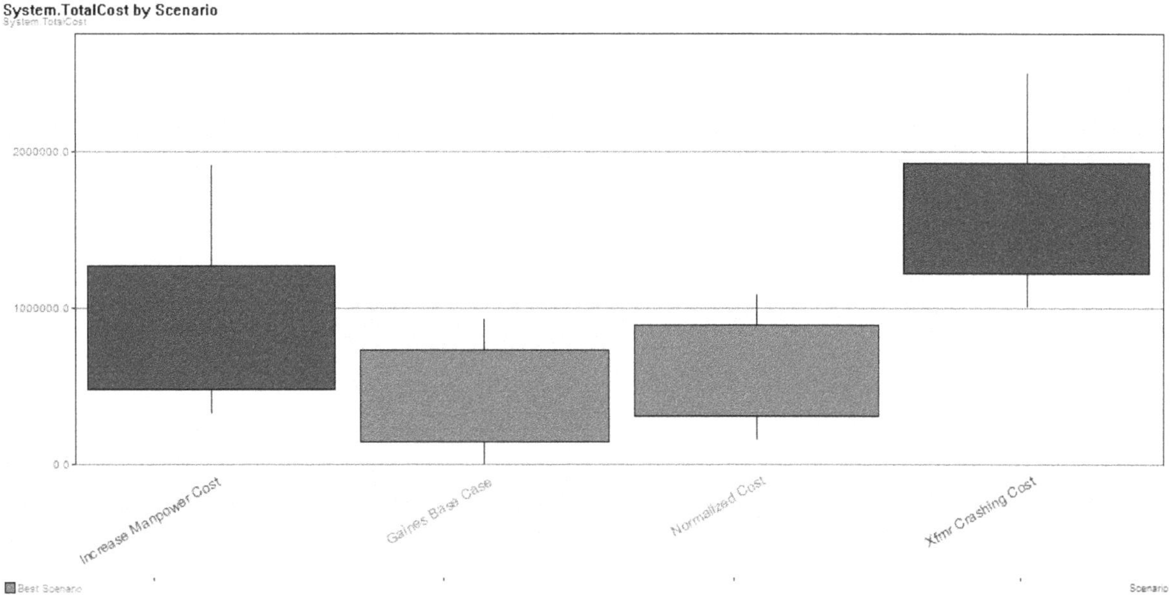

Figure 25. Transformer installation total cost by scenario data

System.Total Cost by Scenario	Maximum Outliers	Minimum Outliers	Second Rank Number	Third Rank Number	Median
Increase Manpower Cost	$1,908,575.11	$322,560.00	$479,953.08	$1,268,540.01	$874,246.55
Gaines Base Case	$925,899.24	$0.00	$146,929.58	$730,204.98	$438,567.28
Normalized Cost	$1,087,179.24	$161,280.00	$308,209.58	$891,484.98	$599,847.28
Xfmr Crashing Cost	$2,501,595.66	$1,008,000.00	$1,218,184.84	$1,928,964.26	$1,573,574.55

Figure 26. Transformer installation cost variables

COST VARIABLE:	MEDIAN:	Q1:	Q3:	MINIMUM (OUTLIERS):	MAXIMUM (OUTLIERS):	INTER-QUARTILE RANGE:
INCREASE MANPOWER (DOUBLE)	$874,246.55	$361,908.00	$1,748,566.00	$322,560.00	$1,908,575.11	$1,386,658.00
BASE CASE	$438,567.28	$36,732.40	$876,975.68	$0.00	$925,899.24	$840,243.27
NORMALIZED	$599,847.28	$198,012.39	$1,038,255.67	$161,280.00	$1,087,179.24	$840,243.28
XFMR CRASHING	$1,573,574.55	$1,060,546.21	$2,358,437.81	$1,008,000.00	$2,501,595.66	$1,297,891.60

(maximum limit) to aid with the decision analysis. The labor cost experiences a comparable change (18.93% base case ~68.73% xfmr crashing) and provides valuable insight with the construction horizon. A surface graph is utilized here to depict the trending cost of each scenario (Figure 27).

The visualization assists decision-makers with crucial funding pronouncements for the summer critical capital projects. The cost minimization problem is simpler to evaluate with accurate and detailed information.

The Value-Added System Cost Scenarios

The transformer crashing technique is infeasible in the initial planning stages of the summer critical projects. However, as the June 1st deadline quickly approaches in the construction cycle, the option becomes more appealing, especially during the heavy spring storm periods when the demand work deplete the required resources to perform the assigned tasks. Figure 28 illustrates the importance

Figure 27. Transformer installation surface graph total cost by scenario

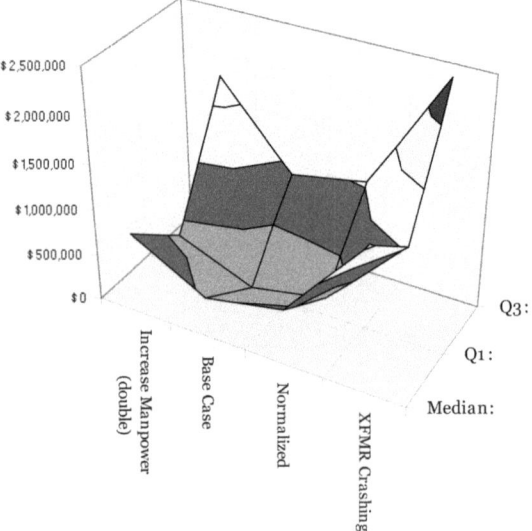

of the "wait-idle" trade-off decisions applicable to the cost paradigm.

Environmental concerns are typically included in the capital project decision-making process, especially during the fishing and/or hunting seasons. The potential conflict between outdoor enthusiasts and the heavy equipment construction cycle is a viable issue the investor-owned utility must take account of during the essential planning stages. There are numerous outdoor-related activities (hiking, biking, canoeing, rock climbing, camping, skiing, snowboarding, spectator sports, etc.) American's enjoy on an annual basis. However, wildlife-related activities may possibly disrupt or delay the construction latest start date to meet major milestones, and more importantly, the summer critical end date (when the equipment is expected to be in-service and fully operation by June 1st). The time constraint assists the astute planner, engineer, and supply-chain decision-makers to develop the construction strategy well in advance – typically in the first quarter of the prior year – or earlier. Resource utilization, permitting, and workaround plans (contingencies) must be formulated and thoroughly scrutinized for a successful capital project implementation. Moreover, the major equipment must be ordered in conjunction with the green primary and secondary routs (to include contingencies) because of the extensive lead times. The field audits and thorough site Walkdowns must be conducted *more than once* to evaluate area specific details (elevations, road access, right-of-way, vegetation, etc.). The detailed analysis must augment any pre-assessment feasibility study and include and possible overlap/conflicts with the hunting seasons (most plans fall short in this area). An Upper Midwestern utility capital project was summarily delayed for twenty days because of fishing encroachments. As a result, the situation created excessive resource idle time. Considerably more people are engaged in wildlife-related activities than ever. An annual "snap-shot" of the participant activities is illustrated in Figure 29.

Figure 28. Transformer value-added cost parameters

Xfmr System Cost	Value Added Cost:	Total System Cost:	Gap Analysis (wait/idle tradeoff):
Gaines Base Case	$381,362.86	$438,567.28	$57,204.42
Xfmr Crashing Cost	$547,371.61	$1,573,574.55	$1,026,202.94

Figure 29. Selected participants in wildlife-related recreation activities: 2006[10] (in thousands (25,431 represent 25,431,000)) for persons 16 years old and over engaging in activity at least once in 2006

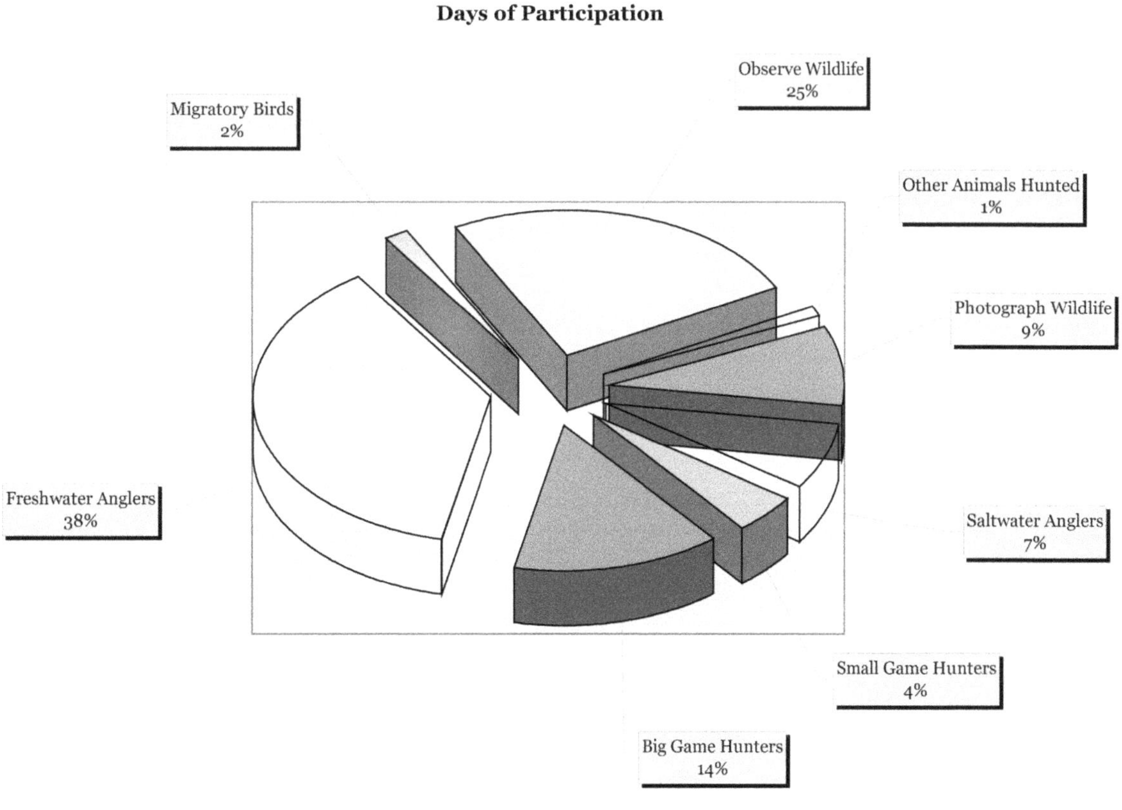

Days of Participation

Thirty-eight percent of the outdoor-related activities (in terms of *days of participation*) are those associated with the freshwater anglers, closely followed by the observation of wildlife category (25%), and the big game hunters (14%).

Bird-watching (2%) and hunting other animals 91%) were on the opposite end of the scale. From a sheer *number of participants* perspective, the freshwater anglers dominated the outdoor-related activities category with twenty-nine percent of the total, followed by those participating in the observation of wildlife category (25%), and wildlife photography (14%). The big game hunting category rounded out the top four at thirteen percent of the number of participants. The electric utility company's goodwill to the community far outweighs the cost in fines/penalties associated with inadequate forecasting of various outdoor-related activities associated with the construction horizon. In order to minimize conflicts, a thorough review of the production permits, seasonal activity schedules, regulatory policies, and local ordinances are necessary for a successful capital project implementation. Figure 30 displays the graphical portion of the base case vs. the transformer crashing scenarios.

The gap analysis exhibits the absolute minimum wait cost of a project to proceed versus the other cost extreme (fire drill) of crashing the venture before the June 1st summer critical deadline. The economies of scale are attained once several lessons learned cases are archived and thoroughly reviewed.

CONCLUSION

It is imperative for decision-makers to resolve uncertainty and prioritize capital projects with perfect (or near perfect) information. Several lean and green variables are considered for large construction ventures. This chapter demonstrated the importance of decision analysis from a cost perspective in the equipment installation sphere. The system capacity margin, coincident peak load, and operational capability are equally important facets of the electrical grid upgrades. The robustness of the grid and its ability to quickly recover from a disturbance are key inputs for system planning. The reliability of the infrastructure is dependent on human interaction and responsive, robust equipment. Renewable energy system integration introduces a high degree of complexity to an already strained electrical grid. The increased cost of alternative energy merits additional research in this area. Nonetheless, the sustainable energy resource (once synchronized and dispatchable to a certain extent) is compatible with the current system. The installation process with green energy

Figure 30. Transformer installation value-added cost by scenario

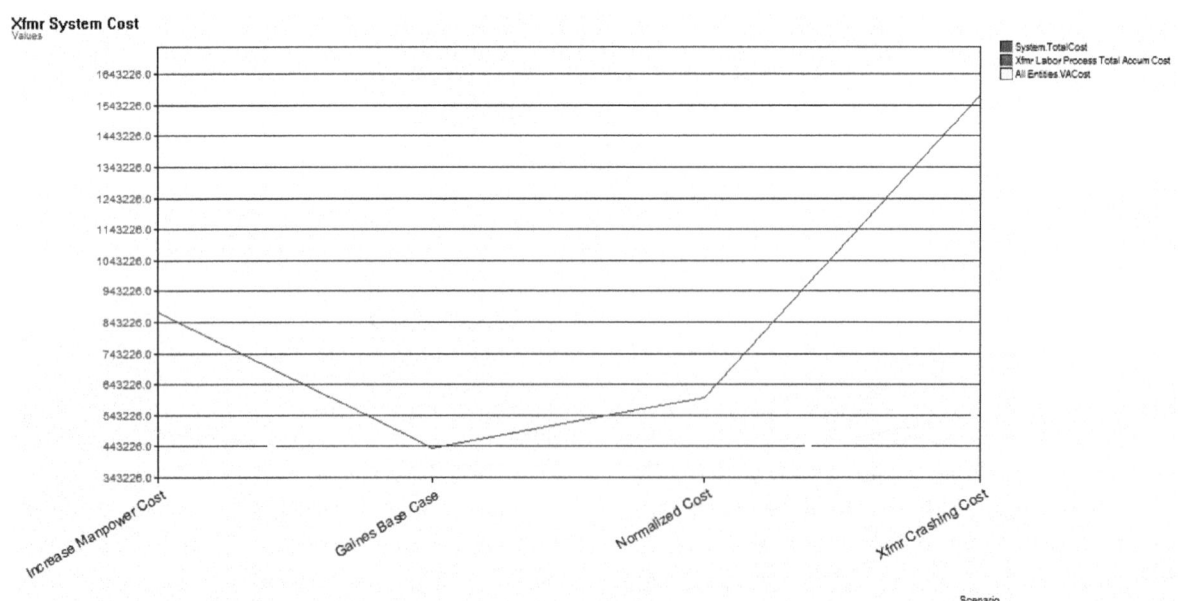

techniques necessitate a thorough evaluation of the supply-chain sourcing in the utility industry. The green technologies must play a prominent role in component manufacturing, end-user emissions, and ultimately the recycling phases of the venture. The sensitivities of the environmental issues relating to the threatened and endangered species are critical elements in the construction arena as well. A thorough assessment of the land and its surroundings by the appropriate group (compliance, legal, and environmental management) is compulsory well before the first shovel breaks ground for the capital-intensive project. The ability for an organization to make use of Lean Quality tools to reduce the equipment installation cycle time, cost, and inventories while minimizing the carbon footprint within the construction envelop is an essential progression towards green stewardship. The cost analysis mechanism encompasses the stimulus-response technique to evaluate a variety of scenarios in a simulated environment. It plays a decisive role in the construction management and budgetary development of capital-intensive projects. The data presented provides a guideline and cost boundaries based on established scenarios. Responsible electric utilities and process industries will benefit from detailed information in construction projects and become stewards for the environment.

REFERENCES

Besterfield, D. H. (1999). *Total quality management* (2nd ed.). Upper Saddle River, NJ: Prentice Hall.

Breyfogle, F. W. III (2000). *Managing Six Sigma.* New York, NY: John Wiley & Sons.

Feigenbaum, A. V. (1991). *Total quality control* (3rd ed.). New York, NY: McGraw-Hill.

Harry, M., & Schroeder, R. (2000). *Six Sigma, the breakthrough management strategy.* New York, NY: Currency/Doubleday.

Juran, J. M. (1999). *Juran's quality handbook* (5th ed.). New York, NY: McGraw-Hill.

Lamprecht, J. L. (2005). *Applied data analysis for process improvement.* Milwaukee, WI: ASQ Quality Press.

Mann, D., & Domb, E. (1997). *40 inventive (business) principles with examples.* Retrieved 8 November, 2008, from http://www.triz-journal.com

Schermerhorn, J. R. Jr. (1993). *Management for productivity* (4th ed.). New York, NY: John Wiley & Sons.

Scholtes, P. R., Joiner, B. L., & Streibel, B. J. (1996). *The team handbook* (2nd ed.). Madison, WI: Oriel, Inc.

Thomas, K. (1975). *The handbook of industrial and organizational psychology* (Dunnette, M., Ed.). Chicago, IL: Rand McNally.

Wexler, M. (2010). Riding on the global warming express. *National Wildlife, 48*(5).

ENDNOTES

[1.] Source: Energy Information Administration, *Electric Power Annual 2007,* (published 21 January 2009).

[2.] Source: 2009 NERC Disturbance Report (published 19 June 2010).

[3.] Source: Edison Electric Institute, Washington, D.C. *Statistical Yearbook of the Electric Power Industry,* Annual.
Notes: Capability represents maximum kilowatt output with all power sources available. Capacity margin is the difference between capability and peak load.

4. TRIZ is a Russian language acronym for *Teoriya Resheniya Izobreatatelskikh Zadatch.* Translated into English it means "The Theory of Inventive Problem Solving."

5. Source: U.S. Army Corps of Engineers, *Waterborne Commerce of the United States,* Annual (2007).

6. Source: Association of American Railroads, Washington, D.C., *Freight Commodity Statistics,* Annual (2008).

7. Source: U.S. Environmental Protection Agency, *National Emissions Inventory (NEI) Air Pollution Emission Trends Data;* (1970-2002).

8. Source: U.S. Census Bureau, 2007 *Service Annual Survey, Truck Transportation, Messenger Services, and Warehousing.* (Published March 2009).

9. Source: U.S. Fish and Wildlife Service, *Endangered Species Bulletin*, Bimonthly (May 2009).

10. Source: U.S. Fish and Wildlife Service, 2006 *National Survey of Fishing, Hunting, and Wildlife Associated Recreation,* October 2007.

Chapter 4
Generator Modeling Practices for Renewable Energy Implementation

ABSTRACT

In order to integrate a renewable energy resource into the current electrical grid infrastructure, a comprehensive assessment of the demand-side usage patterns is compulsory. It includes such economic variables as tourism within the area, fuel usage, industrial/agricultural output, labor productivity, employment by sector, household income, and the electricity prices by group. A few of the key metrics as inputs for the integrated resource planning are described in this chapter. A thorough assessment and detailed analysis is also required for each generator request to investigate the proposed electrical impact on the grid.

OVERVIEW

The key characteristic for the successful integration of new generation within the electrical grid is planning. The complexities involved with the inter-related components of the power system requires detailed analysis and models to quantify specific attributes in the electric footprint. The models are then evaluated and thoroughly assessed to determine how the system reacts to real-world conditions with various change. Hence, the models are an important asset in the planning process. The system analysis that formulates the basic upgrade of the generations and transmission entails integrated econometric linear-programming based models, which includes statistical analysis, optimization, simulation, and the calculus of variations. The major function of the system models is to provide detailed analysis of a proposed modification and prescribe specific alternative solutions – typically addressed by multi-criteria decision methodologies (Kendall & Kendall, 1994). The optimization process in generation capacity planning (expansion) and

DOI: 10.4018/978-1-4666-2839-7.ch004

a utility integrated resource plan, for example, is a non-linear process, integer, stochastic, and multi-objective optimization problem designed to minimize cost and maximize reliability. The integrated resource planning considers a full-range of feasibility supply-side and electrical demand-side options and evaluates each against a common set of objectives. The paradigm provides an opportunity for system planners to address complex issues in an inclusive, structured, and seemingly transparent method. The techniques utilized tend to be more diagnostic than prognostic in most cases where the *health* of the system is described but not necessarily *how to cure it*. Nevertheless, the primary objective of integrated resource planning from an electrical utility perspective is to develop a least-cost alternative (plan) that can meet the customer-energy service requirements and environmental improvements. Planning is an important tenet in the renewable energy generation implementation process. It was espoused by one of the early leaders in America who once said:

By failing to prepare, you are preparing to fail. (Benjamin Franklin ~ 1706 - 1790)

When the selected generator system integration plan is agreed upon by the stakeholders, an intensive evaluation is conducted based on the electrical impact within the proposed footprint. The plan, or request, then enters a queue for deeper analysis of the system, for impact studies typically at the regional Independent Service Operator facility. The major problems encountered at this level are "request queue overload," where most plans are insufficient financially and deficient in resources. The queue quickly becomes backlogged as higher-priority transmission and distribution (T&D) projects are evaluated over the generation requests. The integrated resource plan requires detailed analysis based on realistic scenarios of the electrical footprint (economic conditions, electricity prices by group, tourism, etc.) as well as the proposed generator parameters.

The restructuring of the system impact study firstly lien the input of the acceptable plans to evaluate, and secondly, with the implementation of a "time-phase" analysis system that relieves the burden experienced within the queue itself. Both propositions are cost-effective measures designed to increase request throughput and reduce cycle time.

ELECTRICITY DEMAND

In order to integrate a renewable energy resource into the current electrical grid infrastructure, a thorough assessment of the demand-side usage patterns is compulsory. It includes such economic variables tourism within the area, fuel usage, industrial/agricultural output, labor productivity, employment by sector, household income, and the electricity prices by group. A few of the key metrics as inputs for the integrated resource planning is described in this section. Several electrical entities exclude critical elements in the resource planning activities particularly with the five and ten year forecasts. A large number of the planning departments are quite content with the utilization of the "ten percent" rule (add/subtract ten percent from last year's activities) when formulating the annual budget. Yet other utilities solely depend on the regional entities and /or Independent Service Operators to provide the crucial economic details in the local electrical footprint. These "guesses" and assumptions are typically in error and profoundly exacerbate the electric demand predicament on a much grander scale. Nevertheless, the demand-side predictions are based on the price of electricity and its competing commodity (i.e. gas) as well as the efficiency of the installed home appliances over a period of time. The volatility of the natural gas prices enhances the probability factor of electric energy utilization (Awerbuch, 1993). Moreover, the forecast must include a method to detect how the economic activity from the industrial and business sectors within the electrical jurisdiction can thrive to create sufficient disposable income to permit the customers to purchase electricity. These economic conditions from the demand-side areas are essential elements with the introduction of generation expansion from the supply-side of the equation. How people spend their time, utilize their money, and employ various technologies must be included in the model. The economic indicators provide a key metric for the performance of the area of interest for the project implementation (Budget of the U.S. Government, FY2013). Integrated resource planning efforts must eliminate the straight-line increase of postulations over a five or ten year period from a linear model and provide the detailed analysis of each individual product class (residential, industrial, commercial, and street lighting) with solid reasons as to why that element may or may not increase over time. The complexity of the electric infrastructure requires this type of evaluation in order to supplement the system with additional generation. One of the categories excluded from a typical resource planning report is tourism. Several regions around the nation are visited by tourist annually and increase the electricity use proportionally. The top five states visited by overseas travelers, for example, from 2000 – 2008 are illustrated in the Table 1.

The international tourism input is one of the projections utilized as a forecast driver of electricity usage and peak demand. An astute power supply resource planner/forecaster in Florida for example, will recognize a decrease of international tourism in this metric from its peak year in 2000 (6,026,000) and prepare accordingly. Even so, the data reflects an uptick recorded in the most recent year 2008 (5,246,000) where the tourism may return to peak levels. Conversely, more oversea travelers are visiting the State of New York (8,413,000 in 2008) than any other state in the nation. The peak demand for the electric service is expected to accommodate these end-users without any monetary outages or long-term blackouts. Interestingly enough, the Hawaiian Islands experienced a decrease in tourism of international travelers over the same time period. The domestic tourism is included in a separate category and counted towards the electricity usage as well. The processes are calculated for each customer group and aggregated to determine system-wide sales. The tourism category alone has the potential to evaluate the detailed consumption of the electric commodity. The time-of-use data compiled by the interconnected metering has the potential to make available detailed analysis during these peak periods. The devices are designed to go beyond the traditional two-way communication of data collecting between the customer and utility to improve restoration times of outages, as well as accurate readings. It is expected to develop real-time usage patterns of the electrical footprint to determine the peak loads for potential generation expansion. These smart devices are also expected to evaluate the load profiles for network calculations and power quality monitoring at the site of the end-user (Pertti et al., 2007). Additional key electricity metrics for the integrated resource planning model is the average

Table 1. Top five states visited by overseas travelers: 2000 – 2008[1] (5,922=5,922,000)

Tourism: Top Five States	2000	2005	2007	2008
New York	5,922	6,092	7,908	8,413
California	6,364	4,791	5,185	5,296
Florida	6,026	4,379	4,683	5,246
Nevada	2,364	1,821	1,768	2,103
Hawaiian Islands	2,727	2,255	1,864	1,825

annual expenditures of the end-user. An end-use model for household electricity exploitation includes separate estimates for the following devices:

- Lighting
- Space Heating
- Air Conditioning
- Fans
- Water heating
- Cooking
- Entertainment

If elected, the deployment of a "customer portal" provides essential data to the electric companies. The value-added service is designed to improve energy efficiency with enhanced data services related to electricity (appliance monitoring, real-time pricing, and demand-side management) (Gunther, 2007). More goods and services are purchased by the typical American worker, for example, as the disposable household income increases. The income is generated by employers in various sectors of the economy. The annual payroll of employees (by service-related industries)[2], has increased precipitously in specific regions of the country. A payroll increase of 8.22% in the professional, scientific, and professional services sector was experienced in 2006. This group includes accountants, scientific research, engineering, computer systems analyst, veterinary and legal service occupations. Additionally during the same time period, the payroll increase of 9.82% was experienced by the administrative, support, waste management, and remediation service sectors. This assemblage comprises of the telephone call centers, collection agencies, security services, waste management, travel arrangement, and temporary service personnel. Lastly, the accommodations and food service sector experienced a 6.70% increase in payroll during the same period. This group includes the hotel /motel personnel, recreation vehicle park employees, boarding houses, and drinking (alco-

holic beverages) establishments. The increase in wages usually equates to an increase in spending. Recently, electronic device purchases and other forms of electric exploitation are the preferred method of electric consumption. The astute power supply utility planner must utilize the data in this metric and plan as a result of the trend. Details of the data are broken down to the utility spending levels. The average annual expenditures increases in accordance to income levels as illustrated in Table 2.

The trend depicts higher income earners ($150k and over) are more inclined to include the various electrical devices and services in their households. The lower income consumers (< $70k) may possess a variety of devices with only sporadic electric use. Typically the forecast entails the population growth statistic classified into dwelling types (single-family, multi-family, mobile home, etc.) for the residential category. It is then classified to the heating (electric/non-electric) use for further analysis. The incorporation of "smart" appliances and other consumer devices (i.e. a remotely controlled air conditioning unit that shuts down during high energy usage) are generally found in the home of high wage-earners. Opportunities exist for other service business platforms to create value in the model (Smedlund, 2012). Yet, in the demand-side resource arena, the customer demand may be curtailed during periods of peak loads via inexpensive purchase power programs to cycle

Table 2. Average annual expenditures of all consumer units by income level: 2007[3]

Income Level	Housing Utility/Fuels
Less than $70k	2,894
$70k ~ $79,999	4,129
$80k ~ $99,999	4,256
$100k and Over	5,234
$100k ~ $119,999	4,657
$120k ~ $149,999	5,077
$150k and Over	5,809

the large electrical devices (water heaters, air conditioners, advanced cooling and heating technologies, etc.) through the local utilities no matter what the income level. Nevertheless, the electrical energy usage translates into utility revenues as one of the many variables encompassed within the integrated resource planning sphere. The revenue for selected states per class is illustrated in Table 3.

Revenues from electricity sales to retail customers, for example, will decrease significantly from year-to-year based on consumer confidence. Recently, several utilities experienced losses in this category based on the current economic conditions. The important financial metric (return on equity, internal generation of funds, interest coverage ratio, etc.) is often a key factor in the decision process for renewable energy implementation. The revenue of the electric utilities assists with the critical decision of managing the load via other means. The overall goal of the process is to avoid the construction of generation facilities that would be operated for only a few hours per year. A secondary goal of the load administration paradigm (peak clipping, valley filling, load shifting, strategic load growth, and flexible load shaping) is to avoid expensive power purchases when the customer load can be shifted (curtailed) at a lower cost, during peak periods. These efforts are achieved through direct load control, price alter-

Table 3. Selected electric energy price by class of service and state: 2007[4] (revenue (in cents) per kilowatt – hour (kWh))

State	Residential	Commercial	Industrial	Total
Hawaii	24.12	21.91	18.38	21.29
Connecticut	19.11	15.39	12.92	16.45
New York	17.10	15.92	8.71	15.22
Massachusetts	16.23	15.20	13.03	15.16
Maine	16.52	12.94	14.11	14.59
New Hampshire	14.88	13.91	12.27	13.98
Alaska	15.18	12.19	12.63	13.28
Rhode Island	14.05	12.67	12.04	13.12
Vermont	14.15	12.29	8.92	12.04
Texas	12.34	9.87	7.79	10.11
Colorado	9.25	7.62	5.97	7.76
Minnesota	9.18	7.48	5.69	7.44
Oregon	8.19	7.20	5.06	7.02
Arkansas	8.73	6.91	5.25	6.96
North Dakota	7.30	6.58	5.24	6.42
Utah	8.15	6.54	4.52	6.41
Nebraska	7.59	6.39	4.78	6.28
Kentucky	7.34	6.76	4.47	5.84
Wyoming	7.75	6.25	4.10	5.29
Idaho	6.36	5.14	3.87	5.07
Average	12.18	10.46	8.49	10.49
Contiguous U.S. Average:	11.27	9.68	7.63	9.66

natives (interruptible rates), and thermal energy storage techniques. The most important costs to the utility are the rate reductions and billing credits involved with the promotion of the special programs. The major focus of these programs from a consumer perspective are the utilization of well-organized devices /methods to include water heater wraps, building insulation, proficient refrigerators, high-efficient lighting, and "smart" heating, ventilation, and cooling processes. The previous loads were considered as "grid friendly" resistive types (incandescent lights, resistive heating, etc.). The newer, energy efficient types (variable speed /frequency drives, AC compressors, compact fluorescent lights) have an undesirable effect on grid dynamic stability. The transmission and distribution system upgrades are essential to formulate compatibility between the load and generation. Oftentimes, these critical component modifications are disregarded when new generator expansion is introduced.

The avoided cost analysis, from a system performance perspective, is all-inclusive quantification of the demand-side options (Busch & Eto, 1996). It is the economic expenditures (capital and operating expenses to supply generation capacity and fuel, transmission, storage, distribution, and customer service) to meet the end-user requirements based on a set of resources. The avoided cost, or "avoidable costs," includes the avoided capability costs, energy cost, and ecological externalities (a cost/benefit from the production or consumption that is not accounted for in market prices. For example, the damage to human health cost from air pollutants) related with the electric service. Resources deemed as less than the avoid cost are thus cost-effective and necessitates implementation within the integrated planning criteria. In order to model and estimate the avoided cost, the identification of a set of options (energy-efficient appliances, direct load control, time-of-use rates and /or interruptible loads, and energy storage) within the demand-side management sphere is required. The estimation of the energy reductions

in peak demand and savings are developed with the use of the expected penetration levels of the individual programs and end-use technologies. The aggregate efforts of the bundled programs are assessed in a block-by-block methodology that results in a seasonal load curve. The avoided cost of the demand-side management activities is then determined with the traditional production costing methods as compared with the resource plan. The situation will create various scenarios (expected cost, variance, and penalty factors) for the objective function. The penalties form a direct relationship with the consumer as either a surplus capacity component (inefficient utilization of installed capacity) or an unmet demand (shortage of electricity during peak periods). The incisive utility planner is aware of these conditions and assesses each proposition with extreme caution. Ultimately, the customer is the determining factor for any new expansion endeavors within the electrical footprint.

BEST PRACTICES AND FORECAST MODELING

Planning originates with the load forecast (Kahn, 1991). The best practice in forecast modeling includes a graduated assessment of techniques from good to the very best in the industry. Several electric utility companies employ some form of the techniques (or a combination) to develop the integrated resource plan. The traditional method currently in use involves the formulation of a "base case." The forecaster typically prepares the base case and several alternative forecasts to compare against an assumption of variables. The high and low assumptions of economic growth, fuel prices, population, or other combinations of the parameters are assessed and compared to the base case and compiled in an annual report. Nonetheless, detailed economic analysis of the geographical service area is a critical input for the forecast model. This is a necessary step in order to obtain accurate information for the electrical footprint. In the *good* trend forecasting model, the assumption of electricity utilization per customer will maintain the same rate in the future. This growth rate is developed from historical sales / peak demand data to estimate future electricity consumption (Nichols & Von Hippel, 2002). The sales records are categorized by customer class and geographical area and the demand records are included in a load profile compiled over a substantial period of time. Straightforward statistical methods are utilized in the trend forecast model. The *better* analysis package, econometric forecasting, comprise of the considerable interactions between the chronological economic variables, sales data, and peak demand. These economic variables include fuel usage, tourism ~ domestic and international, industrial/agricultural output, household income, employment by sector, labor productivity, and electricity prices by class. The selected production indexes by U.S. industries are shown in Table 4.

The data depicts an increased trend in the electric power generation, transmission & distribution category over the period and fluctuations in the natural gas distribution sector. The projections of the variables are easily acquired from financial institutions, bureau of economics, etc. to drive the forecast of electricity utilization and /or peak demand. The progressions are characteristically executed independently for each cus-

tomer class, and aggregated to extrapolate organizational sales. One of the *best* forecasting methods is the end-use model. The technique develops approximations of electricity requirements with comprehensive analysis of its utilization. The evaluation is reasonably exhaustive and presents additional information for the utility planners. The model affords integrated predictions of electric and peak power demands. The end-use forecasting technique is flexible enough to provide the foundation for the energy-efficient options - (smart air conditioning units, intelligent thermostats, efficient lighting technologies, etc.) that can sense various signals (chiefly comprised of cost, frequency, and voltage stability) and communicate bi-directionally via the grid infrastructure with the utility without compromising system reliability - by making changes to the parameters and comparing the plan to the base case. The implementation of operational intelligence as a solution to real-time analytics within the grid infrastructure will assist in the evaluation of the demand-side data. The methodology consumes large amounts of metered data and actively performs analytics in search of anomalies, trends, or patterns that indicate problems or opportunities for key stakeholders (Bohan et al., 2011). In the local area forecasting model, comprehensive experience with the prior development patterns forms the foundation of this technique. It is utilized in the sub-transmission and distribution areas of the

Table 4. Selected industrial production indexes by industry: 1990 – 2008[5]

Selected Industrial Production Indexes by Industry: 1990-2008								
	1990	2000	2003	2004	2005	2006	2007	2008
Electric Power Generation, Transmission & Distribution	76.9	97.2	102.1	104.2	107.2	107.8	110.5	110.2
Manufacturing, Durable Goods	52.8	105.2	102.7	107.0	112.8	117.8	120.2	116.3
Natural Gas Distribution	85.6	98.6	100.9	98.8	97.1	91.3	98.1	100.2
Manufacturing, Non-Durable Goods	87.7	102.2	100.1	102.0	104.8	105.7	106.7	103.7

(2002=100)

business. Nevertheless, a combination of all three of the major forecasting methods (trend, econometric, and end-use) is considered as *best practices* within the industry.

The introduction of the proposed generation on the supply-side involves specific attributes for the integrated resource plan. The capital and operating cost of the generator is typically the first and foremost concern for the utility. Questions such as the procurement and operating costs, as well as the fuel expenditures, are usually bounded as one of the first attributes. The reliability issue and its associated history is another consideration of the power supply technology – both internationally and domestically. The control room of the future must be able to utilize efficient predictive tools to monitor and dispatch the wind and solar generation for reliable delivery of electricity (Jones & Cheung, 2012). Another attribute involves the efficiency of the proposed technology, specifically the net amount of electricity produced per unit of input fuel. As for the transmission & distribution system upgrades required to accommodate new expansion, the efficiency equates to the percentage of power /energy lost during transmission. The quantities and quality of fuel for generation is another attribute of the proposed power supply. Also, the operational life of the technology as well as the decommissioning costs is major attributes to consider for generation expansion projects. The

expected net value of the unit after its useful life must be measured. Additionally, the environmental impacts of the proposed power supply technology are of a major concern for the socially responsible electric utility company. The quantities of air pollutants as well as liquid/solid wastes per unit of electricity produced are significant attributes for supply-side management. Moreover, the ecological impact of the unit during the construction and decommissioning phases must be considered. The attribute includes the land use requirements for the proposed technology as well. Finally, the proposed plant capacity of the unit is thoroughly analyzed for its size and available generation capacity throughout the year. These major attributes (as well as other parameters) are utilized in the supply-side model to evaluate the effectiveness of the technology.

The criteria used to assess an integrated resource plan involve several detailed categories from both the supply and demand sides. The implementation of a renewable energy resource is evaluated in several stages. The resource plan is designed to evaluate the feasibility of the technology as it relates to the current infrastructure (Nichols & Von Hippel, 2002). The best practices of the integrated resource planning area are illustrated in Table 5.

The integrated resource plan is a "living document" subjected to many revisions through-

Table 5. Best practices of the integrated resource plan

Financial Criteria:	Performance Criteria:	Energy Security:	Environmental:	Other Criteria:
• Overall plan cost (capital, fuel, and other costs) • Utility net income • Return on equity • Internal generation of funds • Interest coverage ratio	• Customers served • Loss of load probability (LOLP) • Reserve margin • Efficiency of energy use (supply/demand side)	• Diversity of Supply (fraction of each fuel used) • Use of renewable resources • Use of domestic resources	• Amount of emissions produced over the life of the plan • Impact on wildlife/ biodiversity • Liquid/solid waste production	• Cultural impacts of the plan • Social implications • Political acceptance • Feasibility of the plan • Impact on economic sectors (positive/ negative) • Employment impacts of the plan • Aesthetic issues (recreation/ tourism)

out its lifecycle. Alternative plans may emerge based on electricity demand, technological advances, fuels prices, and regulatory policies. Nevertheless, the plans are vigilantly reviewed for feasibility and subsequently evaluated based on a set of criteria – quantitative as well as qualitative. A systematic evaluation approach is best to select the top alternatives. The integrated resource plan is a multi-criteria /multi-attribute analysis problem commonly solved via sophisticated mathematical algorithms. Ultimately, the integrated resource plan is a combination of the supply and demand-side resources. Once implemented, the plan is closely monitored to assess its effectiveness. The demand-side management program, for example, will evaluate the energy consumption and associated processes (marketing, quality control measures, field delivery of the program, etc) to test the efficacy of the plan.

SUPPLY-SIDE EVALUATION AND PROBLEM FORMULATION

The electric utility business is one of the most capital-intensive industries in the world based on the large investment in generating units. It is akin to the paper mill organization with many interlinked processes where investment decisions affect practically every aspect of the industry (Everett et al., 2010). The generation options for the electric utility company includes new power plant construction, purchasing electricity from independent power producers/interconnection, extending the lives of older generating units, or increasing the capacity /efficiency of the older units (Glover, 2001). Typically, detailed plans (alternatives) are evaluated in terms of reliability and cost aimed to satisfy environmental, energy security, and performance criteria of the proposed generator. The reliability aspects of the objective must entail sufficient generation available to meet the load. The network configurations required to accommodate the proposed genera-

tion is classified in the regulated transmission planning category. However, the complexity of new generation expansion typically encompasses both entities as major investment decisions are determined by the market instead of the formally vertically structured, centralized utility. There are numerous tradeoffs and risks involved in the overall generation planning arena. New expansion financial investment has shifted from the once regulated customer to the de-regulated market forces where generation investments are generally uncertain. Consequently, the generation requests increased precipitously over the past few years and subsequently imposed a burden on the regional independent service operator system impact study queue. The planning for the doubling of electricity every ten years by the vertically integrated utility has become more complicated since the early 1970s. In a de-regulated environment where any financially-backed legitimate investor can request a generator (usually in the same location as others) for system implementation generally creates project uncertainties throughout the electrical footprint. The exceedingly long lead-times for the transmission line permit (6~10 years) poses additional problems within the planning horizon. An example of the situation illustrates the generator requests in one particular southwest region of the country exceeds 40,000MW. A Midwestern region has received requests closer to the 60,000MW range. Both regions do not current have the available transmission and distribution networks to accommodate the new expansion projects, yet the system impact study is required to move the project forward. A thorough assessment and detailed analysis is required for each generator request to investigate the proposed electrical impact on the grid. Computer simulations analyze complex mathematical algorithms with hundred of thousands of non-linear equations as well as the unknowns, and formulate models of the network power flows in the transmission and distributing system of the proposed generator implementation. The current electrical capacity is included in the objective function. These calculations encompass power plant start-up/production costs, no-load costs, and operating reserve (Carlson et al., 2012). The solutions may consume anywhere from a few seconds to several days, depending on the accuracy of the data and model integrity. Nevertheless, the power supply integrated resource problem is structured to evaluate the least-cost alternative. The problem is formulated to evaluate the attributes, uncertainties, and options. A decision database is consequently developed by computing attributes for a larger set of scenarios. Decision support processes and scenario-based configurations are utilized in the petroleum industry as well (Meyer, 2011). The exploitation of the tradeoff concept is subjugated to identify a comprehensive decision set once the inferior alternatives are rejected. A final strategy is subsequently developed once the plans in the decision set are thoroughly analyzed. If a utility evaluates cost ($/kWh), for example, to implement various power supply sources (wind turbine generator, remote substation, storage units, and photovoltaic array) in the electrical footprint, it will encounter hundreds of alternatives for the decision problem based on the uncertainty factors of resource availability, economic conditions, and the forecasted load growth. The expert, or group decision problem is utilized to evaluate the societal risk for complex system analysis (French, 2012). The objectives of the example are simple: minimize the cost and emissions while maximizing the system reliability. The incremental steps involved in each unit (i.e. the wind area square footage and capability relationship between the minimum /maximum outputs) are a component of the decision set to determine the optimal combination of the technology capacities. This concept applies to every unit introduced into the current system. The model output of the simulation, therefore, is expected to produce the cost of the energy production ($/kWh), SO_2 emissions (tons/yr), PV cost ($/ft^2), and the load service as major attributes of the plans previously prioritized by the decision-maker's priority matrices.

The plans are analyzed and re-assessed based on these priority weightings of the decision-makers to produce acceptable results. Moreover, the system integration involves the thorough assessment of the current grid conditions (as-is) migrated to the future (to-be) condition. The best alternatives are compared in the optimal energy mix for the final implementation strategy. The utilization of a decision circuit, a method to exploit conditional independence by combining the benefits of decision trees and influence diagrams, can calculate the solution of multiple decision problems in a single representation (Bhattacharjya & Schachter, 2012). The technique is well-suited for real-world decision problems. The large solution set for a complex system requires advanced search methods to achieve the optimal effectiveness of the stated objectives. One such method utilizes a knee-set algorithm. As a multi-objective optimization problem, the goal of the decision-maker is to identify a Pareto set of optimal solutions within a tradeoff region of acceptable plans (alternatives) (Miettinen, 1999). The computation of selected points generated by the algorithm creates a bulge, or knee in the Pareto front. This area represents the optimal compromise in a multi-objective optimization problem readily acceptable by the decision-maker. The value of every plan is calculated and prioritized based on the measure of its composite distance to the best solution. The variance of this composite distance is estimated with an error model (generally assigned as 10% of the expected values because of inconsistent priority judgments). The identification of the decision set by search of the plans with the potential to overlap the minimum distance solutions employs a 90% confidence interval within the program. The development of a global decision set is approximately determined as the union of the conditional decision sets (Mattson et al., 2007). In the meta-heuristic approach, the inclusion of a combination of genetic algorithms, simulated annealing, tabu/scatter search techniques and their derivatives, are the chief forces behind the model.

The process improves the solution set by utilizing an adaptive memory approach (which solutions work well before and consequently recombined into a new, better solution) (Glover et al., 1996). The search avoids local solution traps as well as noisy (uncertain) model data. Nevertheless, the search algorithms are designed specifically for the decision-maker to act upon acceptable alternatives based on the new expansion criteria.

Once the plan (alternative) is vetted, reviewed, and selected by the decision-makers, it is subsequently sent to the regional independent service operator as a request for generation. The plan, along with other entities requesting some form of generation, T&D expansion, long-term megawatt reservations, etc. is thoroughly assessed by the study group. These system impact studies are currently based on a First-In-First-Out precedence queue, and consequently become overloaded. The queue generally has higher-priority T&D projects for study before *any* generation request is systematically processed. Moreover, the generator requests typically exclude the "to-be" conditions (as previously outlined), for the new expansion. Hence, the required front-end analysis of the requests are shifted to the impact study queue without the detailed due diligence by the requesting entity. Because of the situation, the backlog of the serial impact studies has grown exponentially. In fact, one particular Midwest region experienced, on average, ten generator requests per month over a forty month period. The requesting entities typically exclude the detailed analysis of the affected large and small generators in the electrical footprint as well. The implementation of the supply-side technologies and the associated alternative plans are systematically evaluated against a plethora of scenarios in the simulation. The intelligent decision-maker must analyze the affects of the entire business landscape once the plan is accepted. The process is designed to eliminate re-work, additional administrative steps, as well as minimize costs. The situation is summarized beautifully below:

One of the tests of leadership is to recognize a problem before it becomes an emergency. (Arnold Glasow ~ b. 1905)

The decision-maker is expected to embrace effective analytical skills as well as leadership qualities beyond the bottom line of the company. The assessments of the maximum values (via simulation) within the proposed project are essential elements a decision-maker possesses in his/her arsenal (Kubiak, 2012). It is a non-intrusive method to evaluate an impending situation with a versatility limited only by the end-users imagination. The queue problem is simulated and thoroughly analyzed in this chapter as well as solved in terms of costs, resources, and cycle-time reduction.

The Queue

The queue theory simply states the following conditions in a random haphazard paradigm:

- The average number of requests arriving per unit time is constant denoted by λ
- The number of requests arriving during any two non-overlapping time intervals are independent or…
- The chance (δt is $\lambda \delta t + 0(\delta t)$) that a request will arrive during any specified time interval of infinitesimal duration (*dt* is $\lambda\, dt$)

The typical requests have service times within the exponential distribution sphere where the chance of the customer's service time will exceed any specific duration as depicted below:

$$t = e^{-t}$$

The key characteristic of this service time distribution includes how the customer is served. The time of service for the current request is completed before the next request is served (Cooper,

1981). The system time (a_n) is denoted by the following equation:

$$a_n = b_n + c_n$$

where: b_n is the waiting time (queuing time) and c_n is the service time.

The arrival times (t_n) and service times c_n are typically written in the following form (Papoulis, 1991):

$$M|G|\infty$$

where M equates to arrivals (for Markov ~ Poisson arrivals or *memoryless*) and G equates to service times (exponentially distributed service times). General, which indicates the arrivals or service times are arbitrary.

D equates to Deterministic and indicates the arrivals are constant.

In a single server queue M|G|1 as in the top portion of Figure 1, the arrival times t_i are Poisson distributed and the service times are c_i. The queue details are graphically illustrated in the bottom portion of Figure 1. When a request arrives in the system, it receives service as the others wait in their respective queues. The request advances in the system when the service is complete. The request exits the system at $t=\tau_i$. The next request enters the system at t_i occupies the last slot in the queue and remains unchanged as other requests arrive. Idle periods are experienced when the system is not busy (no requests in the queue). The busy time is a random variable equal to the time interval from $t=\tau_i$ when the service period begins to $t=\tau_i + busy_i$. The queue holds subsequent requests until the system is not busy and ready to receive the request. Timing of the busy/idle cycle is important for multiple processes with limited resources.

Figure 1. The server queue

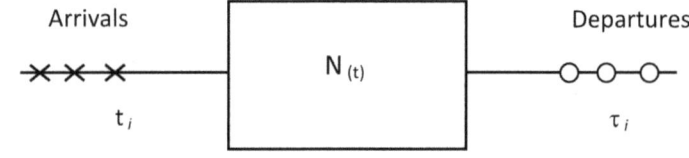

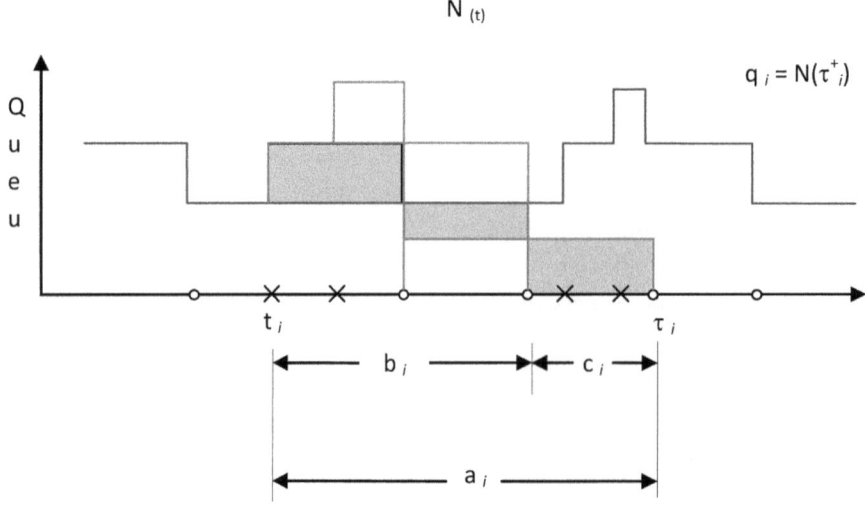

$$b_i = \tau_{i-1} - t_i$$

$$a_i = \tau_i - t_i$$

$$= b_i + c_i$$

where a_i is the system time, b_i is the waiting time, and c_i is the service time.

Priority service within a queue is sometimes necessary to reduce overburden conditions. This application diminish the value of the "first-come-first-served" (FIFO) rule (Gross & Harris, 1997). The variant of the basic queue problem include the spacing of the arrival times per customers requesting service as well as the resources available to service the requests. As an unstable protocol, the "nearest-to-go" system assigns the highest priority request closest to the destination. If a renewable energy project successfully pass the front-end screening process (without internal re-work), it is prepared to proceed with further detailed analysis within a *time-phased* study period (re-evaluation of the request at different time progressions of the project to validate constraints within the system sphere). The core of the analysis is based on system engineering principles that include the following (Martin, 1997):

- Process development lifecycle and configuration management
- Process risk, operational level quality assurance and evaluation
- Metrics for quality assurance and process evaluation
- Metrics for cost estimation and operational effectiveness evaluation
- Total quality management

Another variety of the service request involves the "lowest-wait time" in the queue, a stable protocol similar to the "shortest-in-the-system" paradigm and gives priority to the request introduced last in the system. The number of requests in the system is characterized as bounded by its parameters and independent of time. The condition is referred to as stability and usually encompasses a network of protocols and a defined traffic pattern within the system. Currently, the regional independent service operators are analyzing practically every generator request in relation to the effective impact on the electric grid. The method not only creates initial queue overload in the system, it also develops an unstable environment under the "first-in-first-out" protocol. The FIFO criteria is the most popular protocol because it is easy to implement and simplistic, yet greedy (work-conserving) in terms of storage and releasing a request. It gives service priority to requests that entered the queue first. This type of arrangement in a simulation model mimics the realistic conditions of the system impact evaluation. In fact, the greedy protocols are those forwarding a request across a process when there is at least one request waiting in the queue. Various schemes are analyzed for stability to include the combinational "nearest-to-go-longest-in-the-system" (NTG-LIS) as well as the "farthest-from-the-source" (FFS) protocols. However, based on the consistency of the generation impact study requests, the simple resource-leveling technique and precedent evaluation, coupled with the "lowest-wait-time" in the queue protocol achieved the cost and time minimization objectives in the simulation model. The "last-in-first-out" (LIFO) scheme is marginally successful in the generator request simulation when arranged with a priority queue combination for overload control. In the absence of a backlog condition, the response time possessed a higher variance than that of the FIFO and subsequently created more abandonments than the scheme. When the mean delay of the LIFO protocol is at overload and high, the favorable increased variance allow more requests to enter the system than FIFO, yet failed several congestion evaluations at critical junctions in the simulation model. At the macro level, the evaluation of every request overburdens the system from the outset and misaligns the required resources performing the tasks.

The TRIZ[6] principle #2 "taking out" is outlined below (Mann & Domb, 1997):

- Separate an interfering part or property from an object, or single out the only necessary part (or property) of an object.
  - Lean manufacturing by elimination of all non-value added activities
  - External laboratory testing
  - Segregation of non-conformant product, material, or equipment
  - Cluster analysis by distilling qualitative customer feedback into quantitative data

The utilization of proficient and cost-effective approaches to incorporate technology into the organization is mandatory to advance the program's effectiveness. Presentation of a combination of related technologies must transform the incoming data into information, and information into knowledge. The process analyst must be able to construct implications about the significance of the data for the decision-makers *before* the queue overload condition existed. This activity is absent in the current process. A front-end screening of the request is unavoidable for the customer to reduce the cycle-time of the system and ultimately the costs. It is necessary to restrict the traffic congestion before queue entry of the impact request evaluation. The generation requests from the customer must provide pre-assessments with the proposed system integration (Grady, 1994). A thorough due diligence (to include detailed checklists of the generator integration parameters) of the request is an important aspect in the pre-screening specialty for the serious inquiry. All resources, financial obligations, applications, cost allocations for system upgrades, etc. must be completed before the system impact study commences. This measure is designed to maximize the overall throughput of the system.

THE GENERAL ELECTRICAL SYSTEM STUDIES

The Power System Simulator for Engineering (PSS/E) is generally conducted to study the optimal timing, size, and location of generation in a region over future years. The program is also utilized to identify priority investments in the main transmission interconnections between sub-regions to optimize power generation requirements over the study horizon. The computer programs and structured data files are designed to evaluate the basic functions of a power system performance simulation to include the following:

- Fault Analysis (short circuit) - Equivalent Construction
- Optimal Power Flow - Open Access & Pricing Calculations
- Voltage Analysis - Dynamic Simulations
- Data Handling, Updating, Manipulation

The optimal power module (OPF) is intended to improve efficiency and throughput of the power system performance studied by adding intelligence to the load flow solution processes (minimizes costs/maximizes performance). The dynamic simulation module examines system disturbances (generator tripping, motor starting, faults, and loss of field) and is an invaluable asset for generator system integration activities. Several of the algorithms utilized in the simulator for dynamic studies, for example, are used in the actual device installed in the field (Ghorai & Reddy, 2008). Because of this, engineers can accurately identify the response rates of a completed system to various transient events. The system studies are derived from detailed computer simulations that represent large geographical areas and include major transmission/generation components individually modeled within the system. The integration of a wind turbine generator to a transmission system, for example, is a challenging task that can have a negative impact on the electrical grid. Voltage fluctuations are created at the transmission level if a turbine trips offline for any reason. Additionally, a large in-rush current is required initially with standard induction turbines start-up activities to minimize voltage fluctuations. Moreover, because of the unpredictable nature of the wind, the variability of the wind speed may impose large and sudden power output changes within the system. Finally, the voltage flicker/harmonics generated by wind turbines may amplify any existing grid flicker and harmonics to unacceptable levels. These reciprocal conditions may affect the operation of the wind turbines as well and possibly cause severe damage to the units. The electrical system modeling of the generators – when performed correctly – seeks to minimize the negative impacts of system integration (Bergen & Vittal, 2000). The typical PSS/E program may be modified in accordance with specific procedures to include generation and transmission maximization model as an input. The adaptation allows the user to evaluate various scenarios such as the isolated operation of each power system, regional operations of the system, and the market conditions with constraints (interconnection capacity and the maximum amount of import capacity from each region). The configured simulation aids in the analysis of network topography, demand, and production/exchange data (Wood & Wollenberg, 1996). The evaluation of the steady-state load flows are also calculated and permit the performance of contingency analysis (a vital attribute in electrical system evaluation). Security criterion based on the voltage profiles and line congestions (thermal loadings) are analyzed as part of the study as well. Finally, the identification of potential network bottlenecks are described in detail with possible transmission system relief solutions. The typical simulation model is designed to evaluate the system reliability in terms of loss-of-load expectation of the generation and loads in a control area (Anderson & Fouand, 2003). The diagram illustrated in panel A of Figure 2 depicts a small region with two generators.

The case is analyzed by the electrical planning experts to assess the impacts of both large and small generators, thermal overloads, flow/network analysis, and voltage constraints when a new generator or long-term megawatt request is introduced in the system. The electrical footprint is comprised of several interconnected zones where generation reliability is expected throughout the region. The interconnected zones are defined based on the limiting interfaces within the control area. An hourly load profile is developed from the generating units within the simulation and adjusted for planned maintenance as well as forced outages to mimic real-world conditions. If a region's available generation is less than load, the simulation program will attempt to distribute support from areas that contain a generation surplus for that hour subject to the transfer limits connecting the regions. The area is considered deficient for the hour (a loss-of-load state) if the generation cannot be delivered. The random forced outages, planned maintenance, and electrical process data are compiled for every hour of the day, month, season, and year to formulate a robust reliability index. The simulation is then conducted hundreds of times with various arbitrary outages on the generators as well as the transmission interfaces to assess the system reliability. The case histories permit the engineering planners to accurately assess random events (i.e. equipment failures) and deterministic policies which administer system operations.

The contingencies are analyzed in detail in the simulated model once a new generator, for example, is introduced in the system. The fault analysis program is conducted to determine the extent of a problem and its associated interconnections. There are numerous actions (faults) computed and summarized in panel B of Figure 2 within a specific five generator, three region case study. The systematic opening of key components in the system was computed and analyzed from various points in the electrical footprint. A sample of the actions is depicted in Table 6. There were

Figure 2. The two generator model and contingency actions

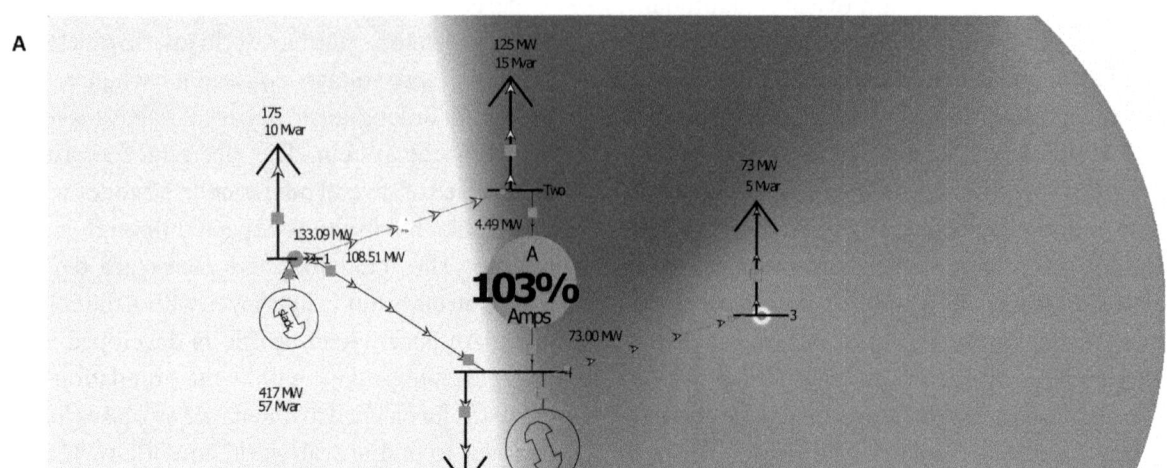

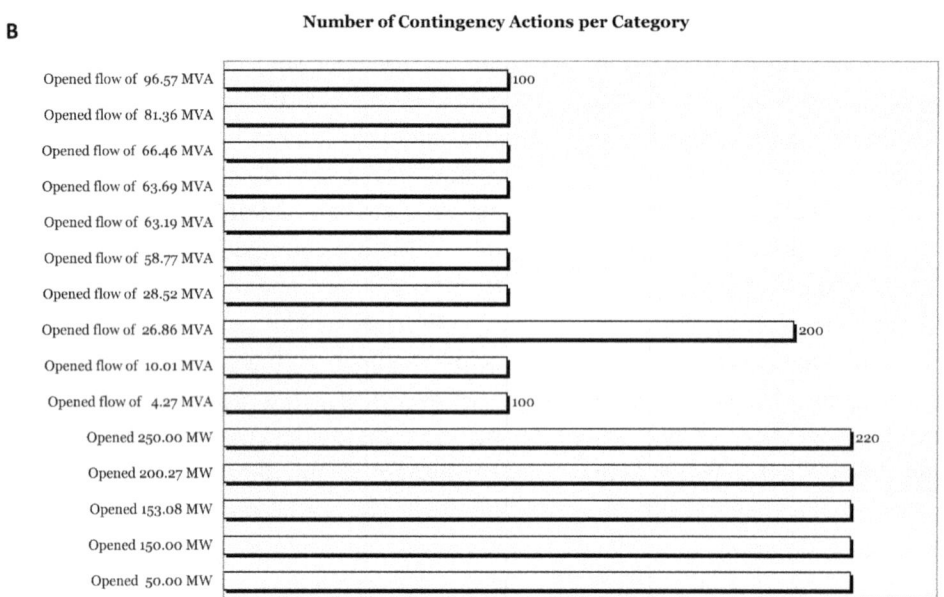

200 faults involved, for example, with the open flow of 26.86 MVA as summarized in the bar chart (panel B of Figure 2) and 220 faults in the categories of open megawatt flow of 50, 150, 153. o8, 200.27 as well as 250. How each of the faults affects the new expansion or parallel activities (if the wind turbine generator was introduced in this area) are of vital importance for all stakeholders occupied in the process. The engineering plan-

ners are expected to evaluate specific outages and conditions that affect the region from practically every perspective. The simulation program assists these experts with this arduous task including scenarios such as long-term effects of megawatts on the transmission system of new generation and projected load growth.

Table 7 illustrates a sample of the violations that occurred during the contingencies. It includes

Table 6. Fault samples for evaluation

Sample of Actions
OPEN Gen 1_138.0 (1) #1
OPEN Gen 2_138.0 (2) #1
OPEN Line 1_138.0 (1) TO 2_138.0 (2) CKT 1
OPEN Line 1_138.0 (1) TO 3_138.0 (3) CKT 1
OPEN Gen 1_138.0 (1) #1

the 120-kV limits, the actual values, and the percentage over the limit (panel A of Figure 3). The contingencies are ranked by order of severity and can be utilized for detailed analysis within the branches.

Table 8 illustrates a summary of the violations that occurred on the transmission lines during the contingencies. It depicts the number of violations (89), for example, from bus #1 to #2, the maximum loading (222.76%), and the aggregate percent overload (2904.11). The simulation has the ability to describe the events that occurred within the faulted areas through selected scenarios. These

important evaluations are critical in the assessment of new generation implemented in the system.

Table 9 illustrates a summary of the limit violations that occurred on the branches and busses during the contingencies. It clearly displays the actual value and the associated limits, as well as the overload percentages on the branches. The element 2 in the Hilltop-Deer Park area, for example, with a 230.47% overload is one of the many branches that may warrant further study for generation and /or projected load growth. The system impact studies are meticulous and detailed. The analysis must follow a strict protocol to obtain valid data for additional generation system implementation.

Another analysis may involve the request of 100 MW of long-term, firm transmission network service from one specific regional source to the control area over a twenty-year period. The regional Independent Service Operator will evaluate the impact study request with seasonal models, summer/winter peak data for selected future years, and a base out year. The simulation is utilized to dispatch generators in the region and adjust each

Table 7. Samples violations with overload rank and limits

Sample Violations					
RANK Overload	Label	Category	Value	Limit	Over Limit Percent
7	G_0000022U1&G_0000077U1&L_0000022-0000066C1&L_0000077-0000055C1	Branch MVA	267.31	120	222.76
6	G_0000011U1&G_0000077U1&L_0000022-0000033C1&L_0000033-0000044C1	Branch MVA	235.33	120	196.11
5.2	G_0000011U1&G_0000066U1&L_0000022-0000033C1&L_0000033-0000044C1	Branch MVA	235.3	120	196.08
4.5	G_0000011U1&G_0000044U1&L_0000022-0000033C1&L_0000033-0000044C1	Branch MVA	235.26	120	196.05
7.9	G_0000022U1&G_0000044U1&L_0000011-0000033C1&L_0000077-0000055C1	Branch MVA	210.11	120	175.09
3.8	G_0000011U1&G_0000022U1&L_0000022-0000033C1&L_0000033-0000044C1	Branch MVA	209.39	120	174.49
7.9	G_0000022U1&G_0000066U1&L_0000011-0000033C1&L_0000077-0000055C1	Branch MVA	202.71	120	168.92

Figure 3. The limit violations and generator loss model

A

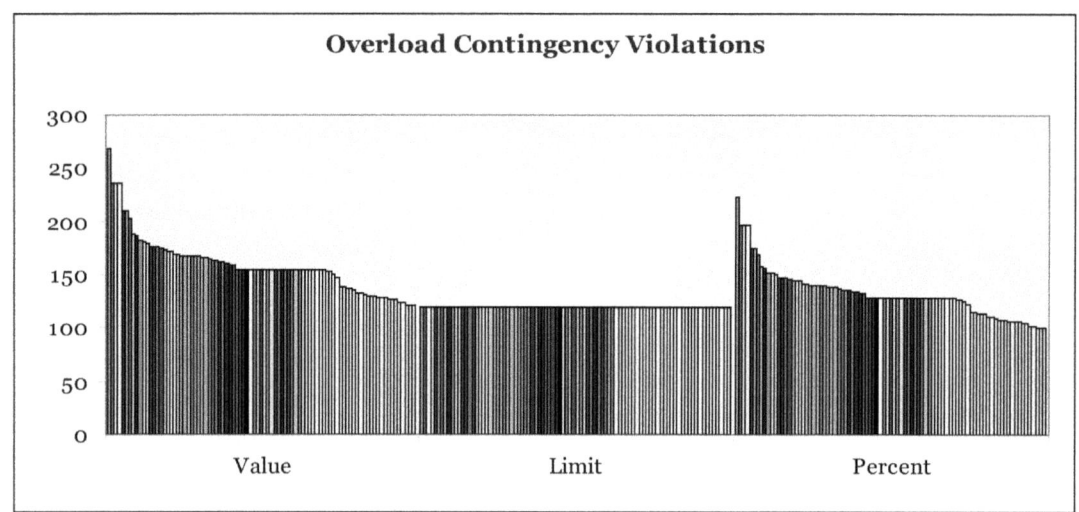

B

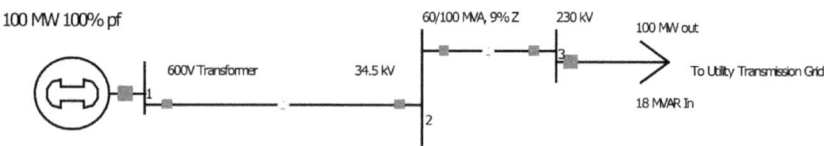

Table 8. Samples violations with overload rank and limits

Line Records								
From Number	**From Name**	**To Number**	**To Name**	**Circuit**	**Xfmr**	**Aggr. Percent Overload**	**Violations**	**Max % Loading Cont.**
1	1	2	2	1	NO	2904.11	89	222.76
1	1	3	3	1	NO	3068.87	90	196.11
2	2	3	3	1	NO	4314.24	132	227.77
2	2	4	4	1	NO	3424.4	107	237.73
2	2	5	5	1	NO	11369.13	207	275.71
2	2	6	6	1	NO	5499.06	127	244.43
3	3	4	4	1	NO	0	0	
4	4	5	5	1	NO	9553.08	164	463.34
7	7	5	5	1	NO	4276.28	188	214.98
6	6	7	7	1	NO	838.56	38	180.33
6	6	7	7	2	NO	838.56	38	180.33

Table 9. Limit violations with specific elements

Limit Violations						
Category	Element	Value	Limit	Percent	Area Name Assoc.	Nom kV Assoc.
Branch MVA	2 (2) -> 3 (3) CKT 1 at 2	124.72	100	124.72	Hilltop-Hilltop	138
Branch MVA	2 (2) -> 4 (4) CKT 1 at 2	136.91	100	136.91	Hilltop-Hilltop	138
Branch MVA	2 (2) -> 6 (6) CKT 1 at 2	460.94	200	230.47	Hilltop-Deer Park	138
Branch MVA	4 (4) -> 5 (5) CKT 1 at 4	278	60	463.34	Hilltop-Hilltop	138
Branch MVA	6 (6) -> 7 (7) CKT 1 at 6	226.95	200	113.47	Deer Park-Pond Lake	138
Bus Low Volts	5 (5)	0.57	0.9		Jackson Circle	138
Bus Low Volts	6 (6)	0.89	0.9		Deer Park	138
Bus Low Volts	7 (7)	0.57	0.9		Pond Lake	138

unit to reflect the expected operating conditions. The simulation is performed to preserve the rights of nearby resources. The results of the impact study are summarized in Table 10.

An additional example the PSS/E program may evaluate are the losses for a typical 100 MW wind farm illustrated in panel B of Figure 3 (Ghorai & Reddy, 2008). The simulated model includes the losses in the calculations as well as the required reactive compensation to satisfy the point of interconnection criteria (18 MVAR total losses). These data are illustrated in Table 11. The program is expected to evaluate the operational capabilities of the wind turbines with a variable power factor range (i.e. 95% capacitive to 95% inductive) and determine if any additional compensation is required.

THE PRELIMINARY RESULTS

In order to perform the impact studies effectively, the proper queue management protocol is utilized for maximum efficiency. The essence of the overload problem entails detailed analysis on the system level for a viable solution (Martin, 1997). The systematic approach utilized to evaluate the overload queue conditions are based on funda-

mental business process experience in the utility industry. Once the process problem is defined and diagnosed, prescribed steps are applied via continuous improvements to mitigate the fault (Sage & Rouse, 2009). The "As-Built" current model evaluates every request without any pre-assessment activity on the part of the customer. Only a simple cursory process is utilized as part of the front-end analysis. The FIFO protocol is prevalent in this model to simulate the actual generator request inputs. The current model was compared with a modified version of the process aptly named as "The Serial" version, where rework reassessment0 of the requests are thoroughly entrenched in the scheme. Once the initial generation request is received, various changes in the load and electrical footprint are experienced over time. The model is designed to reassess the electrical condition after a significant amount of time passes to simulate real-world conditions. The "Parallel Model" evaluates the generation requests after a thorough review by the customer performed at the front-end of the process. The model utilizes probability assessments embedded within the decision blocks as a tool for screening each generation request. Based on the analysis of selected independent service operators and assumptions, the review rejects 33% of the requests, which are

Table 10. Impact study high-level results

Thermal Overloads:	Flow-Based Analysis:	Network Analysis:	Voltage Analysis:
• Mitigation of thermal constraints of limiting elements/ projects in queue • Utilized various peak year scenarios (2013 /2018).	• This service will not be restricted based on the flow-based analysis.	• Zero megawatts of transfer are available from source to the region without reinforcements.	• **2013:** Results illustrates no base case voltages are outside of the acceptable range criteria
• Service cannot be granted at this time	• The start date of the requested service is not within the 12 month horizon.	• Project evaluation alternatives are utilized to mitigate the loadings of the limiting elements (resolution of the summer emergency ratings)	• **2013:** However, the 100 MW transfer combined with the loss of a major transformer results in post-contingency voltages outside of the acceptable range criteria and affects busses in the surrounding system: o 12 ~ 161-kV busses o 61 ~ 69-kV busses
			• **2018:** No base case or post-contingent voltages outside of the acceptable range.

Table 11. Generator loss calculation

Turbine:	Generator Step-Up (600 V Transformer):	Collector Grid Losses:	Collector Grid Charging:	Power Transformer Losses (60/100 MVA):
0 MVAR (inductive)	-7.0 MVAR (inductive)	-4.0 MVAR (inductive)	2.0 MVAR (capacitive)	-9.0 MVAR (inductive)

subsequently re-submitted after the customer corrects the problem. Following the initial pre-assessment, the request is then reevaluated. The decision block in the next stage rejects 10% of the submittals based on the industry standards. Service is granted after the thorough analysis of the network assessment, flow evaluation, voltage contingency review, and thermal overload investigation. An additional resource (the technical analyst), is embedded in the model to perform any re-work activities if the major process wait times exceeds a pre-established limit. All three of the significant models are based on the FIFO queue scheme to accurately evaluate the preferred performance among the schemes. The best simulation was then compared with the "lowest-wait time" in the queue to emulate a generation request near its completion cycle. The situation reflects actual field conditions where some requests has the necessary funding, resources, etc. and will typically move

ahead (leap frog) of others in the system. The request is prioritized with the available resources and service is subsequently granted *only* after the comprehensive studies are completed. The parallel Model is illustrated in Figure 4.

The Queue Waiting Costs

The generator requests were compared with the three variations of the models from a cost perspective. The results of the analysis are illustrated in panel A of Figure 5. The average queue waiting cost of the main processes produced distinct results – most notably in the network analysis category where a 72.20% improvement was experienced between the current model and the parallel model. The serial model reduced the cost by roughly half when compared with the current model in the same category. Further investigation revealed cost improvements in the flow-based analysis (61.59%),

Figure 4. The parallel model for queue evaluation

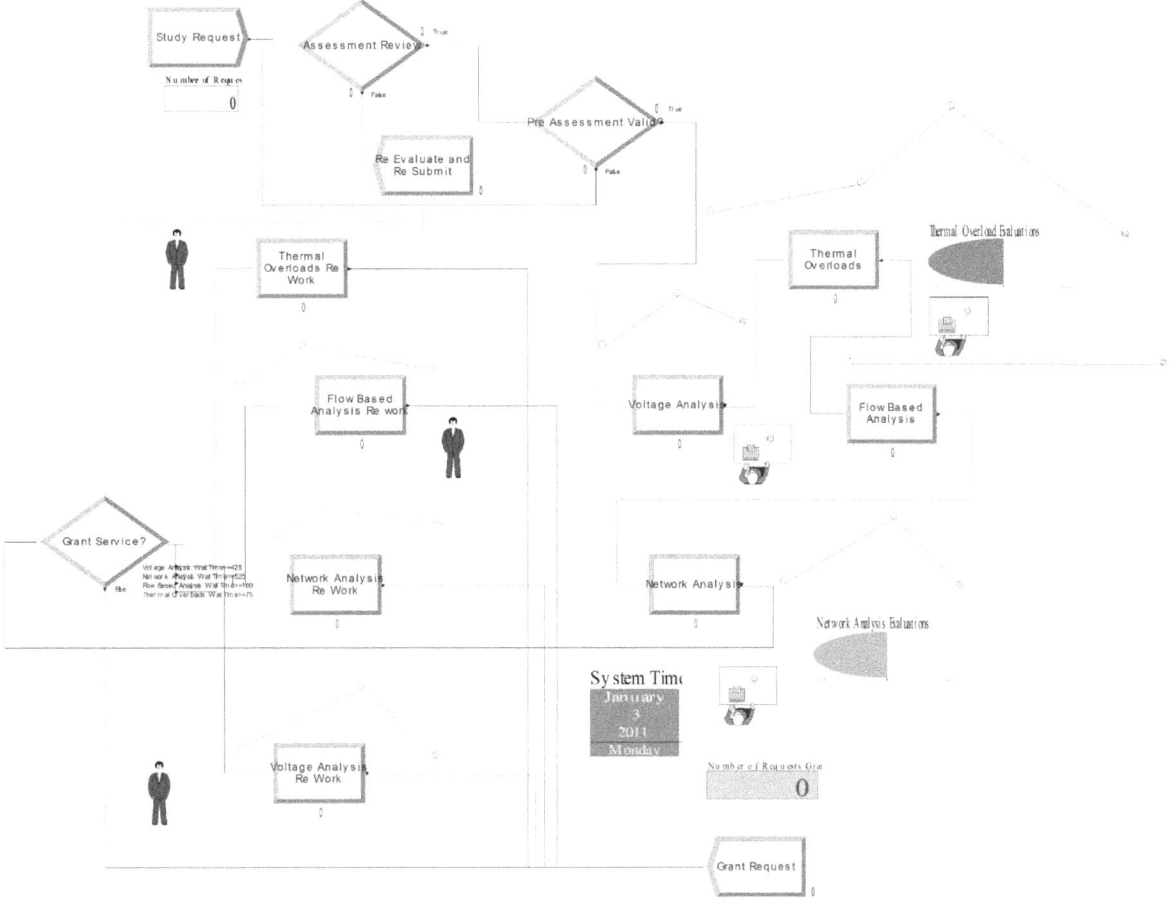

and voltage analysis (39.48%) categories as well. The current process excelled in only the thermal overload category by 13%. Most engineers in the profession fail to realize the costs associated with queue waiting times. These expenditures have the potential to increase exponentially if left unimpeded, and must be harnessed via proper cost containment strategies.

The Number of Requests Waiting in the Queue

The number of requests waiting for service in the thermal overload and network analysis queues for the current process model exceeded both the parallel and series simulation models. These data is

illustrated in panel B of Figure 5. The category improvements were in the magnitude of 79.29% and 62.10% for the respective queues. Minimization of the average request waiting time in the system is a sub-goal of the operation. The reduction of generator requests in this metric alone facilitates an efficient and highly effective operation in the system impact study realm.

The parallel model experienced a modest improvement over the current model in the network analysis category (35.42%). These data is illustrated in panel C of Figure 5. The metric is designed to evaluate the total accumulated time a request is in the system. The improvement in this area is aimed to streamline the entire process and optimize efficiencies. Interestingly enough, this was

Figure 5. The preliminary results for the parallel model comparison - part I: total average accumulated time

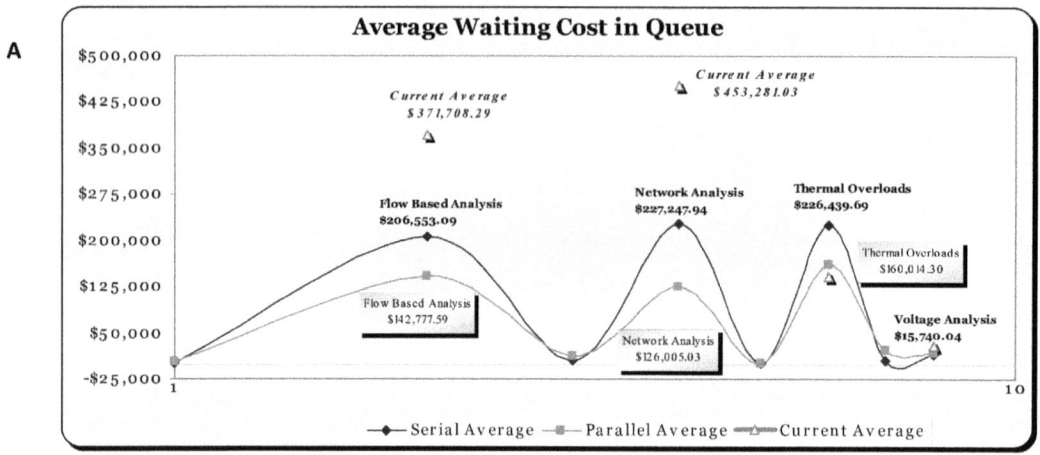

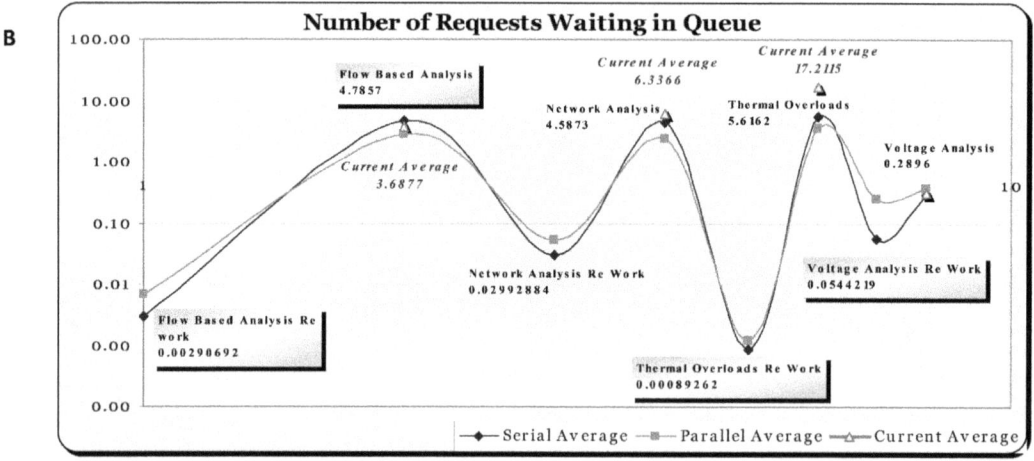

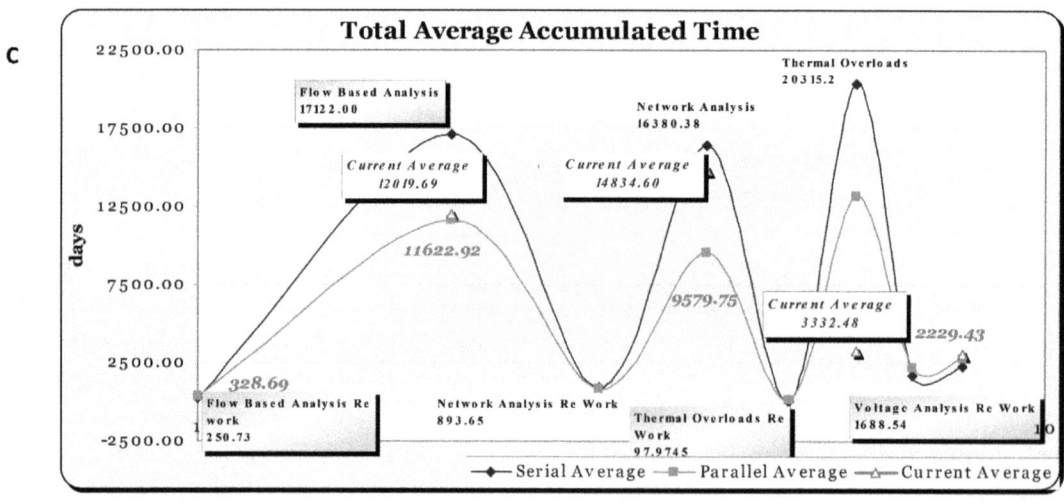

the only category that displayed any appreciable improvement over the current methodologies. The serial model demonstrated no improvement in all categories when compared with the current model. The accrued time in the system for a generator request is an important attribute in the effort to reform the process while reducing the backlog. Appropriate restructuring of the resources and related queues are cost/time reducing initiatives to complete indispensible electrical analysis of the proposed generator requests as proficiently as possible.

The Average Queue Waiting Time

The average queue waiting time (in days) is calculated and compared with all three models. How long a generator request waits in the queue for processing is an integral component of the time the unit spends in the entire system. The minimization of this important element is designed to create increased throughput and efficient cycle times that are currently lacking in the system. The parallel model exhibited marked improvements over the current model in the flow-based analysis (61.59%), network analysis (72.20%), and voltage analysis (39.48%) categories. These data is illustrated in panel A of Figure 6. The thermal overload category displayed a sign of degradation (by 13.04%) when it was aptly compared with the current model. Notwithstanding, the serial model exhibited improvement over the parallel model of the major categories that entails the re-work activities. It is a clear indication of the necessity of the reassessment process involved within the impact study analysis envelop. The task is typically performed by technical analyst/engineer to re-evaluate the electrical footprint since the initial request submittal. Parallel flows, adjacent construction activities, and additional large & small generators impact influence the synergy of the initial generator request. Moreover, the non-value added activity conducted within the re-work process is planned to identify any flaws (mis-steps) overlooked in the main processes. It

espouses the TRIZ principle #34 of "Discarding and Recovering" as outlined below (Mann, D. and Domb, E., 1997):

- Conversely, restore consumable parts of an object or system directly in operation.
  - Re-work of a non-conforming product
  - Re-energize continuous improvement initiatives

The Average Number of Requests Processed

This category is an essential attribute towards system optimization. These data is illustrated in panel B of Figure 6. The increased throughput of the four major processes in both the serial and parallel models over the current model is true positive indicators of a robust system. The data exhibited marked improvements in the flow-based analysis (60.0%), network analysis (51.05%), thermal overloads (75.23%), and the voltage analysis (59.48%) categories. The resources are performing above expectations when compared with the current model. More requests are processed, on average, utilizing the FIFO protocol and with the implementation of a stringent review scheme. Nevertheless, the generator request must be thoroughly capable after detailed analyses are performed.

This metric evaluated the actual number of generator requests (per resource) during the prescribed impact study period. These data is illustrated in panel C of Figure 6. The parallel model experienced a profound increase in generation requests handled when compared with the current model. Specifically, the analyst resource increase its productivity by 61.28% and the engineering resource experienced similar productivity gains (60.98%). The assigned technical analyst in the parallel model revealed a slight increase (4.14%) over the serial model. Nevertheless, the resources in the parallel model performed well above the status quo (more than doubled) production levels embedded within the current model.

Figure 6. The preliminary results for the parallel model comparison ~part II: the total number of requests seized

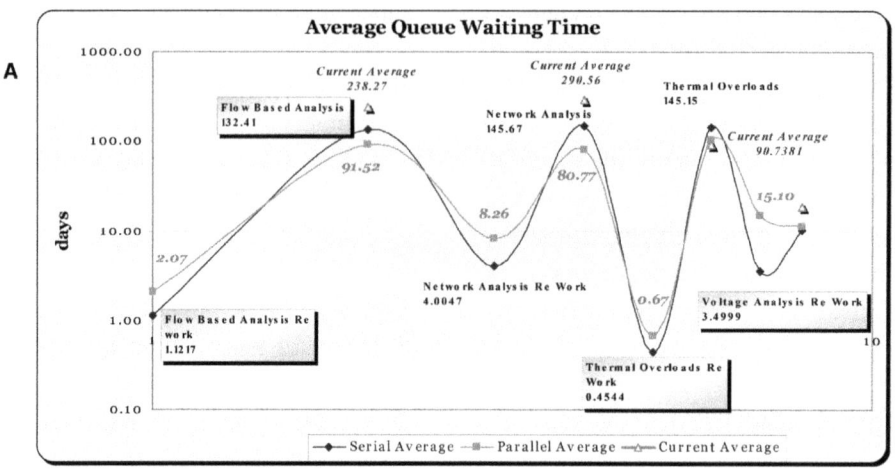

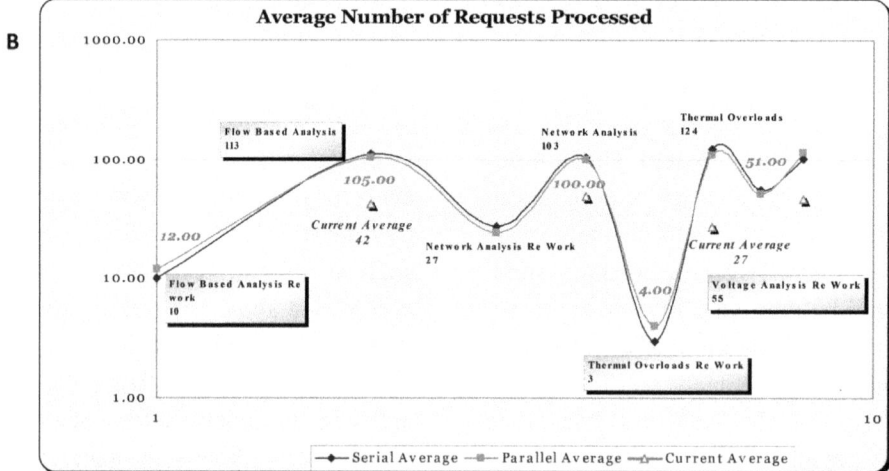

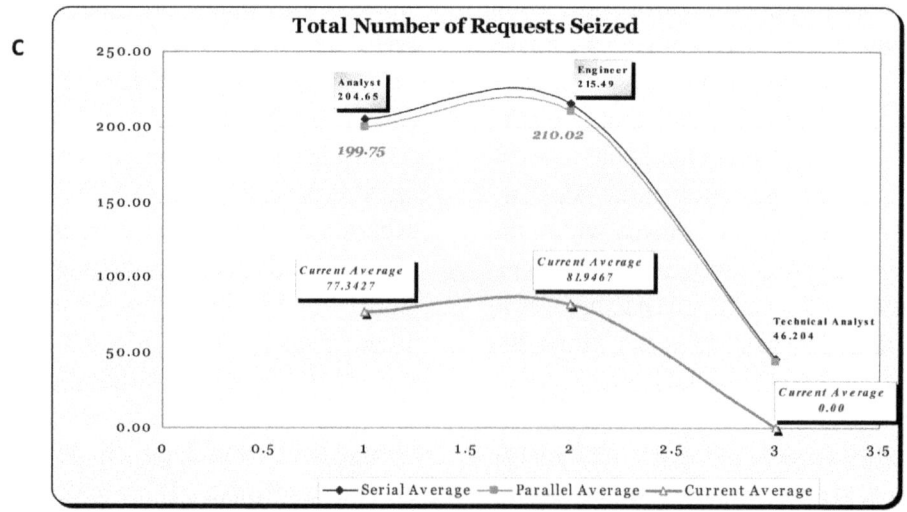

THE FINAL RESULTS

The "lowest wait time" in the queue protocol was developed as the optimal model to grant generator requests as a result of the impact studies. The model excelled in several key performance areas that relieved an overburden system. Moreover, when the model was compared with the best simulated model (parallel version), the data demonstrated profound improvements in the critical cost and time metrics of the queue. The bottlenecks are precipitously reduced, resources are freed, and most importantly, the quality of the generator requests is legitimized with a strong sense of reliability. Consequently, the simulation achieved a high-level of effectiveness by meeting and/or exceeding the system objectives (Clymer, J. 2009). On the macro level, the reduced cycle time of the impact studies directly affects the renewable energy generator system integration through improved efficiencies of time as well as cost. The proficiencies provide unyielding data for the proposed project and any alternative actions poised within the plan. The four key metrics of the lowest wait model are discussed in the following paragraphs.

The Average Accumulated Time

The lowest wait model demonstrated profound improvements in over 62% of the assigned categories when compared with the parallel model and over 75% enhancement when compared with the current process. Specifically, the model experienced improvements in the main categories of flow-base analysis (parallel~7.90%, current~10.94%), network analysis (parallel~15.88%, current~45.68%), thermal analysis (parallel~7.34%), and voltage analysis (parallel~4.33%, current~14.55%). These data is illustrated in panel A of Figure 7.

Gains were also noted in the voltage analysis re-work area (8.87%). However, the request accumulated time of the lowest wait model experienced a slight degradation with the thermal overload and network analysis categories and a significant loss in the network analysis when compared with the parallel model. The re-work (re-evaluate) activities are designed as a cluster trigger to analyze the system after a period of time at specific milestones to ensure the boundaries of the proposed generator is within the electrical horizon. Because of the resource precedent of the re-work models (both parallel and lowest wait) and tradeoffs, the thermal overload resources were activated near the end of the current process. Consequently, the data suffered significant losses in this category when compared with the lowest wait model in particular. Nonetheless, the overall system gains of the model more than make up for the deficiencies in both costs and time.

The Queue Time

This category experienced improvements in 75% of the metrics when the lowest wait model was compared with the current model and 62.5% enhancement when compared with the parallel model. These data is illustrated in panel B of Figure 7. The notable enrichments were in the flow-based analysis (parallel~17.96%, current~77.19%), and voltage analysis (parallel~6.43%, current~43.38%). Other gains were experienced in the re-work area of the voltage analysis area 95.56%). The overall waiting time a generator request spends in the queue is drastically reduced with the introduction of the lowest wait time model. The vetted item receives the required service in a timely, more organized manner where electrical details are not overlooked. Moreover, the systematic approach to the impact studies ensures a quality product as an output of the process. The losses experienced in the re-work areas are negligible when compared with the overall system gains in the time element. The data clearly illustrates a strong need to restructure and manage the current system impact study queue.

Figure 7. The final results for the lowest wait time model comparison: part I

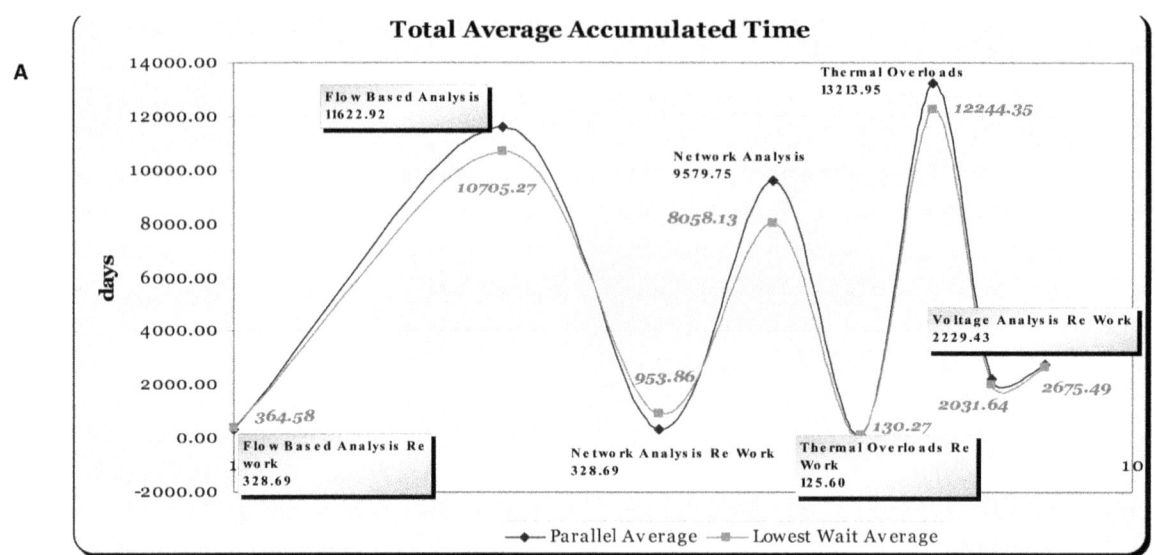

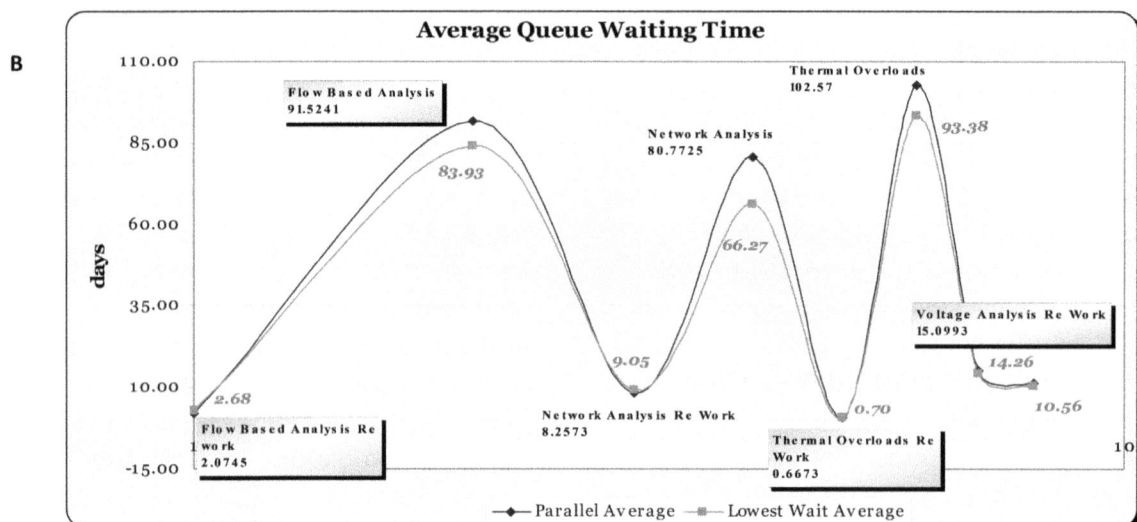

The Queue Waiting Cost

The lowest wait model experienced profound cost reductions in virtually every category. These declines simply equate to cost savings (avoidance) to the financial bottom line. The goal of the avoided cost is to assist the decision-makers with opportunities of where *not* to spend. The paradigm translates to aggregate or gross benefits of the system. The incremental changes implanted in a process typically have a huge effect on the bottom line. When the flow –based analysis metric in the lowest wait time model was compared with the parallel model ($11,845 cost avoidance) and the current model ($240, 776 cost avoidance), - from a sheer cost perspective - noticeable changes were shown in the data. These data is illustrated in panel A of Figure 8.

Other positive indicators are depicted in the network analysis (parallel~$22,631, current~$349,907), and voltage analysis (parallel~$1,132, current~$12,616) categories. By

Figure 8. The final results for the lowest wait time model comparison: part II

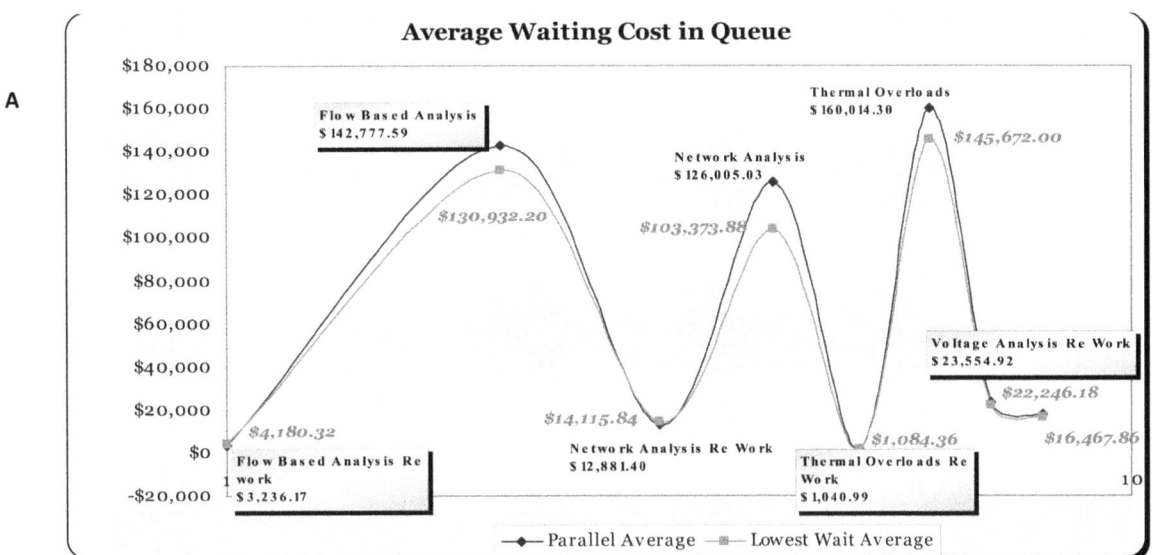

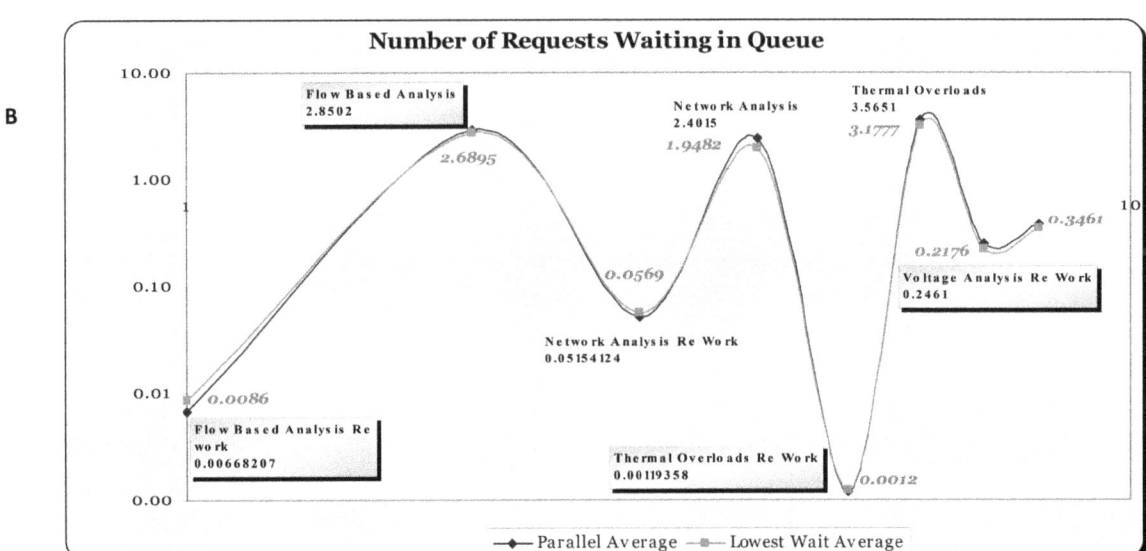

modifying the impact study request queue time it will provide approximately $600,000 in cost avoidance dollars. However, since the simulation investigated a *current process,* these dollars are *presently wasted.* The decision-makers must have the courage to act as visionaries during the planning stages of generator implementation in a complex electrical grid. The potential for bottlenecks in the system impact studies evaluation

process was obvious from the outset. The act of waiting for an event to fail is extremely costly. One of the greatest visionaries of the twentieth century once said:

Leaders are visionaries with a poorly developed sense of fear and no concept of the odds against them. (Dr. Robert Jarvik ~b. 1956)

The Number of Requests Waiting in the Queue

The lowest wait time model provided substantial gains (overall 62.5%) in the study categories of the number of requests waiting in the queue when compared with the parallel model and improvements in the order of 75% when compared with the current process. The most significant improvements were in the network analysis (parallel~18.88%, current~69.25%), and the thermal overload (parallel~18.88%, current~69.25%), categories. These data is illustrated in panel B of Figure 8. There are simply fewer generator requests in these queues as the worker productivity increased when the lowest wait time scheme was implemented. The queue adjustment optimizes the overall productivity function (earned value) and improves response time.

SUMMARY

Planning is an essential component for any type of renewable energy system integration. It begins with the customer (load) forecast and ultimately circles back to the customer. The generation system integration plan, if implemented correctly, is a seamless process with virtually no loss of load situations (high reliability) and minimum implementation cost. As the demand for the electric commodity increases, so does the cost to the customer. The integral balance between the demand-side activities and the supply-side generation invokes an area within the least-cost region for the customer and profitability for the utility company. Yet, the essential planning must encompass feasible supply-side alternatives to effortlessly integrate the renewable generation to meet increased demand where applicable. Moreover, extensive upgrades of the transmission and distribution infrastructure are required to accommodate planned new generation in the electric grid. These activities necessitate extensive analysis of the current system to appropriately translate and integrate the proposed generation to a future system. The overall system is expected to be cost effective, exhibit a high degree of reliability, flexible, and produce zero emissions from the perspective of the utility, customer, regulators, and the general public. The detailed analysis entails how the new generator technology impacts the electric system thermal overload capacity, flow-based assessments, network evaluation, and voltage boundary simulation investigations. Ideally these concurrent processes are conducted during the transmission and distribution upgrades and systematically implemented on a first-come, first-served basis. However, this situation is not always the case as the regional independent service operators experience queue overload from the recent increase in volume of the generator requests. The overburden conditions deplete resources, increase costs, and degrade the quality component in the electrical planning sphere. The inclusion of a front-end screening process to evaluate the validity of the generation requests coupled with a queue protocol restructuring reduced the backlog as well as the waiting cost in a simulated study. The effort proved that the current model is processing fewer requests at an exceedingly higher cost in the queue. The ineffective methodology of inspecting every incoming request is proven to be an expensive proposition for the entities involved. When the process simulation model is utilized as an evaluation tool to determine quality, cost, and system vulnerabilities in the planning arena, it will benefit the entire electrical analysis paradigm. The detailed evaluation of the generation integration plan is a paramount activity and must eclipse any rapid advancement of renewable energy implementation in order to satisfy an agenda item on the requestor's list.

REFERENCES

Anderson, P. M., & Fouand, A. A. (2003). *Power system control and stability* (2nd ed.). Piscataway, NJ: IEEE Press.

Awerbuch, S. (1993). The surprising role of risk in utility integrated resource planning. *The Electricity Journal, 6,* 20–33. doi:10.1016/1040-6190(93)90048-P

Bergen, A., & Vittal, A. (2000). *Power system analysis* (2nd ed.). Upper Saddle River, NJ: Prentice Hall.

Bhattacharjya, D., & Schachter, R. (2012). Formulating asymmetric decision problems as decision circuits. *Decision Analysis, 9*(2), 138–145. doi:10.1287/deca.1110.0226

Bohan, B., Ferrar, B., & Luigs, M. (September 2011). Addressing the big data concern in the utilities sector. *Utility Automation Engineering T&D Magazine.* Retrieved 15 May, 2012, from http://www.elp.com/elp/en-us/index

Budget of the U.S. Government. (FY 2013). *Performance and management section,* Analytical Perspectives Volume. Retrieved 29 May, 2012, from http://www.whitehouse.gov/omb/budget/analytical-perspectives

Busch, J. F., & Eto, J. (1996). Estimates of avoided costs for electric utility demand-side planning. *Energy Source, 18.*

Carlson, B. (2012). MISO unlocks billions in savings through the application of operations research for energy and ancillary service markets. *Interfaces, 42*(1), 58–73. doi:10.1287/inte.1110.0601

Clymer, J. R. (2009). *Simulation-based engineering of complex systems.* New York, NY: John Wiley.

Cooper, R. B. (1981). *Introduction to queuing theory* (2nd ed.). New York, NY: John Wiley.

Everett, G., Philpott, A., Vatn, K., & Gjessing, R. (2010). Norske Skog improves global profitability using operations research. *Interfaces, 40*(1), 58–70. doi:10.1287/inte.1090.0471

French, S. (2012). Expert judgment, meta-analysis, and participating risk analysis. *Decision Analysis, 9*(2), 119–127. doi:10.1287/deca.1120.0234

Ghorai, M., & Reddy, N. (2008). *Meeting the IESO interconnection requirements for Ontarian wind farms.* Presented at the European Wind Energy Conference, Brussels, Belgium. Retrieved 23 August, 2010, from http://www.ewec2008.info

Glover, F., Kelly, J. P., & Laguna, M. (1996). New advances and applications of combining simulation and optimization. In J. M. Charnes (Ed.), *Proceedings of the 1996 Winter Simulation Conference,* (pp. 144-152).

Glover, J. D., & Sarma, M. S. (2001). *Power system analysis and design* (3rd ed.). Toronto, Canada: Thompson Learning.

Grady, J. O. (1994). *System integration.* Boca Raton, FL: CRC Press.

Gross, D., & Harris, C. M. (1997). *Fundamentals of queuing theory* (3rd ed.). New York, NY: John Wiley.

Gunther, E. (2007). Field and device technologies: Customer portals, home area networks and connected devices. *GRID INTEROP 2007 Forum Proceedings* (Gunther-1), Richland, WA.

Jones, L., & Cheung, K. (March 2012). You're going to need a smarter crystal ball. *Electric Light and Power Magazine.* Retrieved 1 June, 2012, from http://www.elp.com/elp/en-us/index

Kahn, E. (1991). *Electric utility planning and regulation,* 2nd ed. Washington, DC: American Council for an Energy-Efficient Economy.

Kendall, K. E., & Kendall, J. E. (1994). *Systems analysis and design* (3rd ed.). Upper Saddle River, NJ: Prentice Hall.

Kubiak, T. M. (June 2012). The significance of simulation. *Quality Progress*, pp. 44-45.

Mann, D., & Domb, E. (1997). *40 inventive (business) principles with examples*. Retrieved 8 November, 2008, from http://www.triz-journal.com

Martin, J. N. (1997). *System engineering guidebook: A process for developing systems and products*. Boca Raton, FL: CRC Press.

Mattson, C. A., Mullur, A. A., & Messac, A. (2004). Smart pareto filter: Obtaining a minimal representation of multi-objective design space. *Engineering Optimization*, *36*, 721–740. doi:10.1080/0305215042000274942

Meyer, M., Van Deventer, L., Wykes, C., & Cawood, E. (2011). Innovation decision support in a petrochemical production environment. *Interfaces*, *41*(1), 79–92. doi:10.1287/inte.1100.0528

Miettinen, K. (1999). *Nonlinear multi-objective optimization*. Kluwer Academic Press.

Nichols, D., & Von Hippel, D. (2002). *Best practices guide; Integrated resources planning for electricity*. Retrieved 24 September, 2010, from http://www.info.usaid.gov

Papoulis, A. (1991). *Probability, random variables, and stochastic processes* (3rd ed.). New York, NY: McGraw-Hill, Inc.

Pertti, J., Antti, M., Kimmo, K., et al. (2007). *Using advanced AMR system in low voltage distribution network management*. Presented at the 19th International Conference on Electrical Distribution. Vienna, Austria. Retrieved 23 February, 2010, from http://www.cired2007.be/pdf/ustopselectedreportss6.pdf

Sage, A. P., & Rouse, W. B. (2009). *Handbook of system engineering* (2nd ed.). New York, NY: John Wiley.

Smedlund, A. (2012). Value co-creation in service platforms business models. *Service Science*, *4*(1), 79–88. doi:10.1287/serv.1110.0001

Wood, A. J., & Wollenberg, B. F. (1996). *Power generation, operations, and control* (2nd ed.). New York, NY: John Wiley.

ENDNOTES

1. Source: U.S. Department of Commerce; International Trade Administration; Office of Travel and Tourism Industries and Bureau of Economic Analysis (BEA). Released May 2009

2. Source: U.S. Census Bureau, *County Business Patterns*. 2006

3. Source: U.S. Bureau of Labor Statistics, *Consumer Expenditures in 2007*, News, USDL-08-1746. (25 November 2008).

4. Source: U.S. Energy Information Administration, *Electric Sales and Revenue 2007*, (published January 2009).

5. Source: Board of Governors of the Federal Reserve System, *Industrial Production and Capacity Utilization* Statistical release G.17, monthly.

6. TRIZ is a Russian language acronym for *Teoriya Resheniya Izobreatatelskikh Zadatch*. Translated into English it means "The Theory of Inventive Problem Solving."

Chapter 5
Ordinal Capital Project Ranking Evaluation and the Quality Component

ABSTRACT

There is a general sense of developing "new" capital projects in a virgin arena and deemed as high-quality than performing the necessary upgrades on an existing project. Based on the increased need to mobilize the necessary resources, each project in the reliability sector must be prioritized for safety reasons and to avoid potential equipment damage to expedite the work. Properly planned capital projects are skillfully integrated within the portfolio envelope. Conditions such as the projects in the reliability sector create blackouts, increase costs, and develop a poor safety environment if left unattended.

INTRODUCTION: THE QUALITY COMPONENT

The quality control techniques and methodologies utilized in the electrical equipment installation arena are typically categorized as a supplementary service. The data analysis is generic and applicable to a wide range of conditions when implemented within a Quality Management System (Aft, 1998). It is analyzed to determine adverse affects on the quality of the service and identify preventive or corrective actions. In the electric utility industry, waste reduction and material integration cycle times are two of the key attributes for equipment

installation, specifically within the renewable energy sector. The three main categories for the quality control techniques are described below:

- Diagnostic Techniques
- Process Control
- Acceptance Sampling

When these techniques are put into practice and applied religiously in a quality plan, the components of the Lean Paradigm will automatically drive the capital project cost downward. Moreover, when the quality plan is utilized as a foundation for utility best practices across the organization

DOI: 10.4018/978-1-4666-2839-7.ch005

(planning, finance, project management office, engineering, maintenance, procurement, line shops, etc.), a drastic reduction in cost/waste is experienced in every aspect of the service. The Lean Paradigm is a systematic approach to provide a path that specifies value, evaluates the best viable sequence while continually improving the process. It espouses a business process of doing more with less. Management of a system of processes within an organization (commonly referred to as the process approach) is a continuing effort over the individual processes in the quality sphere. The combination and interaction of these processes is advantageous as a value-added service, meeting the stated requirements, evaluating the process performance effectiveness, and continuous improvement aspects based on the object measurement (ANSI/ISO/ASQ Q9001-2000, 2001). The Quality Assurance Plan is a component of Project Management Standard for the electric utility company. The document must encompass the major operational tenets of 1) Management Responsibility to include objectives, authority, and review, 2) Resource Management that entails a provision of resources, competence, awareness and training, 3) Process Realization to include the operational control aspect, customer communication, and management of the monitoring & measuring devices, and finally 4) Measurement, Analysis & Improvement mechanisms for an internal audit, control of the nonconforming process, analysis of data, improvement, and corrective/preventive action.

An example of the diagnostic technique germane to the electrical utility business utilizes cause-and-effect diagrams as well as Pareto Analysis for equipment installation scheduling delays. The technique is designed to pinpoint potential problems in a process that directly affect quality outcomes. Even the simplest processes, such as visitors entering a facility can be improved immensely when the method is utilized to categorize problems (Conklin, 2012). The continuous upgrading effort within a substation/line rehabilitation for renewable energy generation implementation invariably creates a plethora of problems in an organization. The Pareto diagram can successfully identify the primary causes of a transformer delay from a list of problems encountered by field personnel as described below:

- Endangered species (pine snakes) were discovered once the ground was broken.
  - Notified the department of environmental protection to re-file the permit applications *(Preventable ~ Rework)*.
  - **Mitigation Strategy:** Seek another parcel in parallel with permitting work effort to avoid massive construction delays *(Corrective Action)*.
- Local construction management resource issues
  - Relieves over-burden line shop resources *(Corrective Action)*.
  - Manage contractors, safety issues, materials, etc. *(Preventive ~ reduces bottlenecks)*.
- Communication
  - The milestone need date (from the project manager to the engineering department) created a two month lag time for the electrical drawings and scope documents *(Preventive)*.
  - Resource turnover of internal/external engineers with no oversight of project control and created an annual engineering project backlog *(Preventive)*.
  - Verbal undocumented updates on critical equipment installation processes *(Corrective Action)*.

The lesser problems escalated to massive project delays in the regional transformer installation example. The analysis revealed – in this particular case, the material procurement – as the primary weak link in the supply-chain. Consequently, the entire venture collapsed and alternative measures were set into motion to correct the practically irreversible scheduling delay quandary. When these problems are graphically illustrated in a Pareto diagram, the scheduling delays become quite obvious. Approximately sixteen percent of the delays were attributed to material procurement issues, seven percent to the material itself as well as engineering design clarifications, and one percent resided within the transformer mobile availability category. An appropriate quality plan, if implemented would mitigate the delay issues and reduce the related expenditures.

The process control techniques espouse the specification characteristics inherent within the progression to meet the quality objectives (Feigenbaum, 1991). It accounts for random variation outside of the control bands (upper and lower control limits) as an aid to identify problems. It is a diagnostic aid designed to segregate changes in the parameters to assist informed decision-makers with the identification and elimination of the root cause of a problem to improve quality and productivity (Zou et al., 2011). As a key indicator for the effectiveness of a quality plan, corrective action is a chief element for the total quality control planning arena. Corrective action, within these programs, is the undeviating rectification of a quality problem that entails product defects, part deviation, process errors, apparatus malfunctions, as well as various paucities that occurs in the equipment installation process. Consequently, the corrective action is designed to prevent nonconformities from *recurring*. Conversely, preventive action provides a means to prevent nonconformities from *occurring*. The selection of an improvement method is essential during this stage in the evaluation process. The problem is easily converted into a magnitude of variability. The positive and negative points are clarified and compared to eliminate the variability. The quality tools employed at this phase include check sheets, histograms, Pareto charts,

and graphs. Control charts are typically engaged to detect variability changes in a process to correct actions. The generated data is subsequently plotted and compared with the control limits of the process. Out-of-control data is commonly utilized to make adjustments and improve the overall process. The scheme aids the decision-maker with distinguishing between random (inherit is a stable progression) and assignable causes (specific) of the process variation. Even in machine-vision systems, where image capturing devices are utilized, the statistical process control augments the data progression and interpretation (Megahed et al., 2011). When used as a financial control mechanism for capital-intense projects, the improvement in forecasting alone is beyond reproach. Each process generates a specific cost average based on the capability factor of the scheme. The minimum/maximum averages are depicted as well to formulate additional control bands within the process. From a scheduling perspective, the process control technique, the total time per entity specifically identifies the potential bottlenecks and time consumed embedded in the process. The perceptive electric utility legal department and project management office may elect to develop the construction contract language to include incentive and penalty factors based on the optimized model of the process control bandwidths.

The acceptance sampling technique aids the decision-maker with the inspection of product lots from suppliers (Evans & Lindsay, 2005). There are numerous defects discovered during the electrical equipment installation that creates additional plant-in-service delays. The sampling techniques are suitable in these cases where a decision on the product quality level is executed but not economically feasible to inspect the entire lot (i.e. insulators, conduit, conductors and fittings, steel structure material, etc.). A variety of sampling plans are available specific to the application. When the sampling plan is implemented correctly, the benefit of reduced time, labor, and cost avoidance is realized as opposed to inspection of the entire lot. It is a cost-effective measure to gain insightful data and characteristics of a specific population (ISO/TR 10017, 2003).

Sampling techniques were engaged with the selection of maintenance problems in a specific Midwestern electric utility company. The computerized maintenance system, for example, is utilized by the company mainly as an equipment monitoring and inspection tool. There were four (4) major activities employed by the system during the study period:

- Monthly Substation Inspections
- Quarterly Battery Checks
- Transformer Oil Analysis Testing
- Circuit Breaker Test Operations

Although the system capabilities were far greater in scope than the company's usage, every single function was not expected to perform to a specific utility configuration platform. However, several layers of functionality remain untapped in the Computerized Maintenance System and were definitely beneficial (financially, economically, and efficiently) to the company's strategic objectives. The basic route types alone could be expanded to include the following critical parameters unique in the utility industry:

- Environmental Report
- Lines - Cathodic Protection Inspection
- Relay - Carrier Checks, Functional Testing, and Re-Calibration
- Substation - Annual Infrared Scan and Planned Maintenance Live Line Washing
- Test Equipment Calibration
- Transformer Electrical Testing

When joined into an effective, highly-utilized work management system, the characteristics of the program could not only assign scheduled maintenance, it would be tasked to automatically assign work packages for demand maintenance as well. The functionality of improved exception tracking existed that records the work order status, video inputs for faster assessments of a situation, and monitoring point data validation to diagnose the information as it was being processed in the system. The company's limited use of the system by field personnel, equipment specialist, and administrators created an inefficient and ineffective platform designed for failure. A sample of the maintenance issues are captured in Table 1.

The equipment information, lists, routes, monitoring sets, and exceptions were only a few major areas the utility company failed to fully utilize. A properly implemented computerized maintenance system must recommend actual replacement and Life Cycle costs for specific equipment items. The major function employed by the company and the percentage of overdue items in the evaluation period is depicted Table 2.

Approximately two-thirds of the scheduled activities were on the overdue routes with the minimally required test/inspection data. The worst offender was code "D", the annual breaker test-operate criteria (41.55%) where Chard Road data was last recorded at 1508 May 12, 2003 and scheduled for another test on May 11, 2004 which never occurred. Although a twenty-five day grace period from the due date is allowed for the substation monthly inspections, this area is exceptionally high (25.32%) with the overdue criteria. The data is captured and depicted in a histogram (Pareto analysis) illustrated in panel A of Figure 1. The four major activities of the system were consistently overdue and in some cases, the process was completely terminated altogether (financial, contractual, manpower, time). A random verification of certain substations in each region revealed a gross under-utilization of the routes in practically every area. Moreover, the accumulated days between the last performance and the scheduled performance on several routes were suspect (Table 1). Many tests were abandoned in mid-year or minimally performed by the contrac-

Table 1. Sample maintenance evaluation

Quarterly Battery Checks (every 90 days)					
Substation Name	Last Time Used	Scheduled Update	Days between use	Lapse Days	Overdue? T/F
Medlock	4/24/2003	7/23/2003	89.40	792.00	TRUE
Barfield	11/13/2003	2/11/2004	89.36	589.00	TRUE
Jay Lord	3/26/2004	6/22/2004	87.61	457.00	TRUE
Henson	5/17/2004	8/15/2004	89.43	403.00	TRUE
Fort	9/21/2004	12/20/2004	89.64	276.00	TRUE
Kendall	5/20/2005	8/18/2005	89.10	35.00	FALSE
Tocsins	8/10/2005	11/8/2005	89.43	-47.00	FALSE
Monthly Substation Inspection (every 30 days)					
Substation Name	Last Time Used	Scheduled Update	Days between use	Lapse Days	Overdue? T/F
Twin Star	5/10/2005	6/9/2005	29.32	105.00	TRUE
Thelon	6/8/2005	7/8/2005	29.62	76.00	TRUE
Bradley Sub	6/23/2005	7/23/2005	29.39	61.00	TRUE
Kendall	7/5/2005	8/4/2005	29.44	49.00	TRUE
Aphelion	8/9/2005	9/8/2005	29.53	14.00	FALSE
Lea Haw	8/23/2005	9/22/2005	29.39	0.00	FALSE
Breaker Test Operational Check (every 365 days)					
Substation Name	Last Time Used	Scheduled Update	Days between use	Lapse Days	Overdue? T/F
Chard Road	5/12/2003	5/11/2004	364.37	863.37	TRUE
Getty Hydro	9/25/2003	9/24/2004	364.37	727.37	TRUE
Serb Plant	10/22/2003	10/21/2004	364.58	700.58	TRUE
Battelle	3/26/2004	3/26/2005	364.62	544.62	TRUE
Project 12	5/5/2004	5/5/2005	364.40	504.40	TRUE
Yellow Lake	11/16/2004	11/16/2005	364.37	309.37	FALSE
F. B. Wesson	12/31/2004	12/31/2005	364.55	264.55	FALSE
Transformer Oil Analysis Test (every 180 days)					
Substation Name	Last Time Used	Scheduled Update	Days between use	Lapse Days	Overdue? T/F
Verge	2/20/2002	8/19/2002	180.00	1310.00	TRUE
Camps (138-kV)	7/25/2002	1/21/2003	180.00	1155.00	TRUE
Bat Casks	8/12/2002	2/8/2003	180.00	1137.00	TRUE
Kenton	8/26/2003	2/22/2004	179.51	757.50	TRUE
Thelon	8/29/2003	2/25/2004	179.39	754.39	TRUE
Tilt Station	9/2/2003	2/29/2004	179.52	750.52	TRUE
White Side	9/4/2003	3/2/2004	179.52	748.52	TRUE

Table 2. Activity problem set

Code	Route-Based Activity Categories	# Overdue	Total Items	Priority Rank	% Overdue
A	Monthly Substation Inspections	39	70	2	25.32%
B	Quarterly Battery Checks	33	71	3	21.42%
C	Transformer Oil Analysis Test	18	18	4	11.68%
D	Breaker Test Operations	64	72	1	41.55%
		Σ=	154	231		

tor based on financial disputes concerning the actual work. Increased usage of these routes must be substantially documented in a timely manner to secure accurate and precise data. The issue of data quality and integrity is inferred when the routes were overdue by such large margins. Future systems must comply with stringent data gathering policies set forth by the company. Although the computerized maintenance system has the robust capabilities to acquire and log the required information, the frequency of the inputs were lacking and the major areas were *not* utilized in the company. The bar coding function for lines and substations, for example, were capable of recording every single inventory item in the substation and/or associated line. However, the compatible function was not used. Several vital equipment fields were not populated and employed for nameplate data. This section alone can assist in the cross-checking paradigms of other databases (planning, asset management, accounting, etc.) for equipment verification. When properly linked, the computerized maintenance system can become a powerful informational database, especially if these unused fields were activated and in frequent use. The expandability of the system to integrate into a crucial repository for assets was virtually untapped. However, if a new system were in place, each line item /asset must be implemented into a relational database to include many of these idle routes and their associated derivatives. The major unused sections were described below:

- Manage List
- Equipment
  - Manufacturers (fields were not fully populated)
  - Vendors (fields not populated)
- Routes - Substations O&M
  - Breaker external inspection
  - Transformer electrical test
  - Portable oil kV test
  - Breaker test-ops
- Routes - Relay/Lab
  - Breaker oil analysis testing (incomplete special instruction field)
  - Relay/calibration function
  - Transformer oil analysis testing
  - Bar code function for lines/substations
- Equipment Information (unpopulated fields)
  - Recommended life cycle
  - Replacement cost
  - Responsible individual/functional group
  - Generate bar code unselected
  - Status
- Monitoring sets
  - Transmission line (120-kV/138-kV/345-kV)
  - Station power transformer
  - Switch (air break/disconnect/ground/ motor operated/spring operated ground)
  - Relay-old (all)
  - Relay-new(all)

Figure 1. Graphical analysis of the anomalies

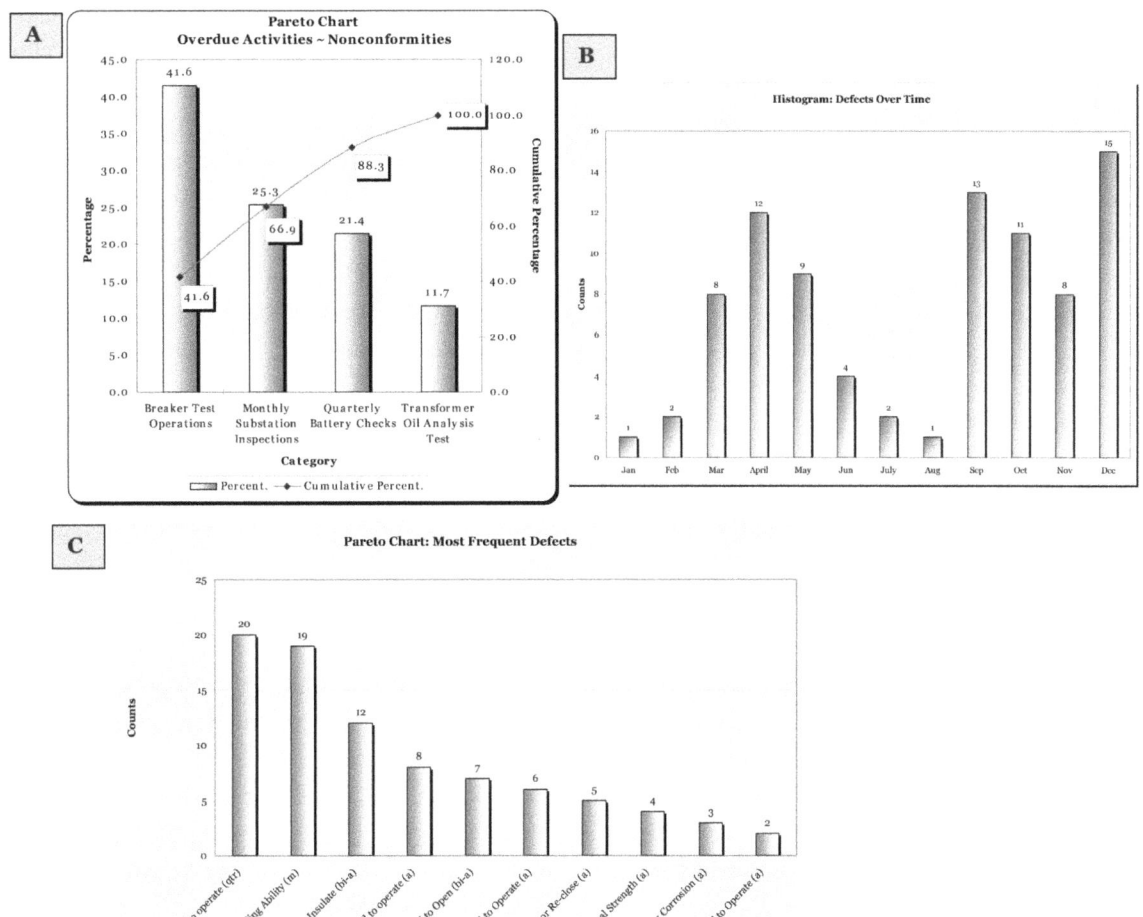

- ○ Potential Transformer
- ○ Generator (station power)
- ○ CCVT/combination CTVT/CCPD
- ○ Current transformer
- ○ Compressor
- ○ Circuit switcher
- ○ Charger
- • Exceptions - Work Orders-Projects-Work Plans
  - ○ Work (types/priorities/conditions/status/resources/reimburse codes/accounts) were coded ~ not used
  - ○ Repair priorities

The details of a failure mode, its criticality, and detection methodology are important attributes in the maintenance sphere. The frequency of inspection of the oil circuit breaker, for example, is just as significant. The quality tool Pareto chart is utilized again to evaluate the compiled problems derived from a technical inspection checklist of the maintenance area (panel C of Figure 1). The Histogram is utilized to evaluate the frequency of occurrences over a period of time. It is designed to observe the overall variations in a process. Most of the incidences were discovered during the spring and fall of the calendar year in accordance to the summer critical performance criteria where the equipment is in the "must use" state. Only the minimally required checks were conducted during this period. These data is depicted in panel B of Figure 1.

The quality trilogy encompasses the principles of quality planning, quality control, and quality improvement (Juran, 1999). The quality actions must be planned in order to achieve exceptional quality management. Moreover, the defined quality must be controlled and the performance improved on a project-by-project basis with tools such as Pareto analysis. The investigation assists with identifying the "big problems" within the quality scheme. Eventually, the activity achieves penetration to a superior level, which is again controlled to prevent future deterioration. The quarterly checks of the oil circuit breaker reveals high occurrences in the *trip free fail to operate* category. Monthly checks produced nineteen events in the *loss of sealing ability* category for the same period. A systematic quality control method to reduce the frequency of incidences was nondescript in the company. Replacement of the entire oil circuit breaker unit once it became too problematic was the only detectable corrective measure. A good quality management system requires the entities to communicate the results as well as keep documents of the improvements achieved. Additionally, projects must be conveyed to solve problems, report progress, provide training, and quality goal setting for improvements.

The Failure Mode and Effects Analysis (FEMA) must include initial actions to reduce the problem occurrences (McCalley et al., 2003). The problem detection methodology and recommended actions is required to take the condition one step farther and hold an entity or individual responsible to complete the measures (Figure 2). The target completion date, actions taken, results of the severity, frequency reduction, occurrences must be documented as a component of the quality improvement efforts. Although the FEMA tool can successfully uncover potential failure modes of a process, it cannot reveal complex deficien-

cies (smaller failures that have a compounding effect) (Nanda, 2012). Another approach utilizes reliability prediction in the early design phase by emphasizing quantitative data from a parent product (Sanchez & Pan, 2011). An exceptional reliability centered maintenance plan incorporates automated templates for the personnel to evaluate the tasks and provide the technical justification for the repairs (Bently, 1999). The template also provides logical maintenance activities to include the key data to track and trend (utilized as triggers for other tasks to establish a baseline) and implementation considerations. The physical assets are subjected to a variety of stresses in the field which creates deterioration by lowering its resistance to externals pressures. When the resistance decreases to a point where the asset can no longer perform, it will fail as well. The stress exposure is measured in a variety of ways to include the operating cycles of the asset, the output, operating times, and calendar/running time. This will create a pattern for the asset to represent the aging and deterioration. The well-established bathtub curve (with a high incidence of failure in the beginning life of an asset known as infant mortality) is followed by a constant or gradually increasing failure probability then a wear-out zone, is the model for electrical transmission equipment. The detection of early failure rate problems induces longer burn-in times for specific products and subsequently reduces overall costs (Ye et al., 2011). Nonetheless, the constant failure probability at all stages in the lifecycle is always a concern (Schlabbach & Berka, 2001).

The utilization of fishbone diagrams and process mapping techniques as a high-level tool to align resources with the utility company's mission statement is a challenging feat. Various quality and project management mechanisms were employed to solicit stakeholder buy-in, especially with high-visibility projects. The sequence of event paradigm on a critical path diagram was utilized in this case to enhance the senses of urgency of the document management turnover process from one electric company to another.

Figure 2. Failure effects of the oil circuit breaker

Failure Mode (criticality)	Components	Failure Cause	Failure Effect	Detection	Maintenance Activity	Frequency
Loss of sealing ability (medium)	Nitrile seals	Moisture ingress leading to dielectric failure. For small section seals, the deterioration is pressure dependent	Loss of sealing ability leading to increased demand on, and the early failure of air system equipment. Insulation media loss.	Visual Inspection	Replacement	1 month for inspection
	Nebar/cork gaskets/joints	Reduction in thickness due to "hammer action" moisture ingress		Visual Inspection	Replacement	1~10 years
Fail to Close or re-close (medium)	Lubrication	Loss of lubrication oil; lubrication degradation	Failure of one of the main purposes of breaker	Check oil level	Re-lubrication	1 year
	Mechannical Part	Motor Failure; compressor seizure; loose connection; contact wear; switch failure; wrong setting		Operational test	Repair mechanical parts; replacement	1 year
	Insulation Oil	Oil degradation, contamination; moisture accumulation		Insulation oil test	Oil filtering, oil replacement	300 hours
	Control Circuit	Close coil fail		Operational test	Repair replace	1 year
Fail to Open (high)	Lubrication	Lubrication degradation	System instability. Major failure. High cost of repair	Inspection	Re-lubrication	5 years
	Mechanical	Mechanism out of adjustment; wrong setting		Check key measurements	Adjustment	1 year
		Weld of shaft crack; glass fiber rods shearing		Visual Inspection	Repair, replacement	1 year
	Insulation	Oil contaminated		Insulation oil test	Oil filtering, oil replacement	Each interval inspection
	Control Circuit	Trip coil failure		Functional test	Repair, replacement	1 month for test
Fail to Insulate (high)	Insulation Oil	Oil loss, containmination, degradation, moisture ingress	Servere damage	Visual Inspection; insulation oil test	Sealing; refilling	Each interval inspection
	Bushing	External bushing insulation failure		Power factor test	Cleaning, greasing; replacement	5 years
Pressure Switch fail to operate (high)	Mechanical parts	Loose connection; subcomponenet failure; out of adjustment; mechanical clog (or crack); contact fouling	Closing with insufficient pressure may result in damage	Operational checking of pressure switch	Tighten; repair; adjust; replacement	1 year
Auxiliary contacts fail to operate (medium)	Mechanical parts	Loose connection; mechanism out of adjustment; linkage binding; subcomponent worn; cracked shaft; contact fouling; contamination	This failure can prohibit proper automatic and manual operation	Physical check of wire termination points/check of auxiliary switch	Mechanical maintenance	Each local breaker exercise
Loss of mechanical strength (medium)	Porcelain to metal joints-cermets, oxide jacking	Frost or rust oxide jacking	Fracture and/or destruction of porcelain. Chemical aging of cermets	Visual inspection; operational test	Cleaning and lubricating; Replacement	Each breaker maintenace
Coating corrosion (medium)	Paint and other coatings	Corrosion, aggravating items; seals, joints, bushings and tanks	Sealing failure; loss of mechanical strength; insulation failure	Visual Inspections	Recoat	1 year for visual inspections
	Housing	Corrosion			Replacement	
	Oil Storage tank	Corrosion			Inspection replacement	
Governor fail to operate (medium)	Governor	Loose connection/sub-component failure/out of adjustment/contact fouling	Can lead to excessive run time and compressor failure	Physical check of wire termination points/check of governor	Mechanical maintenance	1 year
Heater Failure (medium)	Heater	Heater element failure/thermostat failure	Air valves and poor response	Check heater operability	Repair, replacement	1 month for check
Internal Cabinet leak (medium)	Internal cabinet	Deterioration of weather seal/compressor oil leak	Overall breaker operation	Visual Inspection	Sealing maintenance	1 year
Trip free fail to operate (medium)	Mechanical parts	Mechanism out of adjustment/subcomponent sticking/loose connection/ switch or relay failure	Damage of equipment	Check key measurements/ operational check	Adjustment / replace/ tighten	100 ops/year

(Leftmost spanning label: Oil Circuit Breakers)

The basic understanding of the available quality tools and their associated uses aided with the implementation of several tenets of the program. The brainstorming strategy sessions, requirement analysis, and graphical process mapping techniques clearly defined the problem to senior management. The business requirement was to establish a document management record center for the company and develop a cost-effective method & processes to procure, identify, categorize, classify, enrich, and store documents in a future repository. The day-to-day operations of the company's core stakeholders (management, internal users, contractors, etc.) required constant handling of public utility documents in the course of its business. These important documents were previously managed, controlled, and maintained under contract by the investor-owned utility at an enormous annual lump sum cost. The implicit requirement for the complete turnover of the company-owned documents (approximately 750,000 in various formats) from the investor-owned utility company by the end of the service agreement was imperative to achieve the strategic goal and develop a sound fiscal management policy.

The identification of document types/counts (engineering drawings, vendor documents, bill of materials, circuit lists, equipment data, real estate, patrol maps, etc.) in each department was required for the company handling, conversion, and storage accommodations in every format (electronic, microfilm, paper). Check sheets was also compulsory for document retrieval functionality (precision, recall, locating), document link management, and overall document quality inspection for anomalies (speckles, clouds, etc). The establishment of a comparable document management system for the records center was a parallel objective of the solution. The company faced enormous challenges that included the following:

- Investor-owned utility records management system was not easily shared or integrated into the company's business applications.
- User interaction that required numerous inquiries and offline research for hidden data.
- The document review/completion process involved multiple, time consuming offline procedures.

Moreover, the technology transfer between both companies was non-existent. The major problems encountered involved technology transfer from one medium to another. Several older substation drawings (dated from 1905 to 1930) were carefully handled with special gloves to preserve the integrity when scanning into the newer system. The culture of the investor-owned utility was reluctant to embrace the newer technological advances and rapid recall of the documents from any field location (Schaar & Wilson, 2010). The overall high-level strategy and quality mechanisms are illustrated in Figure 3.

The implementation of a Three-Prong Approach (document processing/conversion of microfilm, paper, Mylar, direct electronic transfer to the company content/document manager, long-term paper storage and destruction) as a putrefaction strategy for the company-wide document retainage. The major objective encompassed every current document holder of the company-owned materials to include, but not limited to, land management, forestry, system control, system protection, environmental, and maintenance departments in the transitional phases. The team (competitively selected based on their expertise in their appropriate field) developed policies and procedures between both companies (physical acquisition, specifications, asset definition, process design, scheduling, drafting standards, joint substation issues, pilot programs, etc.) to facilitate the document turnover process and encompass accountability. The activity included the document split-out (turnover start), contractual agreements, pilot program, hardware and software implementation, debugging, system evaluation, storage area acquisition, file cabinet amortization, and personnel training. The following year scheduled activities encompassed full file turnover, processing/indexing, and storage/destruction of specific documents. Quality measures were utilized in each phase/milestone and thoroughly reviewed. The Ishikawa (or fishbone, cause and effect) diagram was utilized to systematically represent and analyze the true requirements for process implementation. The diagram organized the major and minor contributing leading effects, assisted with the definition of a problem, and identified probable causes by narrowing down the possible ones. The key to the diagram assisted the team members with systematic generation of ideas and provide the correct feedback for the direction of causation. The document management fishbone diagram is illustrated in Figure 4.

The ultimate diffusion of data in the technology transfer sphere was a contact activity, especially as it pertained to the physical acquisition of the documents. The implementation of an independent document management center had an immediate impact on the bottom line with the abolition of the enormous annual administration fees. The estimated cost avoidance of $7.5k with the elimination of the microfiche documents alone was realized within the first year of use. The increased user efficiency of the web-based portal

Figure 3. Document management high-level strategy

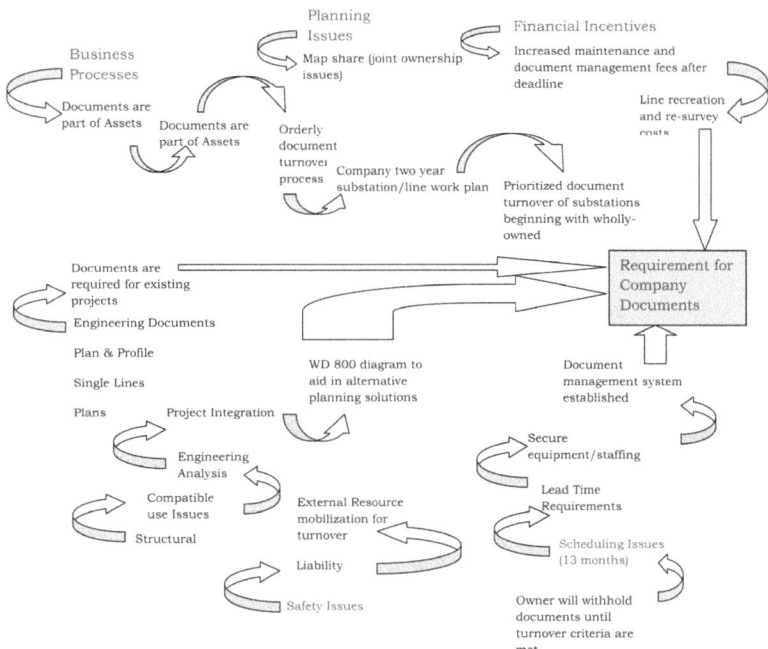

and improved adherence to regulatory requirements were additional benefits of the company-owned document management system. Moreover, the ability to implement and administer the system internally, and share knowledge across disciplines was supplemental benefits of the program. The entire process consumed sixty-nine weeks for document ingestion to the company. The program was modeled and simulated to optimize the available resources before the end of the service contract. The document management system utilized enforceable business rules and security characteristics unique to utility company.

THE QUALITY MANAGEMENT SYSTEM

The Quality Management System is required in the electrical installation processes and equipment upgrades within the substation and transmission line rehabilitation work efforts. Detailed process mapping and organizational interaction is essential for a successful renewable energy generator system

implementation. However, top management must commit to the quality initiatives throughout the entire process. This total commitment must include a viable quality policy as well as establishing achievable quality objectives for the work processes and the associated work environment (ISO 10005, 2005). The quality plan must encompass tenets of the sample outlined in Table 3.

Top management is expected to provide input concerning the strategic processes of the capital-intensive project. Continual improvement efforts for future projects are also provided by the organization's senior management (ISO 10006, 2003). The quality culture is expected to derive from effective leadership to ensure the infrastructure is in place to meet the stated objectives. This includes total quality commitment by all levels of the organization.

Knowledge alone is not enough to get the desired results. You must have the more elusive ability to teach and to motivate. This defines a leader. If you can't teach and you can't motivate, you can't lead. (John Wooden)

Figure 4. Fishbone diagram of the document management ingestion

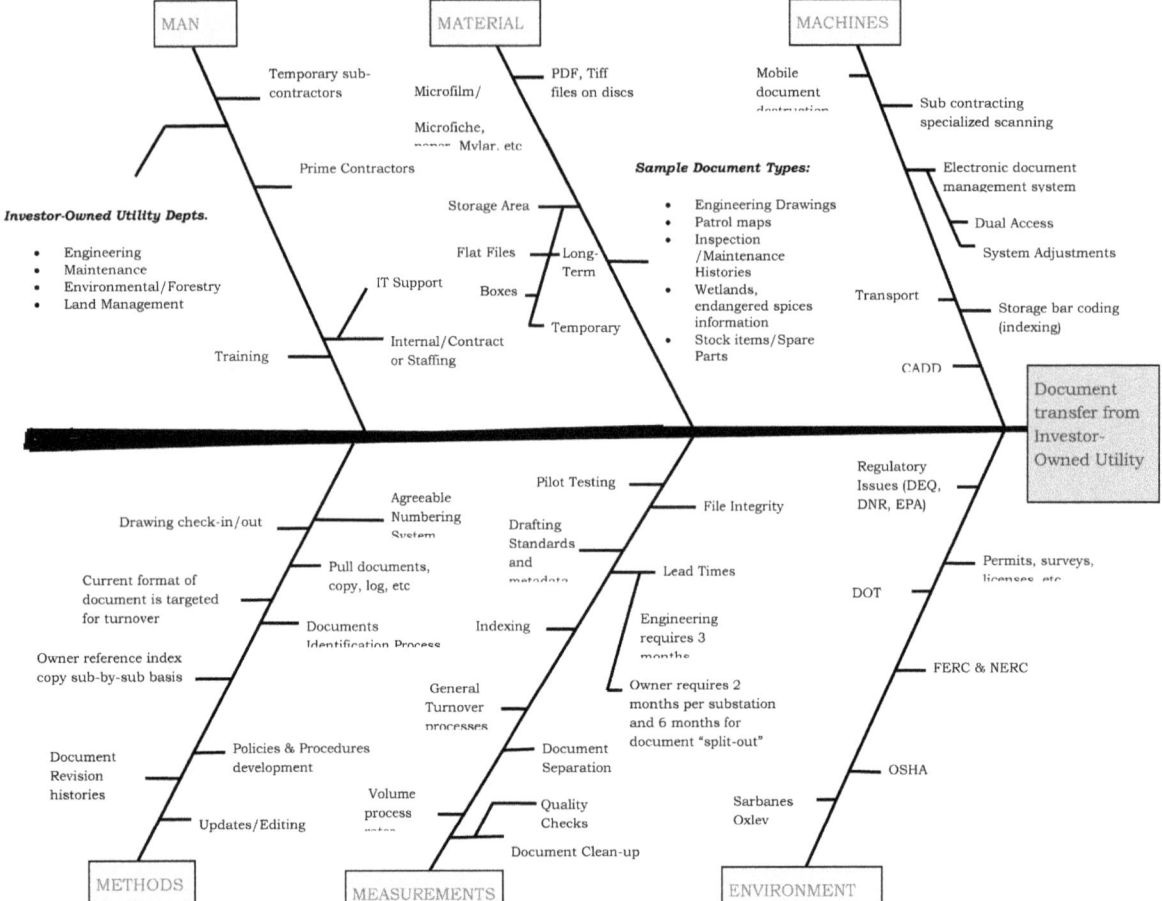

Ultimately, the quality objectives of the organization are the responsibility of top management. This group must drive the essential quality components on a daily basis. One of the quality gurus, Tom Peters, emphasized the leadership technique of "Manage by Walking Around," (MBWA) to actively engage in the listening and training aspects of the paradigm. The manager is able to provide assistance and *on-the-spot* interpretation of quality values when affianced in these techniques. Moreover, it demonstrates humaneness and a sense of caring when managing by walking (not wandering) around. An example of the MBWA technique involved a perceptive top manager in an eastern electric utility and a tiger team tasked to investigate overdue (both over budget and behind schedule) capital projects with regional managers. The top manager was more than willing to listen to the field problems and reinforce the quality plan. Discussions were focused on resource and material issues with regional problematic substations. A recurring case concerned a particular incorrect cable at the substation as a direct result of the quality control initiatives and methodologies. It was discovered, ironically after top management conducted a regional capital project meeting with the line shop, substation design/services, and the supply-chain personnel, (the forces, events, entities, and people affecting the situation at hand) on the MBWA technique. During the field discussion on inter-department coordination and staging/ material preparation for this particular substation, the cable was ascertained as too short to connect to the equipment. The noticeable discrepancies and inconsistencies with the available information, initiated by the top manager, were corrected immediately via the appropriate supply-chain channels. Further discoveries revealed that the die sets were non-compatible in accordance with the design and corrected as well. The profound lessons-learned from the field audit encompassed the following tenets of the quality assurance paradigm:

- Material Integration was common in all three (3) effects
  ○ Resource Management and Performance
  ○ Based on material logistics, delivery times, and system integration into the substation
  ○ The Project Manager must reduce delays with proper communications (progress reports, department coordination) and via monitoring/controlling techniques
- Engineering design problems *must* be resolved during parallel operations
  ○ Perform Quality Checks for accuracy (i.e. phasing, general scope, alignments, Bill of Materials, etc). Utilize "Checklists" where applicable
  ○ System Configurations
- Material Preparation and Staging
  ○ The project manger *must* "Manage-By-Walking-Around" to ensure adequate part compatibility, free of defects, etc.
  ○ Provide frequent project site updates/ vendor performance reports
  ○ Provide documented quality inspections, corrective & preventative actions.
- Thoroughly investigate previous "Lessons Learned" packages.

The top manager re-introduced each regional project manager to the quality standards handbook and implemented at least one quality problem/ resolution during subsequent Capital Project Planning Meetings.

The incorporation of the Plan, Do, Check, and Act (PDCA) circular paradigm in all of the processes is the touchstone for an effective quality management system (ANSI/ISO/ASQ Q9001-2000, 2001). The objectives must be carefully *planned* to deliver the results in accordance with the established quality policies. *Do* by executing the processes. *Check* by verification and monitoring of the process against the quality policies and report results. *Act* takes actions to continually improve the process. The model ensures adherence to the quality continuous improvement efforts established by the organization.

THE SINGLE CAPACITY PROJECT

The substation and transmission line maintenance of specific units will escalate to serious problems if left unattended. Several of these known problems can only be mitigated through methodical substitution of entire units. Consequently, these same projects are elevated to a high enough level where the competition for the budgeted dollars is fierce. There is a genuine need to upgrade and reinforce the electrical system via the correct project priority channels. However, the majority of the projects are not properly ranked, scheduled, or budgeted with a systematic process. A sample of a circuit breaker replacement program from a regional independent service operator capital project catalog forms the basis of an appropriately implemented priority ranking scheme. Several of these units were over-burdened and required systematic replacement in the electrical footprint. Detailed inspection of selected oil circuit breakers revealed several hierarchy problems to include interrupter concerns with butt style contacts (slower operating breaker) and the air cells filled with oil.

Table 3. Quality plan outline

Requirement:	Descriptions	Details:
1. Scope **a. Inclusions** **b. Exclusions**	How will the organization meet customer and regulatory requirements?	• What is the reference documents utilized for the quality management system? • What are the conditions of conformity? • What specific quality processes are included/excluded?
2. Quality Objectives	Established by top management at relevant levels/functions within the organization	• Must be measurable and consistent with the quality policy • The foundation for quality policy, suitability, and organizational compliance requirements
3. Responsibilities	The quality management system is supported by a committed top management team impressed throughout the organization.	• Policy, objectives, management reviews, and resource availability
4. Documentation	Documented statements of a quality policy and objectives, a manual on quality, and document control.	
5. Records	Must be established and maintained to provide evidence of quality conformity.	• Continued operation of the process generates further record that confirms the process is meeting the requirements
6. Resources	Process and project implementation must be performed by trained, competent personnel fully aware of the quality component.	• Competence is the capability to display use of knowledge, skills, and behaviors to accomplish the desired results
7. Project Inputs	Requirement analysis	
8. Customer Communication	The stakeholders are entitled to feedback, process information, amendments, and complaints.	• Inquires are addressed via project meetings • All communications must include the project manager
9. Design & Development	Control of the verification and validation of the process and project implementation with effective communication between interfaces.	• Inputs: Functional & Performance Requirements • Outputs: Process and project acceptance criteria
10. Procurement	Purchased products must conform to the specified requirements set by the organization	• The quality controls and constraints must be appropriately applied of incoming equipment for conformity
11. Installation & Commissioning	Specific processes and procedures defined for equipment installation	
12. Special Processes	Corrective Action Status	• Statistical trend surveys
13. Configuration Management	Utilization of the current version of the quality configuration management mechanism	• Configuration control, design changes, and traceability
14. Customer Property	The organization must identify, verify, protect, and safeguard customer property incorporated in the process for project implementation.	
15. Product Handling	Detailed procedures are established for each component of the process	
16. Non-conformities	Ensures process conformity of the project implementation with the quality management system.	• Material non-conformities must be handled in accordance with the procedures set within the quality plan and supply-chain
17. Monitoring & Measurement	Establish the frequency of project progress reporting.	• The progress of the project is recorded via timesheets and updated weekly on the master schedule

continued on following page

Table 3. Continued

Requirement:	Descriptions	Details:
18. Internal Audit	Specify which phases to ensure process conformity of project implementation.	• Audit schedule at the design, execution, and/or close-out phases.
Author _____ date:_ Quality Mgr _____ date: _ Document # _____ date: _		

There were also profound decreases in the pre-charge levels of ten percent per year as well as complicated interrupter designs which submitted to intense labor demands when inspected internally (yet no performance issues with the latter condition). Many of the problems were attributed to adverse conditions that included poor designs (improper/inadequate design that require frequent maintenance), slow or worn mechanisms (in need of refurbishment), and bad actors (frequent correction or on-condition maintenance is required based on a variety of problems). One particular project required the replacement of the entire gas insulated bus and breaker configuration because of unsuccessful attempts to patch the system. The $5.8M dollar effort utilized an open air type system as the preferred replacement. Excessive run time can damage the compressors (some breakers have no run time meters) and must be checked monthly for air leaks. The data ought to be documented and run time meters installed where applicable. Comparing inspection results with the predictive maintenance data is a component of the *best practices* methodology. When an identical inspection is performed on an oil circuit breaker, data must be gathered on the condition of the mechanisms and piece-parts. This is a technique of corroborating the accuracy of insights from predictive maintenance tasks.

Most of the problematic oil circuit breakers were replaced with the SF_6 type of units, which posed a detrimental environmental effect. One of the chemical elements that reside within high voltage equipment in an electrical substation is Sulfur Hexafluoride gas (SF_6). Because of its unique compound properties, the gas is used as an electrical energy insulant and arc quenching medium frequently found in high voltage electrical equipment utilized by utilities and their major industrial customers. Unfortunately, the same properties that compose SF_6 so effectively for use in electrical equipment also make it a potent greenhouse gas (GHG) with significant global warming potential if it is released into the atmosphere. In fact, SF_6 gas is more potent than CO_2 (23,900 times) with an atmospheric life of 3,200 years (Mueller, 2005). Consequently, the gas is now targeted for emission reduction and, as an environmental priority, warrants replacement technology in the electrical industry.

The environment is an integral component of *any* quality and strategic plan when evaluating new technology. The gradual destruction of the protective ozone layer with the release of chlorine into the upper echelons of the atmosphere are potential dangers posed by the chlorofluorocarbons (CFC) used in the electrical industry. The CFC atoms are extremely stable within the confines of the technology; however the radiation levels in the Earth's upper stratosphere creates the decomposition by converting O_3 to O_2 (Molina & Rowland, 1974). The ozone layer essentially blocks the harmful high-frequency ultraviolet light from the Sun generally attributed to the increase in skin cancer. As a contributor to the greenhouse gas effect, the SF_6 gas must be handled carefully with the used amount maintained as low as possible (less than 10kg for a 145-kV breaker). Additionally, the leakage rate must be minimized to less than 0.5% per year for the live tank circuit breakers to produce low SF_6 emissions. The leakage is generally developed around the flanges over a period of time (20 to 30 years) and considered as a normal characteristic of the technology. Deep curtailments are compulsory to mitigate the effects of greenhouse gas emissions in climate change (UK HM Treasury, 2006). Moreover, an independent alternative analysis and evaluation strategies of the replacement equipment for the gas circuit breakers of targeted utilities and process (primarily gas) industries is required to reduce the emission production. An exploration of numerous potential solutions and evaluation of each before accepting the preferred methodology is also a priority. A thorough investigation of the high voltage ($\geq$765-kV) air blast breaker as well as the low voltage ($\leq$ 38-kV) vacuum type breaker must be included in the in-depth alternative analysis. Moreover, evaluations of the service leak frequency, disposal, and maintenance best practices merit the necessary scrutiny to facilitate the agenda. The circuit breaker program is in dire need to substitute the SF_6 units to a more environmentally-friendly device. Incisive planning and strict adherence to the quality management system would have produced the cost avoidance of several millions of dollars. However, because of the intense competition for the annual budgeted dollars coupled with the *fix-it-now* summer critical capital project criteria as a corrective measure in rework activities. The planned upgrades are essential for renewable energy system integration. Yet, the point of interconnection for a wind turbine generator, for example, must not contain perilous electrical equipment that will continue to devastate the environment. The contradiction of implementing renewable energy units to substations with *upgraded* equipment that is a known contributor to greenhouse gas emissions is unconscionable for the truly green electric utility.

Nineteen major units were included in the sample study of similar circuit breakers in the capital upgrade projects. All were competing for the same capital dollars at the highest priority levels. Nevertheless, the results of the study ranked the projects quite differently than the original cost allocation/general need priorities typically experienced in the industry. The derated equipment was aptly modeled with appropriate replacement capacities for future in-service dates. All of the projects escalated to a higher priority category based on the equipment condition assessments and preferred simultaneous funding prospects. Randomized and opportunistic outage schedules to facilitate the work effort were the normal replacement procedure employed in the region. The established priority arrangement evaluated the nineteen projects and logically ranked the units in accordance with set criteria. In the ranking scheme, the projects with insufficient supporting data received a *no priority* designation no matter the cost, age, and condition of the equipment. The importance of the documentation for funding justification is an indispensable element in the quality plan, as illustrated in Table 4.

Approximately half of the projects evaluated received the *great priority* label based on its current equipment capacity levels (the threshold was set to greater than 110%). Moreover, almost half of the projects received the same label for priorities based on the lowest capacity levels (the greatest priority level attached to projects less than 59.99% capacity levels), of which more than half were common with the other high-threshold group. The decision strategy evaluated the derated condition and determined which circuit breaker provided the lowest capacity for the highest currently installed capacity, regardless of cost. Ties were broken by means of the installation year, where the oldest unit received the highest priority. In the case of an additional tie, the priority by rank (highest capacity) project was the determining factor for the tie-breaker. If on a rare occasion a subsequent tie was formed, it was broken by the priority rank (lowest capacity) project. Naturally, the *great priority* projects should have received the highest ranking in the queue. The top five priorities at both capacity levels are ranked accordingly. The number one ranked Clarement "A" 288 circuit breaker produced the largest margin (57.70%) in the model logic although it is classified as fourth at both capacity levels. Conversely, the project with the smallest margin (Artganti 6W8) ranks on the bottom of the precedence scale based on the logic. It was the youngest circuit breaker (installed in 1988) in the study scheduled for replacement. Sadly, the equipment is scheduled for replacement with the *same* type of problematic circuit breakers as before. Because of the lack of documental justification, the oldest equipment in the pool (Alpune 188 installed in 1946) is scheduled to receive no funding. The precedence results ranked the Spaulton "A" 36M9 circuit breaker still within the top five for the work effort. The perceptive decision-maker will recommend an outage for the replacement of the 36M9 along with the Narrow 377 circuit breaker to strategically substitute both units before the summer critical period. The issues at the Champson substation rank only one of the circuit breakers in the overall top five. The decision-maker may elect to prioritize six projects at the substation level, beginning with circuit breaker 288 and ending with the 588 unit, with different time-phased outages to maximize efficiencies. The cost effectiveness of the overall benefits practiced in the circuit replacement program with parallel operations, strategic planning, and Lean (just-in-time) methodologies would diminish the overall cost and resources to facilitate the necessary upgrades. It is designed to deliberately reinstate the problematic components with the correct (environmentally-friendly) equipment, the oldest, as well as the units that will provide the best performance (based on margins) for the work effort. The timing of the outage schedules can be prioritized as well to coincide with "ready-to-install" capital projects that will minimize idle time while maintaining efficiencies. It is most desired to minimize the output response within the circuit breaker replacement program. The units in the study for each documented replacement is based on the component wear, overload conditions, and eventual equipment deterioration. The *smaller-the-better* characteristic in the loss function aims to reduce the result where the ideal target is zero (Taguchi et al., 2005). It is here where the quality loss is quantified as it relates to specification performance, product distinctiveness, and the overall financial loss of the installed equipment. The loss function is outlined below:

$$L=ky^2$$

$$k=A_o/y^2_o$$

$$L=k*(\partial^2+y^2)$$

The increasing parabolic function includes the cost utility as $L(y)=ky^2$ where:

$L=$ loss

$k=$ proportionality constant

$A_o=$ Replacement cost

$y_o=$ unit capacity

$\partial^2=$ reduction of the variability around the average

$y=$ average

When the capacity (y_o) is at the optimal level (52.6) or exceeding the threshold limit (110), the replacement cost (A_o) is constant ($160k) for each circuit breaker.

Thus,

$$k= 160,000/(52.6)^2 = 57.829$$

And the loss function: $L(y) = 57.829\ (y)^2$

The computed losses are in the magnitude of $316,839.06 (high threshold limit) to $316,388.08 (lowest capacity) with *only* a $450.98 range.

Table 4. Circuit breaker upgrade priority ranking

	Substation and Breaker:	Cost:	Type & Replacement:	Year Installed:	Percent Capacity:	Modeled Replacement Percent Capacity:	Priority by Current Capacity:	Priority by Rank *(Highest Capacity)*:	Priority by Lowest Capacity:	Priority By Rank *(Lowest Capacity)*:	Margin Difference:	Final Priority Rank:
1	Mossy 345-kV	$5,800,000					No Priority		No Priority		0.00%	
2	Mossy A 13B7	$160,000	Oil	SF6	1-Jan-1966	106.80%	54.30%	*High Priority*		*Great Priority*	3	52.50%	*6*
3	Mossy B 13M9	$160,000	Oil	SF6	1-Jan-1973	102.40%	52.60%	*High Priority*		*Great Priority*	1	49.80%	*8*
4	Champson A 148	$160,000	Oil	SF6	1-Jan-1964	111.70%	64.30%	*Great Priority*	5	*High Priority*		47.40%	*10*
5	Champson B 188	$160,000	Oil	SF6	1-Jan-1964	108.30%	62.30%	*High Priority*		*High Priority*		46.00%	*13*
6	Champson C 288	$160,000	Oil	SF6	1-Jan-1964	111.20%	54.70%	*Great Priority*	7	*Great Priority*	5	56.50%	*2*
7	Champson D 388	$160,000	Oil	SF6	1-Jan-1960	111.20%	63.40%	*Great Priority*	6	*High Priority*		47.80%	*9*
8	Champson E 500	$160,000	Oil	Kit	1-Jan-1960	109.20%	62.30%	*High Priority*		*High Priority*		46.90%	*11*
9	Champson F 588	$160,000	Oil	Kit	1-Jan-1960	107.60%	63.50%	*High Priority*		*High Priority*		44.10%	*14*
10	Spaulton A 36M9	$160,000	Oil	SF6	1-Jan-1968	113.30%	58.20%	*Great Priority*	1	*Great Priority*	8	55.10%	*5*
11	Spaulton B 36B7	$160,000	Oil	SF6	1-Jan-1968	103.40%	53.10%	*High Priority*		*Great Priority*	2	50.30%	*7*
12	Narrow 377	$160,000	Oil	SF6	1-Jan-1968	113.00%	56.80%	*Great Priority*	2	*Great Priority*	7	56.20%	*4*
13	Clarement A 288	$160,000	Oil	SF6	1-Jan-1968	112.30%	54.60%	*Great Priority*	4	*Great Priority*	4	57.70%	*1*
14	Clarement B 388	$160,000	Oil	SF6	1-Jan-1968	111.20%	54.70%	*Great Priority*	8	*Great Priority*	6	56.50%	*3*
15	Artganti 6W8	$160,000	SF6	SF6	1-Jan-1988	104.10%	74.90%	*High Priority*		*High Priority*		29.20%	*15*
16	Semphall 499	$160,000	Oil	SF6	1-Jan-1957	112.30%	66.00%	*Great Priority*	3	*High Priority*		46.30%	*12*
17	Alpune 188	$160,000	Oil	1-Jan-1946			No Priority		No Priority		0.00%	
18	Chrastien	$1,100,000					No Priority		No Priority		0.00%	
19	Beckdile	$1,100,000					No Priority		No Priority		0.00%	

Therefore, the cost avoidance for each circuit breaker is approximately *double* the replacement expenditures. Moreover, the entire replacement program – from the quality perspective – for the documented units equates to $4.745M. Since rework effort is highly recommended to replace the SF_6 equipment with environmentally-friendly units, the cost avoidance dollars swells to approximately $10M. The quality component is an essential analysis mechanism for the incisive strategic decision-maker involved with the renewable energy system integration efforts.

Pre-assembly of the materials, for example, enhances efficiency and improves the quality throughout the construction cycle. Lean quality techniques are specifically designed to reduce waste and advance efficiencies in an organization (Womack & Jones, 2003). Lean thinking paves the way to do more with less – resources, materials, space, and time. The astute manager will incorporate the Lean paradigm in the construction cycle to include the following characteristics:

- Human activity that absorbs resources yet creates no value
- Transportation of materials and resources with no purpose
- Downstream resource wait time for upstream activity completion
- Services that fail to meet the customer requirements

Top management in the organization is keenly aware of the system integration time and cost issues that envelops renewable energy. Yet this same group of people are unaware of the cost of poor quality as a percentage of material, resources, etcetera the system integration affords the program. The Lean processes merged within the quality plan will not only heighten the continuous improvement initiatives, it will also *minimize* time and reduce the overall costs. Project goals are definitely met with alternative strategies and methods (Pietroforte, 1997). The various managerial approaches must entail a certain level of flexibility while maintaining the quality component throughout the development process. The incorporation of some form of the Lean paradigm as a set of techniques and principles will provide a different viewpoint of the equipment installation process. It is an improvement of the initial reliability aspect intended to eliminate the design weakness (Hall et al., 2010). The Lean theory is designed to accomplish improvements in three common areas 1) design of the process system as a level of managerial action 2) control of the process to achieve the desired goals and, 3) leadership of the improvement of the process system. As a fundamentally customer-focused principle, Lean is ideal for the integration of renewable energy to the electrical infrastructure. However, equipment upgrades are inevitable in order to facilitate the system integration capital project.

In heavy construction ventures, communication is the key component that delivers a quality product (Park & Meier, 2007). Each department is obliged to provide detailed information and data on the progression of the project. Nonetheless, these data are rarely documented, archived, and utilized in lesson-learned databases for future project implementation. The project time estimates typically employ a variation of subjective ad hoc techniques where the completion time is highly dependent on the estimator's personal experience (McCrary et al., 2007). The high variability inherit with the method reduces the quality component throughout the capital project implementation process.

One quality method that involves all employees in the organization for equipment improvement resides in the maintenance arena. It is an effective method classified as *total productive maintenance* utilized to improve reliability, quality, and the overall production (Duffuaa et al., 1999). The equipment performance loss is monitored and documented to include outages, reduced outputs (derated), idling and momentary outages, and its related impact on quality. Tenets of this program espouse the concept of improving the overall equipment effectiveness, time availability, and the performance efficiency based on the quality rates. Moreover, the initiative goals are to incorporate a systematic approach for reliability, maintainability, and the lifecycle cost of the equipment. It includes all levels of management and workers (operations, supply-chain, maintenance, engineering, equipment management, and administration) through small group activities and team performance to improve equipment effectiveness. These personnel must provide detailed documentation – as well as automated transcriptions – to augment any known equipment problems experienced during operations. The action ensures quality throughout the lifecycle while creating value for future system upgrades. An example of a typical outage that affected ancillary equipment is described in Box 1[1].

The stability reduction issues are a major concern from an equipment perspective. A marginal line in a parallel path may suffer quality-wise and operationally if it experiences continued stress. As part of a Reliability-Centered Maintenance program, the equipment is periodically checked and monitored as problems are documented to identify any corrective measures (Moubray, 1997). Detailed documentation ensures effective maintenance in the quality sphere. It assists with the elimination of equipment *bad actors* (lemons) or poor designs within the electrical substation. The equipment well-documented maintenance frequency may escalate specific units to a higher priority on the replacement schedule. Periodic triggers based on the economic lifecycle cost of critical equipment within the electrical footprint must be in place to evaluate *repair-vs.-buy* decisions. A burned down transmission line in one area will produce negative consequences in another if the equipment is problematic. The detailed documentation will provide the necessary leverage to replace and upgrade such equipment.

The Ordinal Capital Project Ranking Evaluation

The circuit breaker replacement program for the selected electric utility was ultimately prioritized on the regional level. There were several other circuit breaker replacement projects channeled to different categories. The essence of the problem was to correctly recognize and prioritize the circuit breaker replacement program based on the general need, ratings and /or equipment limits, reliability, and upgrade. Ultimately, the small replacement program was ingested in a larger prioritization matrix, mis-labeled, and exposed to defunding actions. The purpose of the ordinal ranking overview is to assess the feasibility of the capital endeavor when compared with the given alternatives (other projects competing for funding opportunities in this case). The ordinal rank (prioritization) forms the numeral which displays the series order for funding decisions. Two of the elements from the TRIZ[2] principle #17, specifically the "Another Dimension" are outlined below (Mann & Domb, 1997):

- To move an object in two- or three-dimensional space
  - Interrelationship diagram
  - L,T,Y,X, C types of matrices
  - Distributed accountability and influence (i.e. Quality audits conducted by the department and advises on technical details, yet everyone is responsible for quality)
- Use 'another side' of a given area.
  - Extensive two-way communication

Box 1.

Start Date - Time: 5/13/2009 1:02:00 PM EDT
Region: NPCC
Affected Utility: Hydro-Quebec Trans Energie
Balancing Authority ID: HQT **Type:** UO
Cause: Fire-Vegetation / Trees
Event Description:
DESCRIPTION: On May 13 at 13:02 a 735 kV line tripped. The automatic reclosing operated normally but the line tripped again. After a few minutes, the Operator re-energized the line and it tripped again. A patrol of the line was requested. The patrol found a fire on several trees close to the right of way. The fire was extinguished and the trees were cut. The line was back at 21:08. This event had no effect on the generation dispatch, transmission, or customers. The stability limit on the corridor was reduced. It is suspected that an electric arc was the cause of the line trip. The investigation is still ongoing.
TYPE OF EMERGENCY: Other
ACTION(s) TAKEN: Other
Category: Fires

◦ Organizational assessment by observing the business from the other side either directly or utilizing external consultants

The advantage of the process is encompassed within a decision structure (a Y-Shaped Chart) with events, numerical rankings, and capital project consequences. The cardinal ranking, or the difference of the alternative's expected value ranges, evaluates the most plausible option of the capital project under evaluation (the next logical sequence in the process). The model logic of the chart is divided into five (5) major project sections and three (3) areas of various levels of need. The objective of the entire process is to examine the top five priority rankings of the Independent Service Operator's capital projects and validate with group analysis using standard utility categorization criteria. The results of the study formulate the input for the cardinal rankings of specific capital projects within the ordinal portfolio. The general pre-planning capital project metrics is shown in the pie-chart of Figure 5.

METHODOLOGY AND RESULTS

The Y-Shaped Chart utilizes a single element from each of the five major categories of the specific requests to formulate the general category. The summation of each request is tallied and given a symbol in accordance with the level of need (relative strength), effort, and urgency. An evaluation is then performed (calculated) to determine the total Level of Effort (LOE) and associated urgency. It is here where the ordinal rankings of the top five project categories are first derived (see Figure 9). The utility industry standard categorization system is implemented for comparison purposes. The capital projects are ranked within their respective sectors and evaluated against industry norms. The final step prioritizes and validates both sets of data and compiles the group in terminal ordinal project

ranking (see Figure 10). It is expected to utilize the list for cost-effective capital project prioritization to optimize resources, the schedule, and overall venture strategy. The chart is shown in Figure 6.

Sample Calculations

Growth (2) + Upgrades (5) = 7 > 5 == Large LOE & Intense Urgency

Safety Hazards (2) + Category Sum (9) = 11 < 15 == Weak Total LOE & Indifferent Total Urgency

The Project Subset Evaluation

The Capacity Category is reserved for developments that involve new loads and system reinforcement associated with specific ventures required to improve, relieve, or correct an existing voltage or thermal condition. The utilization of *design thinking*, entrenched in ethnographic techniques, will place the decision-makers in each stage of the improvement process (Gattiker, 2012). The method is beyond the traditional data gathering and problem-solving practices as the decision-makers mentally transform to end-users. Nevertheless, the data in this sector clearly depicts areas of growth and demand within the electrical footprint. Further inspection of the subset pie chart data confirms several strengths in relationships of the capital projects. The "new expansion" category consumes nearly half (43%) of the Capacity Project total shown in panel A of Figure 7. The increased urgency and high Level-of-Effort will place the request near the top of the capital project priority list. The "upgrades" category is in the second tier of the Capacity Project total with equally high urgency and effort levels. Nevertheless, when referenced with the "Y" Shaped Matrix, the need is not as great when compared with the "new expansion" projects. There is a general sense of developing "new" capital projects in a virgin arena and deemed as high-quality than performing the

Figure 5. Project pre-planning general metrics

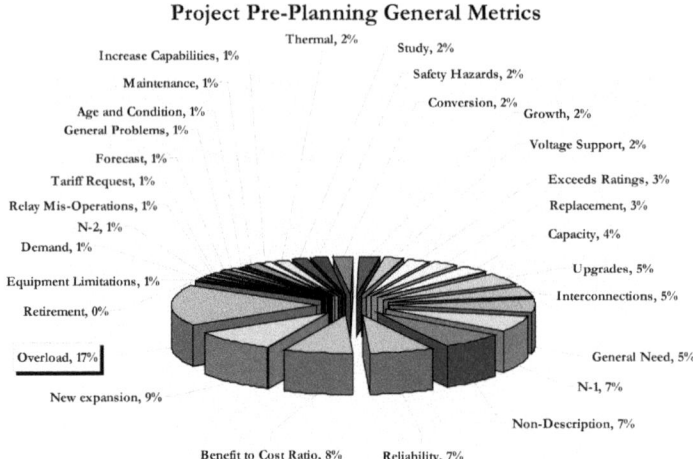

necessary upgrades on an existing project. The "increase capabilities" and "capacity" categories are interchangeable in some regions. The same argument can be made for the "demand" and "growth" categories. Nonetheless, when taken as a sum of its parts in each category as described,

it comprises of the balance (36%) in the Capacity Project total.

The Reliability Category is reserved for developments that involve improvement and reinforcement for the reliability of the infrastructure asset. These data depicts equipment that exceeds its

Figure 6. Project pre-planning y-chart

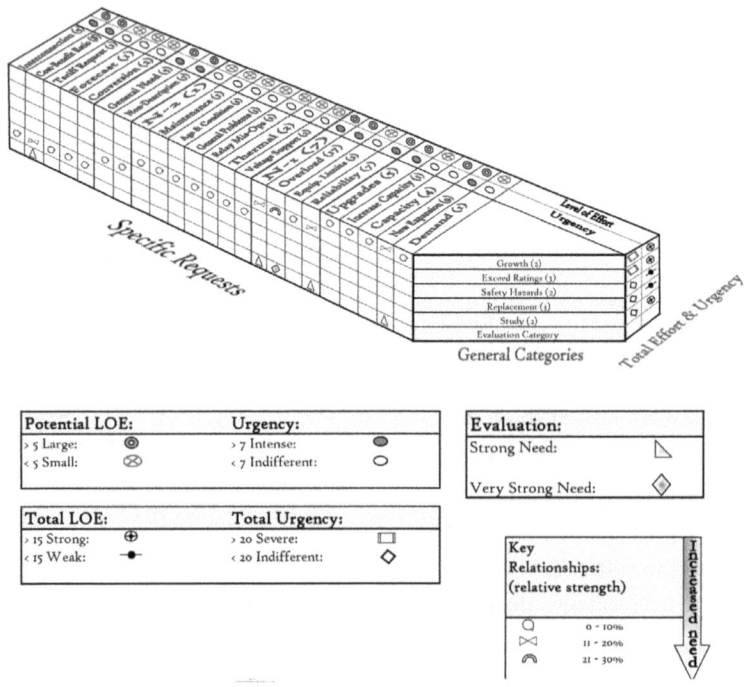

Figure 7. (a) Capacity project exploded metrics; (b) reliability project exploded metrics

Ordinal Capacity Project Percentages **A**

Upgrades
21%

Capacity
17%

Demand
3%

Growth
10%

New expansion
43%

Increase Capabilities
6%

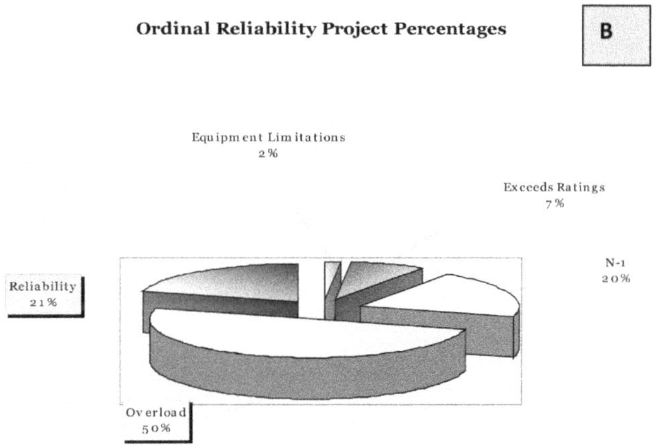

Ordinal Reliability Project Percentages **B**

Equipment Limitations
2%

Exceeds Ratings
7%

N-1
20%

Reliability
21%

Overload
50%

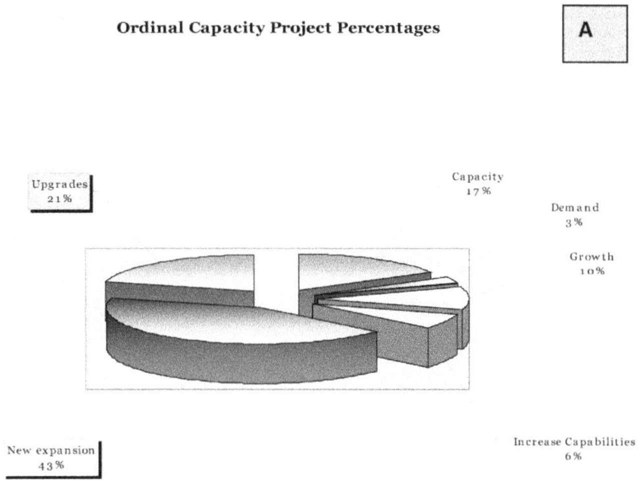

normal ratings and jeopardizes the reliability of the system. The "overload" category consumes half (50%) of the Reliability Project total as portrayed in panel B of Figure 7. There is an urgent need and an extremely high Level-of-Effort for the projects in this area. In fact, various managerial tactics and/ or failed strategies from the condition projects area produced the problematic ventures in the "overload" category. The "Y" Shaped Matrix ac-

curately confirmed the serious conditions in three (3) categories within this sector (N-1, Overload, and Reliability). All of these categories require an ardent sense of urgency as well as an increased Level-of-Effort to accomplish the tasks. Based on the increased need to mobilize the necessary resources, each project in the reliability sector (91%) *must* be prioritized for safety reasons and to avoid potential equipment damage to expedite

Figure 8. (a) Condition project exploded metrics; (b) "other" project exploded metrics

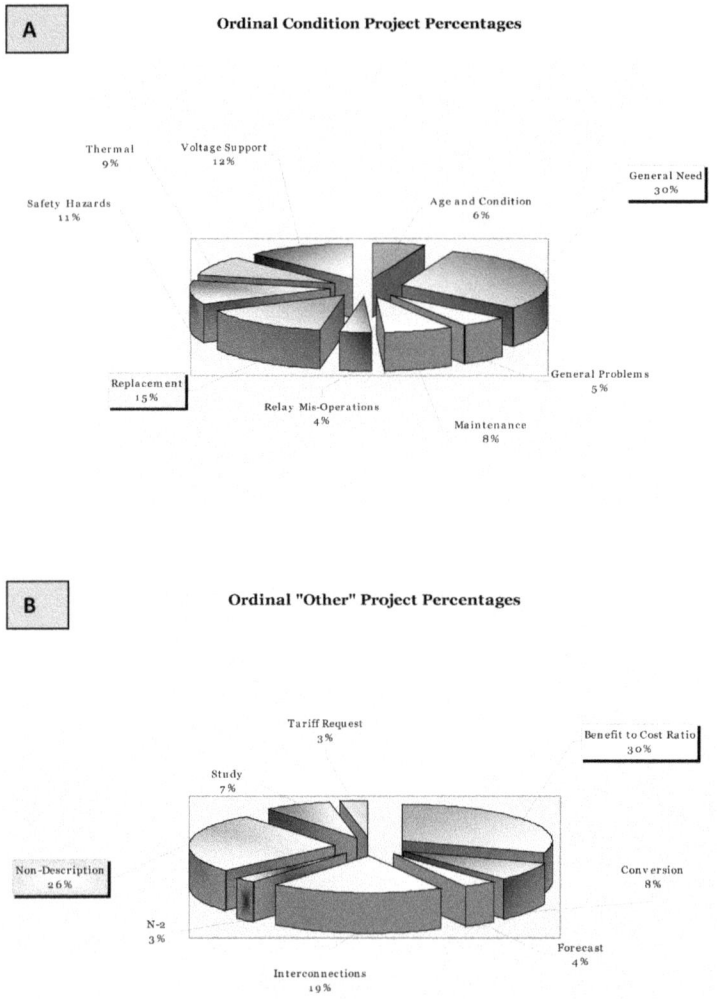

the work. Properly planned capital projects are skillfully integrated within the portfolio envelope. Conditions such as the projects in the reliability sector create blackouts, increase costs, and develop a poor safety environment if left unattended (IEEE Standard 902, 1998). Mitigation strategies *must* be in place within the lower echelons of the project prioritization scheme *before* any endeavor attains such status. A properly administered quality plan by top management will alleviate any future problems encountered in this category.

The Condition Category is reserved for developments that involve the replacement of equipment because of inability to obtain parts or obsolesce. These data depicts problematic equipment that has served its useful life in the system. The "General Need" category consumes nearly one-third (30%) of the Condition Project total as shown in panel A of Figure 8. When conjoined with the "General Problems" category, the projects increase to 35% without citing specifics. The terms are open to interpretation based on the utility's experience with the individual project. An urgent need is not required and the Level-of-Effort for the projects is extremely low. Nevertheless, these projects gradually increase to a higher level of need if not systematically prioritized within its own sector. The nine (9) categories in the Condition Project arena signify various states of the existing field equipment. A utility *must* provide the necessary resources in order to mitigate any potential problems. The "Replacement" category comprise of 15% of the projects in this sector. The other third of the pie chart represents safety and potential regional blackout issues if left unchecked (Safety, Thermal, and Voltage Support). Although the relative need is low, the urgency and Level-of-Effort to complete these projects are exceedingly high in accordance with the "Y" Shaped Matrix. Each project *must* be prioritized within this sector and evaluated against the alternatives for optimal cost, quality, and safety issues.

The "Benefit-to-Cost Ratio" category consumes 30% of the "Other" Project total as shown in panel B of Figure 8. There is a Level-of-Effort required as well as a moderate need for this category. Although the total urgency is ranked as "indifferent" on the "Y" Shaped Matrix, the work in this category exhibits a very strong sense of capital planning and fiscal responsibility. Ideally, the project requests are derived from this

Figure 9.

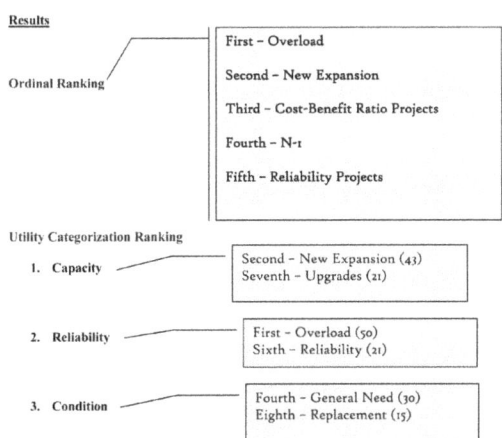

Figure 10.

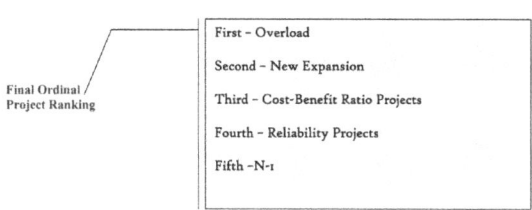

area as the resources and time allows. The "Non-Description" category for the projects are unclear and without merit. Twenty-six percent of these projects are categorized as a viable venture, yet warrants no urgency or effort. The capital portfolio managers must provide specific requirements and justification in order for project recognition. The other requests (Interconnections, Tariff Requests, Study, N-2, Forecast, and Conversion) are normal project investigative areas considered as routine for the Regional Independent Service Operator.

The "Non-Description" and "General Need" categories are technically invalid due to a lack of project detail. Therefore, "N-1" is equivalent to "Non-Description" (6.84%) of the project total for the 5th place position and "Reliability" (7.07%) of the project total will replace the "General Need" (5.45%) in the 4th position to validate the ordinal ranking.

CONCLUSION

A cursory inspection of the data generates unyielding and intuitive results. The "Overload" condition in the "Exceeds Rating" category exhibits a very strong need to fund the projects. The sense of urgency and level-of-effort are equally as high. Nevertheless, the model logic ranks the projects in a category quite differently.

- Although the N-1 & Reliability categories exhibit a strong need, both are equivalent to "New Expansion" and "C-B Ratio" in the "Growth" and "Study" areas respectfully. However, upon further inspection, the "New Expansion" category demands a higher ranking based on the model logic, closely followed by the "C-B Ratio" projects.

- The apparent weakest of the top five in the evaluation category (C-B Ratio) exhibit indifferent urgency and strong LOE. Nevertheless, the projects in this category are ranked third in accordance with the model logic.

- The majority of the specific requests are within the "Exceeds Ratings/Reliability" categories with an urgent need. These problems prioritized the entire list based on neglect and/or specific focus on the two other main categories (Growth/Study). If re-positioned properly in the capacity, reliability, and condition sphere as proved in the model logic (along with careful resource allocation), the "Overload" category is essentially eliminated.

The techniques can be utilized in high-quality industries (manufacturing, energy, heavy construction, and petrochemical) to evaluate the details of the process. Resource planning and strict attention to detail are essential elements within the overall scheme.

REFERENCES

Aft, L. S. (1998). *Fundamentals of industrial quality control.* Boca Raton, FL: CRC Press.

ANSI/ISO/ASQ Q9001-2000. (2001). *Quality management systems – Requirements.* Milwaukee, WI: Quality Press, American National Standard, American Society for Quality (ASQ)..

Bently, J. (1999). *Introduction to reliability and quality engineering* (2nd ed.). Upper Saddle River, NJ: Prentice-Hall.

Conklin, J. D. (April 2012). Next in line: Always look ahead to the next project for maximum quality gains. *Quality Progress*, pp. 44-47.

Duffuaa, S., Raouf, A., & Campbell, J. (1999). *Planning and control maintenance systems: Modeling and analysis.* New York, NY: John Wiley and Sons, Inc.

Evans, J. R., & Lindsay, W. M. (2005). *The management and control of quality* (6th ed.). Florence, KY: South-Western College Publishers.

Feigenbaum, A. V. (1991). *Total quality control* (3rd ed.). New York, NY: McGraw-Hill, Inc.

Gattiker, T. (March 2012). Rethinking design. *Quality Progress*, pp. 33-37.

Hall, J., Ellner, P. M., & Mosleh, A. (2010). Reliability growth management methods and statistical methods for discrete-use system. *Technometrics*, *52*(4), 379–389. doi:10.1198/TECH.2010.08068

IEEE Standard 902. (1998). *IEEE guide for maintenance, operation, and safety of industrial and commercial power systems.* Piscataway, NJ: The Institute of Electrical and Electronic Engineers, Inc.

ISO Standard 10005. (2005). *Quality management – Guidelines for quality plans.* New York, NY: American National Standards Institute.

ISO Standard 10006. (2007). *Quality management – Guidelines to quality in project management.* New York, NY: American National Standards Institute.

ISO/TR Standard 10017. (2003). *Guidance on statistical techniques for ISO 9001:2000.* New York, NY: American National Standards Institute.

Juran, J. M. (1999). *Quality control handbook* (5th ed.). New York, NY: McGraw-Hill, Inc.

Mann, D., & Domb, E. (1997). *40 inventive (business) principles with examples.* Retrieved 8 November, 2008, from http://www.triz-journal.com

McCalley, J., Vooris, T. V., Jiang, Y., & Meliopoulis, A. P. (2003). *Risk-based maintenance allocation and scheduling for bulk transmission system equipment*. Retrieved 3 January, 2006 from http://www.pserc.org

McCrary, S. W. (2007). Validation of project time decision-support tools and processes. *Journal of Information Technology, 23*(2).

Megahed, F., Woodall, W. H., & Camelio, J. H. (2011). A review and perspective on control charting with image data. *Journal of Technology, 43*(2), 83–98.

Molina, M. J., & Rowland, F. S. (1974). Stratospheric sink for chlorofluoromethanes: Chlorine atom-catalyzed destruction of ozone. *Nature, 249*(5460), 810–812. doi:10.1038/249810a0

Moubray, J. (1997). *Reliability centered maintenance* (2nd ed.). New York, NY: Industrial Press, Inc.

Mueller, R. (2005). 10 steps to help reduce SF_6 emissions in T&D. *Utility Automation and Engineering/T&D*. Retrieved 4 June, 2009, from http://www.elp.com/index

Nanda, V. (February 2010). Preempting problems. *Six Sigma Forum Magazine*. Retrieved 1 April, 2011, from http://asq.org/knowledge-center/index.html

Park, B., & Meier, R. (2007). Reality-based construction project management: A constraint-based 4D simulation environment. *Journal of Information Technology, 23*(1).

Pietroforte, R. (1997). Communication and governance in the building process. *Construction Management and Economics, 23*, 71–82. doi:10.1080/014461997373123

Sanchez, L., & Pan, R. (2011). An enhanced parenting process: Predicting reliability in product's design phase. *Quality Engineering, 23*, 378–387. doi:10.1080/08982112.2011.603110

Schaar, T., & Wilson, R. D. (2010). *Quality basics simplify complex engineering document management challenge*. Making the Case for Quality: The Knowledge Center for the American Society for Quality. Retrieved 15 May, 2010, from http://asq.org/knowledge-center/index.html

Schlabbach, J., & Berka, T. (2001). *Reliability centered maintenance of M.V. circuit breakers. Power Tech Proceedings* (*Vol. 4*). IEEE Porto.

Taguchi, G., Chowdhury, S., & Wu, Y. (2005). *Taguchi's quality engineering handbook*. Hoboken, NJ: John Wiley and Sons, Inc.

UK HM Treasury. (2006). *Stern review on the economics of climate change*. Retrieved 20 June, 2009, from http://hm-treasury.gov.uk/sternreview-index/html

Ye, Z., Tang, L., & Xie, M. (2011). A burn-in scheme based on the percentiles of the residual life. *Journal of Quality Technology, 43*(4), 334–345.

Zou, C., Jiang, W., & Tsung, F. (2011). A LASSO-based diagnostic framework for multivariate statistical process control. *Technometrics, 53*(3), 297–309. doi:10.1198/TECH.2011.10034

ENDNOTES

[1.] Source: 2009 NERC Disturbance Report (published 15 June 2010).

[2.] TRIZ is a Russian language acronym for *Teoriya Resheniya Izobreatatelskikh Zadatch*. Translated into English it means "The Theory of Inventive Problem Solving."

Chapter 6
The Substation Below–Grade Analysis

ABSTRACT

The preparation of the substation layout for equipment installation is a vital element in the success of major capital projects. Equally important are the key integration processes involved in the equipment commissioning and plant-in-service phases of the endeavor. The integration of man, material, machinery, and methods require an intricate balancing act to maintain a specified budget, as well as timely component installation.

OVERVIEW

The below-grade process of a wind turbine generator implementation begins once the renewable energy integration challenges and criteria (meteorology, transmission, land, public relations, civic engineering development, permitting, and environmental studies) are satisfied. A comprehensive review of the geotechnical studies, surveys, and engineering heavy permits consume only a small portion of the pre-construction phase of the capital-intensive project. A more thorough analysis of the site work is expected to consume a greater share of time for the below-grade assessment. The negotiations with turbine equipment providers (and long lead-time procurement issues) are included in this phase and implemented in the master schedule. The development of the environmental construction compliance plans as well as the mobilization arrangement is encompassed in the pre-construction activity set. Most importantly, the prime contractors are secured to perform the work. Ideally, *all* of the applicable permits are approved and the project has positive community support. Moreover, the landowner issues – specifically the *liability during construction* portion – of the agreements are stipulated and clarified to all parties involved. All of the studies are expected to produce positive results after thorough assessments of the construction site.

DOI: 10.4018/978-1-4666-2839-7.ch006

The more opinions you have, the less you see. (Wim Wenders ~ b. 1945 -)

Nevertheless, problems occur during the early stages (before the first bulldozer arrive on site) of the pre-construction activities. These problems create unnecessary costs and expenditures when valuable resources are misappropriated. Examples of site specific field problems are outlined below:

- A 230/12.5-kV – 14MVA unit at a transmission substation on the east coast was delayed for two months because of a non-compatible engineering design. The electric utility applied for the local environmental/land permits in conjunction with the pre-construction phase for the particular parcel. Furthermore, the $5.8M capacity project for new load was behind schedule when work crews assessed a habitat suitable for endangered species. There were no contingency plans established and the utility was forced to evaluate an alternative site (sunk costs: $316k).

- A line re-configuration and engineering field re-design of the specific substation component layout was required for a Midwestern utility work effort. The system reinforcement project was further delayed with a change in prime contractors during the engineering re-design construction work stoppage of six weeks. The project was re-bid and a pre-construction schedule re-established with the new contractor (sunk costs: $150k).

The incisive civil engineering program manager is expected to thoroughly evaluate the construction site and all associated assessments. During the conceptual layout phase, permit layout development, and the engineering heavy permit phase, the pre-construction group is deemed as active participants. This involves the principle number five of the TRIZ model as stated below (Mann & Domb, 1997):

- **Merging:** Make operations contiguous or parallel; bring them together in time.
- It is designed to eliminate the need of crisis management.

Therefore, in order to determine the mechanical strength of the wind farm foundations, for example, the pre-construction team is co-mingled with the resources and components of the geotechnical study. The team must be in agreement with the condition of the turbine foundation and improvement methods. One U.S. Eastern wind farm site proposed turbine locations in several mine spoil areas due to previous strip mining activity in the area. These proposed foundations were on various base soils (natural medium dense to very dense overburden soils, bedrock, and/or mine spoil) and improved via deep dynamic compaction techniques. The pre-construction team is considered proactive in mobilizing the resources to accomplish this task. The geotechnical assessment report typically contains a summary of the findings as well as the design and construction foundation recommendations. As part of the re-evaluation process, the astute program manager must review the selected soil samples index test lab results for geological tangible data (not assumptions). The entity must appraise the submitted rock core samples for compression strength (crushing in psi) data as well as the test boring logs. The test is designed to measure strain for deformation in a borehole and illustrates visual measurements of the features. Moreover, the bore test examines joint orientations that include strikes, dips, rock names, and engineering properties. The pre-construction manager must be familiar with the sub-surface exploration program tenets, to include test pits, trenches, short tunnels, and drilling of vertical/horizontal or oblique bore holes. The securitization of the resources and the specialized equipment is an important step during the early phases of the process to prevent project delays. The overturning of steel transmission-line towers owing to a foundation failure is a direct consequence of inadequate planning and soil assessments. In the transmission tower design, forces of side thrust and lateral movement must be considered for the foundations. The typically bolted towers require passive pressure around the foundation with mandatory compacted backfill during construction for stability and to prevent sliding (Subramanvian & Vasanthi, 1990). The soil assessment is an imperative component within the pre-construction sphere. Verifiable processes developed within the pre-construction envelope are designed to reduce cost and produce a timely wind turbine generator implementation.

The Lean Quality System is designed firstly, as a level of managerial action; secondly as a control mechanism of the process to achieve the desired goals, and lastly to provide leadership of the improvement within the processes. The evaluation methods utilized to assess the complex potential problems consist of several alternatives to include interventions, scenarios, and detailed analysis with respect to the multitude of quantitative and qualitative data (Rossi & Freeman, 1993). The wise civil engineering manager will ensure that the resources are mobilized and the integrity of the assessment work is beyond reproach before construction commences.

BUILDING FROM THE GROUND UP

The foundation of any substation capital project is an essential element for the current and future electrical capacities. How this is accomplished is just as important as to why, when, and where. The processes involved on the first day of construction greatly affect system conditions and any future upgrades to the electrical grid. Poor or erode soil conditions as well as inadequate grounding will have devastating effects (safety, power quality, outages, etc.) in the electrical footprint. In order to incrementally improve reliability within the infrastructure, small steps are first introduced to reduce system outages. For example, improving the grounding on a transmission line will decrease

the number of outages on the shielded line. It begins with the soil resistivity and ends with the strength of the structure. The naturally occurrence of lightning wreaks havoc on the electrical system as described in the events in Box 1[1].

The lightning strikes in these events created several outages well over sixteen hours, loss of over 125 MW in load, and tripped a 451 MW generating unit. Precautionary measures must include provisions for prevention/reduction at the design and construction levels. There are several methods *integrated into the process* designed to reduce the number of outages due to lightning. The most effective method is the utilization of an overhead ground wire attached to the structure such that the shield angle is thirty degrees or less (RUS Bulletin 1724E-203, 1994). The addition of a bayonet attached to the structure (based on its strength) to adjoin an overhead ground wire may be included in the design engineering/pre-construction phase. Moreover, lightning arrestors grounded properly, are utilized to decrease flashovers. The civil engineering program manager must align the work effort with the design engineering resources to review the Earth resistivity information for the site to assist with the location/quantities of the lightning arrestors. The devices will not increase the loadings on the structures. These activities are expected well before an outage to the substation occurs and must be *integrated into the processes*. Additional ground rods must be driven deeper to obtain lower resistivity soil layers. The lightning dissipation device is another method utilized to reduce outages. Grounding is a key component for the successful performance of these devices and highly resistive soil is essential. The dissipation device lowers the voltage differential between the ground surface and cloud charge below flashover levels. The surrounding air is then ionized by sharp points on the devices, allowing the safe transfer of the charge to the grounding system. A combination of all of the methods must be included within the substation to minimize lightning outages. Gallop-

Box 1.

Start Date - Time: 7/2/2009 10:44:00 PM EDT **Cause:** Weather-Lightning **Event Description:** DESCRIPTION: On July 2 at 22:44, a 345 kV line and associated SPS tripped. An electrical storm with lightning was in the area at time of trip. The line was synchronized on July 3 at 01:12 and all other switching was completed at 01:25. TYPE OF EMERGENCY: Major Distribution System Interruption ACTION(s) TAKEN: Repaired/Restored; Other **Start Date - Time:** 7/13/2009 12:29:00 PM MDT **Cause:** Weather-**Lightning Strike/Equipment Failure** **Event Description:** DESCRIPTION: On July 13 at 12:29, the utility's system separated and the suspected cause was a lighting strike. The islanding resulted in 10 MW of contracted load being shed plus approx. 70 MW of load. TYPE OF EMERGENCY: Major Transmission System Interruption **Start Date - Time:** 8/14/2009 9:06:00 PM EDT **Cause:** Weather-**Thunderstorm, Lightning** **Event Description:** DESCRIPTION: On Aug. 14, the utility reported that during a severe thunderstorm, lightning struck four spans west of a 345 kV line. The line did not relay at that time. On Aug. 14 at 21:06, the static wire failed mechanically and fell across the 345 kV insulator strings resulting in a high impedance fault. Eventually, the static wire failed and came down on the 138 kV line taking that line out and a distribution transformer – ultimately affecting 8,391 customers and 45.7 MW of load. When the 138 kV bus tie breaker opened on the 132-60 line dropped the distribution transformer and associated equipment. This also resulted in the tripping of a 451 MW generating unit. The event lasted 01:56. The 345 kV line repairs were completed on Aug. 18 at 19:57. TYPE OF EMERGENCY: Other ACTION(s) TAKEN: Repaired/Restored; Other

ing conductors are causes for outages as well created by a phase-to-phase fault. An example of this type of outage is described in Box 2.

Several methods are utilized to reduce this type of outage to include the following:

- Increase the separation between the conductors by raising/lowering the structure crossarms.
- Reduce the span length with additional structures.
- Install/re-conductor with T2 type after a thorough evaluation of the structure strength, guying, and associated mechanical factors.

- Include mid-span spacers designed to eliminate conductor "slap" and detuning pendulums to dissipate the low frequency energy.

The vortex-induced (Aeolian) oscillations will cause conductor fatigue. Eventually, the stresses conductor may break and cause an outage. These vortex-induced oscillations are caused by sustained high winds in excess of 22 mph to 37 mph. The oscillations generates narrow-band random waves arriving at the cable supports (Emil & Scanlan, 1986). The cable is not perfectly flexible and creates oscillatory bending stress near the supports resulting in fatigue. The situation is

Box 2.

Cause: Weather-**Galloping Conductors** **Event Description:** DESCRIPTION: On Dec. 21 at 12:40 the utility separated its interconnection tie. The suspected cause of the trip is galloping conductors resulting from winter/wind conditions in the service area. Affected transmission and the interconnection link were restored 18 minutes after the trip. TYPE OF EMERGENCY: Major Transmission System Interruption ACTION(s) TAKEN: Repaired/Restored; Other

prevented with cushioned supports to alleviate the bending stresses. Moreover, the tuned-mass damper (Stockbridge Damper in Figure 1) is a counter-vibrating mass with a wideband frequency to suppress the last half-wave (nearest the support) generated by the cable oscillation.

The incremental design steps and construction procedures are *integrated into the process* in order to minimize the potential for electrical power outages.

The Soil

The soil evaluation at the selected construction site is a vital tenet for the success of the renewable energy system integration. The soil surrounding the transmission tower, for example, must not only support the unit, it is expected to resist a great amount of upward pull as well as side thrust. This is one of the deciding factors for the foundation footing selection. Nonetheless, the process begins with the execution of a subsurface exploration program to assess the construction of foundations to support the propose wind turbines. This program typically consists of test borings

Figure 1. The Stockbridge damper

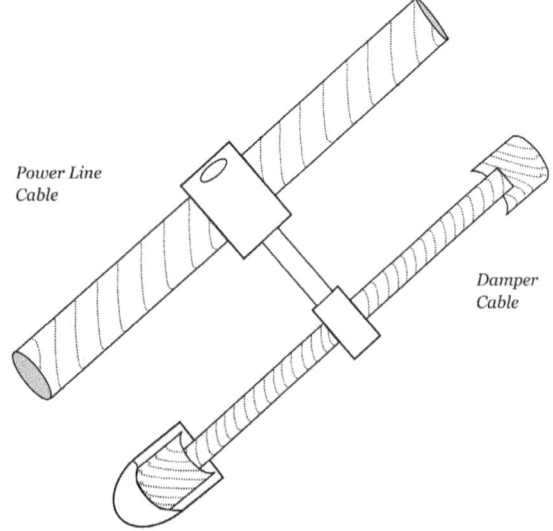

Power Line
Cable

Damper
Cable

conducted near the approximate location of the turbine and test pit excavations. The quantities are dependent on the number of proposed units (i.e. 35 proposed wind turbines equate to 35+ test borings). A description of the terrain and topology (i.e. mountainous, access paths, etc.) is expected in the assessment. The detail of overburden thickness (i.e. near surface to 85 feet thick) of the proposed area must be documented as well. The subsurface exploration program may encompass electrical resistivity testing at each of the wind turbine locations and multi-channel analysis of the surface wave profile along certain sections. The soil resistivity testing is an aid to identify locations, depth to bedrock, and other geological phenomena. The resistivity has a direct impact on the corrosion of underground pipelines. Lower resistivity equates to higher corrosion activity and dictates if protective equipment is used. Moreover, the soil resistivity directly affects the grounding system design where the area of the lowest resistance is desired to achieve the most economical benefit. The electromagnetic method is utilized to evaluate contaminated sites (i.e. industrial hazards, waste) as applied to subsurface mapping (Steward, M. and Gay,, M. 1986). The technique is designed to reveal *any* potential problems from previous disposal within the site. Moreover, rocks and sediments conduct electricity as a consequence of ions in the solution of pore fluid, the charge layer on clay, and conduction via metallic materials (rare) (Jordan, T. and Constanti, D. 1995). The soil conducts electricity in some locations and is a poor conductor in others. The soil resistivity is chiefly predisposed by the type (clay, shale, etc.), moisture content, electrolyte amount (dissolved salts, minerals), and temperature. The various soil characteristics from three distinct areas (Centralia[2], Mansker[3], and Rimer[4] Series) of the country are illustrated in Table 1.

The grounding system depends on the accuracy of the soil resistivity test for correct performance. When the grounding grid is satisfactorily placed into service, the results equate to no costly change orders for ground *enhancements*. Once again, the quality is *integrated into the process* to avoid outages and re-work. Temperature and moisture content are stabilized at greater distances below the surface of the earth. The grounding rod must be driven at considerable depths to achieve the best results (ideally the water table). The soil resistivity testing consists of inserting four equally spaced and in-line electrodes into the test area. This technique is commonly referred to as the 4-point method (or Wenner Four Point Method) (Lyncole, 2005). The 2-point method is simply the resistance measured between two points as the name implies. Nevertheless, the 4-point method provides a higher degree of accuracy. The basic operation utilizes a ground tester instrument where a constant current is injected through the ground via the source and the outer probes. The resulting current flowing through the ground testing area (the resistor) creates a potential difference, which is then measured by the two inner probes. The resistance is calculated using Ohm's Law (R=V/I) and displayed on the face of the instrument. The soil resistivity is measured in ohms-meter (the resistance of a volume of ground of one cubic meter). The conversion factor of the displayed resistance is multiplied by the constant 1.915 and the probe spacing. The formula is depicted as:

$$\rho \text{ (Ohms-m)} = 1.915 \times R \times A$$

Where:

ρ = soil resistivity in Ohm-m

R = display output in Ohms

A = distance between electrodes in feet

1.915 Constant

The detailed formula for the soil resistivity testing is illustrated below:

$$\rho_a = 4\pi * a * R / 1 + (2a / \sqrt{a^2 + 4b^2}) - (a/ \sqrt{a^2 + b^2})$$

Assume b = 0

$$\rho_a = 2\pi * a * R$$

The lower the ground resistance, the safer it is within the substation for all resources (personnel, equipment). The accepted industry standards are developed that stipulates transmission substations must be designed not to exceed 1-ohm. Some regions in the country find it difficult to obtain 100-ohms or less whereas other areas can easily achieve 5-ohms or less (the maximum recommended resistance in the distribution substation).

The test pit evaluation is one of the most accurate investigative methods of the subsurface exploration program (Chen, F. 2000). The technical engineer can examine and acquire samples of the subsoil strata, layer and lens, and stratification factors in detail. Nonetheless, there are strict limits involved with these activities as outlined below:

- When the twelve foot depth of the test is limited to the reach of the backhoe
- When the high water table averts excavation
- When the standard penetration resistance test is required
- When unstable soil prevents the technical engineer from entering the test pit subjected to the Occupational Safety and Health Administration regulations

Accidents occur in practically every industry and the construction sector is certainly not immune to it. A synopsis of a construction fatality while conducting an assessment is described below:[5]

Table 1. Selected U.S. soil characteristics

Series:	General Characteristics	Water Capacity	Permeability (inches/hr)	Shallow Excavations (6-8 ft)	Shrink-Swell Potential	Erosion Factor (k)	Soil Reaction	Construction	Average Temperature (°F)
Centralia	These soils are fine-loamy, mixed, mesic utic haploxeralps. Depths: 0-17" Loam (USDA Texture) 17"-26" Dark brown clay loam (10 yr 4/3), yellowish brown (10 yr 5/4) dry, moderate fine and medium sub-angular blocky structure; hard firm, slightly sticky and slightly plastic. 17"-38" Silty clay loam, clay loam. 49"-60" Grayish brown (10 yr 5/2) clay loam, light brownish gray (10 yr 6/2) dry, common medium prominent yellowish red (5 yr 5/8) mottles.	0.19-0.21	0.6 – 2.0	Severe: slope	0-17" Low 17"-38" Moderate 38"-60" Moderate	0-17" 0.32 17"-38" 0.28 38"-60" 0.32	0-17" 5.1~6.5 17"-38" 5.1~6.0 38"-60" 4.5~6.0	Road fill - poor; low strength Sand – improbable; excess fines Gravel – improbable; excess fines Topsoil – poor; slope	Jan. ~ 39°F July ~ 65°F Dec. ~ 41°F
Mansker	Clay loam (0-3% slopes); calcareous loamy soils, sloping and strongly sloping. Highly calcareous clay loam of the up-land over fragmental clay loam caliche that is more than 3 ft thick and comprise from 10-50% of the soil mass, by volume.	0.10 and 0.18	0.2 – 0.5	Moderate	Moderate	---	---	Road fill - fair Road Sub-grade- poor Sand – poor (localized pockets) Gravel – not suitable Topsoil – good	Jan. ~ 30°F July ~ 80°F Dec. ~ 32°F
Rimer	Loamy, mixed, mesic aquic arenic hapudalfs (0-4% slopes). Consist of poorly drained soils on till plains/ outwash plains and in glacial drainage ways. Clay, silty clay loam (USDA tex-ture)…35%~55% clay.	0.08-0.12	Less than 0.2	Severe: cut-backs cave (the walls of excava-tion tend to cave in or slough); wetness	High	0.32	6.1~8.4	Road fill - poor; low strength; shrink-swell Sand – improbable; excess fines Topsoil – fail; too sandy	Jan. ~ 22°F July ~ 76°F Dec. ~ 27°F

Technical Engineer Dies after Being Struck and Crushed by a Bulldozer/ Earth Compactor--Pennsylvania

Summary

A 31-year-old technical engineer (the victim) died after being struck and crushed by a 12-ton bulldozer/compactor at a private landfill while he was attempting to obtain a soil sample for analysis. A landfill employee was bulldozing a new access road across the landfill to a new dumping area. The steel compactor-type wheels on the bulldozer were 36 inches high, 30 inches wide, and equipped with 3-inch-round, 1-foot-long cleats that compacted the soil as the road was being bulldozed. As the equipment operator began to bulldoze the road in the direction away from the landfill office and entrance, the victim arrived at the landfill in his pickup truck. The victim drove his truck to the new road, parked on the side, and with his back to the bulldozer, prepared to collect a soil sample from the new road. To compact the ground further, the equipment operator began to back the bulldozer over the area he had just traversed. As the bulldozer continued to back up, the victim was caught and fatally crushed between the right side rear compactor wheel and the body of the bulldozer. NIOSH investigators concluded that, to prevent similar incidents, employers should:

- Ensure that safety features incorporated into the design of machinery is operable at all times. At the time of the incident, the bulldozer's back- up warning alarm was inoperable. Equipment back-up alarms are designed to alert pedestrians and other equipment operators that equipment is moving in reverse.
- Instruct workers to ensure that equipment operators are aware of their presence before workers perform any tasks near operating equipment. Workers performing tasks in the vicinity of operating heavy equipment should alert the equipment operator of the location of their work area prior to beginning work. This should eliminate inadvertent contact between the operating equipment and workers in the vicinity.
- Consider equipping machinery with devices that will eliminate blind spots behind the machinery. The bulldozer involved in this incident was equipped with a rear view mirror at the front left side of the cab. The mirror afforded the operator a clear view of the area around the left side of the machine; however, the operator had a limited view of the area to the right rear side of the bulldozer. A rear view mirror could be mounted on the front right side of the cab to allow the operator to view the area around the right side of the bulldozer. An additional mirror could also be mounted on the left rear of the cab to afford the operator a view directly behind the bulldozer.

Most civil engineers rank the field test data higher than the laboratory test of the soil samples. A drill rig (percussion or rotary) is utilized to bore the engineer's required depths (typically 50 feet for each turbine location and lesser for the foundation). The various lab tests performed (swell, consolidation, direct shear, tri-axial shear, and compaction) are designed to determine minimum and maximum values of the sample to obtain the average for the design. The accuracy of these tests must be thoroughly verified in the pre-construction phase of the project. A combination of the information from the field drilling logs, field penetration data, visual examination of the samples, and technical expertise in the area is required of the civil engineering program manager. Ultimately, the main purpose of the soil tests is to determine the values of shear, consolidation, and the resistivity of the element.

Grounding

Correct grounding of an electrical substation provides a safe and reliable system for personnel and equipment. A dedicated Earth driven grounding system is comprised of ground rods and/or a buried counterpoise approved by the design engineer (RUS 1728F – 810, 1998). There are several assumptions within the capital project sphere that includes (although not limited to) the following circumstances:

- Environmental permitting is not required
- Subsurface soil conditions are normal
- Grading is not required
- The grounding and ground grid design are adequate

The grounding and soil issues are usually on the list of assumptions. However, there is absolutely no substitute for a poorly designed grounding system. Improper grounding results in diminished lightning arrestor protection for the transformer, inadequate relay coordination, an increase in the ground potential rise and mesh voltages, and undesirable current flow in unintentional ground paths. Moreover, the potential of shock hazards to substation personnel and equipment damage are directly related to poor grounding. One particular Midwestern electric utility recently budgeted $2.38M to upgrade the static and grounding system for a 138-kV substation. By comparison, the typical cost is $1/square foot for the initial installation of an average size substation or approximately $42k. The Lean Quality techniques emphasize the minimization of re-work and non-value activities in a process. In fact, one of the quality gurus espoused the concept of ideal perfection (Besterfield, D. et al., 1999).

The performance standard is zero defects (Crosby, P.B., 1961)

Nonetheless, periodically verifying the grounding system integrity is an important maintenance activity (FIST, 2005). Corrosion, mechanical damage, and loose connections affect the equipment conductor grounding. Overtime, a poorly installed grounding system is compromised with equipment additions and/or removals. Also, construction activities in the area may have a significant effect on the grounding system. These related area construction actions are rarely confirmed in juxtaposition of existing substation operation. A ground grid acceptance test must be included in the installation process. The test is designed to detect faults (installation and maintenance) associated with system grounds (Edwards, F., 1993). Examples of the fault discoveries utilizing the test in one particular region are described below:

- The grounding grid was not installed as designed. Furthermore, the grid segments were not installed.
- Transformer and capacitor neutral connections were deteriorated.
- Damage to the existing grounding grid and Supervisory Control and Data Acquisition (SCADA) installation due to construction activities. The equipment ground risers were severed which created isolated grounding sections.
- Transmission line and substation shield wire terminations were deteriorated.
- The deterioration created excessive current flows in unintended ground paths (conduit, inter-phase piping, and control wire cable trays) in one substation.

If left undetected, the problems are compounded exponentially and eventually create outages, costly equipment damage, and personnel injury. Work crews must be fully aware of the potential hazards involved in the substation. A synopsis of a work fatality within the substation is described below:[6]

Laborer Dies from Electrical Injuries Sustained in an Electrical Distribution System Substation in Virginia

Summary

A laborer died 15 days after a 10.5-foot-long galvanized pipe he was carrying contacted an energized 12,500-volt jumper wire at an electrical distribution system substation. One end of the jumper wire was attached to a step-down transformer at a position approximately 11 feet above ground level. The other end of the jumper wire was attached to an overhead powerline. The victim was part of a two-person crew assigned to pull wire through a newly installed underground conduit. The victim positioned a truck containing reels of wire, a reel rack, and a galvanized pipe that was going to be used as a reel rack spindle, inside the substation approximately 8 feet from a transformer. While his co-worker (the crew leader) was working on a separate task approximately 40 feet away, the victim apparently lifted the pipe from the back of the truck and turned toward the transformer with the pipe in a vertical position. The pipe contacted the jumper wire, and the current passed through the pipe and the victim to ground, injuring the victim. NIOSH investigators concluded that in order to prevent future similar occurrences, employers should:

- Evaluate their existing safety program to ensure that appropriate procedures to reduce worker exposures to hazards, especially electrical hazards, have been developed and implemented. Employers should ensure that existing safety programs include specific written procedures and guidelines for workers to follow pertaining to hazardous exposures likely to be encountered. Particular emphasis should be given to electrical hazards (e.g., energized jumper wires), and the need for and proper use of personal protective equipment. For example, employers should require that any employee entering a substation be provided and required to wear hardhat, gloves, and boots suitable for the maximum voltage of the equipment in the installation. Adherence to employers' safe working procedures and guidelines should be enforced at all times.

- Provide employees with adequate training to ensure that they can recognize potential hazardous exposures. OSHA standard 1926.21(b) (2) (2) states that "the employer shall instruct each employee in the recognition and avoidance of unsafe conditions and the regulations applicable to his work environment to control to eliminate any hazards or other exposure to illness or injury." Employers should provide employees with adequate training to ensure that they can recognize potential hazardous exposures. When new company procedures or guidelines are developed or existing ones modified, employers should ensure that workers are provided with appropriate supplemental training.

- Conduct initial jobsite surveys to identify hazards associated with each jobsite and develop job specific methods of controlling these hazards. Employers should conduct jobsite surveys to identify potential worker hazards so that appropriate preventive measures (e.g., subsequent training to employees specific to identified site hazards), to control these hazards can be applied prior to the start of any work. Two characteristics of this jobsite combined to produce a very serious hazard: 1) an energized jumper wire located 11 feet above ground level, and 2) the use of a conductive 10.5-foot galvanized pipe in the vicinity of the energized jumper wire. Such potential hazards can be minimized by ensuring that employees maintain a safe distance from energized conductors, by providing employees with non-conductive tools and materials, and/ or by de-energizing or covering electrical conductors with insulating material (e.g., line hoses). (Note: The conductive galvanized pipe involved in this incident was to be used as a spindle to support the reels of wire on the reel rack. A spindle made of wood, fiberglass, plastic, or other nonconductive material, may have been substituted for the galvanized pipe.) Additionally, when work needs to be completed within a substation, employers should consider isolating the substation and de-energizing all circuits in the installation before work begins. To minimize disruption of service to customers, employers could schedule such work inside substations during periods when customers are minimally affected or consider providing service through alternate paths in area electrical networks.

The grounding grid plays a vital role in the electrical system operation. Occasionally, transformers are grounded to provide a source for zero-sequence current, and may be arranged to convert a three-wire (zigzag grounding bank, ground wye-delta), ungrounded circuit into a four-wire grounded circuit (Short, 2006). Telephone interferences are another reason to ground a distribution transformer in particular because of the current flow to the grounding grid. The grounding bank displaces a portion of the current from ground to the phase conductors to decrease the neutral current that interferes with the telecommunication components. Additionally, grounding transformers are utilized during abnormal situations within the electrical footprint. When transformers are arranged in a delta-grounded wye-connection, the system will back feed the circuit if a line-to-ground fault occurs. The sequence of events will ultimately prevent an over-voltage in such circumstances. Substation transformers are especially overloaded during the summer peak periods. The trade-off is typically between the loss of life/equipment replacement cost when the units operate at elevated temperatures. The life expectancy of a transformer decreases exponentially when the hottest spot of the conductor temperature exceeds its maximum nameplate rating. The insulation is degraded at this critical point and halves the life for every 46.4°F increase in operating temperature. The loading specifications and guidelines for the station transformers are provided for safe and efficient operation (Tillman, 2001). It is critical for the pre-construction team to properly install the grounding grid to ensure the associated equipment is operated within its prescribed ranges, and avoid the upper limits/tolerances that may degrade its useful life. The substation is expected to expeditiously clear faults based on an efficient grounding system. Consequently, the electrical reliability of the system is improved from good substation grounding (Switzer, 1996). The grounding techniques include provisions for bonding metallic cable trays to improve the fault current path as well as providing a direct route to Earth (free of bends) for the surge arrestors. Moreover, the importance of grounding the foundation ensures correct operation of the equipment. The effective grounding provides potential faults a low resistance path to the concrete, grounding electrodes, reinforcement bars (rebar), and the Earth. The connections (compression, exothermically welded, bolted, wedge) between the grounding grids must provide a low path of resistance as well. For personnel safety, eliminate the touch potential hazard with Equipotentiality mesh safety mat to equalize voltage between the worker and the equipment. These are some of the essential measures *integrated into the process* for an efficient grounding grid implementation.

GENERAL PROCESS BACKGROUND

The pre-construction phase is blusterous and full of concurrent activities. The prime civil contractor and program manager will conduct site surveys to stake out the exact location of the wind turbines, electrical lines, applicable access roads, and facility access entrance ways. The activity includes the aerial acquisition of the contour interval topographic survey data, the Operations & Maintenance (O&M) building laydown areas, and the electrical interconnection of the switchyard. The geological and engineering evaluation typically consists of the observance/analysis of the rock structure and overburden features. It is an imperative step to assist contractors with the removal of the rocks. The overall terrain must be physically examined and a surface map prepared to indicate the geological features. Aerial photographs/core drilling is utilized to assist with the tasks involved to obtain samples. These explorations are expected to produce geologic data for development purposes (favorable access routes/ site clearing/groundwater/weather conditions) affecting excavation operations. The detailed geotechnical investigation immediately follows the survey data to identify subsurface conditions. This is an important stage of the process as it dictates the design specifications of the foundations, access roads, underground trenching, and the electrical grounding system. The site-specific construction details are developed soon afterwards to encompass the topography, environmental conditions, title information, and the geotechnical data. The construction specifications are expected to adhere to applicable federal, state, and local codes as well as good utility industry practice. The construction environmental compliance plan provides guidance for all work crews to adhere to the sensitive ecological and/or cultural resources of the project. The sediment and erosion control measures are implemented at this stage as well. Several additional pre-construction walkovers of the affected areas are conducted by the prime/ sub-prime contractors, landowners and agency representatives a week before the schedule construction to identify sensitive resources to avoid, limits of clearing, location of drainage features and the layout of the sediment/erosion control

measures. Any modifications are agreed upon by the stakeholders before the construction activities begin. The layout areas, operations & maintenance building, and substation/interconnection points must be fenced and gated for safety and security reasons. Temporary warning fences are usually erected in areas where public safety risks are imminent and site personnel are unavailable to control public access (electrical collection system trenches, excavation foundation holes) where applicable. A high visibility plastic mesh fence is utilized around unfinished turbine bases and excavations to warn work crews of the potential danger. All of these steps are *integrated within the process* to ensure the highest efficiencies and quality in the electrical substation implementation.

The site and equipment factors of the intense capital project involve tactful negotiations with the prime contractor's asset allocations. System optimization necessitates equipment selection is based on the evaluation of all features/excavation requirements. The prime contractor may purchase high-cost equipment specifically for the project. However, the astute civil engineering program manager is advised to review the equipment cost alternatives to maintain the capital budget. Certain pieces of rock handling equipment, for example, can be leased to avoid excessive expenditures during the early stages of the project. Once the sediment and erosion control measures are installed, the clearing and grading processes follows where applicable. This progression typically occurs at the appropriate existing and /or access roads staging areas, turbine/facility locations, interconnection switchyard areas, electrical substation locations, and the transmission line corridor. The general operations planning sequence within a worst case scenario sphere involves rock removal and blasting strategies germane to the project. Typically the first step is the removal of overburden rock and other weather material. The drilling and blasting follows as the next step in the excavation. The percussion and rotary drills are industry standards for projects of large scopes and size. Local blasting laws for rock excavation are in place to protect populated areas and existing structures from the following:

- Vibration
- Excessive noise
- Dust/debris generated by blasting

The exclusive utilization of the bulldozer, one of the first major pieces of equipment on the construction site, is secured for land clearing, stripping, backfilling, ditching, side-haul cuts, spreading, and dozing rocks (O'Brien et al. 1996). Moreover, the clearing of vegetation is performed only as necessary. The excavations produced by the clearing activities are backfilled with compacted Earth and/or aggregate on site. Compaction is the process of increasing the unit weight of the soil mass via static or dynamic forces resulting in the expulsion of air and water. The task is carried out by a rubber-type compactor, vibratory compactor, or steel-wheel roller compactor types. The excavators include power shovels, backhoes, draglines, and / or clamshells. The debris disposal must utilize an approved facility specifically designed to handle such waste. When the project requires accessible access roads for the wind turbine components and cranes within the steeper and more topographically dissect lands on the construction site, the blasting of rock to reach the minimum gradient is necessary. A tractor mounted ripper or rock hammer may suffice to perform the task where applicable. Nevertheless, detonation is the preferred method required for foundation construction in areas where bedrock exist above the grade level, and must be conducted in accordance with the established blasting control plan. The task is designed and performed by a contracting specialist with significant experience in the field. The specialist is expected occasionally monitor seismograph activity of the explosions as appropriate, and limit the task to normal daylight working hours (typically 8:00 am to 5:00 pm). The shot design must be structured with correct placement and stemming of the explosive with applicable blast

blankets to minimize fly rock occurrences. The road construction is generally a three step process and consists of the following procedures:

1. Clear and grade
2. Laydown aggregate the gravel (base) with dump or live-bottom trucks
3. Proof roll and compact the gravel

The material placed in the road or foundation areas must be compacted to at least ninety percent of the maximum. The heavy equipment scheduled for the work effort includes front-end loaders (to dig, scoop, lift, carry, and dump into the hauling units, bin hoppers, conveyors, and stockpiles) and the hauling units. The multi-purpose front-end loaders can be utilized to transport, spread, excavate, and compact the fill material. The hauling units consist of on-highway rear dump truck tractors and off-highway haulers (rear-dump truck, bottom dump tractor-trailer, and/or side-dump tractor trailer) units. The civil engineering program manager must encompass the elements of the Lean Quality paradigm for system optimization. Examples of these components are dump truck cycle times, the days required of the units, as well as the available days. In order to minimize delays (and costs) for this one element within the entire below-scheme, the civil engineering program manager must evaluate the details of the operation. The hauler operating conditions are illustrated in Table 2.

The loam contains 7-27% clay, 28-50% silt, and less than 52% sand. The cycle time and the travel time general calculations are described below:

- Cycle Time = Loading Time + Haul Time + Dump-and-Turn-Time + Return Time + Spot-and-Delay Time
- Travel Time$_{(minimum)}$ = Road Length$_{(feet)}$/ Average Speed x 88[7]

Other factors attributed to the haul unit intermittent delays are one-way haul roads, blind

Table 2. Hauler operating conditions

	Material Hauled:	Loading Area:	Rolling Resistance:	Supervision:	Weather:
Favorable Conditions	Topsoil. Loam/clay mixture, compacted coal, "tight" (no rock) Earth	• Unrestricted in length/width • Dry & smooth (under constant maintenance)	Under 4%	Constant at both Loading and Dumping areas	• Ambient temperatures (40°F ~ 60°F) • Little rainfall or other weather delays
Average Conditions	Clay with some moisture, soft or well-ripped shale, loose sand with some binder, mixtures of different Earth, sand-gravel mixture	• Some restrictions in length/width • Dry with some loose material	4% ~ 8%	Intermittent at both Loading and Dumping areas	• Ambient temperatures (60°F ~ 100°F) • Moderate rainfall or other weather delays
Unfavorable Conditions	Heavy (dense) or wet clay, loose sand with no binder, coarse gravel (no fines), caliche or un-ripped shale.	• Tight restrictions in length/width • Wet, slippery, and/or soft • Units load uphill or on a side slope	Above 8%	None (no spotters at either end)	• Ambient temperatures in excess of 100°F or 0°F • Excessive rainfall or other weather delays

corners, cross traffic, bridges, railroad crossings, delay at passing points, and multiple curves and/or switchbacks. Continuous delays are a direct result of inexperienced management/supervision, unskilled operators, wet/slippery haul roads, long downgrade hauls, and/or extremely high/variable rolling resistance. The excavation of each foundation is typically ten feet with a width of fifty or sixty feet in a spread foot design. The top of the foundation is a pedestal either flushes with the ground or extended six to eight inches above the ground (Fuller, 1983).

As part of the quality control methodology for equipment installation, the evaluation of the fiber optic communication equipment installation for the substation is required to validate and carefully weigh the priority of the work plan. Analysis of a Midwestern substation revealed that a failure of an analog system resulted in the loss of remote control abilities and synchronization of two generating units ensuing in the subsequent loss of megawatt delivery in the area. The system modernization to digital microwave/fiber optics communications necessitated upgrading existing relays and

communications paths. The philosophy must be systematically integrated into the installation processes of the renewable energy implementation. Review of the geotechnical data for thermal resistivity of the excavated soils and observation of the open trenching technique within one of the substation is essential for the placement of electrical collection system cables and fiber optic communication lines. The collection and communications systems are constructed parallel to the roads running from and between the generators to the project substations and the Operations & Maintenance (O&M) building. Communications between the equipment and centralized computers located in the O&M building for the project was facilitated by the underground fiber optic cable system co-located with the electrical collection system. The channel is excavated by a *Trenching Machine* (used to cut openings in the Earth to lay underground cables) to a depth of 4 feet and width of 2 feet. The extent of the open trench at any given time must be minimized to merely those distances essential to conduct the work. The electric cables and fiber optic communications cable are placed into the trench and connected to the wind turbines, transformers, and associated equipment, followed by the backfilling of the channel with the excavated trench material. Select backfill is typically placed around the electric cables to facilitate heat dissipation. The backfilling process occurs after the cable integrity is tested and confirmed. The documentation of the cable usage and associated material cost is necessary for the historical inventory and project management (lessons learned) databases for quality compliance.

THE SUBSTATION BELOW-GRADE ANALYSIS

Process Analysis

The preparation of the substation layout for equipment installation is a vital element in the success of major capital projects. Equally important are the key integration processes involved in the equipment commissioning and plant-in-service phases of the endeavor. The integration of man, material, machinery, and methods require an intricate balancing act to maintain a specified budget, as well as timely component installation. Consequently, the major objectives are to minimize the costs (labor, material, contingencies) and time (engineering, construction, contingency labor) for successful summer critical project implementation.

The below-grade substation work consists of eleven major processes, three contingency delay factors, and five general resources. These processes are developed in the model from industrial standards, subject matter experts, and financial statistical analysis. The parcel preparation is an essential phase for the entire scheme. As incorporated processes are implemented in the work plan, the associated resource costs are captured as well. The material itself is the main focus in the model and act as a viable resource for value-added costs. If the material is delayed (especially crucial for a summer critical project) the process system cost

increases substantially via mis-aligned and idle resources. The substation below-grade analysis accurately detained these costs and provided ardent recommendations for process improvement.

The Initial Model Set-up

A thorough assessment of the below-grade process sequence was performed to develop a logical work flow pattern in the model. The major aspects of each process is not all inclusive for every below grade project. However, the process is specific to the Modular Substation Model parcel preparation based on the statistical analysis of 4416 maximum hours allowed for the critical path (a sub-set of the master schedule 12840 hours for total project completion). The entire model and general process flow is depicted in Figure 2.

The resources are shown to the left of the diagram at the initial stages of the process whereas the completion block is depicted on the right. The sequential steps of the main process include the individual resource and the associated work time to perform the task(s). Waiting queues were established at specific points to maintain continuity as well as to develop statistical data. The contingency delay area captures the material/resource via an exceed condition in the system.

The Entities

The general parameters of the four major entities are shown in Table 3.

The model is based on the material cost and its affect on the resources managing it. A positive rate ($295) was assigned to the material as value-added (VA) cost in the system with associated holding cost/hour ($50), non-valued added (NVA) cost ($65), waiting cost ($25), and "other" (transport/transfer) costs. The standard rates for the resource labor (engineer, construction, and testing) are captured in Table 1 as well to include the related (VA, NVA, Hold, Wait, Other) costs, to produce one unit. Each entity was modeled to a

Figure 2. The modular substation below-grade process map

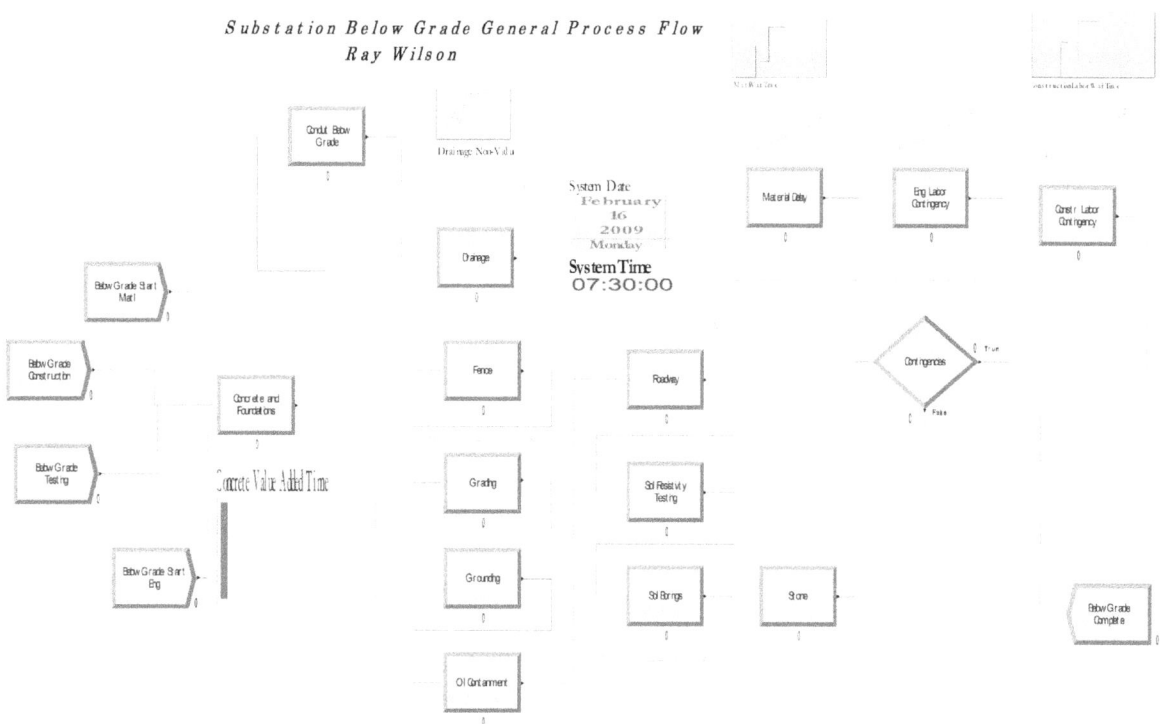

prescribed schedule for the 4416 hour work completion[1]. The appropriate resource busy/hour and idle/hour allocation was also modeled from industrial statistics for this specific project and shown in Table 4.

The scenario analysis model includes the resource schedules.[8]

The Decision Parameter

The contingency delay is solely based upon the non-value added cost (>$985) of the material wait time (Table 5). The cost includes overhead expenditures of retaining the material in the system as well as the associated costs of continuing to hold any resources it owns.

The contingency is realistically based on the resource availability and dedication to the project. The delay parameters include (but are not limited to) the following conditions:

- Material Procurement (p=0.160, $119k worse case). The decision to expedite process (p=0.040, $60K best case).
- Engineering Issues (p=0.148, $111K worse case). The decision to expedite process (p=0.037, $56K best case).
- Construction Resources (p=0.136, $102K worse case). The decision to utilize "work around" process (p=0.034, $52K best case).
- Outage Clearances (p=0.114, $86K worse case). The decision to expedite process (p=0.029, $42K best case).
- Permits
- Weather
- Accidents

The general conditions of the top four delay problems are depicted in the decision tree in Figure 3. The least-cost option of the four is the clearance approved (p=0.222) alternative ($4817) whereas

Table 3. Entity parameters

Entity Parameters	Picture	Holding Cost/hr	Value Added Cost	Non-Value Added Cost	Waiting Costs		Other Costs
Matl below grade	Picture. Widgets	50	295	65	25	15	10
Engineer Labor	Picture. Man	38	125	50	30	15	12
Construction Labor	Picture. Truck	27	95	40	20	18	15
Testing Labor	Picture. Woman	29	75	35	18	17	17

Table 4. Entity utilization

Entity	Type			Busy/hour	Idle/hour	
Matl	Based on Schedule	Below Grade Schedl	Preempt	0.45	0.55	0
Eng Labor	Based on Schedule	Below Grade Schedl	Preempt	0.7	0.3	0
Contstr Labor	Based on Schedule	Below Grade Schedl	Preempt	0.54	0.45	0
Test Labor	Based on Schedule	Below Grade Schedl	Preempt	0.09	0.9	0

Table 5. Decision parameters

Decision Parameters					If		Cost ($)
Contingencies	2-way by Condition	50	Attribute	Material Delay. Wait Time	Entity.NVACost	Matl Wait >	985

the high cost option is on the material procurement branch (p=0.210) alternative ($9601). Therefore, the general costs associated with the top four contingencies in an alternative chance model is approximately $7.5K. This value is reflected in the contingency process of the below grade model and subsequent analysis.

Major Processes

The project parameters are based on the statistical analysis of historical below grade work. The specifics are shown in Table 6.

Many of the process were combined to accommodate the maximum allowable time for the below grade work. A mixture of high and medium priorities was utilized based on the work effort and system requirements. The resource allocation was divided by the single entity (non-value added) working within the process to multiple entities (value-added) performing the tasks. The most likely value for the material delay (seize, delay, release) is approximately 1070 hours (maximum 1284). The delay processes are all classified as non-value added activities as well. A decision-maker may utilize these data to formulate minimum amount of time to complete the below grade work as a benchmark for contract incentives, etc.

All of the scheduled resources are normalized to 4416 hours for the analysis (Table 7).

Figure 3. The delay problem decision tree

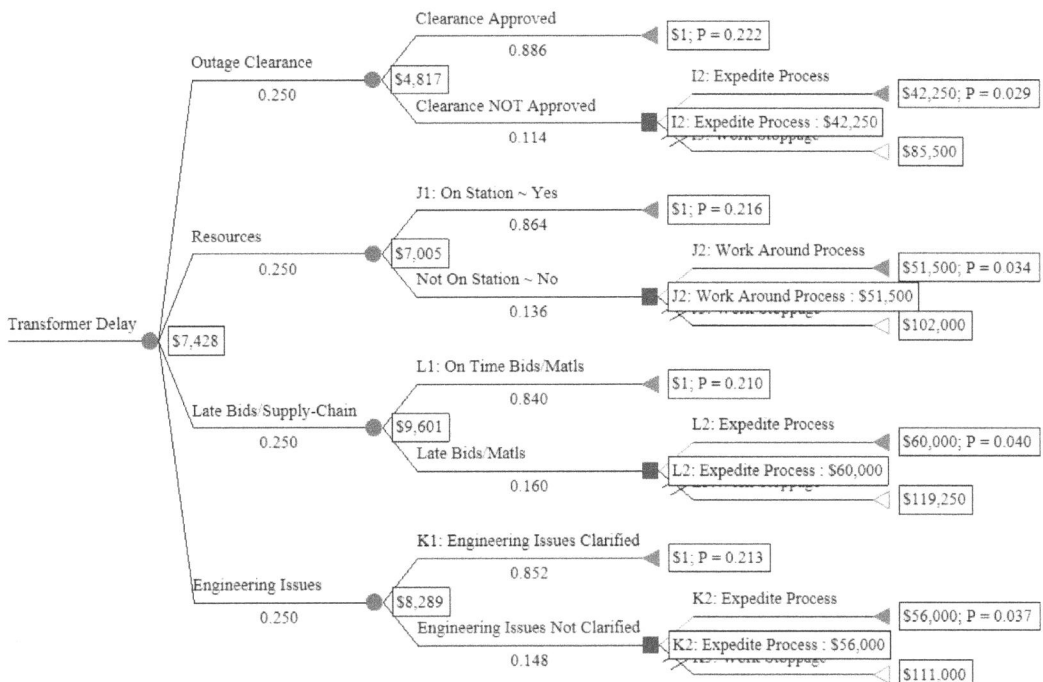

Table 6. Entity allocation

Major Processes	Type	Action	Priority	Units	Allocation	Minimum (hrs)	Value (hrs)	Maximum (hrs)
Conduit Below Grade	Standard	Delay	High(1)	Hours	Value Added	192	240	288
Stone	Standard	Delay	Medium(2)	Hours	Value Added	288	360	432
Material Delay	Standard	Seize Delay Release	Medium(2)	Hours	Wait	854	1070	1284
Eng Labor Contingency	Standard	Seize Delay Release	Medium(2)	Hours	Other	4	6	8
Constr Labor Contingency	Standard	Seize Delay Release	Medium(2)	Hours	Wait	8	12	16
Concrete and Foundations	Standard	Delay	High(1)	Hours	Value Added	1152	1440	1728
Drainage	Standard	Seize Delay Release	Medium(2)	Hours	Non-Value Added	6	8	10
Fence	Standard	Seize Delay Release	Medium(2)	Hours	Non-Value Added	4	6	8
Grading	Standard	Seize Delay Release	High(1)	Hours	Non-Value Added	8	11	14
Grounding	Standard	Delay	High(1)	Hours	Value Added	1152	1440	1728
Oil Containment	Standard	Delay	High(1)	Hours	Value Added	62.4	72	81.6
Roadway	Standard	Seize Delay Release	Medium(2)	Hours	Non-Value Added	2	4	6
Soil Resistivity Testing	Standard	Delay	High(1)	Hours	Value Added	96	120	144
Soil Borings	Standard	Seize Delay Release	Medium(2)	Hours	Non-Value Added	1.5	3	5

KEY PERFORMANCE INDICATORS

The total system cost for the model is approximately $577K (before resource leveling application) to complete the Modular Substation below grade work (30 replications). Further analysis of the potential vulnerabilities within the model (75 replications) and a modified resource schedule produced a more static and stable solution for the total system cost of approximately $594K (modified resource leveling application). The most intriguing part of the evaluation was based on the dramatic decrease in the wait cost ($27K) and slight increase in the idle cost ($3k) between the two scenarios. In the first documented scenario, there are several material items "waiting" in the queue for processing with only a $7K idle cost related to all resources. Conversely, when the resource schedule was slightly modified (no increase in labor) to perform the work effort (sensitivity analysis), the idle time of the engineering, construction, and test labor increased slightly as the below grade work progressed. The non-value added cost remained relatively flat with a small reduction ($4K) in value-added costs. The "Wait Cost" experienced a 91.56% improvement when compared to the Below Grade (BG) Standard. The "Idle Cost" depicts 146.70% degradation when compared to the BG Standard. Both conditions are shown in the Figures 4 and 5 (the smaller piece of pie equates to an improvement over the standard).

Time

The highest indicators are discussed in the documented $577K system cost model for illustrative purposes unless otherwise noted. The construction labor consumed, on average 3450 hours of value-added time in the model and only 29 hours of non-value added time. The wait time consumed 16 hours for the construction labor on average with a maximum value of approximately 3500 hours to complete the below grade project.

Cost

The associated cost of the construction labor was approximately $93K on average in the value-added category and less than $1K of non-value added costs. The wait cost for the construction labor amounted to $460 and $14 in other costs as part of the below grade work effort. The total cost (busy cost + idle cost = usage cost) for the construction labor was approximately $95K.

Work in Process (WIP)

All resources were at least 85% efficient in the work-in-process category phase of the analysis. The raw results are described below:

- **Construction Labor:** 48.5% improvement (measured perspective) from Below Grade (BG) Standards of units work-in-process
- **Engineering Labor:** 28.46% improvement (measured perspective) from the BG Standards of units work-in-process

Table 7. Resource value

Resources				Value (hrs)					
Below Grade Start Eng	Engineer Labor	Constant	Schedule 1	4416	1	Hours	1	Infinite	0.00001
Below Grade Start Matl	Matl below grade	Constant	Schedule 1	4416	1	Hours	1	Infinite	0.00001
Below Grade Construction	Construction Labor	Constant	Schedule 1	4416	1	Hours	1	Infinite	0.00001
Below Grade Testing	Testing Labor	Constant	Schedule 1	4416	1	Hours	1	Infinite	0.00001

Figure 4. The wait and idle cost comparisons

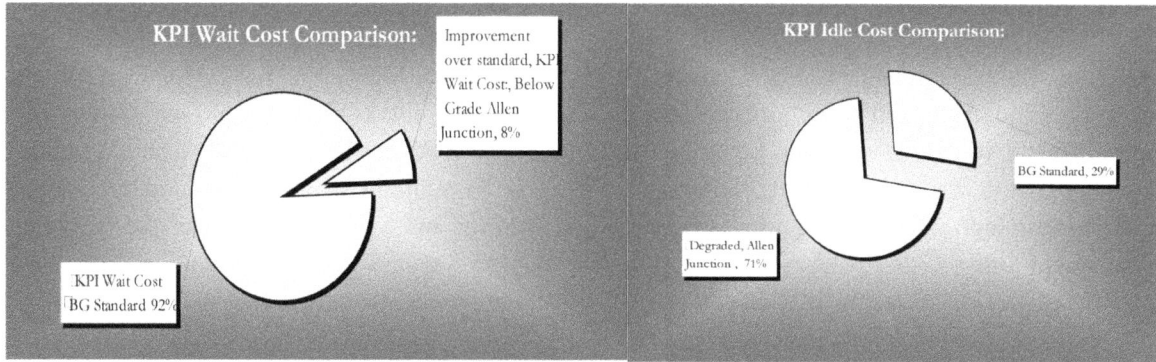

Figure 5. The work-in-process labor comparisons

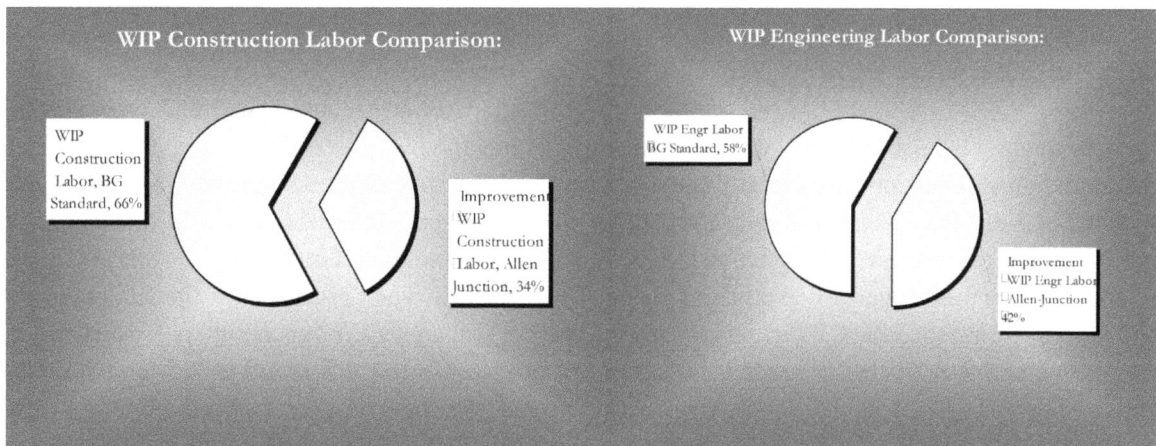

The charts display the minimum improvement percentage from the BG Standard (smaller piece of pie equates to an improvement).

Time per Entity

The "Concrete and Foundations" entity consumed 1435 hours on average (maximum average approximately 1586 hours) and the "Grounding" entity consumed 1462 hours on average (maximum average approximately 1576 hours) of value-added time. The "Grading" entity consumed approximately 11 hours of non-value added time. In the wait time per entity category, the "drainage" entity consumed approximately 6 hours of time (queue time). The total time per entity is an important tenet of the overall report. The upper control limit is derived from the "maximum average" column by the front-end decision maker (contract developer) to formulate the absolute maximum time an entity may perform the task before a penalty is imposed. An example of this presented in Figure 6.

- "The contractor shall not exceed 1586 man-hrs when performing work related to concrete and foundations subject to a 10% penalty of the project total cost."

The time is based on when an entity enters a process until the entity leaves.

Accumulated Value Added and Non Value-Added Times

The "Concrete & Foundations" process accumulated approximately 5740 hours in value-added times. The "Grounding" process accumulated value-added time was approximately 5850 hours. Both of these high indicators represent the total value-added time accrued for all entities in a process. Conversely, the "Grading" process consumed approximately 44 hours of non value-added time (a negative indicator). The below chart indicates a 0.19% slight deviation (virtually even) of the specific Modular Substation process from the below grade standard.

Accumulated Wait Time

The "Drainage" process accumulated approximately 25 hours of wait time for all entities in the system. Since this is a queue based characteristic, the parallel operation is a viable option to reduce the material wait time. Figure 7 depicts the "Grading" process experienced a 96.24% improvement when compared with the below grade standard.

Total Accumulated Time

The total time accrued for all four entities in the process. The highest are in the "Concrete & Foundations" process (5740 hrs) and the "Grounding" process (5850 hrs) as previously discussed above. The "Grounding Grid" is degraded by 279.15% from the BG Standards; however the "Concrete & Foundations" depicts a 15.80% improvement in the process (Figure 8).

Cost per Entity

The highest value-added cost is depicted in the "Grounding" process (approximately $53k) closely followed by "Concrete & Foundations" (approximately $52k) process (a positive attribute). Conversely, the "Grading" process consumed the highest ($402) non-value added cost. The "Drainage" process consumed approximately $216 of non-value added cost in this category. The "Grounding Grid" process was degraded by 26.52% when compared to the BG Standard. The "Concrete & Foundation" process was also degraded by 10.60% however the conduit below grade process demonstrated a 20.49% improve-

Figure 6. The foundation and grounding grid time comparisons

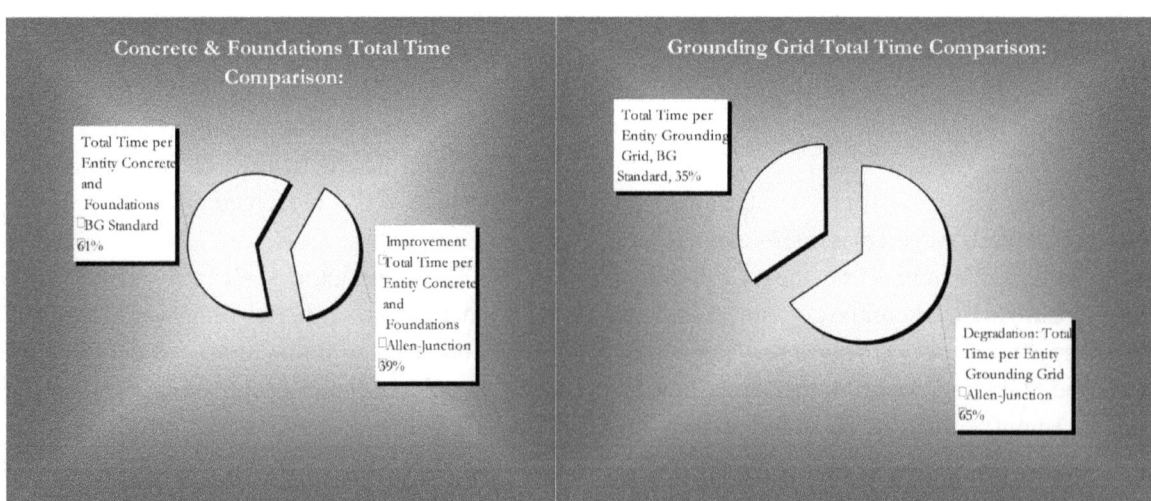

Figure 7. The grading process accumulated wait time comparisons

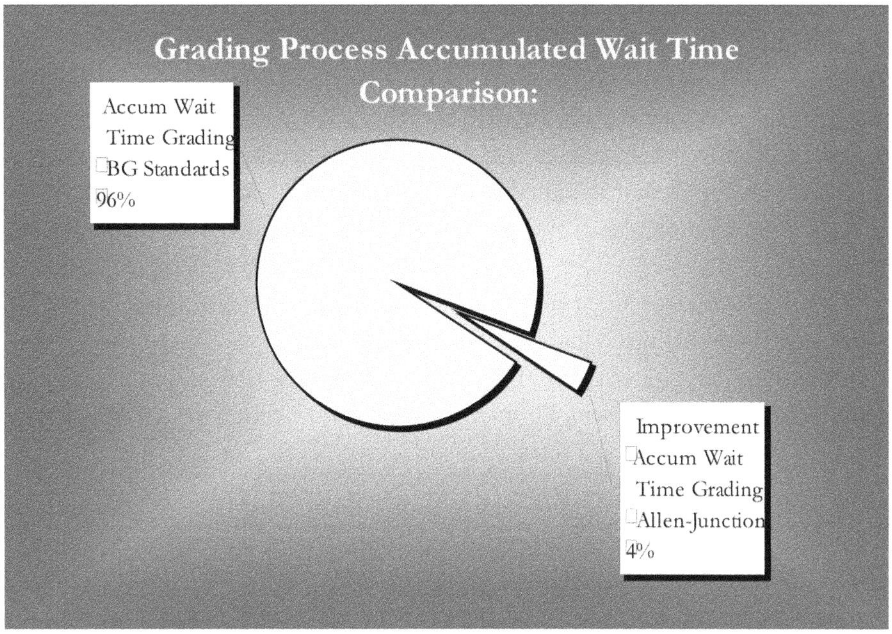

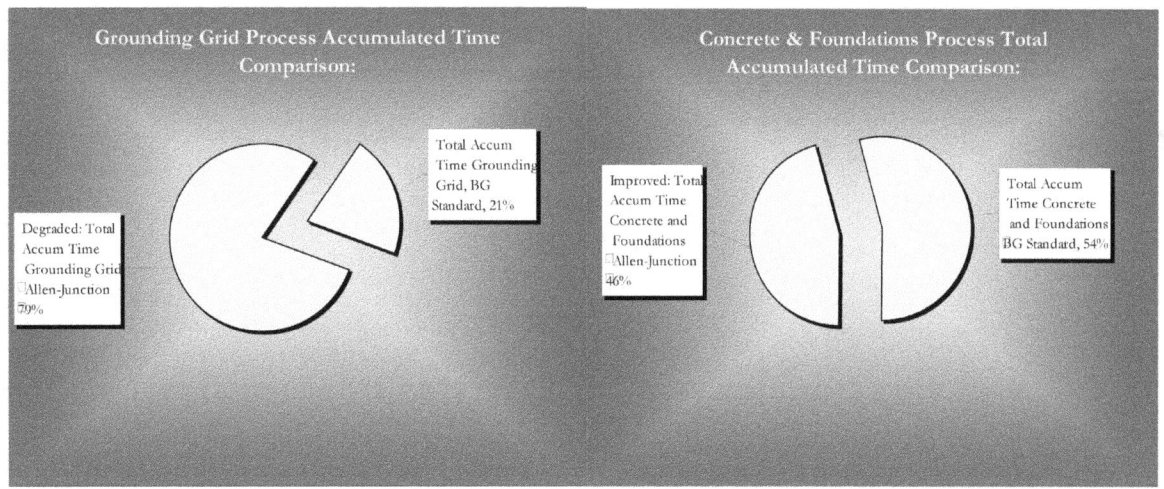

Figure 8. The grounding grid and foundation accumulated time comparisons

ment when compared with the BG Standard (Figure 9).

Total Cost per Entity

One of the most important tenets of the entire report lies within this summary page. It provides the necessary guidance to develop a budget based on the maximum average (or maximum values) as the upper control limit in a statistical process control arena. An example of this to increase the bell curve lies within the "Concrete and Foundations" process where the maximum average value is approximately $57k. Conversely, the minimum average value is approximately $47k (used as the lower control limit). An example of an actual total cost parameter for a substation capital project is depicted in Figure 10.

Cost control methodologies are employed around these parameters over the individual project's lifecycle. Close scrutiny of the monthly budgeted vs. actual cost of the "Concrete & Foun-

dation" work effort allows the decision-maker to monitor/control the element costs *before* an unscheduled event occurs. Subsequent benefits of the control mechanism allows the decision-maker to determine problematic work areas, conflicting resources/demands, and potential cost overruns (Tables 8 and 9 and Figure 11).

The table provides a quick overview of key expenditures within the project's sphere. The statistical limits are shown along with the actual spend data.

When the data is periodically resubmitted in the analysis module, tighter tolerances within the

Figure 9. The cost per entity comparisons

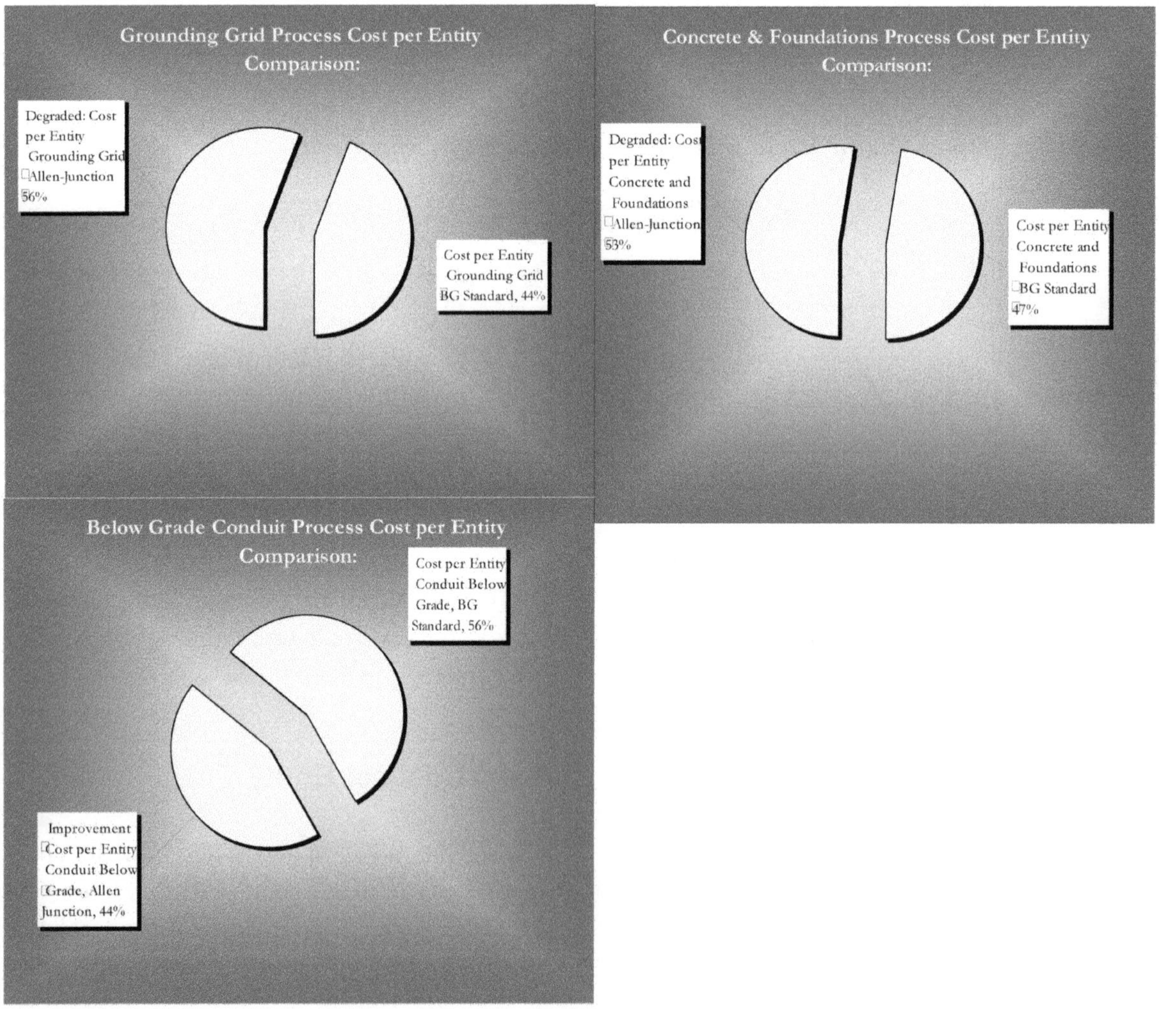

Figure 10. Total project SPC limits for cost control

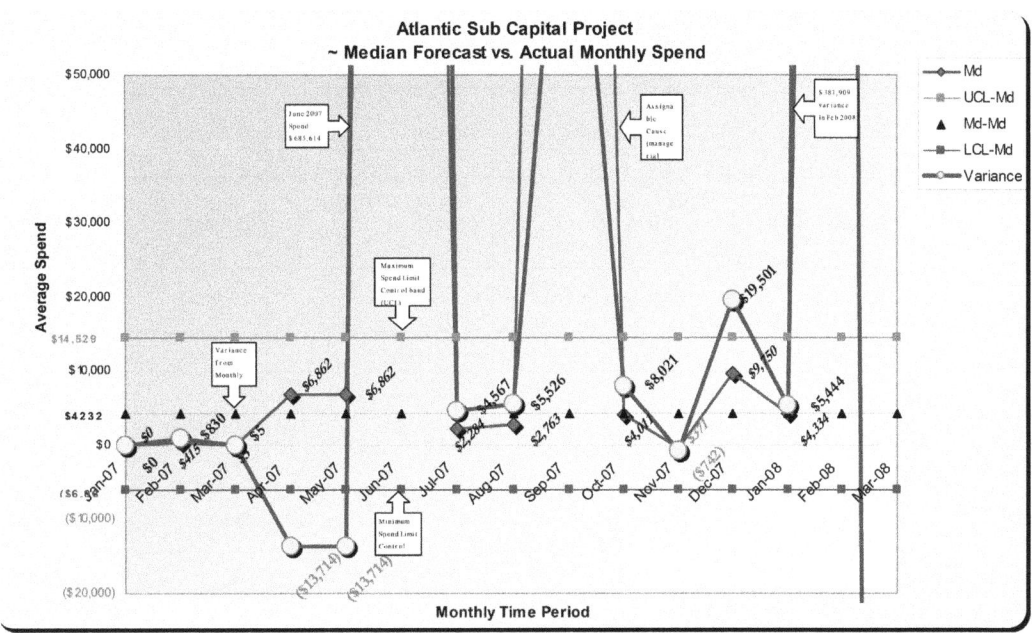

six-sigma realm are revealed to create an historical database for future projects. The cost savings of this process are unparallel. The "Wait Cost per Entity" experienced a 97.2% improvement whereas the "Total Cost per Entity" category revealed a 52.34% improvement. However, the "Grounding Grid" total cost per entity is degraded by 26.52% when compared to the BG Standard.

Accumulated Value-Added/Non Value-Added Cost

The average minimum accumulated value-added cost for the "Oil Containment" is approximately $10k with the average maximum accumulated value-added cost is in the order of $211k. Resource-leveling techniques will certainly improve these costs across the entire spectrum (ideal). Conversely, the average non value-added cost is expected to decrease. Currently, the highest indicator in this category is the "Grading" process ($1.6k). The highest average accumulated wait cost is within the "Drainage" process ($865). Again,

process re-balancing and selective sequencing will alleviate specific conditions in the model. The "Grading" process experienced a 22.73% improvement when compared with the BG Standard; however the ideal value is to achieve zero NVA.

Other-Number In/Out

The average number in and out of the system to produce one complete below grade unit by each entity is four. If the schedule was compressed (2208 total hours), the system would produce zero units with skewed resource capacities (construction labor processing eight units whereas soil Resistivity testing processes zero units). Sensitivity analysis of the system provides the required insight to optimally balance the process within the prescribed constraints. The ability to control specific factors and minimize the increase in prediction variance is essential elements in the process (Monroe, E., et al., 2010). There were three units in the "Material Delay" and zero units processed out of the system. The one-to-one number in/out relationship as it

Table 8. Total project cumulative spend to date

2008	Project Cost Analysis ~ Cumulative Spend				
	Likeliest	Min	Max	Limit	YTD Actuals
Construction Overheads- Supervision	$359,372	$323,434	$395,309	$372,954	$229,519
Construction Overheads- Engineering	$84,708	$76.237	$93,178	$87,909	$53,997
Construction Overheads- A&G	$179,364	$161,428	$197,300	$186,143	$93,095
Contractors- Other	$1,774,000	$1,596,600	$1,951,400	$1,841,051	$633,575
Major Total Construction Expenditures: Altantic Sub Project	$2,397,444	$2,157,699	$2,637,187	$2,488,057	$1,010,186
Actual Spend percentage	42.14%	46.82%	38.31%	40.60%	

Table 9. Project resource worst case scenario

2008	Project Cost Analysis ~ Worse Case Scenario Major Construction Capital				
	Likeliest	Min	Max	Limit	Actuals
Construction Overheads- Supervision	$359,372	$323,434	$395,309	$372,954	$395,315
Construction Overheads- Engineering	$84,708	$76,237	$93,178	$87,909	$93,180
Construction Overheads- A&G	$179,364	$161,428	$197,300	$186,143	$197,305
Contractors- Other	$1,774,000	$1,596,600	$1,951,400	$1,841,051	$1,951,405
Major Total Construction Expenditures: Altantic Sub Project	$2,397,444	$2,157,699	$2,637,187	$2,488,057	$2,637,205
Actual Spend percentage	110.00%	122.22%	100.00%	105.99%	

Exceeded the Most Likely criteria by 10% ($239,761 in savings), the minimum criteria by 22.22% ($479,506), and most importantly, the maximum portion of the range was reached. The acceptable +1 standard deviation was exceeded by 6% ($149,148) as well.

pertained to the other entities maintained system integrity.

Waiting Time and Cost

The average waiting time for the "Fence" queue was approximately three (3) hours closely followed by the "Drainage" queue (2.66 hours). The ideal

Figure 11. Total project graphical cost control targets

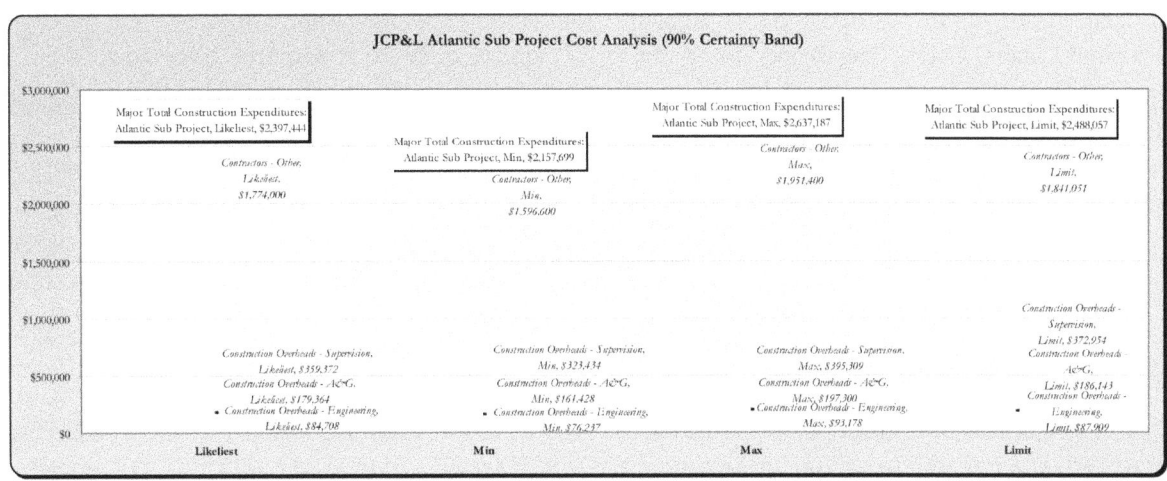

Figure 12. Total foundation and grounding grid cost comparisons

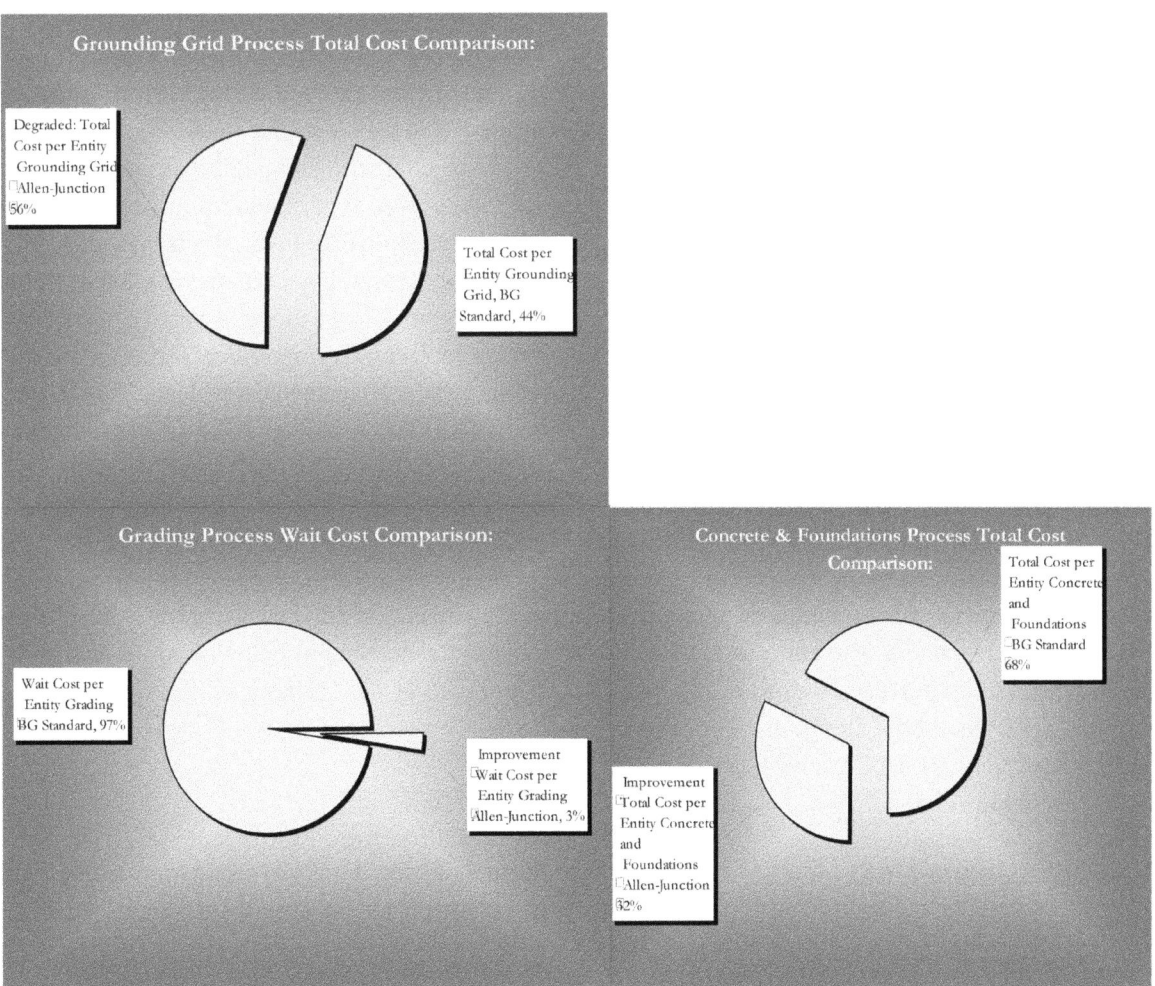

theoretical condition for the category is zero. The associated average wait cost of the entity wait time was approximately $101 (fence) and $97 (drainage) to include the overhead cost of retainage and its related resources it owns.

- Recommend increasing the process priority from medium to high to reduce the fence queue time/cost. Results: The fence waiting time decreased to 2.18 hours and the cost decreased to $78.18.

The "Material Delay" queue possessed the largest (0.230) number of units waiting in the system. Nevertheless, the overall through-put of the other entities was very acceptable for the model.

- Recommend increasing the process priority from medium to high to reduce the queue amount and evenly disperse the units across the contingency process.

Instantaneous Utilization/ Number Busy/Number Scheduled/Number Seized

A quick "snap-shot" of the entire process exhibited the material category averaged 0.1447 units of resource utilization. The number of scheduled resources (0.8217) falls above the acceptable (0.75) range for the industry. Notwithstanding, the schedule utilization (the time period the resource was actually scheduled in the system) displayed the average material (0.1761) units as compared with the other resources. The engineering labor resource seized, on average 20 units to produce on below grade unit out of the system. The "Busy Cost" improved by 91.49% when compared with the BG Standard (Figure 13).

Busy/Idle/Usage Cost

The "Material" busy cost (the product of the average number of busy units) exceeded $287 as the

highest indicator in this category. The "Testing Labor" consumed approximately $3265 in idle cost on average. A resource re-evaluation is in order to reduce the testing labor idle cost.

- Recommend increasing the busy/hour to 0.25 (idle/hour decrease to 0.75) of the labor entity. Results: The average labor idle cost decreased to $2726 in the testing category. The other resource idle costs remained relatively flat.

The test labor degraded by 9.42% whereas the number busy improved by 82.77% as compared with the BG Standard.

Project Summary

The company internal cost estimators woefully underestimated the below grade work effort of $130k by $419k (low end) to $447k (high end). These original estimates neglected vital time/cost statistics as well as pertinent resource parameters. Detailed reviews of deliverables from senior project managers and subject matter experts revealed the conceptual cost estimates were typically low. The project scope/change orders usually increased the original estimates in the order of 5X ($650k) to 6X ($780k) on average to complete the below grade work effort. The analysis and detailed cost estimates by astute financial planners and field operations personnel will potentially provide the company with $213k ($650k-$447k) *minimum* cost avoidance dollars as a front-end evaluation tool for the substation below grade work effort. The summary data is depicted in Figure 14.

REFERENCES

Besterfield, D. H. (1999). *Total quality management* (2nd ed.). Upper Saddle River, NJ: Prentice Hall.

Figure 13. Engineering labor number busy comparisons

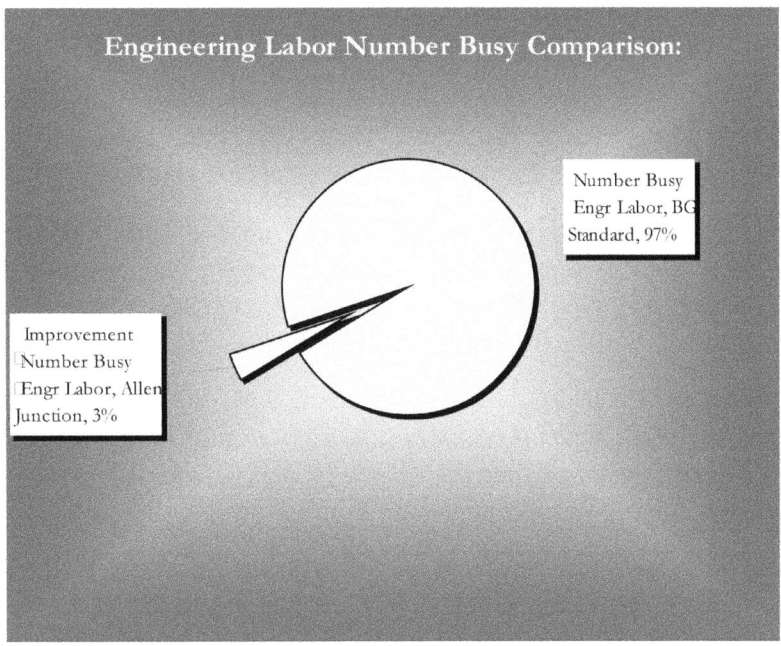

Figure 14. Substation below-grade work effort comparison project summary

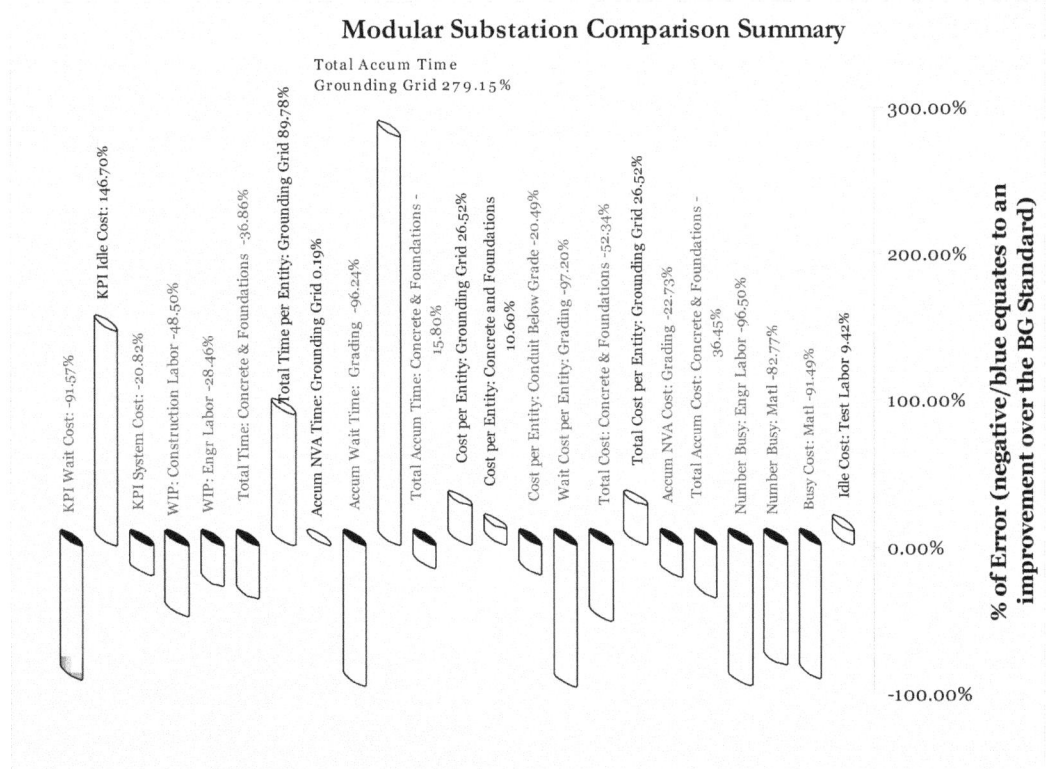

RUS Bulletin 1724E-203. (1994). *Guide for upgrading rural utility service transmission lines.* United States Department of Agriculture Rural Utility Services.

Chen, F. H. (2000). *Soil engineering: Testing, design, and remediation.* Boca Raton, FL: CRC Press.

Edwards, F. W. (1993). *Substation grounding system evaluation.* Conducted by Virginia Power. Presented at the Southeastern Electric Exchange. New Orleans, LA.

Emil, S., & Scanlan, R. (1986). *Wind effects on structures: An introduction to wind engineering* (2nd ed.). New York, NY: John C. Wiley & Sons.

FIST. Facility Instructions, Standards, and Techniques. (2005). *Maintenance scheduling in electrical equipment,* Volume 4-1B. U.S. Department of the Interior, Bureau of Reclamation, Denver, CO.

Fuller, F. M. (1983). *Engineering of pile installations.* New York, NY: McGraw-Hill.

Jordan, T. E., & Constanti, D. J. (June 1995). The use of non-invasive electromagnetic techniques for focusing environmental investigations. *The Professional Geologist,* pp 4-9.

Lyncole Company. (2005). *Soil resistivity testing four point Wenner method.* Application Note LEP-1001. Retrieved 10 October, 2009, from http://www.lyncole.com

Mann, D., & Domb, E. (1997). *40 inventive (business) principles with examples.* Retrieved 8 November, 2008, from http://www.triz-journal.com

Monroe, E., Pan, R., Anderson-Cook, C. M., Montgomery, D. C., & Borror, C. M. (2010). Sensitivity analysis of optimal designs for accelerated life testing. *Journal of Quality Technology, 42*(2), 121–135.

O'Brien, J., Havers, J., & Stubbs, F. (1996). *The standard handbook of heavy construction* (3rd ed.). New York, NY: McGraw-Hill.

Rossi, R. H., & Freeman, H. E. (1993). *Evaluation: A systematic approach.* Beverley Hills, CA: Sage Publications.

RUS 1728-810. (1998). *Electrical transmission specifications and drawings 34.5-kV through 69-kV.* United States Department of Agriculture. Rural Utility Services.

Short, T. A. (2006). *Electric power distribution and systems.* Boca Raton, FL: CRC Press Taylor & Francis Group.

Stewart, M. T., & Gay, M. C. (1986). Evaluation of transient electromagnetic soundings for deep detection of conductive fields. *Ground Water, 24,* 351–356. doi:10.1111/j.1745-6584.1986.tb01011.x

Subramanian, N., & Vasanthi, V. (1990, March). Design of tower foundations. *Indian Concrete Journal,* (March): 135–141.

Switzer, K. W. (1996). Eleven practical tips for grounding substations. *Electrical Construction and Maintenance Magazine.* Retrieved 18 December, 2010, from http://ecmweb.com/mag/electric_eleven_practical_tips/index.html

Tillman, R. F. (2001). Loading power transformers. In Grigsby, L. L. (Ed.), *The electric power engineering handbook.* Boca Raton, FL: CRC Press.

ENDNOTES

1. Source: 2009 NERC Disturbance Report. (published 15 June 2010)
2. Source: USDA Soil Survey of Lewis County Area, Washington State Department of Natural Resources and WSU Agricultural Research Center. (Washington, May 1987)

3. Source: USDA Kansas Agricultural Experimental Station, Soil Survey of Ford County Kansas. (Kansas, 1985)

4. Source: USDA Michigan Department of Agricultural Experimental Station, Soil Survey of Kent County, Michigan. (Michigan, 1980)

5. Source: Centers for Disease Control and Prevention, National Institute for Occupational Safety and Health (NIOSH). Fatality Assessment and Control Evaluation (FACE) Program. (FACE 9307, 1993)

6. Source: Centers for Disease Control and Prevention, National Institute for Occupational Safety and Health (NIOSH). Fatality Assessment and Control Evaluation (FACE) Program. (FACE 90-31, 1990)

7. 88 =5280 feet/60.

8. See Appendix for a detailed discussion of the Scenario Analysis Model.

Chapter 7
Transmission Line Reconductoring Process Analysis

ABSTRACT

The electrical transmission congestions transpire when scheduled or actual load flows across a line/ substation are restricted by the desired levels – either physically or by the electrical line capacity. Several electric companies are aware of the congestion problems and the associated costs, where new transmission upgrades are oftentimes incorporated with new generation in close proximity with the load. Other companies are taking advantage of the benefits Smart Grid technologies propose that captures renewable energy production to defer the transmission upgrades based on the peak load growth. The industrial operations process analysis of the Line Rebuild project is an important cost-reduction mechanism to evaluate system integration. The value-added evaluation process provides the necessary front-end analysis for prudent decision-making of capital improvement endeavors.

OVERVIEW

The three most important structural elements that formulate an electrical distribution system are 1) transmission lines, which transmit high voltage electricity from power generation facilities, 2) substations, which decrease the voltage in the transmission lines to distribution levels, and 3) distribution lines, which convey the electricity to customers.

DOI: 10.4018/978-1-4666-2839-7.ch007

The Transmission Lines

Transmission lines are high-voltage conductors (60-kV or greater) that convey bulk power from remote generation sources to the electrical service area. Overhead high-voltage conductors are approximately an inch in diameter and are comprised of aluminum strands or a mixture of aluminum and steel strands. The lines are isolated electrically by the adjacent air and are not wrapped with insulation material (Fink & Beatty, 2001). The overhead transmission line is the standard technology to convey electrical power. Most damage to the overhead conductors is attributed to fault-current arcs. The conductors strands are significantly weaken from the enormous heat generated by the arc. The protection circuits are designed to prevent such occurrences; however, electrical utilities experience such burn downs (especially with covered lines) when instantaneous tripping schemes were not utilized or applied incorrectly (Barker & Short, 1996). Consequently, long outages may occur due to the high-impedance fault condition which is somewhat difficult to detect once the line plunges to the ground. Generally, the conductor can withstand significant temperatures for a few seconds and maintain its potency. However, if the relaying or fuse fails to clear the fault in a timely manner, the conductor anneals and loses strength. Additionally, connectors and splices are susceptible to weak links in overhead lines based on inferior engineering designs, hostile environment, and/or poor installation (Jondahl et al., 1991). Compression splices (squeeze in) and good component cleaning practices are essential for high-quality contact with the conductors.

Support structures for the overhead transmission lines range from single wood or metal poles and H-frame towers between 50 and 90 feet tall for lower voltage lines (i.e., 60-kV and 120-kV), to metal H-framed or lattice frame towers between 75 and 130 feet tall for higher voltage lines (i.e., 345-kV). The proposed single/double circuit steel lattice towers for a capital project in the western region are designed to convey 500-kV[1]. The average heights of these towers are 120-160 feet (single) and 170-210 feet (double). It is not advantageous for circuits that serve the same substation to be located or double circuited in the same proximity to each other (e.g., in the same corridor or on the same pole). The consequences are less reliability because both circuits could be compromised by the same event (e.g., wind storm or lighting strike). Moreover, safety issues would increase from maintenance activities including working on energized lines because one circuit is designated to carry power. In cases where the circuits are serving separate substations, double circuit configurations would not have the same dependability issues provided that they are each serving other load centers and there is existing redundancy built into the distribution system. In some cases, these circuits can be situated on the same structure.

Transmission lines are designed in accordance with requirements of the National Electrical Safety Code (NESC), and speak to design issues such as the following (Clapp, 2006):

- Clearances between the lines and other characteristics, such as ground or water surfaces, roadways and railways, other conductors or communication wires, and buildings;
- Use of shield wires along the top of the transmission line to protect the conductors from lightning strikes;
- If an energized line plunge to the ground, high-speed relay equipment will sense this condition and actuate breakers that would de-energize the line a tenth of a second to half a second;
- Connection of metallic parts to an electrode in the ground (grounding) in order to protect employees and the public from the hazard of electric potential; and
- Principles for mechanical and structural design, selection of materials, and construction practices to ensure that towers, conductors, and insulators are sufficient to withstand normal and unusual loads, such as ice and wind, to ensure that pole spans are adequate to prevent conductor or structure failure, and to ensure that ample clearances are maintained.

An example of the technical details for specific transmission line requirements are described in appendix II.

The Substation

Since electricity in the transmission lines is generated at voltages greater than what the customer can use, the conductors are first directed through substations, which regulate or reduce the electric voltage to levels that can be conveyed to the customer. The chief task for the substation is to transport power in a proscribed manner while facilitating the necessary switching arrangements and/or connections within the grid. The basic functional requirements for the substation consist of the quantity and type of connections, the system electrical information, the energy and transport path through the system, and the related unavailability costs. Substations generally consist of a control building and steel structures that support the essential electrical equipment, including the terminals to receive transmission lines and transformers that convert power to a dissimilar voltage. The reliability of the system partially depends on the availability of the scheduled generation units and its associated electrical path to the customers. As a quality aspect to minimize outages created by scheduled maintenance and/or failures is to design and configure the equipment with low maintenance requirements within the substation. The fraction of time the electric power is unavailable (typically expressed in hours per year) attributes to the overall bottom line of the utility. It is imperative to maintain the reliability and availability of the equipment in the substation. The general power path consists of three major sections: 1) lines, 2) power transformer, and 3) the switchgear. The dominate cause of maintenance is within the line and transformer areas. The substation maintenance work increases the risk for disturbances (weaker system with workers in the substation, equipment faults, etc.) and can lead to outages to customers. The reduction of these maintenance activities produces less cost for manpower to perform the work, higher personnel safety (fewer or no workers on site since the high voltage system involves risk of injury – electrical shock, falling, etc.), and increased reliability. Substations are generally located in the central part of a service area, defined by the level of customer demand, or load, in order to sanction proficient distribution of power to customers. The substation, typically unmanned with limited road access for transporting equipment, maintenance and/or repair, are contained within secure sites to ensure public safety.

The Distribution Lines

Electricity from substations is subsequently conveyed to the customer through medium-voltage 25-kV distribution lines. Since distribution lines transmit lower voltages, they necessitate less substantial lines and support structures than transmission lines. Distribution conductors are generally half an inch to an inch in diameter and are shored up by T-shaped wood poles approximately 40 feet tall. The distribution line must be designed to withstand, isolate, and quench the effects of lightning strikes, and maintain high insulation between the phases to avoid most flashovers (Harness, 2000). Moreover, the distribution line must be designed and constructed in such a manner as to eliminate the electrocution of raptors. The steel pole construction configurations, where short distances between the bare energized phase conductors and the grounded pole, are specifically dangerous for the bird. Since distribution lines are most prevalent in highly developed areas, and because they operate at lower voltages, distribution lines are more easily and frequently placed underground. Numerous engineering aspects limit the length of underground electric transmission services. As the voltage increases, the engineering constraints and costs significantly increase as well. Consequently, underground distribution lines (12-kV to 24-kV) are not unusual when compared with some underground transmission system (less than 100 miles in one Midwestern region). Nevertheless, the spine of the distribution system is placed overhead and susceptible to catastrophically events. Oftentimes, tree and branch failures create electrical faults at the distribution level by bridging phases – approaching conductors into each other that cause equipment damage which subsequently produce outages. During a ten year period, the electric utilities identified eighty-one major storms that attributed to over $2.7B in damages (Johnson, B., 2005). During normal operating conditions, the industry must mitigate the tree-related outages through aggressive vegetation management programs. There are vegetation-related transmission outages as well, with the potential to cause even a momentary outage (Kretschmer & Hundrieser, 2001). The number of "grow-ins" (outages caused by vegetation growing into lines from inside and /or outside of the right-of-way) and "fall-ins" (outages caused by vegetation falling into lines from inside and/or outside the right-of-way) that causes sustained outages of transmission lines are meticulously compiled and recorded by the North American Electric Reliability Corporation (NERC). Thirty-three percent of the regions reported a vegetation-related outage during the first quarter of 2009[2]. The excessive cost based on storm stress alone justifies an evaluation of tree interaction with the electric system in the structural design process (Simpson & Van Bossuyt, 1996). Because distribution lines provide all electrical customers, they are the largest component of the electrical system.

The integration of renewable generation within the electrical involves intricate analysis of the transmission line upgrades throughout the system. These upgrades are designed to relive congestion, improve reliability, and provide voltage support by thoroughly evaluating the alternatives to accommodate the new expansion efforts.

BACKGROUND

The electrical transmission congestions transpire when scheduled or actual load flows across a line/ substation are restricted by the desired levels – either physically or by the electrical line capacity. Moreover, operational restrictions are developed and enforced to protect the security and reliability of the electric grid[3]. The North American Electric Reliability Corporation (NERC) developed policies and procedures specifically for the orderly relief of congestion for the interconnections. The exclusive utilization of the transmission loading relief procedures (or congestion management) is implemented with other interconnection-wide policies for systematic load curtailment. There are numerous constrained transmission paths throughout the electrical corridors across the regions. Several electric companies are aware of the congestion problems and the associated costs, where new transmission upgrades are oftentimes incorporated with new generation in close proximity with the load. Other companies are taking advantage of the benefits Smart Grid technologies propose that captures renewable energy production to defer the transmission upgrades based on the peak load growth[4]. These benefits include direct transmission congestion relief as well as capacity factors for the renewable generation expansion. Moreover, the transmission reliability limits are relaxed and additional voltage support is provided. The demand-side management efforts are designed to reduce the overall electrical supply costs and improve the reliability of the grid. A sampling of the transmission and distribution congestions projects are illustrated below:

Selected Midwest Region Projects

Project 1: Conduct a study to determine the specific impacts to the SENE binding constraint.

Need: Binding Constraints both Day Ahead (DA) and Realtime (RT) have increased over the past year from 10th to 4th DA and from 11th to 6th RT in the electrical footprint. This impacts not only BEM but other entities that transverse across Illinois.

Alternatives: Secondary parallel line to relieve the congestion. The expectation is the study would provide several alternatives.

Cost: $7,500,000

Project 2: New 765 kV line from Rockport to Greentown

Need: Mitigate thermal and voltage concerns associated with (N-1), (N-1-1) & (N-2) multiple contingencies - Indiana study. Project will reduce transmission congestion and losses, enhance stability performance, and simplify complex operating conditions and provide for alternative connection possibilities for new generation (both renewable and conventional) with increased transmission exit capabilities.

Alternatives: Developed though the study - Possible alternate line routes to accommodate intermediate step down stations

Cost: $1,000,000,000

Selected Western Project

Project 1: California - Oregon AC Intertie Addition Project

Description: Upgrade California-Oregon AC Intertie to improve transfer capability, decrease congestion and improve stability.

Cost: $47,700,000

North Expansion Project (NEP)[5]

Project Description: The NEP construction plan will require the construction of 213 miles of new 230 and 115 kV transmission lines, and three high-voltage "interchanges" and switching sta-

tions at various locations to connect to the existing transmission system. The cornerstone of the NEP projects is the Hitchland Interchange 345/230/115 kV. This will be constructed in the middle of the existing Potter-Finney 345 kV transmission line. From this interchange, the Moore County – Hitchland 230 kV line, and the Hitchland – Prairie 230 kV line will be constructed. 115 kV transmission line constructions from Hitchland will connect the new interchange into the existing transmission network in the Oklahoma panhandle.

Project Cost: Approximately $109 million.

Project Benefits: The project goal is to increase reliability, reduce congestion, and allow additional load serving capability in the areas from Amarillo, Texas to Guymon, Oklahoma. The NEP segments are located in area experiencing significant requests for wind generation interconnections, and the NEP facilities will facilitate the interconnection of additional wind generation in the north Texas panhandle.

A number of the congestion areas identified are classified as *conditional* by the Department of Energy. These are areas where transmission congestion exist, yet is exacerbated if additional generation resources were integrated in the system without increasing the transmission capacity in the region. The cost estimates for the capital-intensive projects typically include the studies, material, and labor. However, these broad-stroke approximations could be drastically reduced if the congestion solutions were based on generation, transmission, distribution, and demand-side options thoroughly evaluated against an assortment of load growth scenarios. The electrical demand-side circumstances must include the energy prices by group, resource development patterns, employment by sector, fuel usage, etc. to establish a robust solution. The demand for electricity from renewable energy resources is expected to expand in some regions and diminish in others as customers migrate to demand-side and energy efficient programs. Since the demand for electricity peaks during the daylight hours, specifically in the summer months, the detailed profiles are essential for new expansion projects. The demand profiles in most regions fail to align with the availability of renewable resources when aggregated as whole, consequently supplemental resources (coal, oil & gas, nuclear, hydroelectric) are required to meet the load. The demand forecast may consist of a short-term outlook for system operations purposes (i.e. 15-30 minutes) to long-term generation planning (i.e. 5 to 20 years) aspects. The variety of the load forecast may include the peak demand (kW), energy (kWh), reactive power (kVAR) and /or the load profile. It is comprised of the appliance load, system load, transmission load, substation/feeder load, or individual customer loads. The intricate balancing act between the resources must be encompassed within the range of scenarios as well.

The wind energy proposal for system integration generally espouses the characteristics outlined below (DOE/GO-102008-2567, 2008):

- Enhancement of electrical transmission system is essential in all electricity-growth scenarios
- Transmission is considered necessary to:
  ○ Relieve congestion in existing system
  ○ Improve system reliability for all customers
  ○ Increase access to lower-cost energy
  ○ Access new and remote generation resources
- Wind Turbine Generation involves additional transmission than some other options as the best winds are often in remote locations

The transmission line costs as well as the system reinforcements increases the total expenditures of the proposed project. The effort entails the upgrade of an existing line to allow the facility to operate at a higher capacity. Details usually include the unsuitable "phase-raising" technology and the replacement of structures (with short outages) to facilitate improvements. Relay replacements and energized transmission line work are also included in the cost estimates as well. In some cases, substation upgrades are required to accommodate the transmission line rating where applicable. If upgrades cannot provide the increased wind penetration levels, new transmission lines must be constructed parallel to the existing lines (removed once the new line is energized). The expensive acquisition of additional right-of-way is required paralleling the existing routes. The general transmission line cost estimates (+/- 20%) and assumptions are outlined below:

- Raw materials, fuel, and labor based on current year unit costs
- Right-of-way, governmental permitting, design materials, construction, overhead construction management, internal utility loads

Congestion relief is an all-encompassing effort to improve the reliability of the system. The Lean Quality processes are usually not included in the cost estimates. Because of this, the project schedules (as well as costs) are significantly affected by resource availability, outage clearances, weather, equipment delivery delays, and permitting issues. The cost of a typical transmission line removal during on-peak conditions is illustrated in Table 1.

The cost increases proportionally with the voltage and amperage to perform the work. The material itself is designated as Aluminum Conductor Steel Reinforced (ACSR) to exploit as a bare overhead transmission line and /or as a primary /secondary distribution cable. Some varieties of the material consist of the steel core wires, which are equipped with a protective covering of Class "A" zinc coating. The concentrically stranded aluminum wires consist of one or more layers of wires helically wrapped around a solid or stranded steel galvanized central core, as another combination of the cable. The correct selection of the new conductor is imperative for the

Table 1. Transmission line removal cost

Cost to remove line out of service for 8 hours.								
Average On-Peak System cost of $17/MWH								
Conductor	Amps	69kV	115kV	138kV	161kV	230kV	345kV	500kV
266 ACSR	440	$7,152	$11,919	$14,303	$16,687	$23,839	$35,758	$51,823
477 ACSR	640	$10,402	$17,337	$20,805	$24,272	$34,674	$52,011	$75,379
795 ACSR	890	$14,466	$24,109	$28,931	$33,753	$48,219	$72,328	$104,824
954 ACSR	995	$16,172	$26,954	$32,345	$37,735	$53,908	$80,861	$117,191
1192 ACSR	1110	$18,058	$30,096	$36,115	$42,135	$60,192	$90,289	$130,853

mechanical strength characteristics as well as the ampacity capability.

The simplest and cost effective technique for line upgrades utilized by the industry is the reconductoring process. In essence, reconductoring is the act of removing the existing conductors and replacing it with a single larger conductor (RUS Bulletin 1724E-203, 1994). This modification technique is employed when the presented structures possess sufficient pole strength as well as ground clearance to accommodate the increase in vertical and transverse loads. The modification is expected to provide for the increased conductor sag as well. The associated hardware and attachments must be thoroughly examined for strength to accommodate the significantly higher line tension of the proposed conductor. Moreover, the revised configuration of the reconductor line will produce dissimilar galloping ellipses. This circumstance must be thoroughly scrutinized in the ruling spans (minimum and maximum) to verify the conductor contact during the galloping condition.

The typical facility study conducted by the Regional Independent Service Operator investigates the transfer capability of an incoming request of a specific megawatt value. The details usually include current capital project alternatives to accommodate the request. These alternatives generally encompass some form of reconductoring or another variety of system upgrades. The reconductoring, for example, of a 556 ACSR section of conductor replaced with a 795 ACSR cable at

167°F is not an uncommon practice for the utility industry. In essence, the line reconductoring in the example increased the ampacity by at least 170 amps as well as decreasing the Ohms/mile by 570. The other option revealed by the simulated study from a thermal perspective is to increase the existing line clearances to support the current operation up to 230°F. The increased rating (to 795 ACSR) must have the associated clearances for 230°F as well. The evaluation is conducted line-by-line, and section-by-section within the electrical footprint. The recommendations from these studies include terminating an existing line in a substation to create an additional line when economically feasible. The transmission line rating upgrade is typically a rudimentary function for capital-intense projects, conducted to relieve an existing condition or to fulfill a requirement for future expansion.

The studies may also reveal pertinent details about the current line characteristics with an adjacent fault. The typical condition is the current line (138-kV) exceeds its summer normal rating during peak conditions. The current line under investigation will subsequently overload if an adjacent fault occurs (the loss of a 345-kV line for example). Hence, the proposal to reconductor at least twenty miles of 336.4 kcmil 26/7 ACSR (530 ampacity) transmission line to the superior 795 kcmil ACSS/TW type of conductor with a 2199 forty hour emergency rating amperage is an overriding decision at an estimated cost of

$2.3M. Depending on the combination of span length, steel core area, and installed tension, the temperature difference between the conductor core and the surface may reach 86°F or more. The annealed aluminum strands in the ACSS/TW conductor is more likely to expand helically in response to modest compressive forces. The main advantage of the ACSS cable – albeit at a higher cost - is to increase the emergency thermal rating by approximately forty percent, lessen the likelihood of vibration fatigue, and reduce the structure height is most cases. The effort would certainly provide accommodation for new generator expansion in the region. However, the current overload condition and system upgrade must be thoroughly evaluated first and foremost.

The alternative for the ordinary round strand of ACSS/AW is the utilization of the trapezoidal aluminum wires (ACSS/TW) which allows for more of the material in the conductor of a given diameter. An example of this is within the diameter comparison of a conductor (Suwannee/ACSS/AW/TW has the same diameter as Drake/ACSS/AW) yet its resistance is approximately 20% less, the thermal rating is roughly 10% higher, and costs about 20% more. The decision-makers are tasked with a variety of line replacement scenarios as part of the transmission upgrade processes. The minimization of electrical losses in overhead lines is important tenets of the replacement procedure. The rate of the losses depends on the conductor resistance and the square of the current. When a large conductor replaces a smaller one, it will reduce the rate of electrical losses. If a 1590 kcmil conductor replaces a 795 kcmil cable, the electrical losses are reduced by 50%. However, the conductor cost is doubled and subsequently increases the structure costs by approximately 35%. Moreover, the emergency line rating established by the local company, increases proportionally as well. The decision to utilize a large conventional conductor versus the smaller type, each with the same thermal capacity, depends on the projected line electrical load magnitude and load factor. Consider, from a financial perspective, a proposed 138-kV line replacement utilizing Drake ACSS/AW versus a similar design with Falcon 1590 kcmil ACSR. The decision-makers will evaluate the electrical load limits (i.e. if it approaches 240 MVA on a regular basis yet averages 180 MVA) along with the present worth cost of the losses over a 40 year life of the line and the construction cost. In this case, the larger conventional conductor and its increased capital cost are effortlessly defensible. However, if the electrical load approaches only 80 MVA on a regular basis and averages only 60 MVA, than the present worth cost of the electrical losses over the life of the line is much less than the present worth cost of construction. In that case, the utilization of the Drake ACSS/AW is the most viable solution. The construction processes play a vital role in decision-making from both and engineering standpoint and a financial perspective. Quality is one of the critical elements in strategic planning and must be implemented in every facet of the decision-making process. The TRIZ[6] principle #23, specifically the "feedback (inverted) Feed forward" is outlined below (Mann & Domb, 1997):

- Introduce feedback (referring back, cross-checking) to improve a process or action.
- If feedback is already used, change its magnitude or influence.
  - Long-term strategic business planning and programming
  - Strategic quality planning and leadership vision
  - Technology road map and trend analysis
  - Predictability within limits as output of the statistical process control
  - Reliability prediction and anticipatory failure determination

Predict and compare. (W.E. Deming)

The cost avoidance potential within the construction sphere of a capital project is an essential key to the lifecycle of the transmission line upgrade efforts as well as the renewable energy generation implementation in the electrical system.

Voltage support and overload conditions are typically the main drivers behind the transmission line upgrade processes. As the *de facto* alternative the majority of utility planners tend to disregard over new expansion projects, the category cultivates gradually more inferior over time until it develops into an emergency effort on the priority list. Nevertheless, the evaluation of transmission line upgrades encompasses three main paradigms as outlined below:

1. Reconductoring
2. New Transmission Line
3. Realignments

There are several derivatives and combinations to the general theme. However, the feasibility of each alternative is assessed for its validity. Reconductoring is typically conducted in stages to facilitate minimum service disruption. The improvements of associated equipment (fiber optic communication, etc.) are performed as necessary in accordance with the scope of work. The reconductoring activities generally provide less voltage support and capacity than a new transmission line or realignment; it relives the electrical system during periods of high load demand. The construction of a new transmission line is yet another alternative usually employed as a cost comparison as well as the overall impact to the system. The acquisition of a new right-of-way and additional access roads increases the project cost and time components. Nevertheless, the new line eliminates generation power flow limitations during n-1 and n-2 scenarios in some cases. The third alternative consists of realignment activities that deviate from the route of the existing transmission line and sited outside of the right-of-way. It is designed to reduce impacts to the land use. Some realignment avoids encroachments to cultural/historical areas and exploits existing structures to avoid transmission line conductors crossing one another. Although the realignment alternative satisfies the electrical requirements of the transmission line implementation, the residential impacts and right-of-way constraints are additional issues utility planners must consider.

The Point of Interconnection

The point of interconnection (POI) for the proposed generation is defined as the physical location on the electrical system at which the requestor owned facilities connect. The change of ownership is characterized at this juncture, and power metering transpires. The proposed generation is designed to maintain electrical delivery within the range of 0.95 leading to 0.95 lagging power factor at the point of interconnection. Some regions require a phasor measuring unit at higher voltage levels (typically 230-kV and above), to monitor the generator's performance. The unit provides high-speed digitized, time-synchronized voltage/current phasors as well as frequency measurements. A number of units capture data at a rate of sixty messages per second to record dynamic performance of the generator and comply with electrical/regional standards (IEEE Standard C37.118, 2005). Performance testing of the generator is required of the owner to characterize the plant responses to system disturbances and control signals. The transmission system must be tested to determine the effects of various types of severe contingencies on system performance to include voltage and angular stability. These tests are performed as a means to study the system for its ability to withstand disturbances beyond those, which would reasonably be expected. Recognition must be given to the occurrence of the Maximum Credible Disturbance criteria in neighboring systems as well and their effect on the system.

Examples of these less probable contingencies to be studied are:

- Unexpected loss of the entire generation capability of any station for any reason.
- Unexpected loss of all lines on a single right-of-way.
- Unexpected loss of a tower line with three or more circuits.
- Unexpected loss of all lines and transformers of one voltage emanating from a substation or switching station.
- Unexpected dropping of a large load or major load center.
- Failure of a fully redundant special protection system to operate when required.
- Operation, partial operation, or mis-operation of a fully redundant special protection system for an event or condition for which it was not intended to operate.

In the event these tests indicate potential for a cascading outage, an evaluation must be conducted to consider 1) the consequences to the system and adjacent interconnection of such a disturbance, 2) the general scope of a capital project(s) to correct the condition, and 3) the operating steps required to minimize the severity of the disturbance. Based on this evaluation, a decision will be made as to whether a capital project should be considered to mitigate the potential risk. The contingency must not cause system instability or cascading outages. In some regions with multi-element outages, it is acceptable to trip a unit(s) to maintain system stability, however all other generating units in the system must maintain transient stability during and after the contingency. A generating unit is considered to be stable if the machine rotor angle and speed curves are well damped by the end of a 15-second simulation. Tripping of units to maintain system stability must be accomplished via local protective relays. The choice of stability corrective measures such as faster relaying, independent pole operation of breakers, additional

transmission, restricted unit output or unit tripping must be made on an individual case basis after considering, probability of occurrence, severity of disturbance, and economics. Selected system voltages and system frequency shall be monitored as part of observing system stability.

The evaluation of the voltage and frequency deviations (as well as the power and voltage schedule) at the point of interconnection is essential data for the system study. These data is utilized to determine the operating transfer limits and network reinforcement for system reliability. The proposed generation must include protective relay systems appropriate for the point of interconnection. The protection scheme is expected to prevent tripping (i.e. distance relays, out-of-step blocking, etc) for stable swings on the interconnection electrical system. The protective relays are designed to detect phase and ground faults on the generator interconnection. Additionally, the owner of the proposed generation must address and formulate mitigation plans for power quality issues prior to interconnections. *Power quality* refers to a wide assortment of electromagnetic occurrence that exemplifies the voltage and current at a given time and at a given location on the power system. This includes degradation (harmonic distortion, flicker), switching transients, and fault-inducted voltage sags (IEEE Standard 1159, 1995). The term *transients* have been used in the analysis of power system dissimilarities for a long time. Its name immediately summons up the conception of an event that is detrimental but momentary in nature. The harmonics may cause increase thermal heating in transformers/reactors, mis-operations of solid state equipment, as well as creating resonant over-voltages (IEEE Standard 519, 1992). The harmonic levels are measured with specialized equipment designated to acquire individual samples of voltage and current waveforms and determine a probability distribution set. The inter-harmonics – distortion calculated in 10 Hz increments – are expected to be below 5% based on the probability distribution within a one-week period. The study of the limits of the associated equipment is expected *before* the new expansion interconnection occurs.

The emerging best practices for the wind turbine generator at the point of interconnection are outlined below (Hawkins & Rothleder, 2006):

- Low Voltage Ride-Through - Ability to Set Ramp Rates
- Power Curtailment Capabilities - Reserve Functions
- Zero-Power Voltage Regulation - Governor Functions
- Guaranteed Power Factor Range at the POI (Voltage Regulation)
- A Specified Level of Monitoring, Metering, and Event Recording

The wind turbine generator is required to remain in service during faults (3ø with normal clearing times 4~9 cycles or 0.15 seconds and 1ø to ground faults with a delayed clearing) as a parameter of the low voltage ride-through criteria. The subsequent post-fault recovery to pre-fault levels is a condition (in most regions) for new wind generation at the interconnection point. The ramp forecasting for hourly energy utilization is another requirement for system operation. Accurate metrological mechanisms are compulsory of the new wind generation to manipulate large ramps within the operational footprint. Because of the variability inherit in the wind technology; a specific level of regulation reserves is required. The operating reserve criteria must be included in the planning scheme contingency analysis parameters. The transmission system is developed to operate at the expected peak and at lower load levels such that the system will maintain voltage stability with the most severe combination of a generating unit and a transmission line removed from service. The system must be evaluated for the most severe combination of two generating units with an additional single transmission line repair *and* for the most severe grouping of one generating

unit plus two transmission lines out of service. Based on this evaluation, operating parameters such as area reactive reserve requirements can be developed. Moreover, a determination can be made as to whether a project should be considered to mitigate the potential risk to include special control schemes.

The inadequate frequency response capability of the wind generation technology (lack of a governor) raises concerns of several control areas. The speed governor system is necessary to regulate the output of the generator as a function of frequency. The speed governor must respond to the area frequency changes to assist with the power system stability. It is expected to have a speed regulation (droop) characteristic adjustable between three to seven percent (five percent is typical). Regional standards must be developed to include the governor requirements or sufficient generation to overcome the deficiency.

THE RECONDUCTORING PROJECT FORMULATED PROBLEM

The reconductoring efforts in one region is now examined from a cost perspective and illustrated in Table 2.

There are five major regions within the example electric utility footprint with a grand total budget of $7.69M for reconductoring projects. The internal labor, external (contract) labor, and materials/equipment are the major categories allocated for funding. Overall, the seventy-three projects are over budget by 10.91% year-to-date (first quarter assessments) and 8.10% over budget in the material and equipment categories. The year-to-date expenditures depict a 15% increase over the overall estimates as well. The material and equipment costs were 15.17% higher, on average, than the budgeted amount. The regional data is illustrated in panel A of Figure 1.

Further scrutiny of the regional data revealed a 65.86% increase in the actual spending above

Table 2. Regional transmission line reconductoring cost

Portfolio Region	Category	Grand Total
Eastern Region	Contractors	$593,300.00
	Labor	$99,151.75
	Materials & Equipment	$199,240.00
Eastern Region Total		$891,691.75
Northern States Region	Contractors	$543,599.00
	Labor	$283,171.56
	Materials & Equipment	$448,874.00
Northern States Region Total		$1,275,644.56
Central States Region	Contractors	$470,500.00
	Labor	$118,422.01
	Materials & Equipment	$247,249.00
Central States Region Total		$836,171.01
Northern Region Upper Branch	Contractors	$2,431,399.00
	Labor	$113,057.54
	Materials & Equipment	$1,861,472.00
	Other	$18,500.00
Northern Region Upper Branch Total		$4,424,428.54
Central Region Middle Branch	Contractors	$144,000.00
	Labor	$6,601.52
	Materials & Equipment	$120,000.00
Central Region Middle Branch Total		$270,601.52
Grand Total		$7,698,537.38

the budgeted amount in the internal labor categories. One particular project budgeted approximately $11.5k for the internal labor where the year-to-date actual expenditures equated to $1.31M (a comprehensible erroneous estimate for the required labor resource). Again, these internal labor estimates that devised the budget are roughly 67% on average below the actual spend for the twenty-three projects in the reconductoring category. Nevertheless, where contractors were assigned reconductoring tasks in this specific portfolio, the year-to-date actual spend was *under budget* by 13.31%. Details revealed, for a specific project, the maximum contractor budget to reconductor a 230-kV transmission line was still under budget by 17.02% at that point in time. An

early assessment of the overall expenditure levels exhibits prudent spending by the contracted labor and indiscriminate spending by the internal labor elements. The contractors were permitted to report and view a monthly budget whereas the internal labor did not. The over spending in the materials and equipment category is predicted to worsen as the reconductoring projects progress through the construction phases. The smallest of the projects material-wise is running above the estimates by 10% at the quarterly snap-shop juncture of the capital expenditures. The initial budgets were erroneous and created a moving forecasting target for the regional managers to predict. The TRIZ principle #2, specifically the "taking out" is outlined below (Mann, D. and Domb, E., 1997):

- Separate an interfering part or property from an object, or single out the only necessary part (or property) of an object.
  ○ Activity-Based Costing methodology instead of allocation accounting
  ○ Individual budgets for different departments
  ○ Business and Quality Goals prioritization
  ○ Lean Manufacture

○ Just-In-Time Inventory Management
○ Prioritize projects through the use of Gap Analysis, Pareto Analysis, etc.
○ Pareto principle of unequal distribution

The utilization of an earned value management system is another mechanism for the organization, planning, performance, measurement, and controlling the capital projects (Kerzner, H.,

Figure 1. Regional transmission line reconductoring cost data

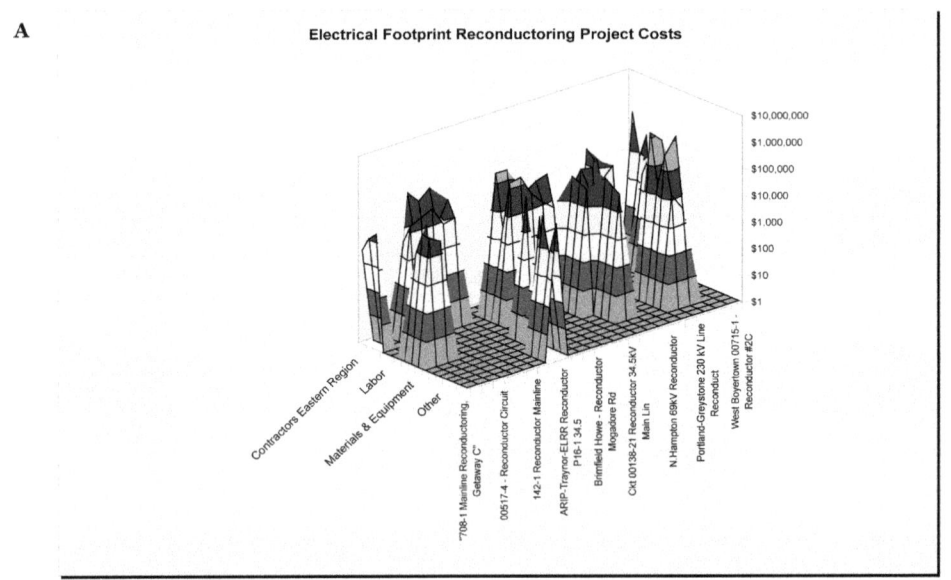

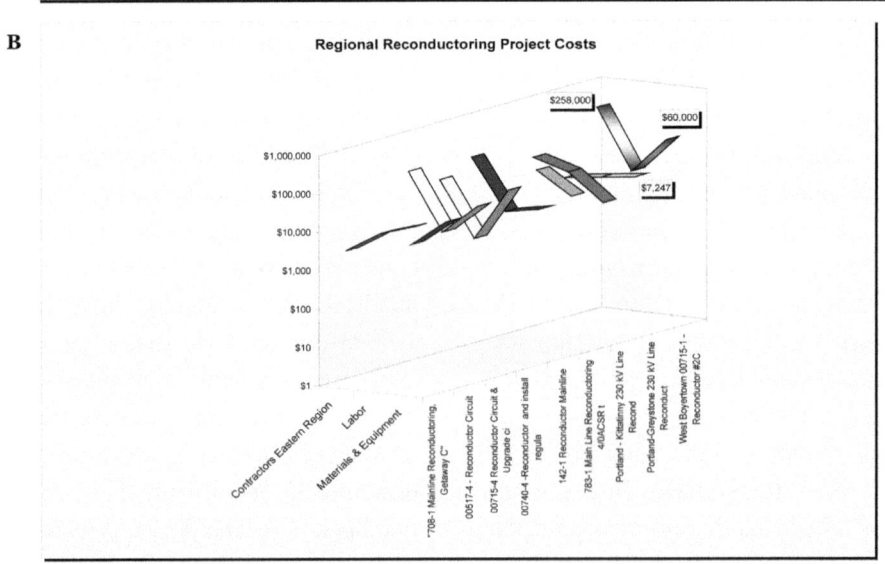

1992). The technical scope of work is logically integrated with the schedule and budget (preferably from a capital project lessons-learned/historical database) to formulate an approved project baseline. Accomplished work and accrued costs during project execution provide critical earned value information to measure performance for comparison to this baseline. The project health is examined by these earned value indicators. The measure is designed to facilitate the formation of a comprehensive, time-phased budget by thorough schedule planning and cost estimation. The time-phased budget plan encompasses all the individual work activities of the project accounts with the expenditures and resources required to accomplish the effort. The project manager – the individual closest to the work effort – must complete a periodic report for the status of the endeavor. This activity is designed to maintain the budget and schedule of the executed project. There are several variations of the theme and software programs available to perform the requisite calculations. However, the astute project manager is expected to gage the scheme's duration and cost based on performance to date. An example of the project status report is illustrated in Table 3.

The original budget of $109k is adequate for this particular project (CPI > 1), scheduled to finish ahead of time (SPI > 1). It is a logical and methodical process a project manager must evaluate, at the very least, on a monthly basis. The contract labor, applicable internal labor, and material estimates must account for the individual task duration embedded within the work breakdown structure of the project. Historical data and simulated modeling are imperative for the cost and time estimates to formulate the project budget. The estimate accuracy improves with the increased quantities of the historical data. Sensitivity analysis is utilized more with simulation modeling where the data is created via *what if* analysis. The high-level individualized project status report will eliminate the excessive spending in specific categories. An incentive system must be in place to reward project managers for main-

taining an accurate budget and schedule for higher-expenditure range endeavors. The system must not punish or remove future funding for projects that are under budget and ahead of schedule as most managers believe. Instead, the reward system must serve as a benchmark for similar capital projects as it resides in the historical and lessons-learned database.

Deeper analysis of one specific project in the Eastern Region, the highest budgeted contracted endeavor ($258k) for example, reveals the lack of a project management earned-value system. The most interesting data illustrated in panel B of Figure 1 is the labor value. The internal labor has exceeded the budget by threefold and the materials/equipment category is over budget by approximately 10%. Yet, the contracted labor is well under budget (approximately 45%) and forecasted to be on schedule with the reconductoring project if the preceding tasks are completed. In a region where the average spending levels are two-thirds above the budget, the resource utilization is clearly imbalanced. The specialty effort of a post-mortem reconductoring project is examined further in the process model.

THE ELECTRICAL TRANSMISSION LINE REBUILD PROJECT

Tell me and I'll forget, show me and I may remember, involve me and I'll understand. (Chinese Proverb)

The Bidding Process

The parameters of the endeavor encompassed a variety of entities performing simultaneously to produce the required output. The major advantage of this post-mortem evaluation were the pre-bid offers of the eight contractors poised to perform the work and the detailed cash flow analysis from the Federal Electric Utility Commission (FERC) codes work breakdown structure elements. The most interesting aspect of the bids were the wide cost range of these elements. An example of the non-standard estimates ranged from $0 to $1,120,469 for foundation work. A second example of these cost elements involved the wide range of the conductor installation estimates ($946,880 to $2,293,300). Ultimately, the inaccurate data were reflected within the sphere of the total bid. The lowest bid for the entire effort was approximately $3.46M whereas the highest bid hovered near the $7M mark. Change orders and contingencies immediately increased the winning bid to $4.85M before the project began.

Table 3. Project earned value status report

Original Budget for the Project:	**$109,000**	
At this date, you should have accomplished this much work:	54%	
At this date, you have accomplished this much work:	66%	
Your actual costs to date:	$44,145	
The Project Manager must perform the following calculations:		
Planned Value (PV)	$58,860	
Earned Value (EV)	$71,940	
Actual Cost (AC)	$44,145	
Schedule Variance (SV)	$13,080	
Schedule Performance Index (SPI)	1.222	
Are you *Ahead* or *Behind* schedule?	***Ahead***	
Cost Variance (CV)	$27,795	
Cost Performance Index, Cost Efficiency (CPI)	1.630	
Are you *Over Budget* or *Under Budget*, compared to your progress?	***Under Budget***	
Assume your progress and spending patterns are expected to continue throughout the remainder of the project.		
Budget at Completion (BAC)	$109,000	
Estimate at Completion ~ Cost (EAC)	$66,886	
Estimate to Completion ~ Cost (ETC)	$22,741	
At Completion Budget Variance (or Variance at Completion)	$42,114	
Do you expect the project to finish *Over Budget* or *Under Budget*?	***Under Budget***	
Next, assume your payment patterns are not expected to continue throughout the remainder of the project. Briefly, do you want to use the original budget for the remaining work?		
Budget at Completion (BAC)	$109,000	
Estimate at Completion ~ Cost (EAC)	$81,205	
Estimate to Completion ~ Cost (ETC)	$37,060	
At Completion Budget Variance (or Variance at Completion)	$27,795	
Do you expect the project to finish *Over Budget* or *Under Budget*?	***Under Budget***	
Next, assume that the work of this project is equally distributed over the entire project.		
Original Schedule for the Project	12	Months
SPI for the current project	1.22	
How long do you expect your project to be? Estimate at Completion (Time)	9.82	Months
At Completion Schedule (Time) Variance	2.18	Months
Do you expect the project to finish *Ahead* or *Behind* schedule?	***Ahead***	

The Initial Model Set-up

The major objectives of the Line Rebuild Model (Figure 2) are to minimize cost in three areas:

1. Labor (Engineering, Testing, Construction)
2. Materials
3. Contingencies (Labor, Material, Engineering)

A secondary objective of the model is to minimize time. The key parameters are outlined below:

4. Engineering (hours)
5. Construction Labor (hours)
6. Contingencies (Material (hours-wait/delay) and Labor (hours))

Process sequencing is extremely important in the project evaluation and assessment. The entire Line Rebuild project was developed to include twenty-five major processes with five contingent resources. The resource utilization and industry standard rates are depicted in Table 4.

A custom weekly master schedule was developed to evaluate each resource/capabilities to provide maximum efficiency for the project based on industry standards (Table 5).

The critical path equated to the maximum of 4936 man-hours. An examination of each major resource is in order to facilitate the process flow. The engineer consumes 965 hours in value-added activities to include:

- Design Engineering

Table 4. Resource utilization

Resources:	Rate:	Utilization:
Engr	$125/hr	6/25=24%
Line Constr	$105/hr	9/25=36%
Sub Constr	$95/hr	5/25=20%
Environ	$75/hr	4/25=16%
ROW	$50/hr	1/25=4%

Table 5. Custom master schedule: capacity

Day/Time:		Efficiency:
1 Mon	0730-1800	0.58
0 Mon	1800-2400	
1 Tue	0830-1730	0.72
0 Tue	1730-2400	
1 Wed	0900-1630	0.86
0 Wed	1630-2400	
1 Thur	0730-1800	0.98
0 Thur	1800-2400	
1 Fri	0730-1430	0.45
0 Fri	1430-2400	

- Relay Recommendations
- Support & Management Functions

The engineering cost estimates (derived from the actual cash flow analysis) were $120,600. The model set-up represents value-added activities where two or more entities were involved. Non-value added activities were classified as a one entity component. The line construction element consumed 1690 hours in activities to include:

- Conductor Installation
- Shield Wire/Motor Operated Air Brake Switch (MOAB)
- Grounding and Clean-up

The line construction cost estimates were in the order of $3.5M. The substation construction element consumed 725 hours in activities to include:

- Survey & Clearing
- Foundations/Structures
- Exit Hardware/Replacement

The cost estimates for this service were in the order of $68.7k. The environmental element consumed 896 hours in non-value added activities to secure permits with an estimated cost of $67,166.

Figure 2. High-level transmission line rebuild process map

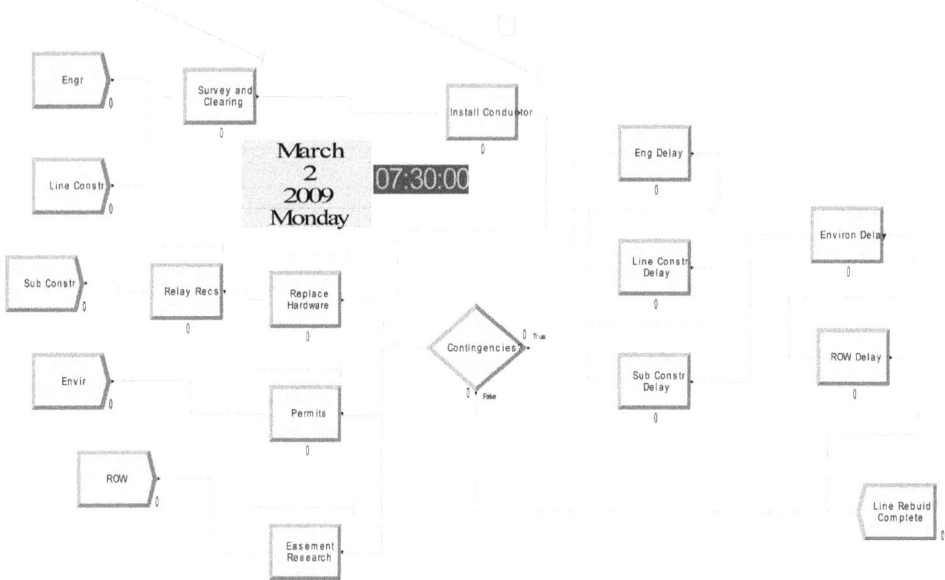

The final element (Right-of-Way) consumed 660 hours of non-value added activities with an estimated cost of $33k for easement research. The contingency components typically are assessed a ten percent rate of the cost estimates. The Line Rebuild Model utilized the non-value added approach with a delay function for each major area. The unique contingency region occupied a special position within the process map, to include a decision block (Table 6). It is a two-way by condition attribute to determine if the non-value added cost exceeds $1500, to impose the delay with specific holding costs (not to be confused with waiting costs). This is an important tenet for the overall Key Performance Indicators.

THE PROCESS FLOW BACKGROUND

The parameters of the model encompassed the five major resources and their related cost functions. An example of these parameters is depicted below and in Table 7.

- $25/hr holding cost
- $125/hr Value-Added (VA) cost
- $75/hr Non-value added (NVA) cost
- $30/hr Waiting cost

The secondary aspect of the resource entity developed its statistical time on the project based on a pre-determined schedule. It was derived from Tables 1 and 2 grounded in the industry standards. An example of the entity/type parameter is outlined below and Table 8:

- Engr Labor
  - Based on Schedule
  - 0.24 Busy/hr
  - 0.72 Idle/hr

The major processes are developed in the model from industry standards and statistical analysis.

The allocations (VA/NVA) are carefully chosen in the model to mimic the actual project construction. As a secondary function, the action category captures the entity based on its specific criteria with its associated priority (low, medium, high).

Table 6. Decision parameters

Decision Parameters					If			Cost ($)
Contingencies?	2-way by Condition	50	Attribute	Variable 1	Entity.NVACost	Entity 1	>	1500

An example of the major process parameters is depicted below and Table 9.

- Install Conductor
  - Seize Delay Release … The process action will *seize* the entity, *delay* it in accordance with the prescribed parameters, and *release* it to the next process.
  - Value-Added Allocation… Minimum (hrs) 820, Value (hrs) 965, Maximum (hrs) 1060

The amount of time scheduled for each resource on the Line Rebuild Project is displayed in the below table as well as the creation/initiation blocks within the process map. An example of this parameter is depicted below and Table 10:

- Environ Labor
  - 896 hrs (value)
  - Infinite (entities until maximum hours are achieved in the replication)

The most important ideology of the technical report is The Key Performance Indicators Summary Sheet. It is designed for executive-level decision-making for high expenditure capital ventures. When utilized in conjunction with the executive summary as well as the decision analysis summation tree, the Key Performance Indicators are a powerful tool for front-end project evaluations. The major objectives are to identify vulnerabilities/bottlenecks and segregate available resources for the project.

The total system cost for the line rebuild project is $423,378 to produce one unit after thirty

replications of the model on the critical path. This figure includes resource/all entity costs displayed in tabular form and pie charts. A major concern for the line rebuild project in its current state is the "wait costs" category ($353k). The resources are waiting for one process to complete before another begins. Further drill down of the data will provide specific details for excessive costs. Another area of consternation in the model lies within the non-value added cost of $30,299. Approximately $5k is consumed in idle costs versus $1805 in busy costs. Ideally, the wait and idle costs are minimal.

The Time Parameter

The Substation Construction category provided the highest value added time (184.67 hrs) for the line rebuild project. Conversely, the environmental labor consumed the most non-value added time (1384.86 hrs). A thorough evaluation of the labor component is warranted to alleviate the potential problem. An ancillary function is required to

Table 7. Entity parameters

Entity Parameters	Picture	Holding Cost/hr	Value Added Cost	Non-Value Added Cost	Waiting Costs
Engineer	Picture. Man	25	125	75	30
Line Construction	Picture. Truck	22	105	58	27
Sub Construction	Picture. Box	20	95	48	23
Environ Labor	Picture. Woman	15	75	38	20
ROW Labor	Picture. Van	18	50	25	17

Table 8. Entity utilization

Entity	Type			Busy/hour	Idle/hour	
Engr Labor	Based on Schedule	Line work	Preempt	0.24	0.72	0
Line Constructor	Based on Schedule	Line work	Preempt	0.36	0.64	0
Sub Constructor	Based on Schedule	Line work	Preempt	0.2	0.78	0
Environmental Labor	Based on Schedule	Line work	Preempt	0.16	0.83	0
ROW Laborer	Based on Schedule	Line work	Preempt	0.04	0.93	0

secure the permits *before* the project start as a viable solution. The average "wait time" for the environmental labor (2246.11 hrs) is exceedingly high for the overall project contribution. When taken in its entirety, the average "total time" of the category was 3631 hrs compounded with the NVA cost ($38/hr) as well as the waiting cost ($20/hr) and holding cost ($15/hr), the environmental labor problem can exacerbated the entire system cost, as the simulated model demonstrated.

The Cost Entity

The model for the line rebuild project exploits the vulnerabilities of the entire process. The category with the highest value-added cost is the substation construction area ($3794.14) on average. Conversely, the environmental labor category consumed ($20,953.67) in non-value added labor costs. As previously discussed in the above paragraph, the time-cost components are intertwined in

Table 9. Entity allocation

Major Processes	Type	Action	Priority	Units	Allocation	Minimum (hrs)	Value (hrs)	Maximum (hrs)
Survey and Clearing	Standard	Seize Delay Release	High(1)	Hours	Value Added	720	800	880
Install Conductor	Standard	Seize Delay Release	High(1)	Hours	Value Added	820	965	1060
Relay Recs	Standard	Seize Delay Release	High(1)	Hours	Value Added	450	500	550
Replace Hardware	Standard	Seize Delay Release	High(1)	Hours	Value Added	110	125	140
Eng Delay	Standard	Delay	Medium(2)	Hours	Non-Value Added	86	96	106
Permits	Standard	Seize Delay Release	High(1)	Hours	Non-Value Added	810	900	990
Easement Research	Standard	Seize Delay Release	High(1)	Hours	Non-Value Added	600	660	750
Line Constr Delay	Standard	Delay	Medium(2)	Hours	Non-Value Added	155	170	185
Environ Delay	Standard	Delay	Medium(2)	Hours	Non-Value Added	80	89	99
Sub Constr Delay	Standard	Delay	Medium(2)	Hours	Non-Value Added	60	72	80
ROW Delay	Standard	Delay	Medium(2)	Hours	Non-Value Added	60	66	72

Table 10. Resource value

Resources				Value (hrs)					
Engr	Engineer	Constant	Line work	965	1	Hours	1	Infinite	0.00000001
Line Constr	Line Construction	Constant	Line work	1690	1	Hours	1	Infinite	0.0001
Sub Constr	Sub Construction	Constant	Line work	725	1	Hours	1	Infinite	0.00001
Envir	Environ Labor	Constant	Line work	896	1	Hours	1	Infinite	0.001
ROW	ROW Labor	Constant	Line work	660	1	Hours	1	Infinite	0.01

the process. The associated wait cost ($33,711.61) for the environmental labor is exceedingly high for the project. When contributed to the *total system cost* ($54,740.20), it gives the impression as the category warrants elimination or re-designs in the process. However, this is only one category of the many facets in the system evaluation.

Process Utilization

The number of units processed per resource range from two (line construction) to seven (ROW Labor) units in the "number in" category after 30 replications of the simulated model. The number of units in/out summation sheet is utilized to evaluate resource usage/level for the vulnerability study. The "work-in-process" summation sheet provides a snap-shot of the resources and average units in process after the final replication. The ROW labor (4.25 avg. units) exhibited the highest amount whereas the Line Construction resource (1.97 avg. units) displayed the lowest number.

Time per Entity

The detailed analysis of the model exploits system weakness in various formats. The Survey and Clearing category represent the highest value-added time per entity (802.53 hrs). Conversely, the Permitting process produced the largest (892.07 hrs) non-value added time in the system. When taken as a whole, the environmental component is a potential problematic area (previously discussed), as compared with the other entities (all less than

172 hrs). The "wait-time per entity" category depicts an exceedingly high (3936.25 hrs) wait time for the relay recommendations which is an input for the total time per entity (4419.96 hrs). The particular potential problem was undetected in earlier analysis and necessitates a thorough exploration.

Accumulated Value-Added Time

The Accumulated Value-Added Time is a practical tool utilized to authenticate specific time groupings. The Survey and Clearing function is the highest (802.53 hrs) accumulated VA Time in the Line Rebuild Project. The Conductor & Installation element is the lowest (0 hrs) VA Time in the model. The Permit category (an earlier identified problematic area) exceeds 892 hrs of non-value added time on the summation sheet. The relay recommendations (3936.25 hrs) were depicted as the highest value in the accumulated wait time summary and the total accumulated time (4419.96 hrs) as well.

Cost per Entity

This is an important summary that captures both value-added & non-value added cost components. The decision-maker can readily identify the key resources/operations in the system that adds value, as well as the non-value added process cost elements. The Line Rebuild Project depicts the Survey & Clearing process as the highest value-added element ($20,544.64 ~also reflected in the accumulated VA cost) and the Permit operations ($13,523.84 ~also reflected in the accumulated NVA cost) as the highest non-value added activity. The second highest NVA cost ($2567.17) was created in the Line Construction Delay category which may cause some concern. The Relay Recommendations generated the highest ($79k ~also reflected in the accumulated wait cost) wait cost per entity on average and contributed to the total cost per entity ($88k ~also reflected in the total accumulated cost). The maximum value is utilized as a budget threshold to monitor potential cost overruns.

Number In/Out

The summary sheet displays the number of incoming units in the process to produce the pre-determined output. It is utilized as a resource-leveling mechanism to determine excessive man-hrs per process. The Line Rebuild Project displays seven incoming units (highest) for the Easement Research and Survey & Clearing categories as compared with one unit on average for the other major processes.

Waiting Time Queue

One method to effectively evaluate the efficiency of a system is to exam the waiting queue. It will determine the potential bottlenecks and process capacity. The line rebuild project displays all of the major queues per process. Based on the process flow of the project, the easement research (3138.17 hrs) and $56k, relay recommendations, (2821. 23 hrs) and $56k, and survey cleaning (2226. 26 hrs) and $49k queue consumed the most time as well as associated costs (reflected in the high-level key performance indicators). The former can be performed as an ancillary activity *before* the project start as a possible solution. The number of units waiting in these queues varies from 1 to 4. An additional resource in each of the potential problem areas will alleviate the bottlenecks.

Resource Usage/Number Busy/Schedule Utilization

The average instantaneous utilization (number busy) for the entire line rebuild project is adjustable in accordance with the simulation model.

Table 11. Core process comparisons

	Engineering Design	Line Construction	Substation Construction	Environmental	Right-of-Way (ROW)	Sub-Totals	Totals with 10% Estimated Contingencies
Estimated Hours	965	1690	725	896	660	4936	5429.6
Total Time per Entity with delays ~ (hrs)	3087.9	2992.37	4620.81	3228.32	65.5	13994.9	-----
Total Cost per Entity with delays ~ ($)	$72,921.68	$74,056.24	$92,527.55	$48,567.59	$982.64	$289,055.70	-----
Estimated Cost	$120,600.00	$130,200.00	$68,750.00	$67,166.00	$33,000.00	$419,716.00	$461,687.60
Percentage of Specialty Effort ~ (estimated hrs)	19.55%	34.24%	14.69%	18.15%	13.37%	**Process Cost Avoidance:**	$172,631.90
Percentage of Specialty Effort ~ (actual hrs)	22.06%	21.38%	33.02%	23.07%	0.47%		
Percentage of Specialty Effort ~ (estimated cost)	26.12%	28.20%	14.89%	14.55%	7.15%		
Percentage of Specialty Effort ~ (actual cost)	25.23%	25.62%	32.01%	16.80%	0.34%		

Figure 3. Time comparison

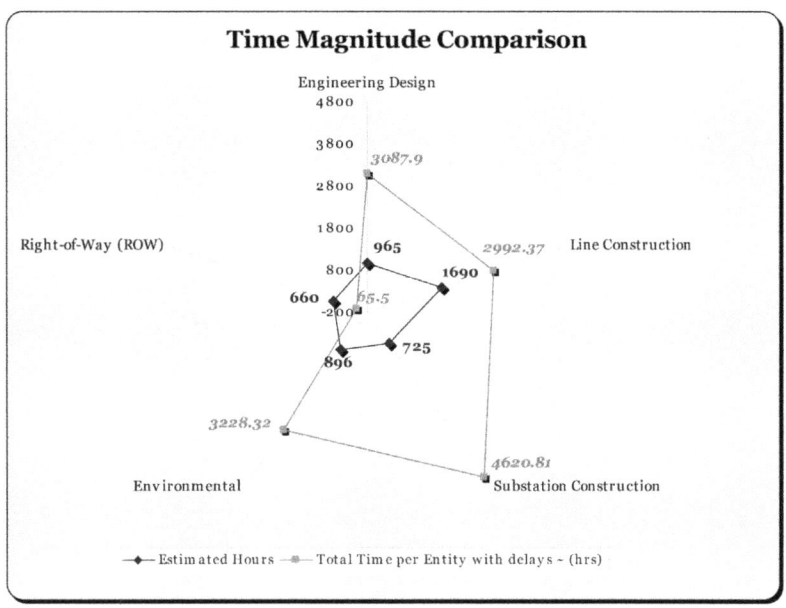

Figure 4. Cost comparison

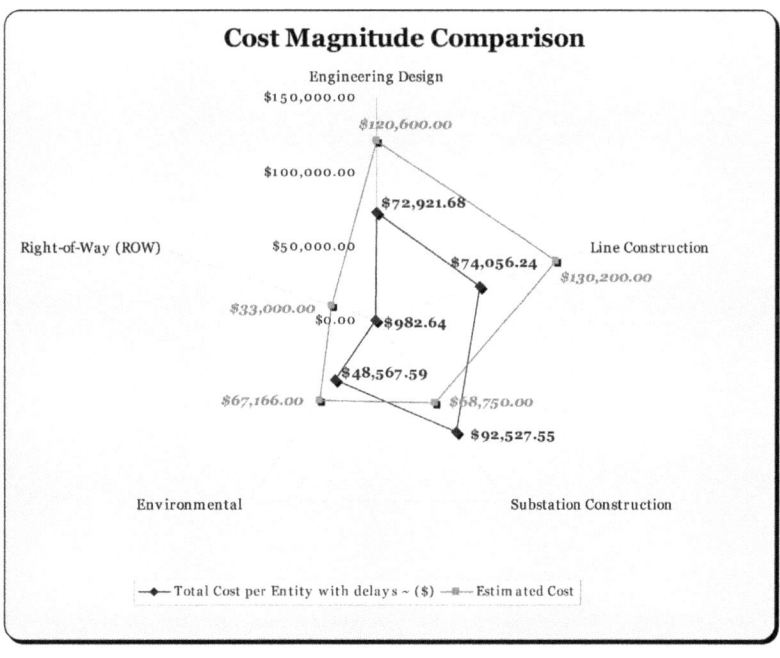

Figure 5. Time percentage comparison

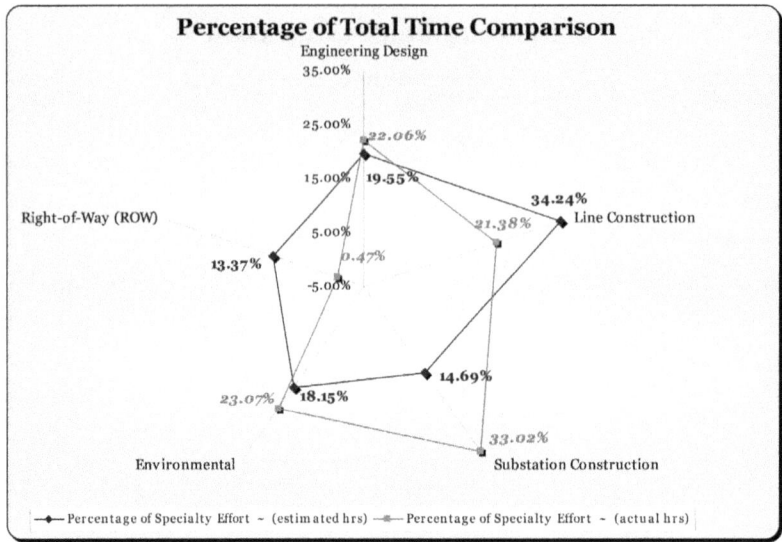

The ROW laborer (busy parameter set 4%) for instance consumes no busy time whereas the Engr labor is busy 0.7238 of the time. Resource leveling techniques and process re-evaluation to *fine-tune* the system (sensitivity analysis) are solutions to the potential problem. Further analysis reveals the number scheduled (0.4589) and the scheduled utilization may increase slightly to accommodate the system processes. An example of this method increased the line construction schedule from (0.4318) to a more acceptable (0.8550) schedule utilization. Moreover, a reduction in the Engr labor from (1.577) to a level on par with the other processes (0.867) will balance the system as well.

Figure 6. Cost percentage comparison

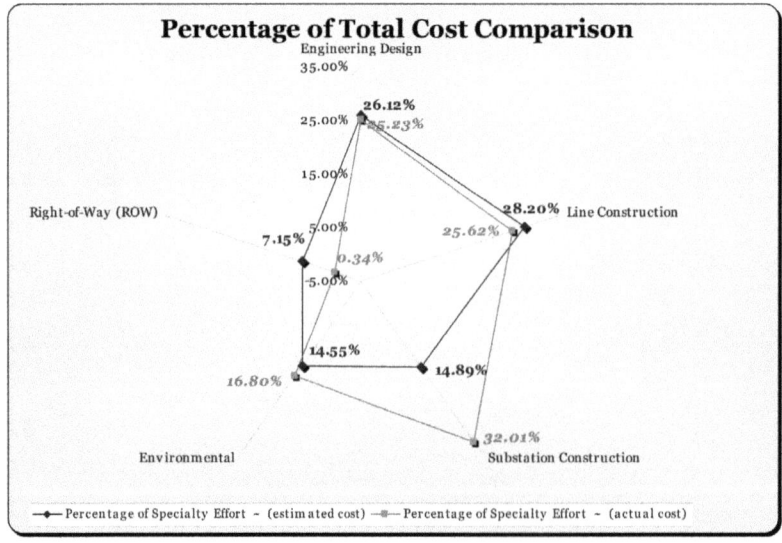

The busy costs associated with each resource (less than $857) are quite-low when compared with the highest idle cost ($2106 for ROW laborer).

Total Number Seized

On average, two units are seized by each resource save ROW labor (0) and Engr labor (2.96) resource leveling techniques and process re-evaluation are utilized to re-balance the system.

RESULTS

The industrial operations process analysis of the Line Rebuild project is an important cost-reduction mechanism to evaluate system integration. The value-added evaluation process provides the necessary front-end analysis for prudent decision-making of capital improvement endeavors. The original estimates are compared with the model processes in the Table 11.

The processes revealed interesting estimation data, methodology, and accuracy as summarized below:

The data clearly depicts lower time estimates with exceedingly high costs. The accuracy of the model is more precise over the original estimates. Once the process sequencing and resource-leveling techniques are implemented, the element estimates are expected to be exact.

The Time Comparison Magnitudes of the Line Rebuild project is captured in the radar diagram (Figure 3).

The estimated hours to complete the work effort are well under-schedule. The ROW tenet only consumed 65.5 hours (compare 660 hrs. of the original estimate) as the lone caveat. In order to produce one unit with the required resources and schedule, the "red" data points are the norm for the Line Rebuild project. The Process Cost Savings alone ($461,687.60-$289,055.70) equates to a capital reserve of $172,631.90. The detailed cost estimates are in the diagram depicted (Figure 4).

Although the delays are embedded within the model to emulate accurate costs, the original estimates far exceeded the actual work effort. The only caveat is within the Substation Construction area where an *under-estimation* event occurred. The Engineering Design effort (especially within the relay recommendation sphere) and the Line Construction tenets were well over the cost estimates. A thorough review is compulsory to avoid such extraneous cost overruns. The time percentage is depicted in the diagram (Figure 5) below:

The original estimates were inaccurate by an order of magnitude to facilitate the overall scheduling problems. The cost diagram (although a bit more precise) failed to address the potential over-budget problems as well (Figure 6 below). In retrospect, the entire bidding process and cost paradigm was based on the "minimum-time/maximum-cost" criteria. The original estimates did not produce the desired financial results.

If the processes were implemented in the Line Rebuild project, the Project Mgmt ($120k), AFUDC category ($440k), and the contingency cost category ($380k) would experience a drastic reduction. In fact, $517k project cost avoidance ($940k actual -$423k model system cost) is realized with the methodology, *before* the implementation of resource leveling. The study proved that the front-end project analysis is a viable tool for intense capital integration developments.

REFERENCES

Barker, P. P., & Short, T. A. (1996). Findings of recent experiments involving natural and triggered lightning. IEEE/Power Engineering Society, Transmission and Distribution Conference, Los Angeles, CA.

RUS Bulletin 1724E-203. (1994). *Guide for upgrading rural service transmission lines.* U.S. Department of Agriculture Rural Utility Services.

Clapp, A. L. (2006). *NESC handbook: A discussion of the national electric safety code* (6th ed.). New York, NY: The Institute of Electrical and Electronic Engineers, Inc.

DOE/GO-102008-2567. (2008). *20% wind energy by 2030: Increasing wind energy's contribution to U.S. electricity supply.* Washington, DC: U.S. Department of Energy, Office of Energy Efficiency and Renewable Energy.

Fink, D. G., & Beaty, W. H. (2001). *Standard handbook for electrical engineers* (14th ed.). New York, NY: McGraw-Hill Inc.

Harness, R. (2000). *Raptor electrocutions and distribution pole types.* Fort Collins, CO: North American Wood Pole Coalition Technical Bulletin.

Hawkins, D., & Rothleder, M. (2006). *Evolving role of wind forecasting in market operation at the CAISO. Power System Conference and Exposition* (pp. 234–238). Piscataway, NJ: The Institute of Electrical and Electronic Engineers.

IEEE Standard 1159. (1995). *Recommended practices for monitoring electric power quality.* Piscataway, NJ: The Institute of Electrical and Electronic Engineers.

IEEE Standard C37.118. (2005). *Enclosed field discharge circuit breakers for rotating electric machinery.* Piscataway, NJ: The Institute of Electrical and Electronic Engineers.

IEEE Standard 519. (1992). *Recommended practices and requirements for harmonic control in electrical power systems.* Piscataway, NJ: The Institute of Electrical and Electronic Engineers.

Johnson, B. W. (2005). *Utility restoration cost recovery.* Washington, DC: Edison Electric Institute.

Jondahl, D. W., Rockfield, L. H., & Cupp, G. M. (1991). *Connector performance of new vs. service aged conductor.* IEEE/Power Engineering Society Transmission and Distribution Conference, Los Angeles, CA.

Kerzner, H. (1992). *Project management: A systems approach to planning, scheduling, and controlling* (4th ed.). New York, NY: Van Norstrand Reinhold.

Kretschmer, R. K., & Hundrieser, K. E. (1996). Reliability: What level and what price? *Public Utility Fortnightly, 139*(20).

Mann, D., & Domb, E. (1997). *40 inventive (business) principles with examples.* Retrieved 8 November, 2008, from http://www.triz-journal.com

Simpson, P., & Van Bossuyt, T. R. (1996). Tree caused electrical outages. *Journal of Arboriculture, 22*(3), 117.

ENDNOTES

1. Source: *Big Eddy-Knight Transmission Project Update,* Fact Sheet, Bonneville Power Administration, DOE/ BP-4131 Portland, OR. (December 2009)
2. Source: *Vegetation-Related Transmission Outage Report,* North American Electric Reliability Corporation, First Quarter 2009 (released May 22, 2009).
3. Source: U.S. Department of Energy/National Electric Transmission Congestion Study/2006.
4. Source: ISO New England, *Overview of Smart Grid – Policies, Initiatives, and Needs* (17 February 2009).
5. Source: *Transmission Projects Supporting Renewable Resources,* Edison Electric Institute, Washington, D.C.
6. TRIZ is a Russian language acronym for *Teoriya Resheniya Izobreatatelskikh Zadatch.* Translated into English it means "The Theory of Inventive Problem Solving."

Chapter 8
Permit Issuance Process Evaluation

ABSTRACT

The most frequent customer complaint is the amount of time it takes to process and issue a permit. The obvious goal is to reduce the time it takes from submittal of an application to the issuance of the necessary permits to be able to start and complete construction. It is an improvement process designed to meet customer expectations while maintaining assurance that the public benefit is achieved. Seven various scheduling and cost configuration scenarios are analyzed to determine the optimal resource leveling condition(s). The foundation is developed for the strategist to assist with the time and cost trade-off evaluation involved in capacity planning for permit processing.

INTRODUCTION

A good leader can't get too far ahead of his followers. (Franklin Roosevelt)

The permit issuance process in the electrical utility industry involves various entities working in concert to provide, review, and implement the document. There are several layers to the process and oftentimes creates unnecessary as well as extensive delays. The different permit types and processing procedures can lead to the long delays and subsequently increase the cost in a capital-intensive project. Some of the permit types and cross-functional document review are listed below:

- Right-of-Way - Deeds
- Vegetation Control - Option Agreements
- Storm Water - County & Local Permits
- Temporary Construction & Access License

The external entities and processes involved with the permit lead to the affected landowners, legal, court filing, and local planning & engineering departments. In fact, many of the system impact studies for wind farm integration conducted by the regional independent service operators include an overall estimated cost for permit fees. However, this figure does not encompass the cost avoidance based on the permitting delays. It is beneficial to evaluate these assumptions in order to develop the

DOI: 10.4018/978-1-4666-2839-7.ch008

true capital project costs. Permitting delays are inevitable in some endeavors and can be mitigated if the correct systems are in place. Redundancy planning and parallel work efforts of non-permit activities are methods utilized to optimize the processing system and avoid enormous delays. The majority of the renewable energy system integration efforts require strategic equipment upgrades within the current U.S. infrastructure. It is expected that the permitting process for the integration work effort to be as seamless as possible to further advance the cause. For example, the development of offshore wind energy in the U.S. currently face challenges with permitting and siting issues (Van Cleve & Copping, 2010). The developers cite the arduous process of permitting one of these sites and a high-level of uncertainty in the arena. There are also challenges related to gaining public acceptance through outreach activities when the offshore wind energy project is proposed. The current process requires streamlining from seven-to-nine years as well as reducing the related costs ($2~3M) based on the environmental studies of operating offshore. The permitting process must be expedited to reduce the associated agency long response times and lower barriers for other developers to enter the arena. The consolidation of similar agency requirements is mandatory to support the streamline process for permit issuance.

The external entities play a vital role in the permit process – from the federal level to the local inspection areas. The capital project is interrelated to the permitting and siting processes and dependent on its outcomes. Specifically, the time required reviewing and issuing the document has the potential to idle expensive resources. Consequently, the planning aspect is essential to reduce the wait and idle times typically associated with capital-intensive ventures. Once the time constraints are identified and reduced to a feasible measure, the associated costs are expected to decrease proportionally. Additionally, the implementation of the managerial soft

skills *"Triple C"* methodologies (communication, coordination, and cooperation) embedded in the process will reduce these costs even further. The team members must be able to frame the problem and analyze solutions. Moreover, team members are expected to communicate these insights and cooperate with each applicable stakeholder to implement vital decisions related to the business situation (Leonhardi, 2011). When several departments are contributing to the process, only one may create a delay within the entire system. By developing specific system control limits, documented best practices, and Lean Quality measures, the requesting company is ensured a timely, cost-effective product from the process. Once the permit is issued, the company is expected to comply with the specific provisions in the document. Many of these stipulations are interrelated with other agencies, permits, and statues germane to the project. The *"Triple C"* methodologies are expected to dominate in these situations.

Essential to the permitting process is communication at *all* levels within the system. Most projects fail because of the lack of inter-departmental communication. For example, the planning commission requires a fourteen day review period of a company-submitted plan and profile drawing for a proposed high-voltage line reconductoring project. *How is the feedback established? What method is used? How are changes implemented?* Theses are only a few of the critical questions that must be asked and developed in a viable communication plan with external entities. Moreover, clear identification of the tasks involved is an effective measure of the resource utilization in the permitting process. The correct *coordination* of various departments performing together is also essential for permit issuance. How well these entities *co-operate* in the course of the permit process is the final leg of the *"Triple C"* triangle. The department of natural resources (DNR), for example, is expected to review and assess the findings and recommendations of regional experts of migratory animals within the proposed project right-of-way.

The discourse between the two entities ought not to delay the permit if the DNR disagrees with the results (there are other remedies to this situation that includes historical data, threatened/endangered species delisting, and the current federal register). The incisive decision-maker has developed mitigation strategies to include the previous and similar regional projects that involved migratory animals with the correct resolution. These efforts are designed to facilitate the permitting process and reduce the delays. The anticipated delay factor by the requesting company must be as proactive as possible to obtain the permit with effective instruments.

The General Process

As a subset of the overall electric utility permit issuance process, consider the regional development agency that has a building department tasked with the review of construction plans and inspections to assure compliance with codes and local ordinances. The most frequent customer complaint is the amount of time it takes to process and issue a permit. The obvious goal is to reduce the time it takes from submittal of an application to the issuance of the necessary permits to be able to start and complete construction. It is an improvement process designed to meet customer expectations while maintaining assurance that the public benefit is achieved. The parameters of the endeavor encompassed a variety of entities performing simultaneously to produce the essential output ~ the issuance of a permit. The objective is to minimize the time required to issue the permit utilizing quality tools and methodologies. A precept of this procedure is embedded within the Continuous Process Improvement technique employed to produce an efficient, streamlined application of the permit issuance progression. The strategic evaluation exploited several other quality techniques to include Lean (particularly the 5S philosophy to eliminate the non-value added activities), Earned Value Management,

Statistical Process Control (for process tracking), and Six Sigma to improve the quality by reducing the variation in the process. Although the preliminary investigation produced satisfactory results in reducing the time to issue a permit, the added component of cost reduction via resource balancing will produce greater efficiencies inherent within the process.

The Initial Model Set-up

The preparation of the process layout for permit issuance is a vital element in the success of timely approvals in the Building Department. Equally important are the key integration processes involved in the resource utilization and contingency phases of the endeavor. The integration of man, material, machinery, and methods require an intricate balancing act to maintain a specified budget, as well as timely permit issuance. Consequently, the most important objectives are to minimize the costs (labor, material, contingencies) and time (engineering, planning, and contingency labor) for successful permit issuance. The major objectives of the Permit Issuance Process Model (Figure 1) are to minimize cost in three areas:

1. Labor (Engineering, Planning, External Resources)
2. Materials (permit)
3. Contingencies (Labor, Material, Engineering)

A secondary objective of the model is to minimize time. The key parameters are outlined below:

4. Engineering (days)
5. Planning/External Labor (days)
6. Contingencies (Material (hours-wait/delay) and Labor (days))

The resources are shown to the left of the diagram at the initial stages of the process whereas the completion block is depicted on the right. The sequential steps of the main process include the

Figure 1. High-level permit process map

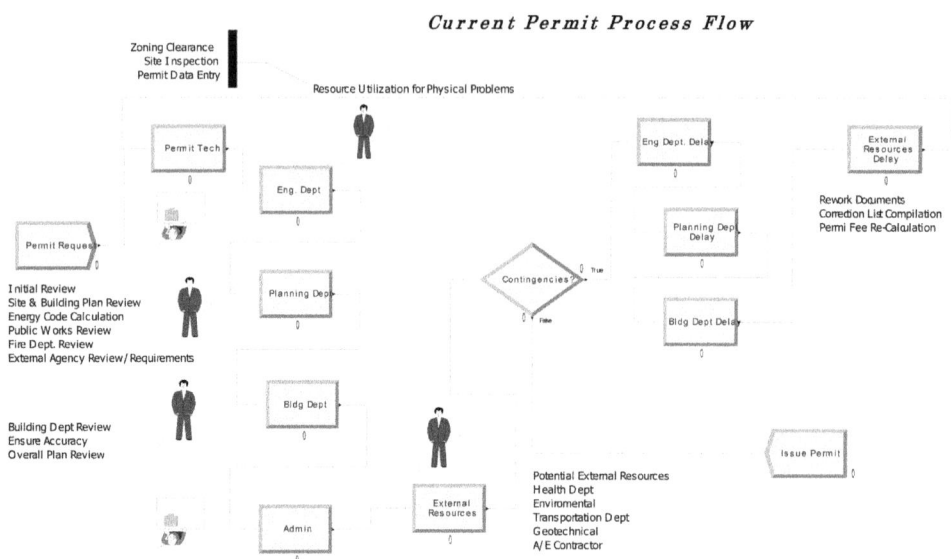

individual resource and the associated work time to perform the task(s). Waiting queues were established at specific points to maintain continuity as well as to develop statistical data. The contingency delay area captures the material/resource via an exceed condition in the system.

Process sequencing is extremely important in the progression evaluation and assessment. It is a key component to determine and exploit the bottlenecks (Theory of Constraints Technique) in a process. The entire Permit Issuance Process was developed to include twenty-five major processes with four contingent resources. Each process encompasses the designated resource(s), accumulated time/cost within the stage, and the "wait/delay" time & cost associated with the entity. These processes are developed in the model from industrial standards, previous project system integration, and financial statistical analysis. The permit issuance preparation is an essential phase for the entire scheme. As incorporated processes are implemented in the work plan, the associated resource costs are captured as well. The material (permit) itself is the main focus in the model and act as a viable resource for value-added costs.

If the material is delayed (especially crucial for expedited requests) the process system cost increases substantially via mis-aligned and idle resources. The high-level permit process analysis accurately detained these costs and provided ardent recommendations for procedure improvement. The resource utilization and industry standard rates are depicted in Table 1.

The model is based on the material (permit) cost/time and its affect on the resources managing it. A positive rate ($295) was assigned to the material as value-added (VA) cost in the system with associated holding cost/hour ($50), non-valued added (NVA) cost ($65), waiting cost ($25), and "other" (transport/transfer) costs. The standard rates for the resource labor (engineer, planning, and external resources) are captured in Table 1 as well to include the related (VA, NVA, Hold, Wait, Other) costs, to produce one unit. Each entity was modeled to a prescribed schedule for the 64 day work completion. The appropriate resource busy/hour and idle/hour allocation was also modeled from industrial statistics for this specific project and captured as capacity.

Table 1. Resource utilization

Resources:	Rate:	Utilization:
Engr	$125/hr	6/25=24%
Planning	$105/hr	4/25=16%
Bldg Personnel	$95/hr	5/25=20%
Permit Tech	$75/hr	4/25=16%
Admin	$50/hr	1/25=4%
External Personnel	$80/hr	5/25=20%

A custom weekly master schedule was developed to evaluate each resource/capabilities to provide maximum efficiency for the process based on industry standards (Table 2).

The contingency delay is solely based upon the non-value added cost (>$985) of the material (permit) wait time. The cost includes overhead expenditures of retaining the material in the system as well as the associated costs of continuing to hold any resources it owns. The contingency is realistically based on the resource availability and dedication to the process.

THE GENERAL EVALUATION APPROACH

Many of the process were combined to accommodate the maximum allowable time for the permit issuance general procedure. A mixture of high and medium priorities was utilized based on the work effort and system requirements. The resource allocation was divided by the single entity (non-value added) working within the process to multiple entities (value-added) performing the tasks. The main elements contributing to the elimination of non-value adding activities are depicted below:

- Excess processing
- Excess production
- Delays
- Transport
- Inventory
- Defects
- Movement

The act to minimize movement (Lean), organize/separate the needed from the unneeded (sort), and to maintain/monitor the process (standardize) encompasses the major tenets of the 5S philosophy (George, M.L., 2003). Consider the *sort* category in the 5S viewpoint for example. It is exploited by the strategist to determine the frequency of use of the material request. Consequently, a label is attached to the permit request for unused portions within a specified period in the overall process. Once the *sort* stage is complete, the strategist can determine the "how many" of an item is required and "where is the best location (task)" in relation to the system. This is considered the *set to order* stage on the process evaluation. The employees can identify the processes that require attention in the *shine* stage of the process and provide the requisite feedback. The *standardize* stage of the 5S philosophy involves employee participation to maintain a schedule for upkeep of the process. The finally stage, *sustain*, permits the collaboration between the employees/management and organization of the process. It is designed to maintain the positive gains experienced in the improvement efforts and

Table 2. Custom master schedule: capacity

Day/Time:		Efficiency:
1 Mon	0730-1800	0.58
0 Mon	1800-2400	
1 Tue	0830-1730	0.72
0 Tue	1730-2400	
1 Wed	0900-1630	0.86
0 Wed	1630-2400	
1 Thur	0730-1800	0.98
0 Thur	1800-2400	
1 Fri	0730-1430	0.45
0 Fri	1430-2400	

avoid the old, detrimental habits encountered in the prior condition (Vanessa, I., 2007).

The most likely value for the material delay (seize, delay, release) is approximately 12 days (maximum 18 days). The delay processes are all classified as non-value added activities as well. A decision-maker may utilize these data to formulate minimum amount of time to complete the issuance process as a benchmark for contract incentives, etc. The preliminary action of the permit issuance process involved critical pre-planning strategies to include the following TRIZ[1] principles (Mann, D., and Domb, E., 1997):

- **Preliminary Action:** Perform, before it is needed, the required change of an object (either fully or partially).
  - Where circumstances permit, perform non-critical path tasks early.
  - Use of "storyboarding" to facilitate creative problem solving (i.e. gathering the data before the "creativity" session(s) typically found in change management activities)
- **Preliminary Action:** Pre-arrange objects such that they can come into action from the most convenient place and without losing time for their delivery.
- **Dynamics:** Allow (or design) the characteristics of an object, external environment, or process to change to be optimal or to find an optimal operating condition.
  - Continuous Process Improvement Techniques
  - Customer Response Teams/Rapid Reaction Force
- **Feedback:** Introduce feedback (referring back, cross-checking) to improve a process or action.
  - **Statistical Process Control (SPC):** Measurements are used to decide when to modify a process (employed with budgets/turn-around time components).
  - Customer surveys/seminars, etc.

- **Feedback:** If feedback is already used, change its magnitude or influence.
  - Multi-Criteria Decision Analysis to compare valid "apples and lettuce".
  - Change a management measure from budget variance to customer satisfaction.
  - Utilize "half-life" as a measure of improvement (i.e. the time taken to half process development time) to encourage large-scale thinking.

The continuous process improvement technique is a strategic approach to reduce cost, advance productivity and quality, and reduce cycle time for an improved output. It is designed to streamline higher-level review processes and introduce efficiencies to permit issuance and performance monitoring. The technique will produce continuous benefits unless suspended or interceded by major organizational impacts (i.e. political, regulatory, or cultural changes). The specific resources involved in the permit issuance process are depicted in Table 3.

The goal of the entire process was to not only reduce the cycle-time of the permit issuance, but also to reduce variation in the process. Several resources must gain approvals from support entities "consecutively" rather than concurrently. Many repetitive of duplicative efforts can be eliminated if the entities simply develop permitting process templates that outline broad steps and provide standard directions on the required items. Process analysis must be conducted by consecutively analyzing each step in the permitting process. A thorough analysis involves reviewing each step separately to determine where improvements can be gained. Inputs for each step in the contracting process must be identified ~ people, procedures, requirements, regulations and directives, required approvals, and systems. Each input provides opportunities for improvements. For example, identifying system improvements can streamline the permit issuance time through such actions as:

- Automated clause insertion,
- Developing system mechanisms to obtain approvals,
- Developing system enhancements for document storage, communication, and retrieval.

The analysis then spotlights on identifying the critical steps in the permitting process. The critical path includes actions that must be in use (versus those that are discretionary) and the identification of the minimum time it will take to complete each. The combined time requirement of the critical path is the minimum time in which a permit can be issued. In order to facilitate improvements, obstructions to streamlining the process are identified and strategies are developed. Events that are not critical must be acknowledged to determine their inevitability. The tasks are to identify activities that delay or interrupt actions along the critical path and to eliminate duplicative activities. The critical path illustrated in panel A of Figure 2 and described in Table 4 (shown on the following pages) depicts the essential core resource data of the Permit Issuance Process.

The Total Cost Evaluation

One of the most important tenets of the entire study lies within data interpretation of cost and time. The study provides the necessary guidance to develop a budget based on the maximum average (or maximum values) as the upper control limit in a statistical process control arena. Cost control methodologies are employed around these parameters over the individual project's lifecycle. Close scrutiny of the monthly budgeted vs. actual cost of the project work effort allows the decision-maker to monitor/control the element costs *before* an unscheduled event occurs. Subsequent benefits of the control mechanism allow the decision-maker to determine problematic work areas, conflicting resources/demands, and potential cost overruns. When the data is periodically resubmitted in the analysis module, tighter tolerances within the six-

sigma realm are revealed to create an historical database for future projects. The cost savings of this process are unparallel.

The Resource Usage/Number Busy/Schedule Utilization

The average instantaneous utilization (number busy) for the entire permit issuance process is adjustable in accordance with the simulation model. The administrative laborer (busy parameter set 4%) for instance consumes no busy time whereas the Engr labor is busy 0.7238 of the time. Resource leveling techniques and process re-evaluation to "fine-tune" the system (sensitivity analysis) are solutions to the potential problem. Further analysis reveals the number scheduled (0.4589) and the scheduled utilization may increase slightly to accommodate the system processes.

Sub-Summary of the High-Level Permit Case

The industrial operations process analysis of the Permit Issuance Process is an important cost-reduction mechanism to evaluate system integration. The value-added evaluation method provides the necessary front-end analysis for

Table 3. Specific permit general resources

Internal Resources:
Permit Technician
Planning Dept
Engineering Dept
Building dept
Administration
External Resources:
Health dept
Environmental
Transportation dept
Geotechnical dept
A/E Contractor

prudent decision-making of process improvement endeavors. The original process is compared with the model processes in the panel B of Figure 2.

The processes revealed interesting estimation data, methodology, and accuracy. The data clearly depicts lower time estimates with exceedingly high costs. The accuracy of the model is more precise over the original process. Once the process sequencing and resource-leveling techniques are implemented, the element processes are expected to be exact. The Time Comparison Magnitudes of the Permit Issuance Process is captured in the radar diagram panel "A" of Figure 3.

The estimated days to complete the work effort are well under-schedule. The Initial Review precept only consumed one-half day (compare 1 day of the original estimate) as the lone caveat. In order to produce one unit with the required resources and schedule, the "red" data points are the norm for the Permit Issuance Process. The time percentage is depicted in the diagram panel "B" of Figure 3.

Although the entire process demonstrates an overall mark improvement (45.31%), the percentage of time comparison data reveals slight differences. Specifically, the initial review effort requires resource-leveling and process sequencing within the internal processes. An increase in the Site Inspection area necessitates a resource utilization review as well. Nevertheless, the cycle time improvements clearly reduce costs, advance efficiencies, and sustain quality.

Performance Metrics Evaluation Tools

Diagnostic metrics *only* inform the organization of the parameters or process being in control. A well-designed performance measurement system integrates only a select number of strategic methods to align strategic goals (Kaplan, R. and Norton, D., 1996). The objective is to tie strategic goals by minimizing the gap between the current state

and the vision. The general paradigm encompasses several phases designed to reduce the turnaround time of permitting when properly implemented in the requesting organization as outlined below:

- **Phase 1:** Understand customer requirements (the voice of the customer).
  - Conduct Quality Functional Deployment to capture requirements and translate to service design.
- **Phase 2:** Assess the current state.
  - Specify Lean measures using a Balance Scorecard (suppliers, human resources, production processes, customers, financial, society). A sample is depicted in Table 5.
  - Perform process mapping and value stream analysis to evaluate the actions, activities, and operations. Classify these as value-added of non-value-added activities.
  - Implement a Lean profile chart and set targets to minimize the gaps.
- **Phase 3:** Analyze the gaps based on the problem.
  - Identify and rank in terms of criticality.
- **Phase 4:** Generate improvement suggestions.
  - Develop cause-and-effect diagrams to assist with the identification from team members.
- **Phase 5:** Evaluate the suggestions and recommendations.
  - Utilize the cost-benefit analysis and lifecycle costing techniques.
- **Phase 6:** Select the preferred alternative.
  - Utilize the Multi-Criteria Decision Methodology.
- **Phase 7:** Implement the decision.
- **Phase 8:** Monitor and Control the Process.
  - Identify the improvement.
  - Maintain the gains.

Figure 2. A) Critical path diagram for permit approval; B) permit process comparisons

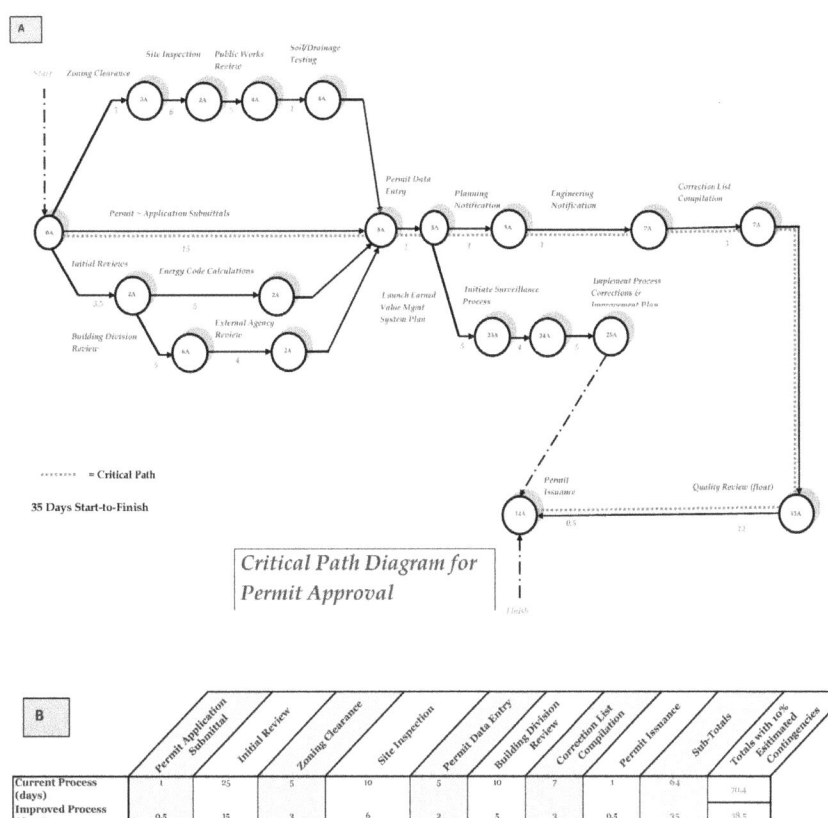

PERMIT EXPECTATIONS AND COMPLIANCE

Once the permit is issued, the company is expected to comply with the stipulations outlined in the document. Expanding the details within permit processes utilizing specific evaluation mechanisms can assist the strategist with viable measures and related data for compliance. Consider the route permit for the high-voltage transmission line venture. The project consist of twenty miles of 115-kV transmission line, 115/69/46-kV substation, and a 115-kV switching station located at the junction of the company's existing 115-kV Lakeview line and the Lakeview 24 line tap. The project is located in Green Township, section 6, township 58N, range 14W. The approval of a route width of 150 feet on either side of the centerline was required to accommodate individual landowner requirements.

The permit conditions are specifically outlined for compliance by the company during construction of the transmission line and related facilities. These conditions are comprises of the following stipulations:

- The company must provide the planning commission with a plan and profile drawing and specifications of the right-of-way

Table 4. Core resource data comparisons

		Current Process	Improved Process	Methodology
0A	**Permit Application Submittal**	1	0.5	
2A	**Initial Review**	25	15	*Concurrent*
	Site Plan Review			
	Building Plan Review			
	Energy Code Calculation Review			*Systems Capability*
	Address Check			
	Public Works Review			*Resource Logistic Planning*
	External Agency Review			*Resource Logistic Planning*
	Fire Dept. Review			*Resource Logistic Planning*
3A	**Zoning Clearance**	5	3	*Concurrent*
	Easements			
	Permit Land Use			
4A	**Site Inspection**	10	6	*Concurrent*
	Physical Problems			
	Special Soil/drainage Testing			
5A	**Permit Data Entry**	5	2	*Proactive Communication*
	Planning Notification			
	Engineering Notification			
6A	**Building Division Review**	10	5	*Concurrent*
	Ensure Accuracy			*Use of Flexible Procedures*
	Meets Code Requirements			
	Overall Plan Review			*Resource Logistic Planning*
7A	**Correction List Compilation**	7	3	*Proactive Communication*
	Permit Fee Re-calculation			
8A	**Permit Issuance**	1	0.5	
	Total Days:	**64**	**35**	

for preparation, construction, clean-up, and restoration for the transmission line.

○ A fourteen day planning commission review period is required to determine if planned construction is consistent with issued permit.

○ No construction will commence until the fourteen day review period has expired or the company is advised, in writing, the plans are consistent and the review is complete.

○ A five day lead time is allowed for submittal of significant change requests. However, no changes shall be made that violates the terms of the permit.

• Construction practices and material specification as described to the environmental

Figure 3. A) Time comparison; B) time percentage comparison

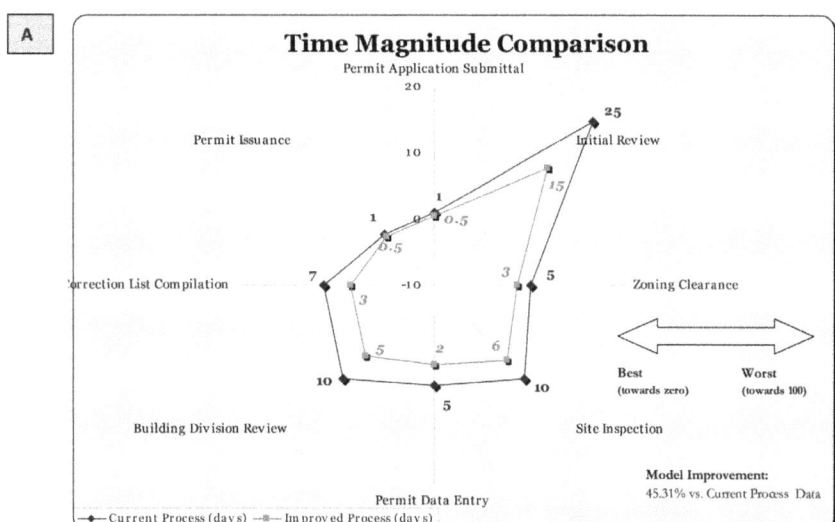

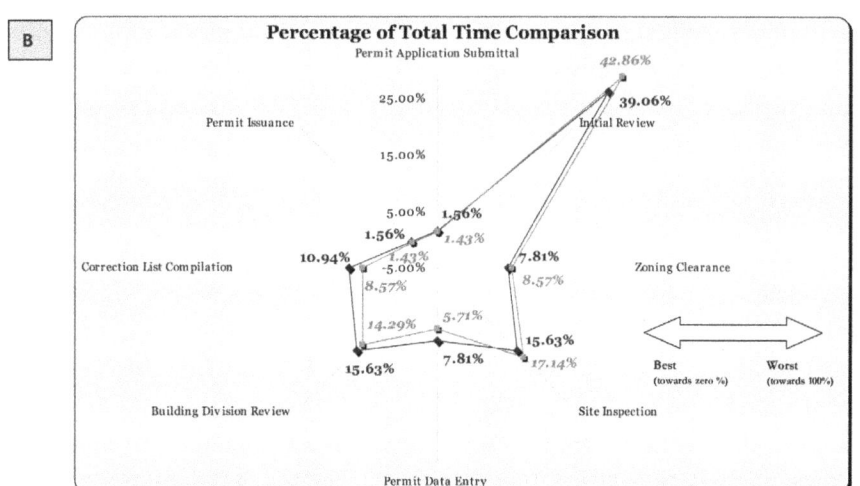

agencies must be strictly followed. A field representative is designated by the planning commission to oversee compliance with the conditions of the permit during construction.

- Waste removal and scrap as a by-product of the construction effort must be disposed of properly at the completion of each task.
- The company is also expected to minimize the amount of trees removed from the right-of-way. Taller tree species that endanger the safe and reli-

able operation of the transmission facility must be removed. Low growing vegetation that will not pose a threat to the transmission facility or impede construction can remain in the easement area.

- The company is expected to work with landowners, the Department of Natural Resources, and the local wildlife programs to restore and maintain the right-of-way to provide a useful and functional habitat for plants, nesting birds and small animals as well as migrating animals. Moreover, the company must minimize habitat fragmentation consistent with inspection and safe maintenance of the right-of-way.

- The company is expected to provide the planning commission with periodic status reports on the progress regarding route finalization, structural design, and the transmission line construction.
- The company must establish a complaint procedure in congruence with the permit prior to the start of construction.
- The company must provide a copy of the permit to the affected landowner upon first contact as part of the notification process.
- At the completion of construction, the company will notify the planning commission three days before the transmission line is placed into service and the date the construction was completed.
- The company must provide the planning commission copies of all of the final *as-built* plans and specifications developed during the project.
- Geo-spatial information, compatible maps and related coordinates must be submitted to the planning commission no later than sixty days after project completion. The data must include all above ground structures related to the transmission lines, switched, and connected substation.
- The electrical performance standards must comply with the established grounding, electric field, communication interference applicable to the National Electric Safety Code and policies, including clearances to ground, right-of-way widths, clearance to buildings, pole erection, and transmission line conductor stringing.
  - The transmission line must be designed, constructed, and operated as such that the maximum steady-state short-circuit current is limited to five

milliamps RMS alternating current between the ground and any non-stationary object within the right-of-way. This includes large motor vehicles and agricultural equipment. The fences are expected to be grounded as well as all fixed metallic objects on or off the right-of-way.
  - The transmission line must be designed, constructed, and operated as such that the electric field measure one meter above ground level immediately below the transmission line must not exceed 8.0 kV/m rms.
  - The transmission line, once in operation, must not cause interference with radio or television, satellite or other communication devices. The company must take prudently feasible action to restore or provide equivalent reception levels in the immediate area just prior to the construction of the transmission line.
- The company must make every effort to avoid impacts to identified historical/archaeological resources during the construction of the high-voltage transmission line on the route. If an impact occurs, the company must consult with state and federal permitting or land management agencies.
- The company is expected to comply with applicable wetland conservation acts and public water policies to include pole placements and impact avoidance measures. A storm water permit from the national pollutant discharge elimination system is necessary if construction activities at the substation and switching station will result in the disturbance of one acre or more of soils. Erosion control measures and best practices must be followed during these activities.

Table 5. Sample balanced scorecard

Suppliers Perspective	Production Perspective	Customer Perspective	Society Perspective
■ The average number of requests ■ Percent of orders on time ■ Percent of requests delivered to point-of-use without storage	■ Performance efficiency ■ Quality rate	■ Service Level ■ Percent of requests delivered on time ■ Percent of completed requests ■ Number of complaints	Environmental
Human Resources Perspective			
Training and Development Measures ■ Availability ■ Number of various areas for employee training	Lead time ■ Processing time ■ Queue time ■ Transportation/set-up time		Safety Issues
Multifunctional team work measures ■ Percent of team communication ■ Percent of skill level for task	Inventory Turnover ■ Average work-in-process ■ Average inventory of finished requests ■ Value-added ratio ■ Non-value added ratio		
Leadership Measures ■ Percent of team leadership ■ Percent of leadership training	Productivity Levels		
Work satisfaction measures ■ Percent of monotony in work ■ Percent of workload ■ Number of employees leaving (turnover, absenteeism, etc)	Request travel distance		
Communication between employees and top management • Percent of suggestions and implementation			
Financial Perspective			
Cost Effectiveness			

- The company must comply with all other applicable state rules and statues required in other permits pertaining to the project.

Consider one explicit section of the permit provisions the company is expected to abide by based on the construction practices. This section consists of developing a waste/scrap material plan, a tree removal plan, and a wildlife assessment plan. The reporting of such measures requires the development of viable plans to monitor and account for the conditions to the planning commission. The strategist may elect to utilize a process to evaluate the worst-case scenario scheduling provision in each plan and assess the resource requirements. The general set-up for the strategy is outlined in Table 6.[2]

All of the plans are expected to meet and/or exceed the compliance in construction standards applicable to the region. The utilization of priority-list scheduling algorithm (similar to the first-in-first-out paradigm) is suitable for some tasks to determine the schedule length. In general, the list is formulated by order of importance. When completed, the task is crossed off the list. Available resources work on the most important unfinished task based on the priority list (then alphabetical order or some other prescribed sequence in case of several free resources). The total schedule days for the assigned resources are determined once all of the tasks in the algorithm are complete. Yet, increasing the resources will make some tasks availability *too early* and others accessible for a late start. If equipment upgrading were allowed in certain assessment tasks, the critical path is reduced but actually decreases the system turnaround time for the plan *instead of speeding it up*. A closer examination of the priority-list scheduling algorithm and related parameters for the waste and scrap material plan, for example, is discussed in the following paragraphs. Tasks w1 and w2 are ready at the start of the process. Task w3 is only ready after task 1 and 2 are complete. Task w4 can be completed anytime and task w5 is ready only after task 3 is complete. Therefore, the critical path is w1, w3 or twenty days as the longest duration. The list processing diagram is depicted in panel A of Figure 4.

If the priority list is w1, w2, w3, w4, and w5, then both engineering and planning departments work on task w1 and w2 respectively.

- The planning department completes w2 in four days and waits for the completion of task w1. The priority list algorithm dictates this available resource to work on the next available unfinished task w4 (5).
- The engineering department completes w1 and starts the next available unfinished task on the list, w3 (12). At this time (t_8), tasks w1 and w2 are completed and tasks w3, w4, are being worked on.
- Planning completes w4 (5) at day nine as the engineering department is still busy with w3 (12). The next available task on the priority list is w5 (2) which are completed by day eleven.
- The plan is not *technically* executable at day eleven as the equipment securitization is incomplete. Therefore, the resource is idle at this time.
- The engineering department completes tasks w3 (12) by day twenty and executes the waste/scrap material plan.

- The tasks are finished in a reasonable amount of processing time with moderate idle time. Scheduling optimization is required to produce a compliance process as timely as possible.

The engineering department total time is twenty days. The planning department total time is eleven days. Therefore, there are thirty-one days total allowed for the schedule using the priority-list algorithm technique.

One optimization method utilizes the decreasing-times processing algorithm (DTA). It prevents the lengthy tasks from being scheduled too late by arranging the tasks in decreasing order of required time (longest tasks first, shortest tasks last). Tasks with equal processing times can be listed in any order. The decreasing-times algorithm for the waste and scrap material plan is illustrated in panel B of Figure 4. Although the idle time increased, the important point is all tasks are completed after nineteen days as the optimal solution (w3, w1, w4, w2, w5). Additional resources drastically increase the idle time and subsequent costs. The worst-case scenario is substantially superior in the DTA when compared with list-processing (COMAP, 2009). The best practices include the construction of all schedules to satisfy the digraph and select the one that best optimizes the prevailing criteria. The tree removal plan is similar to the waste and scrap material plan as depicted in panel C of Figure 4. There are six days of idle time in the list-processing for tree removal plan.

The vegetation management and wildlife assessment plan utilizes the same scheduling philosophy as the waste/scrap material scheme and tree removal plans. Tasks E1 and E2 are performed simultaneously by team members in the vegetation department and the respective environmental organization. Precise allotments of days are allowed to identify the plant and animal natural habitats within the proposed construction zone. Task E5 cannot start until the completion of these assessments. The restoration effort evaluation (E3) and communication criteria (E4) can be performed anytime in the logical sequence of events. Nevertheless, the critical path is identified as E1, E2, and E6, ending at the twenty-eight day point. Once the environmental resources completes the assessment stipulated in E1, the securitization of the inspection personnel immediately follows (task E2 was finished two days prior by the vegetation management department as illustrated in panel "A" of Figure 5. The task E4 is performed by the vegetation management organization at time t_{10} to establish communications with the local landowners, the department of natural resources, and regional wildlife officials. Both company organizations are expected to complete task E4 and E5 by t_{18}.

Since the mutually resources are available at this particular juncture in the schedule, the priority task (E3) is assigned to the environmental department to assess the restoration efforts in the right-of-way. Task E6 is assigned to the vegetation management organization to establish the communication and coordination plans in the general scheme of the permit boundaries. The environmental department is finished at t_{25} after task E3, shortly followed by the vegetation personnel with the completion of task E7, to facilitate the execution of the wildlife assessment plan. The optimal schedule within the parameters of the decreasing-times algorithm criteria is achieved with *no idle time* in the model. The schedule is fairly balanced with the vegetation personnel performing within twenty-nine days total and the environmental organization operating within twenty-five days. The total schedules for the resources are therefore allowed fifty-four days total to complete the allocated tasks. The assignments with the community contact characteristics (E4 and some portions of E3) can be broken up into sub-tasks. However, it will change the dynamics of the scheduling system. The restoration efforts in the right-of-way (E3), for example, can be completed anytime by the assigned resource within the boundaries of the paradigm. Yet the logical sequence of events and the time consumed is during the initial right-of-way clearing before the con-

struction cycle begins. The communication establishment with adjacent landowners is yet another tenuous issue that must be completed in a timely manner. The perfunctory resistance and hostility to the construction efforts from the landowners is anticipated and mitigation plans must be in place as part of these sub-tasks. Moreover, constructive coordination and cooperation efforts is anticipated from the local wildlife agencies with minimal or no resistance to the proposed construction. A viable public relations plan is compulsory as sub-tasks within the overall wildlife assessment plan to facilitate the permit processing/compliance and decrease the turnaround time.

The decreasing-times processing method averts the prolonged responsibilities from being scheduled too late in the system as the priority list is prearranged in diminishing order of the requisite time. When the strategist properly develops the scheme, the worst-case delay scenario is estimated from an overall scheduling perspective when the two algorithms are utilized and compared as depicted in panel "B" of Figure 5. There are precedence relations rules that demand specific tasks precede others and always override the priority list if there is a conflict between the two. Nonetheless, the permit issuance criteria are to minimize the non-value added time/cost as well as to eliminate idle time. Ultimately, the critical path method prevails over most priority list algorithms. However, the best practice is to develop all practicable schedules and select the appropriate technique to optimize the established criteria.

SCENARIO ANALYSIS

A re-visit to the permit processing model is necessary to amplify the constraint time from sixty days to one year. The non-value added decision block is also increased to a $3000 level. The integrity of the model is the same; only these minor adjustments (and queue service levels) were implemented to reflect the annual realistic cycle times. Seven

Table 6. General compliance plans

Task:	Waste and Scrap Material Plan
W1/8	Waste process evaluation (which tasks produce waste)
W2/4	Identify waste/scrap material and proper disposal methods
W3/12	Secure equipment and disposal site(s)
W4/5	Establish communication/coordination plans
W5/2	Execute the waste/scrap material plan
Task:	Tree Removal Plan
T1/8	Tree identification within the plan (evaluation of danger trees)
T2/8	Vegetation evaluation within the plan (identify potential shrubbery)
T3/9	Secure equipment for tree removal
T4/5	Establish communication/coordination plans
T5/2	Execute the tree removal plan
Task:	Wildlife Assessment Plan
E1/12	Identify natural animal habitats/assess fragmentation habitats
E2/10	Identify functional plant habitats/assess fragmentation habitats
E3/7	Evaluate restoration efforts in right-of-way
E4/8	Establish communication with department of natural resources, landowners, and local wildlife departments
E5/6	Secure inspection resources
E6/5	Establish communication/coordination plans
E7/5	Execute wildlife assessment plan

scenarios were analyzed to determine the most cost-effective methodology for permit issuance. It is beneficial for the strategist to evaluate the various leveling techniques and capacity requirements involved for the system throughput.

The schedule utilization is designed to assist the astute decision-maker with the resource capacity determination in a prescribed plan. The scenario progression of the permit issuance process is focused on the optimal resources within the parameters of the schedule – from "Constant Material replica" to the "Five Day Schedule with Additional Technical Assistance and Administrative Resource model." How well the resources are scheduled to produce the maximum output of permits with the minimal amount of re-work is a delicate balancing act faced by all decision-makers.

However, once the optimal point is achieved, the related costs are decreased dramatically. It is therefore advantageous for the analyst, planners, and decision-makers to actively seek the optimal level of a given scenario germane to the available resources and realistic scheduling. As a first step, the material (permit) is allowed for processing by all resources based on a period of constant duration. The maximum average of the data is represented in panel "A" of Figure 6. This scenario illustrates an increased percentage of the material (permit) for subsequent enhancement as well as the resources. Although this situation is ideal, it is unrealistic. Engineering and planning personnel, for example, are typically working on other projects, multi-tasking with the current permit process, or performing non-productive duties essential to the operation of the daily routine. Yet, this is the highest level for all of the resources with the exception of the material itself and the administrative personnel. A second scenario (Brute Force) involves all entities based on the duration of the schedule. In essence, all of the assigned staff is dedicating one hundred percent effort at a constant level to produce the acceptable permit. The Brute Force technique amplifies the availability of the permit for processing by the resources. The proportional schedule utilization is diminished tremendously. The permit technician – an integral member of the processing team – is performing less than six percent within the system. Moreover, all of the assigned resources are operating well below the fifteen percent level to produce a viable permit. The scenario is valuable by providing insight for the baseline of the system for improvement purposes; however, it is costly and inefficient. In order to reduce the gap between the vision, or ideal situation and reality, the master schedule is introduced in the system for the third scenario. No work is performed during the holidays and weekend. Consequently, all of the resources increased precipitously percentagewise. The Material is omnipotent as the resources enhance the product within the allotted time. The scenario is a profound improvement over the Brute Force model, yet inefficiencies abound. The chart illustrates the maximum average of the scheduling utilization to assist the decision-makers with the upper boundaries of the production with the as-

Figure 4. Algorithm results for scheduling plans

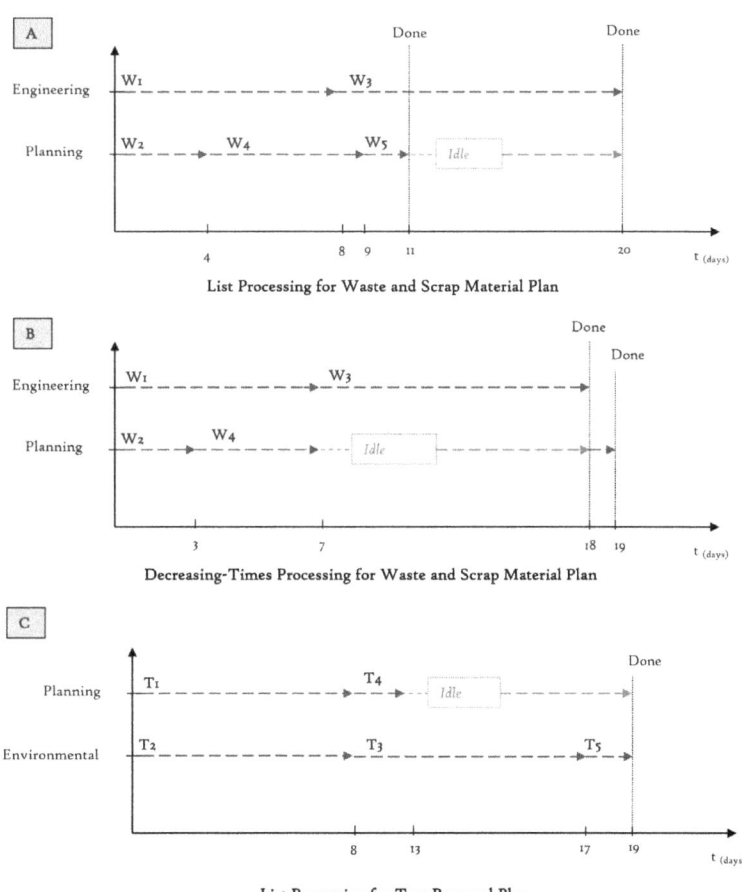

signed resources. Schedule data above this ceiling is deemed as unacceptable and can be reflected in the applicable construction contracts.

The next scenarios evaluate the "Four-Ten Hour Work Weeks" and the robust five day schedule maximums. The increase in production capacity from a scheduling perspective is embedded within the robust "Five Day" model over the "Four-Ten Hour Work Weeks" replica. Minor adjustment of the scenario includes the addition of the permit technician in various assisting roles in the major processes, with some success in the building and technical areas. The additional administrative assistance in the major resources slightly increases the productivity levels (engineering, administrative, permit technician) and yielded the lowest accumulative cost. However, this model will amplify the communication channels and adversely affect the initial coordination processes. Nevertheless, the realistic scenario is the closest approach to the ideal model without modifying the precedence and priority tasks. Overall, the model is the most resource-effective of the intense permitting schedule. The increased, incisive utilization of the administrative personnel and permit technician methodically dispensed in key departments allowed resource-leveling activities in the planning, engineering, and building inspection areas of the entire system. When this model is implemented, the permit turnaround time is expected to be exact.

The average total wait time in days for the permit processing per entity is an interesting metric. It is designed to assist the strategist with constraint determination as well as effective resource utilization in key areas. The scenarios are evaluated and compared against a baseline indicator (Brute Force) and a secondary baseline (Constant Material) to formulate ideal conditions. This "vision" is designed to complete the assigned tasks within the minimal prescribed time. Both of these baseline scenarios produce an acceptable output from system well below three days per entity

as illustrated in panel "B" of Figure 6. In fact, the planning and building departments performed at extremely low turnaround levels (less than a day in one case) to satisfy the ideal conditions. Nevertheless, the baseline scenarios simulate these ideal conditions whereas in reality, the resources and schedule inherit a certain point of *task dedication* to produce such low turnaround levels.

Consider the realistic "Master Calendar" scenario, where the resources actually *ramp-up* to specific efficiency levels, take scheduled breaks, and are allowed weekends/holidays of. The permit queue waiting time is extremely close o ideal conditions in eighty percent of the categories. However, since the permit is an important document within the overall scheme in the construction process, a major delay in the permit queue can have massive cost consequences in the downstream activities (i.e. equipment contracts, resource mobilization, etc). Consequently, the excessive time spent in the "Administration" (40.8 days) and "External Resources" (53.4 days) are queue consumed by the permit is undesirable in the "Master Calendar" scenario. The permit resides in the respective queues during this time *before* it is processes and potential re-work is assigned. The external resources are expected to comply with the incoming policy and quality procedures to ensure an efficient turnaround point for schedule optimization. Moreover, additional cost-effective resources in the administrative department resource the queue waiting time in subsequent scenarios. For example, if the entities were committed to a "Four-Ten Hour Day" schedule to produce a permit from the system, what are the consequences experienced by the permit in the corresponding queues? The administrative personnel process the required elements in less than nine days. Moreover, the external resources attached to the project completed the assigned tasks in approximately nine days. Yet this is not the optimal level for the system to process a permit. This particular scenario illustrates vulner-

Figure 5. Scheduling plan and worst-case scenario comparison

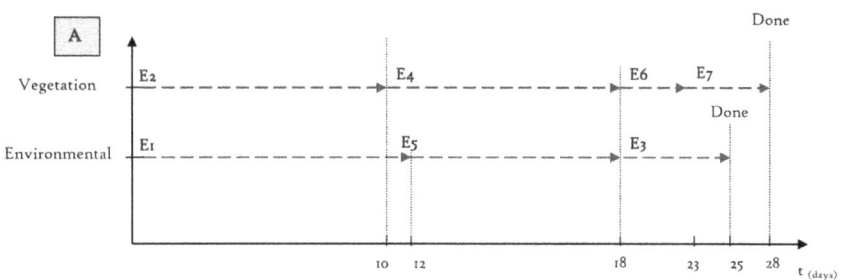

Decreasing-Times Processing for Wildlife Assessment Plan

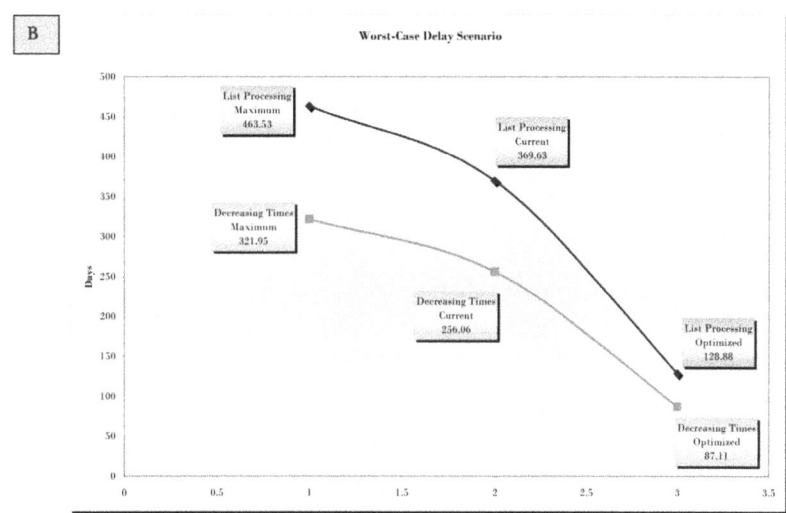

abilities in key processes – namely in the delay and re-work areas. The resources are seemingly focused on the process queues as the delay queues accumulate additional units. Ultimately, this is an inefficient and costly scenario. Incisive strategists are keenly aware of these types of constraints and plan accordingly. These plans ought to include subsequent scenarios to mitigate excessive delay queue processing times with various cost-effective, realistic schemes. An extension of the "Four-Ten Hour Day" schedule scenario that encompasses the queue delay mitigation philosophy is found in the "Five day Schedule" plan and its derivative "Five Day Schedule with Technical Assistance." Both circumstances produced a seventy-five percent improvement in the delay departments at various

points in the system over the "Four-Ten Hour day" scheme. However, the latter scenario is far superior over all viable plans in half of the assigned tasks. The trade-off and compromise of these conditions is inevitable. It is here where the astute decision-maker determines if a two-day delay in the "External Resource" department, for example, between the most cost-effective and time conducive scenarios to optimize the turnaround process. The scenario that effectively reduces the permit queue delay time is the "Five Day Schedule with Technical Assistance and More Administrative Personnel." It provides an overall effective time reduction scheme to process accumulated delays in the system and operates within reasonable time parameters. However, after a thorough assessment of the minimum/maximum and average time sta-tistics (along with related costs), an organization can readily determine the most favorable scenario that eliminates delays and effectively utilizes the resources to its optimal potential. It is then when the reality approaches the ideal conditions.

The absolute minimal processing time to pro-duce a permit is illustrated in panel "A" of Figure 7. The basic question(s) posed by the strategist ought to include the turnaround efficiency compo-nent – How soon can the organization process an effectual permit? What is the nominal amount of time to produce the unit? What is the most profi-cient scenario utilized to process the permit with the current system? The incisive decision-maker is prepared for these questions with viable solu-tions. Moreover, the strategist is able to assess the advantages and disadvantages inherit within each scenario *after* detailed analysis, and thoroughly explain why one plan is preferred over the other.

Consider the baseline scenarios of the "Con-stant Material" and "Brute Force" plans utilized as ideal conditions. The material is based on the constant duration of the incoming permit requests. The highest attribute in the delay queues (the wait time) is processed immediately and the resource queue utilize the first-in-first-out philosophy. This paradigm is the basis for subsequent scenarios. The baseline scenarios in the minimal average time metric depict this situation for each entity and its related tasks. The worst of the conditions is within the "External Resource Delay" catego-ry at approximately six days. These delay catego-ries are very important as the re-work queue ac-cumulate and requires additional work effort. Yet the baseline models successfully produce a permit from the system as each task is performed within six days or less. This is a clear vision of a perfect system free of defects and absolutely no time off for the resources from the processes. The reality scenarios include scheduled time-off, breaks, and holidays for the assigned resources. Also, the resource efficiencies are calculated to ramp-up to the realistic capabilities during specific time

periods in the schedule. The "Master Calendar" scenario is an example of the resource scheduling paradigm. It is designed to allow breaks in the assignment (weekends, holidays, etc.) while completing the tasks. The scenario in the minimal average time metric illustrates a viable plan to accomplish eighty percent of the tasks within the allotted timeframe. However, the administrative and external resource departments are exces-sively high to produce the required output. There-fore, the scenario to a "Four-Ten Hour Day" schedule was developed to assess the minimal average time required to produce the same output. This particular plan excelled in several categories (seventy percent improvement when compared with other categories) at various levels. However, the "Four-Ten Hour Day" scenario displays signs of weakness in the delay and rework categories – areas of importance for efficient throughput. The logical progression is to develop five day time scenarios for comparative purposes. The interesting aspects of the "Five Day Schedule Plan" and the "Five Day Schedule with Technical Assistance" are embedded within the resource utilization philosophy. The additional resource (permit technician in this case) clearly reduced the processing times in the delay categories and only performed poorly in the building resource area by three days. The scenario can realistically accomplish the assigned tasks in twenty days or less per category. The astute decision-maker evaluates the organization's capabilities and limitations based on the permit process incoming requests, backlogs, and future requirements. The minimal average time is the lower control limit (optimistic point) in a statistical process control paradigm. In this case, it can be utilized as a benchmark to award additional benefit expendi-tures for closing out a project in a timely manner, or as an incentive for the external resources to meet or exceed company goals.

The most feasible, cost-effective scenario is the "Five day Schedule with Technical Assistance and More Administrative Personnel." This plan effectively utilizes idle administrative personnel and the permit technician to assist with the potential constraint areas – mainly in the delay departments. The engineering area is the only exception where the process indicated a slight degradation. The technical details are typically addresses by the professional engineer at this level, and rudimentary work performed by the technician. The increased personnel are operating proportionally within deficient resource assignments to assist with schedule demands. The decision-makers are faced with these various time-based scenarios during strategy sessions. The solutions oftentimes involve tradeoff analysis and resource capacity evaluations to implement the plan successfully. The incisive decision-maker has a thorough comprehension of the system capabilities to integrate the preferred scenario and produce a viable permit as efficiently as possible.

The average wait time per entity for the permit processing in the various scenarios is illustrated in panel "B" of Figure 7. Although the general time patterns are similar when compared with the total time per entity and the queue wait time charts, the results are quite different. The metric examines the accumulated wait time when the permit enters a process and is subsequently delayed or when the permit resides in the queue until it exits the holding area. There are associated holding costs per entity based on the resource allocation level, category grade, and efficiency. Nevertheless, the importance of the permit wait time is designed to assist the decision-makers with analytical data to determine the best scenario with the available resources. The permitting process may incur unexpected delays and create a temporary work stoppage at some point in the system. However, the incisive strategist may elect to seasonally adjust the schedule to accommodate the work-life balance employees actively seek in the workforce. For example, the decision-makers may adopt the "Four-Ten Hour Days" scenario for the summer months and still maintain the requisite permit processing levels. Additionally, if any lost time is

incurred during this period, the organization can revert to a more aggressive scheduling scenario to meet demand. Nevertheless, the average wait time for the permit processing is a metric that necessitates the utmost attention by the strategists. The permit wait time in the electric utility industry vary from as little as one-half day to years. A continuous process is obligatory to mitigate these excessive delay times.

Consider the planning department resource delay as a function of the system output. Ideally, the planners can perform the permit effort with less than a two day wait period in the "Constant Material" and "Brute Force" scenarios. However, when these same resources are attached to the "Master Calendar" scenario, the average wait time degrades to approximately eleven days based on resource efficiencies, ramp-up times, vacations, etc. The details required in some permitting applications are thoroughly reviews in various phases by different entities of the document development and creates wait time. Other documents utilize a general boilerplate to assist with the minimization of the delays. Nevertheless, the most viable and realistic scenario for the planning department, from an average wait time perspective, is the "Four-Ten Hour Days" scheme. In fact, if this realistic scenario was adopted, the wait times will decrease in eighty percent of the categories. Moreover, if the external entities were aligned with the permit issuance organizational goals to increase throughput, the entire process would be more cost-effective. Yet the idling of resources based on the current permitting process is a chief concern in the industry. All stakeholders must be thoroughly and actively engaged in order to reduce costs and the associated wait times.

The mitigation strategies for the wait time processes include the Triple "C" Triangle of communication, coordination, and cooperation across the permitting and siting spectrum. For example, inter-agency coordination is necessary to outline and align the specific permitting requirements/ standards. In a virtual team setting, the leadership

Figure 6. Schedule utilization and queue wait time scenarios

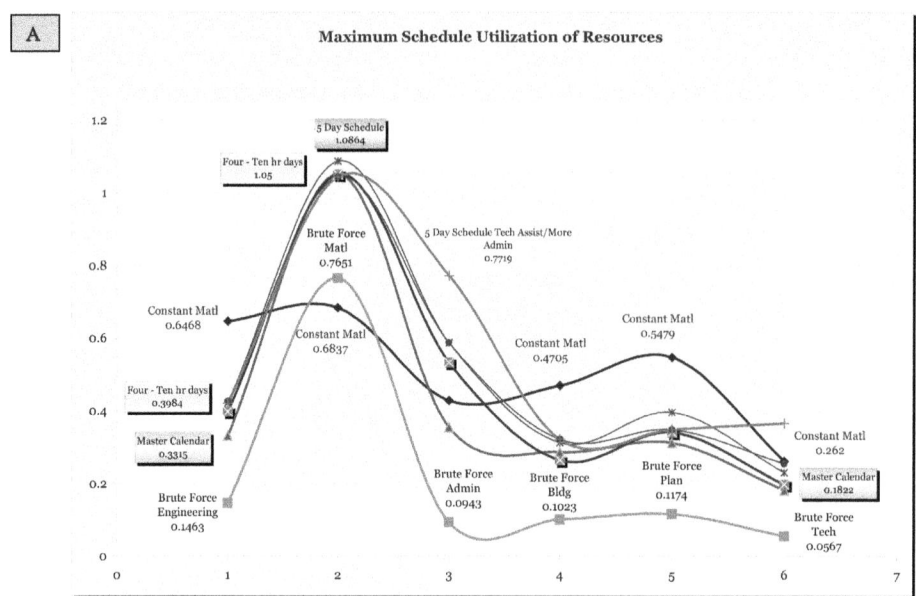

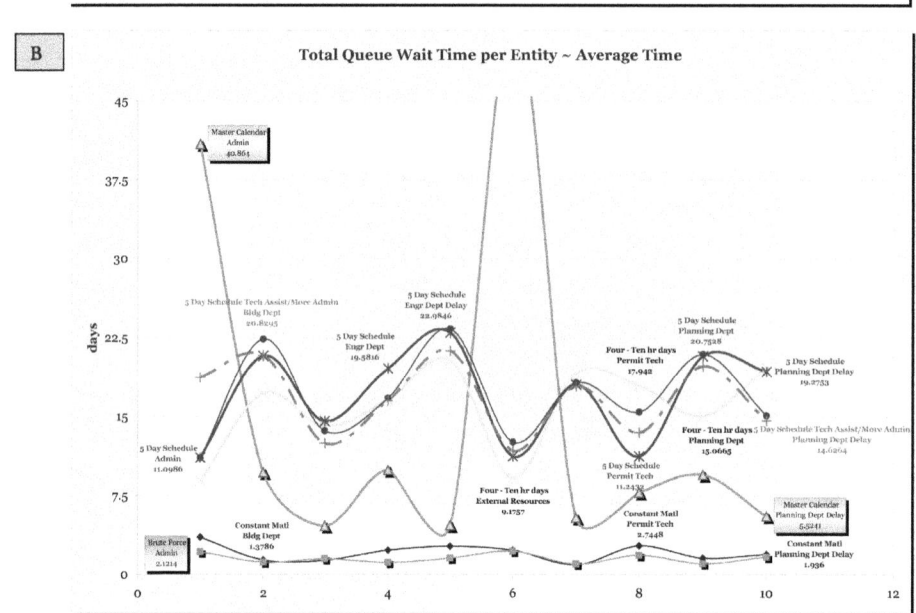

challenges are much greater and complex. Nevertheless, co-located teams can overcome these challenges by developing trust and encouraging openness (Levasseur, 2012). The leader of a high-performing virtual team must actively manage the communication process from initiation to the project close-out phase. The implementation of milestones/deadlines is another essential element to allow the stakeholders to plan accordingly and manage expectations. The incorporation of environmental assessment plan, site evaluation plan, and related plans to satisfy governmental requirements is performed *concurrently* with spatial planning activities to shorten the wait time embedded in the permitting process. The strategic streamlined process is most successful in Europe.

The amplified role opportunities of key members of the permitting team are an important attribute of the cooperation tenet in the Triple "C" Triangle. Moreover, the attachment of incentives to minimize the assessment times in essential areas is beneficial for the entire organization. The final leg of the Triple "C" Triangle involves effective communication throughout the permitting process. The astute strategist will emphasize the direct and indirect benefits of the project scope, organization, and potential adverse impacts, to key members of the team to facilitate the permitting process. These strategic progress sessions are designed to reinvigorate and motivate the stakeholders in regards to the project's performance requirements, resource/time constraints, and the boundaries of the plan. Both internal and external factors are explored to determine the root cause of the delays and promote mitigation strategies to produce an effective permit turnaround in the system. Once these limitations are recognized and properly addresses, the wait times of the permitting process is expected to be non-existent. Management of these *soft-side* qualitative administration skills is essential in practically every aspect of the renewable energy generation integration of the current electrical infrastructure.

The total accrued cost by the entities in the station logic directly related with the activity is represented in this metric. The decision-makers can utilize the data to determine costs associated with each scenario and the best approach to implement the specific process. The scenario progression originates with the "Constant Material" duration model and ends with the "Five Day Schedule with Technical Assistance and More Administrative Personnel" version of the simulation. The iterations between the two models describe the various approaches suitable for resource utilization. However, the trade-off between cost and time is an ever-present factor to produce a permit from the system.

Consider the "Brute Force" scenario as a baseline, where all of the assigned resources are associated with a duration schedule in panel "A" of Figure 8. Strictly from a cost perspective, the model presents the average accumulated costs well above comparable scenarios in the departmental delay (re-work) areas. Yet the "Brute Force" model is the least-cost approach for eighty percent of the main processes. Although it is unrealistic to utilize the resources at full capacity and total dedication to the project, the scenario provides an ideal condition of the potential costs involved with the process. The "Master Schedule" scenario ties the related resources to a realistic calendar with weekends and holidays off. The average accumulated cost for each activity decreased in half of the processes in this scenario. Only the external resources and the administrative categories were outliers in the model. The administrative personnel department is an internal controllable resource that can implement cost containment strategies. However, the external resources are slightly more difficult to control because of the unpredictability of the conditions. Nevertheless, with incoming material and resource protocols, the system is expected to minimize the accrued costs. The "Four-Ten Hour Days" model is very cost-effective on several different levels – specifically in the administration and external resource categories. However, this scenario is relatively ineffective in the re-work/delay areas. The average accumulated costs for the planning department delay, for example, is $114k (minimum average $60k and maximum average is $157k). By comparison, the "Five Day Schedule" model is $86k in this delay category and lower in the respective re-work areas. Moreover, the permit technician and the engineering departments are too costly in the "Four-Ten Hour Days" category when compared with other scenarios.

The resources are focused on one or two tasks per day as the delay factors increase precipitously. The variations of the five day schedule depict a profound cost improvement in seventy percent of the categories over the "Four-Ten Hour Days" scenario. The resources slightly minimize the delay cost and increase throughput. Based on

Figure 7. Total time and total wait time scenarios

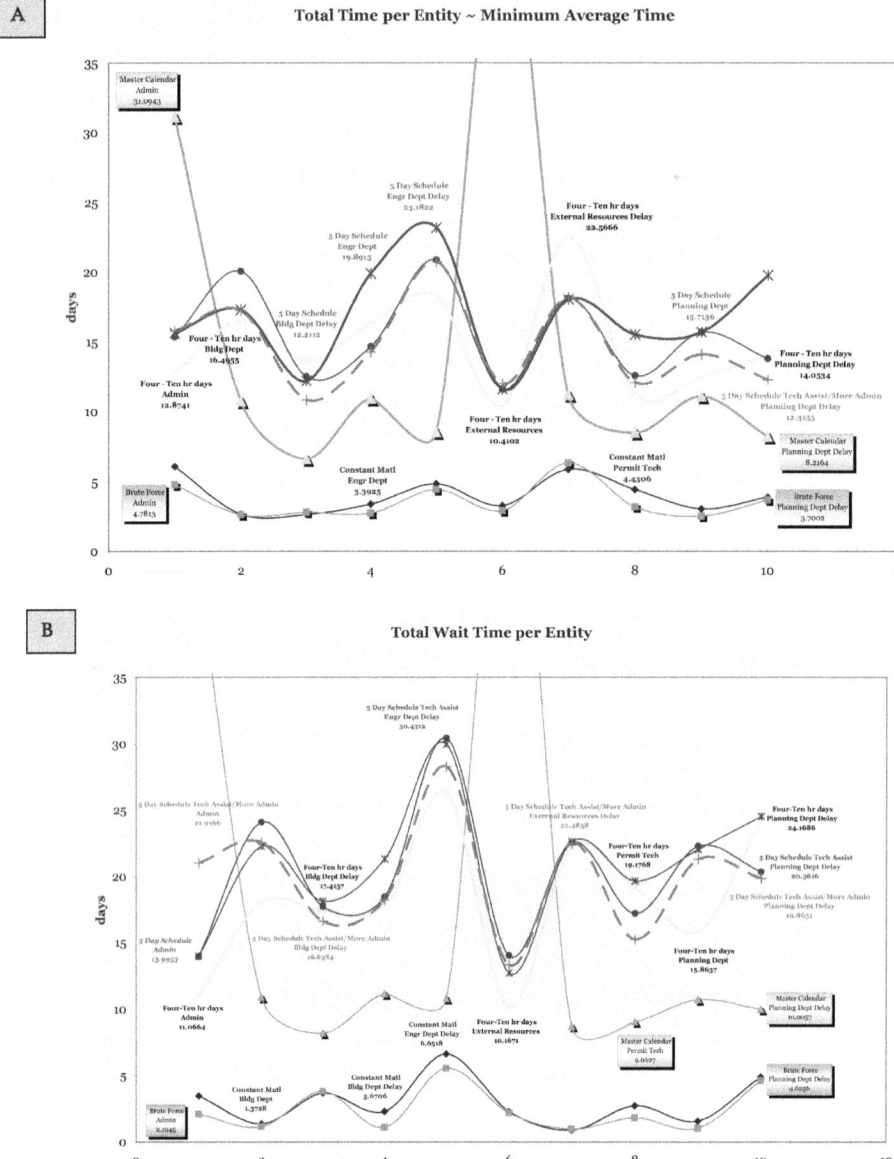

the decision-maker's proclivity for cost reduction in the entire process, the determination to minimize the high cost resources (engineering, planning, building inspection) versus the lower cost resources (administrative, permit technician) is a delicate balancing act. Several cost avoidance strategies are implemented to reduce the delays inherit within the system. Nevertheless, the resource-leveling technique is one of the most effective strategies to minimize cost. The method is effectively captured in the "Five Day Schedule Technical Assistance with More Administrative Personnel" model. The scenario yields the lowest accumulative average cost by shifting the administrative expenditures with the other resources. Moreover, the model drastically reduces the delay

Figure 8. Accumulated and idle cost scenarios

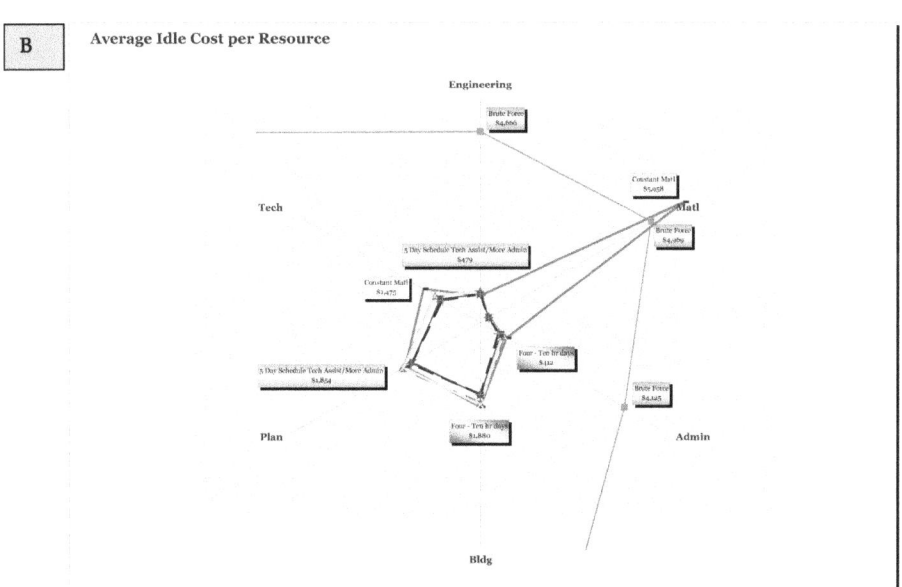

costs when compared with other scenarios. The only caveat in the scheme is the slight increase in expenditures of the administrative personnel in the building and planning departments. Depending on the organization's tolerance for the process to approach the maximum budget threshold, the model is well-suited to deploy a cost-effective solution for processing a permit. Minor adjustments with

the queue waiting times will align the realistic scenario with the sub-goals of the organization as it approach the ideal version of the paradigm. The incisive decision-maker will thoroughly comprehend potential constraints in the process and readily implement cost containment strategies.

The cost of an idle resource is ideally minimal. The efficient system id designed to process the material at a sufficient rate for each resource to properly handle. The "Brute Force" baseline model produced excessively high cost as illustrated in panel "B" of Figure 8. The secondary baseline scenario, "Constant Material," produced high costs in the material area, yet moderate expenditures in the remaining areas. The permit technician, for example, experienced a moderate cost decrease within the various scenarios (from $1,475 in the "Constant Material" model to a mere $877 in the "Five Day Schedule technical Assistance More Administrative Personnel" model). The idle cost reduction was incrementally experienced in every resource category until an optimal level was obtained. The additional administrative assistance (Five Day Schedule Technical assistance More Administrative Personnel model) in key areas provided the required buffer to level the resources in the process. These assignments of the administrative personnel and permit technician allow higher cost employees to pursue indispensable technical elements involved in the permitting

process. The scenario practically achieved the goal of system-wide cost minimization on a resource-by-resource basis. The incisive strategist thoroughly understands these related costs in the permitting process and is proactive with the minimization protocols and procedures well before system implementation.

The average busy cost per resource is illustrated in Figure 9. How well the resourced may cost-effectively respond to the various scenarios is the basis of the evaluation. Both baseline models ("Constant Material" and "Brute Force") produced high busy costs given the schedule duration within the permitting process system. The resources are designed within the parameters of the system to effectively process the incoming permit request in a minimum amount of time. Moreover, the related costs to accomplish the specific assigned tasks are expected to approach an optimal level during the course of the process. When the engineer is busy, for example, the resource cannot perform additional tasks until the current task is

Figure 9. Average busy cost scenarios

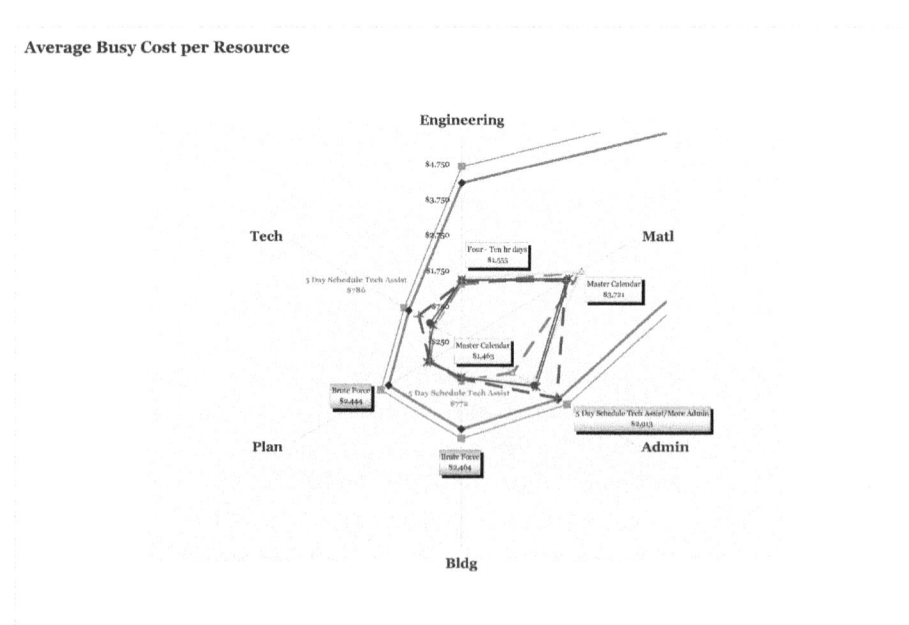

released. The engineer may work on concurrent tasks related to the prioritized task; however the resource is expected to complete the assignment within the prescribed time constrains. The optimal cost-effective point for this same engineer is in the "Master Calendar," where the expenditures equate to $1358. This scenario is slightly better than the others in every category with the exception of the material area. It is here where the decision-makers consider a nominal trade-off between modest increases in the busy cost in one or two categories (permit technician, administrative personnel) in a preferred scenario (Five day Schedule Technical Assistance More Administrative Personnel) versus the larger accumulated cost metric. The evaluation of the same two categories, for instance, produces a $44k difference in the permit technician category between the two scenarios ("Master Calendar" and "Five Day Schedule Technical Assistance More Administrative Personnel"). The administrative category in this same metric illustrates a larger difference between the two scenarios ($228k) that must be considered if the decision-makers prefer the "master Calendar" approach. Although the model excels in eighty percent of the categories, it is not cost-effective in the remaining two which equates to over $375k when compared with the "Five Day Schedule Technical Assistance More Administrative Personnel" model. However, the strategy sessions will encompass various tenets of the advantages and disadvantages of each scenario as compared with the overall scheme. The incisive decision-maker is keenly aware of the cost models and thoroughly comprehend the trade-offs and challenges involved in the processes. How well the strategists respond to these challenges to minimize cost and cycle time is the sub-goal of an effective organization. The excessive costs implicated in the permit process can be diminished with the proper protocols, level resources, and cost-effective high-quality methods implemented in the paradigm. When these conditions are recognized and acted upon, greater achievement will take place in the affected organizations.

REFERENCES

COMAP. (2009). *For all practical purposes: Mathematical literacy in today's world* (8th ed.). New York, N.Y: W.H. Freeman & Company.

George, M. L. (2003). *Lean Six Sigma for service.* New York, NY: McGraw-Hill.

Kaplan, R. S., & Norton, D. P. (1996). Linking the balanced scorecard to strategy. *Management Review, 39*(1), 53–79.

Leonhardi, D. (November/December 2011). Soft skills: The 'killer app' for analytics. *Analytics Magazine,* pp. 34-37

Levasseur, R. E. (2012). People skills: Leading virtual teams – A change management perspective. *Interfaces, 42*(2), 213–216. doi:10.1287/inte.1120.0634

Mann, D., & Domb, E. (1997). *40 inventive (business) principles with examples.* Retrieved 8 November, 2008, from http://www.triz-journal.com

Van Cleve, F. B., & Copping, A. E. (2010). *Offshore wind energy permitting: A survey of U.S. project developers.* Richland, WA: U.S. Department of Energy, Pacific Northwest National Laboratory (PNNL-20024).

Vanessa, I. (2007). Swiss precision. *Industrial Engineer, 39.*

ENDNOTES

1. TRIZ is a Russian language acronym for *Teoriya Resheniya Izobreatatelskikh Zadatch.* Translated into English it means "The Theory of Inventive Problem Solving."
2. Note: The task relates to the item number and time in days (i.e. W1/8 is the first task in the waste and scrap material plan and consumes eight days to complete).

Chapter 9
Electrical Financial Project Analysis

ABSTRACT

A comprehensive contemplation of the essential goals in an organization is essential to evaluate the capital integration problem. The development of cost tracking methods derived as a direct result of the study. The manager's deficiency in visible dashboards of the project cost and schedule data require immediate attention. The financial details includes the previous year's results, the median cost, the year-to-date average cost, the long-term average cost, the upper control limit, the lower control limit, the year-to-date cost, the last-year-to-date cost, the variance from last-year-to-date cost, the year-to-date budget, and the variance from the budget. The implementation of the Project Management Financial & Scheduling Optimization Tool in capital planning is required to control the individual project budget and corresponding schedule. The guidelines included checkpoint methodology, responsibility metrics, historical actual spend data, forecasting logic, and process management flow diagramming.

INTRODUCTION

A detailed understanding of the fundamental business requirements for an organization is required to assess the capital-intense system integration problem. The thorough evaluation of cost overruns within the project management organization based on an annual $730M capital budget within the Asset Management Department was consumed with "out-of-scope" projects well behind schedule. Several resources were displaced in various regions across the three state footprints with no systematic procedure in place to control the cost overruns. The project manager's ad hoc procedures lacked consistency within the organization. The implementation of the Project Management Financial & Scheduling Optimization Tool in capital planning is required to control the individual project budget and corresponding schedule. The guidelines included checkpoint methodology, responsibility metrics, historical actual spend data, forecasting logic, and process management flow diagramming. Project managers not only must rely heavily upon the traditional

DOI: 10.4018/978-1-4666-2839-7.ch009

on-time, on-budget, and on-quality performance measures, but also with the added *on-strategy* dimension central to managing project success and to alleviate leadership challenges. Many leadership tasks relate to developing a vision of the project outcome that is practical, yet capable of mobilizing and motivating team members to accomplish the project's goals and objectives. The leadership vision engages stakeholders who are not actively involved in the project; it also motivates them to sustain their support over the project's timeline. The solution-building negotiation approach to defining the scope of a project, and then clearly communicating this to the project team and other stakeholders defines a strategy for realizing the vision, and translating the strategies into operational plans and results. The core problem was the scarcity of instability of strategy to properly develop and express project vision connected through measurement to tangible business outcomes. The organization linked planned outcomes to their corporate strategy using a measurement framework, referred to as *performance management*, and is common within corporations (Thomas, 1975).

The implementation of these formal stringent guidelines provided clear project vision as well as comprehensible goals and objectives. It also actively involved the executives and sponsors in the project. The alignment of the project manager's resources via systematic tracking, contingency fund management, construction overheads, mitigation strategies, and project status communication produced profound results in the overall operating budget. Additionally, the project manager ensures that the accounting and planning departments are provided with accurate equipment data, costs, and plant-in-service dates. The implemented procedures imposed strict guidelines for the project manager to follow that included a final technical report (significant results, discussion of objectives and deliverables, examples of progress), a financial report (final invoices, cash flow), and resource performance evaluations (technical difficulties, feedback reports, activity verification, lessons-learned). Moreover, the financial and scheduling optimization tool allowed the project manager to utilize essential monitoring and control techniques during transformer installation. After an annual review of the process, the results produced an astounding fifty-five percent reduction in cost overruns in the transformer installation portfolio. The project managers were performing within the guidelines and saving precious budgeted dollars in the company. The Financial & Scheduling Optimization Tool provided a systematic approach to project management and aligned the company's objectives with the on-budget, on-quality, on-schedule, and on-strategy philosophy.

Synopsis

The Financial & Scheduling tool is designed to assist the project lead with effectively managing the master schedule, the monthly budget, and the overall project strategy. The six sigma-based effort produced business successes in:

- Cost reductions
- Productivity Improvements
- Cycle-time reductions
- Service improvements

The tool is an interactive, time-based system (and includes "what-if" analysis) designed from a project manager's perspective. The effort produced consistent monthly forecasting accuracy (80%~95% range) and included the following:

1. Improved transformer installation turn-around time
2. Improved material integration (46.5 days) and construction processes (25.5 days)

3. Reduced costs by $41k on an overall, per project basis
4. Utilized Analysis of the Variances methodology

PROJECT BACKGROUND

The sample projects selected for this endeavor involved substation transformer work based on industrial construction standards and the Project Management Philosophy. The integration of these processes encompassed transformer projects in the "Substation Design Engineering Completions," the "Three-month Look Ahead," and the "Summer Critical" projects. Several projects involved long time spans with evidence of little or no control of the expenditures. The regions of the Company's footprint were well represented in the study. The details (protection device settings, return retired equipment, delivery of distribution equipment, etc) of each of the projects were considered based on the master schedule and financial data. The data integrity of these values was based on the following assumptions:

1. All historical actuals and budget data is high quality without replacement/revisions.
2. All historical scheduling data is in congruence with the project outcomes.
3. The study period encompassed transformer and associated endeavors in a thirty-four month period.

The selected projects are depicted in Table 1. Other transformer projects were evaluated as well. However, the above table encompasses all of the major variables required to complete the "Design of Experiment." The financial portions of the study revealed little to no actual spend data for a specified period. The auspicious situation lead to the exploitation of the interactive financial sensitivity analysis tool designed specifically for the project manager. The fundamental probabilities of the monthly/annual forecast were converted into comprehensible values and included in the data as well. The flexibility of the model allows for Operation & Maintenance, Line Projects, and other endeavors based on the parameters.

Range Financial Chart

The calculation of the measures of financial dispersion is depicted in the range chart (Besterfield, D.H., et al. 1999). It is simply the series of monthly observations of the transformer actual spend data as compared with the budget and produce the difference between the largest & smallest values. Symbolically, it is given by the formula:

$$R = X_h - X_l$$

Median Financial Chart

The median chart is a simplified variable control chart that minimizes calculations of the financial data. The median control limits are determined below:

$$UCL_{Md} = Md_{Md} + A_5 R_{Md}$$

$$LCL_{Md} = Md_{Md} - A_5 R_{Md}$$

Where:

Md_{Md} = Grand median (Median of the medians)

A_5 = factor for determining 3s control limits

R_{Md} = Median of the subgroup ranges

The range control limits are determined from the formulas:

$$UCL_R = A_6 R_{Md}$$

Table 1. Controlled core transformer study projects

Project Name
A. Replace bank 1 and 2 with one 230 - 34.5 kV 125 MVA transformer and Build a 5-breaker at the "A" substation 230 kV ring bus. B. MODULAR Branch Substation, Install (34.5/12.5) 14MVA Modular Substation
C. Install 500/230 kV Transformer Oil Containment and Firewall
D. Replace Bank 2 transformer with 14 MVA add DX circuit
E. Replace No.4 Transformer
F. Install a 138/69 kV Transformer & Reconductor the 69kV Line
G. Substation – replace control system, thyristor valves and cooling system for the 230 kV service
H. Substations No.2 & No.3 138/69kV - Separate the two transformers
I. Replace #1 Transformer
J. Install 230/115 kV Transformer and 115 kV Breakers
K. Substation (add spare 64 MVA transformer)
L. Substation - Un-parallel substation transformers and upgrade low side transformer CT's
M. Install 115-kV Line to point "A" & with CB's and add 230/115-kV Transformer
N. Install New 36-13kV Modular Substation for Transformer Relief
O. Install R/P LA's on #1, #2, #3 Transformers
P. Replace failed Transformer
Q. Install R/P LA's on #3 & #4 Transformer
R. Project Switch house - Convert to New Substation
S. Transformer Replacement
T. Project "B" Add 138/69 kV Transformer
U. Install repaired 230-115 kV transformer from section "C"

$$LCL_R = D_5 R_{Md}$$

Where: D_5 and D_6 are factors for determining 3s control limits R_{Md}

Trend Projections and Regression Analysis

For a linear trend, the estimated transformer actual spend expressed as a function of time is:

Next Month = b_0 + b1* t

Where:

Next Month = trend value for actuals in period t

b_0 = Intercept of the trend line

b_1 = Slope of the trend line

Where: The formula for computing b_1 and b_0 is:

$b_1 = \Sigma t\ \Sigma y_t - (\Sigma t\ \Sigma y_t)/n$

$\Sigma t^2 - (\Sigma t)^2/n$

$b_0 = \bar{y} - b_1 * t$

t = 1 corresponds to the oldest time series value

t = 35 corresponds to the most recent time series value

Where:

y_t = Actual value of the time series in period "t"

n = number of periods

y = Average value of the time series

t = Average value of "t"

The regression analysis is utilized to relate the desired variable to forecast to other variables that are supposed to influence that variable. The range and median charts are shown in Figure 1 and Figure 2.

The chart monitors the average, or centering of the distribution of the data from the process (where is the data clustering?). The financial data for this category comprise of the contactors performing the work effort for a particular project. The general assessment of these data reveals the following conditions:

- Over Control
- Large systematic differences of the material/contractor quality (points near or outside the limits)

There are several methods utilized to correct detected problems in the financial realm. The key is the ability to monitor (track) critical financial categories and recognize potential anomalies, and provide the proper correction methodologies instead of recording the data and passing it to the stakeholders. The corrective action for this particular situation is as follows:

- Investigate material variation
- Eliminate managerial over-adjustment of the process
- Check the control limits
- Evaluate test procedures, inspection methods, and/or frequency

The chart provides and overall "movie" of the entire financial process on a month-to-month basis. A detailed view of the data is shown in Figure 2.

The chart monitors the range or the width of the distribution (how tightly clustered is the data?). Both of the charts are utilized in conjunction of the project financial data to control potential cost overruns and detect impending problems in the field. The general analysis of the above process is as follows:

- Mixture of a process of distinctly different quality (points near or outside the limits)

Figure 1. Median chart for contractor spend category

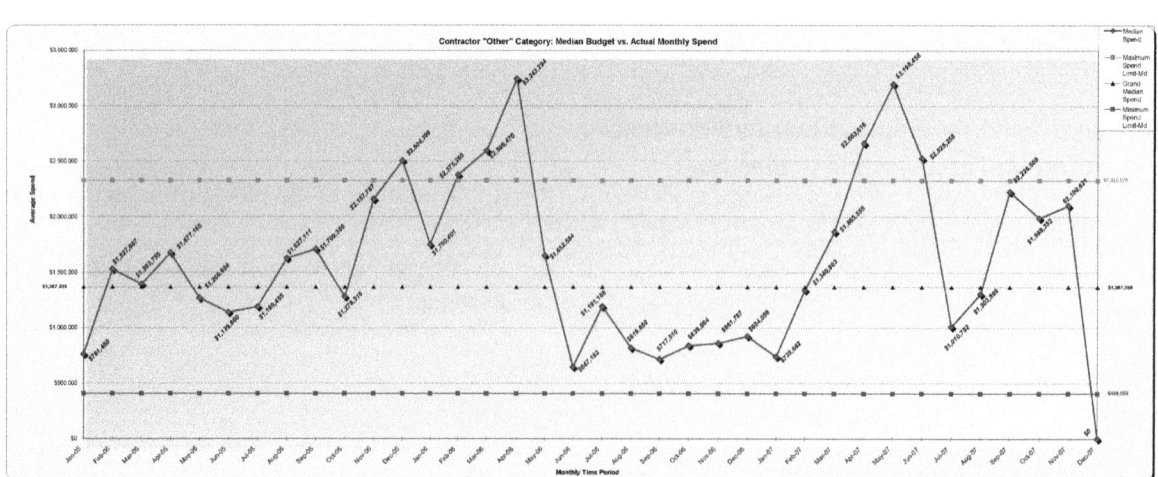

Figure 2. Range chart for contractor spend category

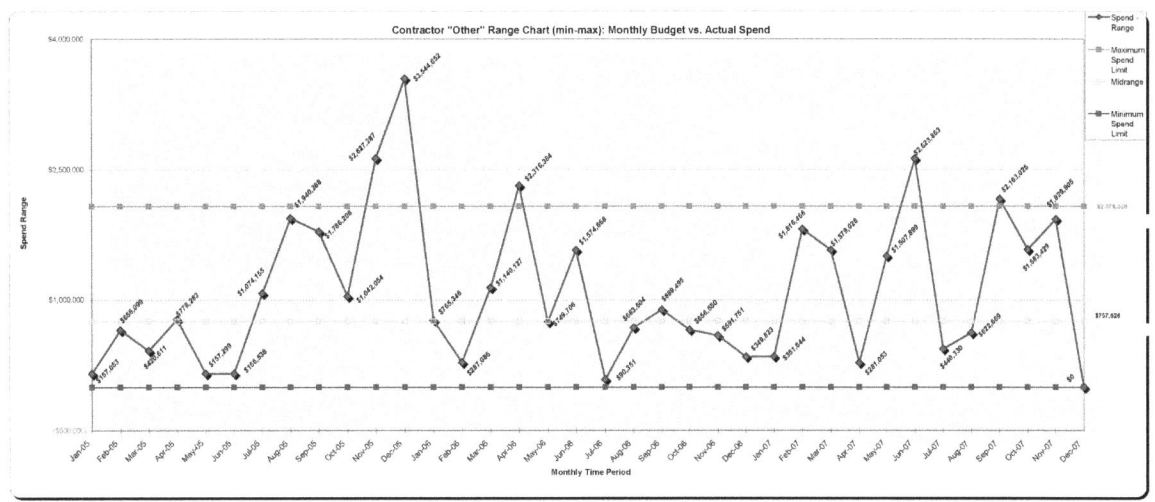

- Detection of dissimilar and diverse ad hoc procedures and/or field policies

The corrective action involves a variety of methods to incorporate the following:

- Investigate material variation
- Eliminate managerial over-adjustment of the process

Oftentimes when a manager obtain an apparently erroneous financial monthly report (data point out of tolerance), an over-adjustment of the process ensues for the next period. Suitable data analysis is designed to compensate for any potential inaccuracies in the process when planned appropriately.

Typical Project Request

Operating Company: Company "A"

Project Name: G. Substation – replace control system, thyristor valves and cooling system for the 230 kV service

Sponsor Need Date: May 1st

Estimated Cost: $2.5M

Project Justification/Reason: There have been more than 18 failures of the Static Var Compensator (SVC) controls over the last two years that have required the dispatch of personnel to the Substation. The SVC is critical to voltage control of the Central 230-kV and 34.5-kV networks. It provides for operability of these systems within the Transmission Planning Criteria requirements. This value was proven during a recent failure. Without the SVC, the 230-kV voltage has approached and exceeded the continuous design of the substation equipment and put it at risk.

The control system has repeatedly failed and become a maintenance issue. The manufacturer consultant was brought in to inspect the control system and thyristors. The conclusion was that the existing controls cannot be repaired and must be replaced. The thyristors are in poor condition and thus impossible to repair and in need of replacement as well. The building where the thyristors are located has inadequate cooling and results in sensitive equipment being exposed to outside conditions and being covered with contaminants.

An additional consultant was hired to perform a power flow and stability analysis to determine whether the SVC can be retired or is needed. The conclusion of this study states that it is needed to prevent wide-area voltage instability on the Bulk Electric System, as well as over-voltage conditions. Both the Dispatch Operations and the consultants agree that the SVC is necessary. The Manufacturer concluded that the SVC is not reparable and must be replaced.

More than 18 failures of the Atlantic SVC controls have occurred over the last two years

The above-proposed project is a classic condition problem in the electrical utility sector. The installation of an SVC in a network can increase the transfer capability, reduce losses, and maintain a smooth profile under varying conditions (Grubaum & Pernot, 1999). Moreover, the device can mitigate active power oscillations via voltage amplitude modulation and provide the required reactive power for dynamic voltage control. The small building is designed to house the thyristor cooling equipment and associated equipment included in the installation process. The work effort is typically ignored until the unit becomes problematic and requires immediate attention. The proactive approach would have eliminated the need after early failure detection (well within the two-year period experienced by this utility company). In retrospect, the utilization of "Beforehand Cushioning" is compulsory. It is a method to *prepare emergency means beforehand*

to compensate for the relatively low reliability of an object, activity, situation, or position (Mann & Domb, 1997).

- Contingency planning
- Establish a worst-case, fall-back position prior to negotiation - 'Best Alternative to a Negotiated Agreement'
- Encourage short, effective meetings by removing the chairs
- Put clauses in contracts requiring arbitration/mediation to avoid litigation
- Begin with 'S' *study* in the Plan, Do, Study, and Act (PDSA) cycle

The electric utility's cost of $2.5M was inescapable because of the insufficient planning/intensive study in this case.

The tools specifically developed for a project manager to track an approved, vetted venture include statistical process control, bar charts, and value hierarchy diagrams. The installation for a substation transformer involves several entities, resource coordination, and contingency analysis for a successful venture. The implementation of the project management philosophy is essential to accomplish even the smallest stage on the entire process. An example of the overall scheme of the transformer installation process is depicted in the

Figure 3. Condition project justification

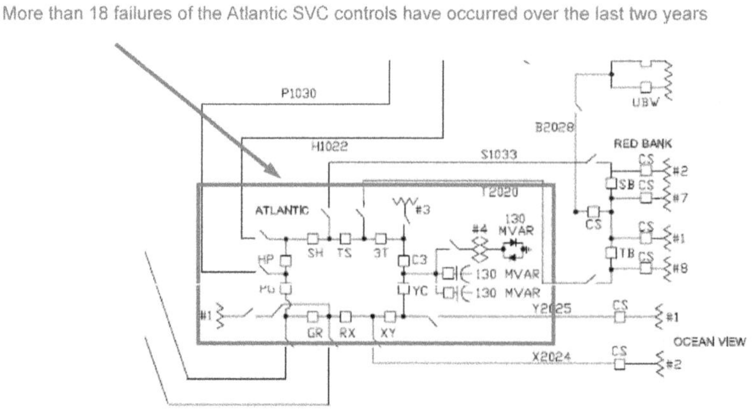

"Surveillance & Witness Assessment" diagram in the Appendix.

The waterfall diagram depicts the delay problem areas as perceived from three (substation design, summer critical projects, and three month transformer project forecast) very different viewpoints. The "Waterfall" diagram is shown in the Appendix. The design of experiment results, with conclusions and recommendations, is depicted in the Appendix.

Problems inevitably occur and impediments of the project timeline are eminent. The project manager with the quality tools systematically evaluates the delays. The transformer study isolated several of the delay predicaments and provided viable solutions. The conversion rate factor equated to 40% (ideal is 90% and above) for the transformer installation process. Although several of the major delay problems were identified in previous chapters (engineering delay, resources, outage clearance, scope change, permits, material, design clarifications, weather, accidents, telephone, equipment availability), additional localized dilemmas impede the progress as well. Some of these regional problems are outlined below:

- Permit problems are unique in one eastern seaboard region and prolong the project delay.
- Communicate the milestones need dates (from project managers to engineering). Typically, a two-month lag date exists in the region for the electronic drawings and scope documents.
- There is a growing need for local construction management to oversee contractors, safety, materials, and day-to-day activities. The utility line shops were deemed as overloaded with a projected backlog of ventures by the end of the fiscal year.
- Updates were verbal and undocumented for several of the modular substation installation projects.

- Known contaminated substation sites created construction contentions. The soil samples revealed satisfactory results.
- The stakeholder need dates vs. the electrical in-service dates were unrealistic between the field crew and the management department. The inconsistency was rampant across several projects in the region.
- Endangered species (pine snakes) were encountered during the modular substation installation. When the ground is broken and reveal the snakes in this case, additional permits are filed with the regional Department of Environmental Protection.
- Incorrect electrical phasing at several modular substation sites was experienced due to poor communication, design clarifications, and lack of resources. Poor clearance on the cross arms of a structure was experienced in another project (utilized "clip-ons" to compensate).

THE VALUE HIERARCHY DIAGRAM

One major factor used to exploit the problem areas is the Value Hierarchy diagram. It was designed to isolate specific parts of the overall transformer installation scheme to minimize the turn-around time. The project manager monitors two critical sections of the venture (cost and schedule) and provides detailed input based on the current data. Sliders are employed to visually display the cost/schedule variable and its relation to the potential delay of the project. A cost example is shown in Figure 4.

As each category improves (slider towards zero), the variable bar and associated cost will decrease accordingly. Conversely, if the supplier relations deteriorate for example, the cost will increase proportionally. The total expenditure is reflected by numerous factors to include a percentage of the processing cost. The maximum range cost of the controls is specifically designed for

Figure 4. Slider controls for material cost

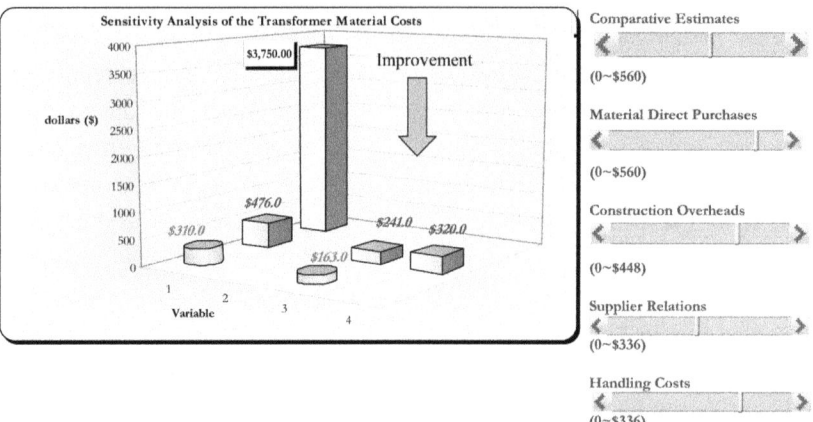

the transformer project within the portfolio based on the particular region. It is imperative for the project managers to expand the estimates and validate the costs accordingly in order to maintain the integrity of the model. Each portion provides a cost input to fifty percent of the overall improvement effort. The transportation cost, internal labor, contingencies and overtime, are expenditures that are rarely documented in the transformer installation process. The Value Hierarchy diagram captures these critical costs and exploits the vulnerabilities associated with the processes. The project manager and related stakeholders are forced to ask the obvious questions if a procedure is required for suitable transformer implementation. A portion of the scheduling analysis is shown in Figure 5.

The scheduling controls are directly related to potential delays in materials, coordination, engineering design, permits, etc. The range is proportion to the project specific parameters. The "Engineering Design Process" (27% of the scheduling input) is an example of the delay (0 to 17 days) procedure. When design clarification is required, the seventeen day maximum is experienced based on the protocols. The exploitation of specific categories will improve the entire transformer implementation process. Tribulations with the "Outage Process" (3% of the scheduling input), as another example, will also amplify the delay factor. Ideally, a cognizant project manager must reduce the timeline near zero (minimize the delays) to achieve a "Lean" process. The Value Hierarchy diagram is shown in the Appendix.

As each category in the Value Hierarchy diagram improves, a corresponding link is displayed in an Overall Transformer Improvement chart. It is a visual display designed for the project manager to readily assess the venture's progress. The detailed parameters linked to the overall display are depicted in Table 2.

The processing, material, and best/worst case costs and schedule are included in the analysis. The detail allows the project manager to examine the overall installation cost from a "bird's-eye" view and capture any potential cost overruns before the problem exacerbates. The Value Hierarchy diagram also provides a localized visual of the entire process to isolate a problem and supply the necessary resources for a correction. These tools were developed as a direct result of the lack of an adequate cost/scheduling tracking mechanism in transformer study. The process was out of the control limits 90.63% of the time in the range chart. In the median chart, the process was out of control 37.5% of the time when compared

Figure 5. Slider controls for scheduling delays

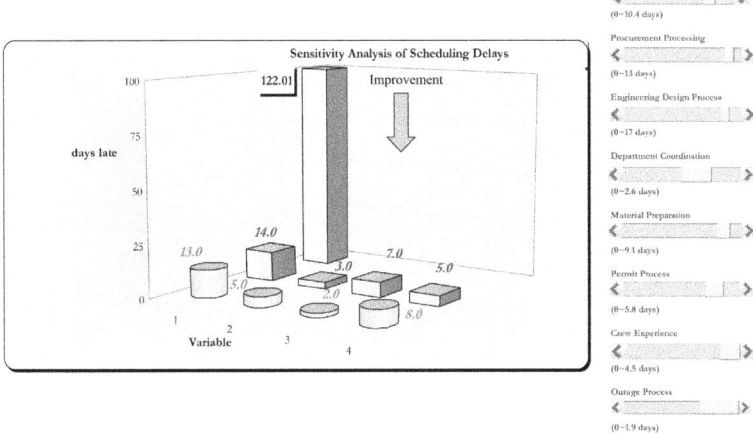

with the actual spending data and the monthly budget. The common practice of the project managers involved "chasing a false budgetary figure" on a monthly basis. If an "under-budget" condition existed for a project one month, the project manager would erroneously continue to remain in this condition for several additional months, depriving resources and mis-managing (over control) the project as a whole. The actual process was not allowed to perform naturally and cross the financial zero to the positive region. The study also discloses, "Over-spending" when the project manager realized the accumulated cash were in various Federal Electric Reliability Commission (FERC) financial buckets. This condition usually occurred near the end of the year and May 1st (summer critical deadline for project implementation). Additional resources were utilized during this time to accomplish the goal and consequently increased the project cost. Another negative attribute that contributed to the overall transformer installation process were the incorrect material estimates. The adage of "garbage-in equated to garbage-out" or (GIGO) was the rule rather than the exception. The cost estimates were found to be in error more than 95%, as accurate data was introduced in the study. Moreover, the "groupthink" condition existed in the engineering/project manager spheres when major equipment was installed in the field (Thomas, K. 1975 and

Figure 6. Overall transformer improvement chart

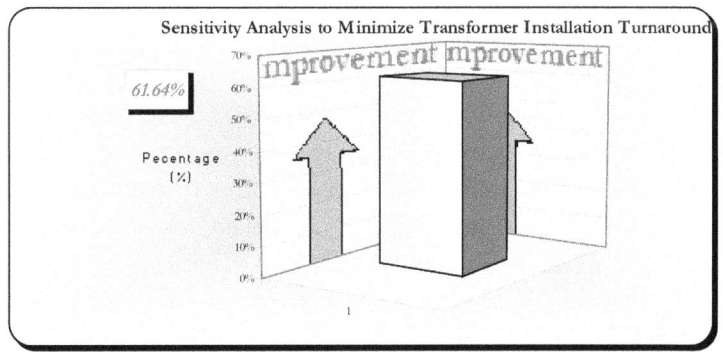

Table 2. Transformer cost parameters

Total Cost Improvement:	4317.000238
40% of Schedule (worst)	73.80487805
40% of Cost (worst)	2590.200143
95% of Schedule (best)	116.8577236
95% of Cost (best)	4101.150226
Best Case I (100%):	**190.6626016**
Best Case II (100%):	**6691.350369**
	6882.01297
	0.616404718

Wisler, D., 2003). The groups were impervious to any wrongdoing in the transformer endeavor. However, the construction crews consistently discovered incorrect drawings and materials, inadequate procedures, and pitiable parcel conditions. These poor circumstances were reflected in the study in the form of cost and schedule. Additional tracking tools and methodologies were introduced to the project manager to provide proper control of the financial and scheduling processes.

PROJECT COST TRACKING METHODOLOGIES

The development of cost tracking methods derived as a direct result of the study. The manager's deficiency in visible dashboards of the project cost and schedule data required immediate attention. The Transformer Project Spend Portfolio chart provided one method to monitor the cost of the projects (Figure 7). Although the initial project estimates were incorrect, the data was utilized to demonstrate the acceptable methodology to track the endeavor's progress. The financial details includes the previous year's results, the median cost, the year-to-date (YTD) average cost, the long-term average cost, the upper control limit, the lower control limit, the year-to-date (YTD)

cost, the last-year-to-date, the variance from last-year-to-date cost, the year-to-date budget, and the variance from the budget.

The chart employs regression analysis to produce the monthly and year-end financial forecasts and trending data as well. The inclusion of similar transformer implementation project historical data provides greater accuracy within the model. The key areas of the tracking chart are the monthly spending and variance from the erroneous budget. The project manager can immediately assess the financial situation of an individual project once the mechanism is implemented and updated. The transformer study proved the effectiveness of the financial tool in a series of "snap-shots" derived from the statistical control charts (median and range). In the utility business, several regions were not engage in the monitoring technology and simply permitted the cost overruns without investigating the specific details of the cause. Moreover, the data logging of pertinent historical data for specific projects was non-existent. Consequently, the project managers were allowed to exercise a "best guess" methodology that exasperated the pseudo financial budget constraints with each endeavor going forward.

The end-of-the year data yields interesting results. The screening objectives immediately discerns the last year actual spend and the current year-to-date actual (more in 2006 at $1.1M

Figure 7. Transformer financial spend chart

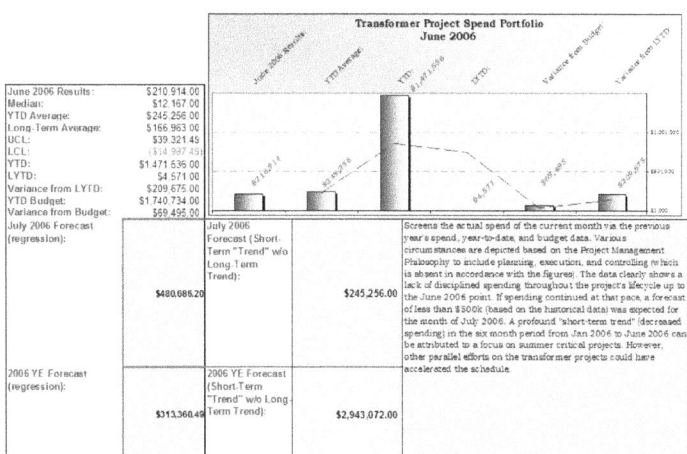

variance). The charts clearly displays an increase in actual spend during this period (24 to 25 in Figure 11) based on year-end dollars. The financial model is extremely accurate in this transformer portfolio (prediction of $3.43K in November for December's forecast $3.95K ~ 87% accuracy with regression analysis). However, the short-term model predicted $2.02K (a 51% accuracy rate) for the month of December. Based on the amount of data in this time period, the long-term year-end forecast prediction in November was $2.62M (72% accurate based on the 2006 budget). When compared with the actual spend (in hindsight), the accuracy of the November forecast increased profoundly ($2.61M/$2.62M = 99.67%). The model maintained its accuracy throughout the test periods despite the erratic behavior of the actual transformer spend. The financial tool is designed to assist the Project Manager with accurate forecasts and cost management objectives throughout the project's lifecycle.

The chart in Figure 8 depicts project progress of the transformer spend patterns. The sporadic spending throughout the year typically increased by year's end as the managerial pressure mounted to exhaust the FERC financial buckets. The data echoed this practice in several regions in the utility industry. The financial mechanism is an excellent tool to compare month-to-month and year-to-year data on a per-project/group basis as well as a reliable forecasting instrument.

The "What-If" Analysis is a powerful tool for the project manager. As a prime coordinator of the work processes for the substation transformer installation, s/he is most familiar with the system integration and cost factors involved on a daily basis. The contingencies and accruals are easily managed with the financial tool. The examination of the "Best Case Scenario," the "Worst Case Scenario," or even in between, the forecast ranges from prior months allows the project manager to accurately predict his/her spend. The historical data (a virtual benchmark) is vital for the accuracy in this model. The financial tool may also be utilized for future budget modeling.

The flexibility of the financial tool introduces various scenarios within the project's sphere. Nonetheless, the transformer study revealed data outliers from the previous year (as well as erroneous forecast data) and large variances from the pseudo budget. Given this environment, the project managers were simply trying to control to a "moving financial target" and created a condition of over control. Since the project managers were essentially "blind" in the financial control of the ventures, the over adjustment of the process was inevitable.

If the financial process is permitted to operate naturally, the data will fall within its prescribed range (UCL=$39,321.49, LCL= -$14,987.49 and Grand Median= $12,167.00 in Figure 10). Interestingly enough, the grand median shifted to $176,202.25 in the latter half of the financial evaluation (Figure 11) as the actual spend was well below the erroneous budget. The managerial over-control is obvious in the process since the data failed to run its natural course. In addition, timed-disbursements at critical points in the project path were more appropriate for financial control in for the entire scheme. The transformer study revealed undisciplined financial control, haphazard project management techniques, and a "fire drill" approach to equipment installation. Nevertheless, a pilot of the financial mechanism in a project equipment implementation produced predictable results, reasonable budget history, and practical "lessons learned" material for future reference.

Regional Cost Forecasting Variance

Tasking the various utility regions with forecasting the entire actual capital project spend versus the budget was a daunting endeavor without any financial mechanisms (Figure 12). Evaluation of the data range was acceptable for such a large process; nonetheless, the regional manager ad hoc forecasting techniques were consistently inaccurate (bottom line in the graph). Stratification of the data near the center of the median (March to June) warrants a detailed investigation of specific capital projects in the portfolio. Perhaps an increased effort in the summer critical projects found its way on the higher priority list and extracted resources from the planned ventures. The managerial decision to payback a particular expenditure at the end of the year ($5.6M) brought the process out of the control limits instead of utilizing planned disbursements. Even though the process was relatively in control 80% of the time, the forecasting data was inaccurate 92.30% of the time for the small region. The financial mechanism will drastically improve the financial forecasting accuracy in this case.

Figure 8. Transformer financial spend chart (November)

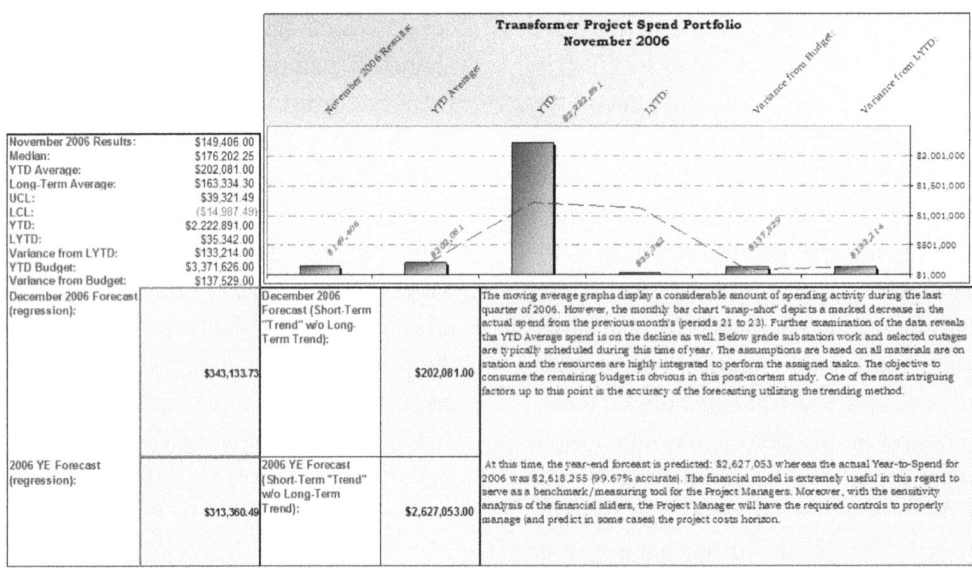

Figure 9. Transformer financial spend chart (February)

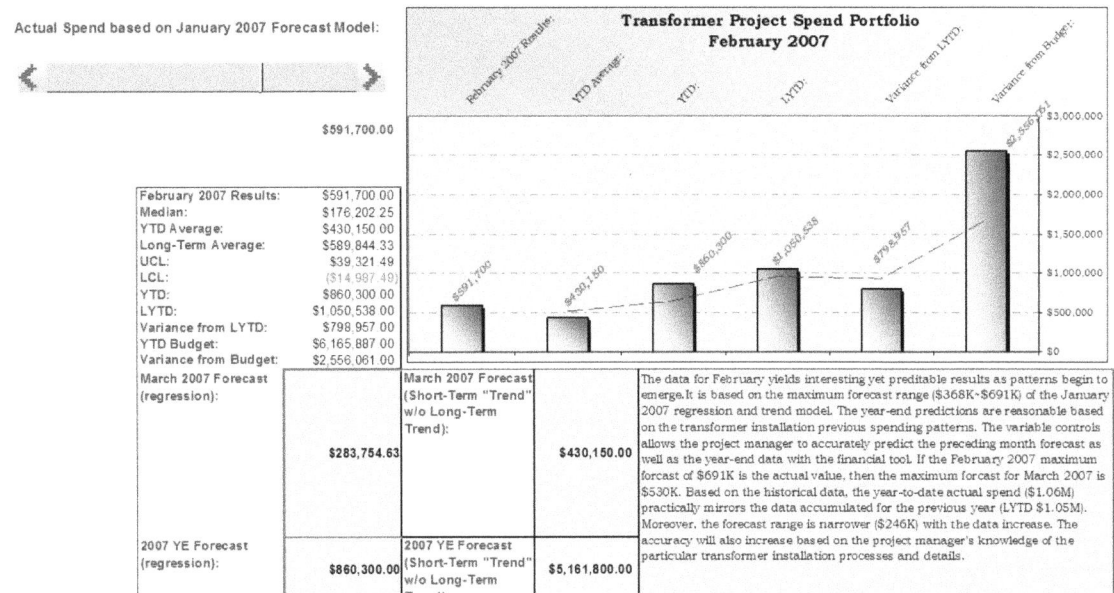

Project Cost Estimates

The regional construction cost estimates are derived from various vendor quotes without regard to its accuracy. The planning engineers typically provide a "best guess" estimate based on the time of day, current weather conditions, or the workload on his /her desk at the time. The correct and acceptable cost estimation for a proposed capital project involves detailed analysis of previous

Figure 10. Transformer financial spend chart (April)

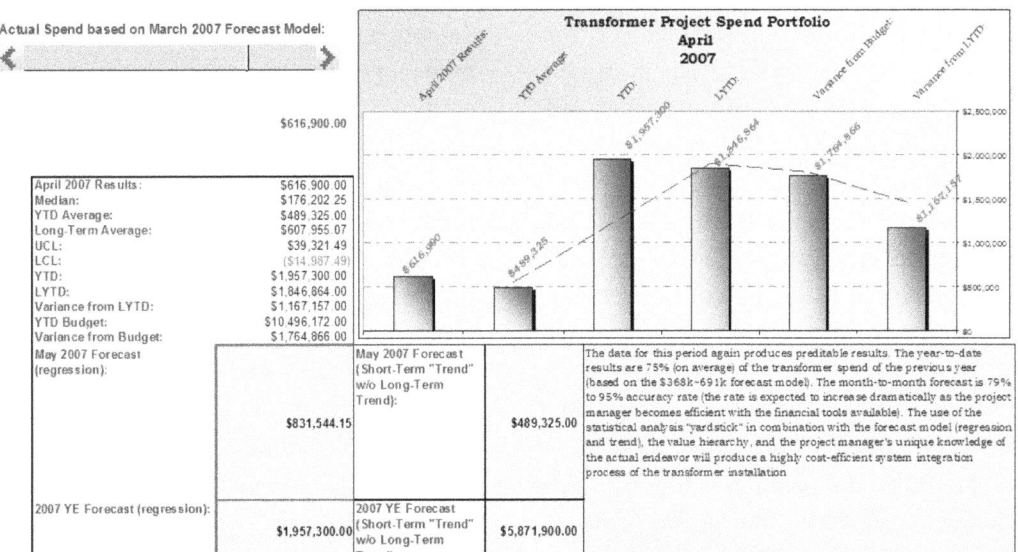

Figure 11. Transformer projects median vs. actual spend data

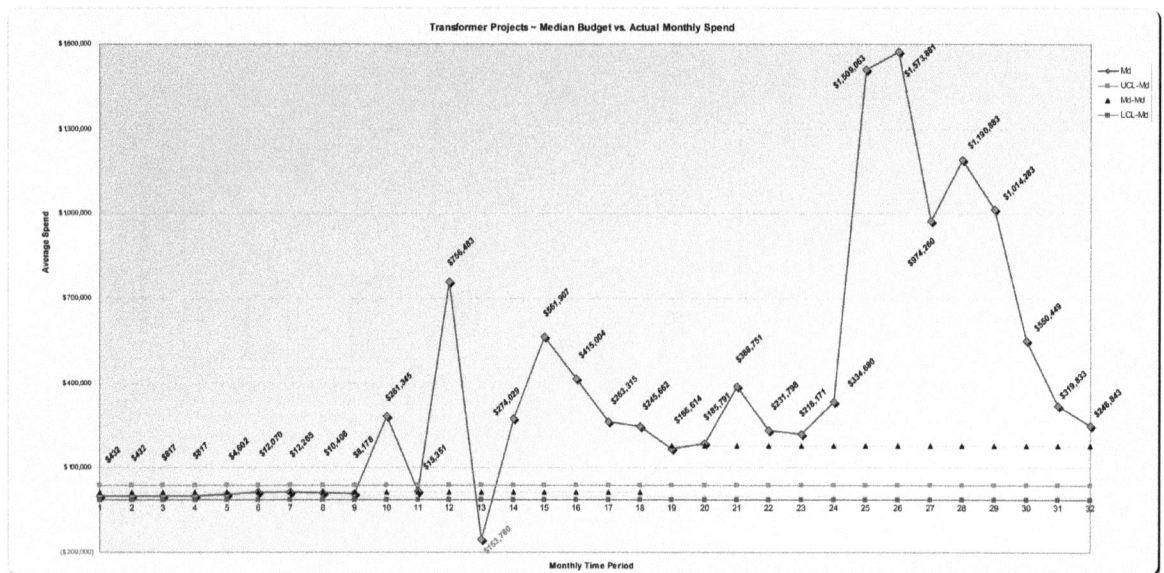

ventures and actual resources "on-the-ground" (manage by walking around concept previously discussed). Each construction project in particular is relatively the *same* yet *different* in the logistics, resources, and materials. One electrical utility company utilizes a "one-size-fits-all" approach when developing cost estimates with a standardized template. The uniqueness of the capital project is automatically void when this method is employed. Some projects require "rock blasting" whereas others do not. A capital project may require extensive soil resistivity testing in an area whereas others do not. Depending on the scheduled internal labor component, a project may require additional external (contracted labor) as well. This cost alone typically inflates the total capital project expenditures, especially during the summer critical period. An example of the detailed costs entailed in the below-grade construction is outlined in Figure 13.

The use of financial simulation tools greatly increases the accuracy of the cost estimates. It provides the necessary spend data range for the project under construction. The project manager may elect to employ the sensitivity analysis com-

ponent provided by the tool in order to decrease/increase the data range as the contingencies (rule of thumb of 10% ~ to approximately 5%) and element cost become more accurate. The minimum and maximum cost estimates are rarely implemented in the large construction capital projects and the contractor change orders usually escalates the expenditures once the contract is awarded. Nevertheless, the element min-max data range provides a critical cost range for the budget development and historical lessons-learned databases for future endeavors. Decision-makers can establish a minimum baseline of a project (for example $96,027 for below-grade construction) based on the project's success and lack of contingencies. In fact, if this method is utilized to its full potential, incentives/penalties can be implemented in specific construction contracts based on similar conditions, regions, and materials. The median cost of the below-grade work in Table 3 above is $113,097 with contingencies (Allowable Funds Used during Construction). The project manager may elect to investigate specific cost elements to obtain more "bang for the buck" based on the technology advances (i.e. battery & charger,

Figure 12. Regional capital projects median forecast vs. actual spend data

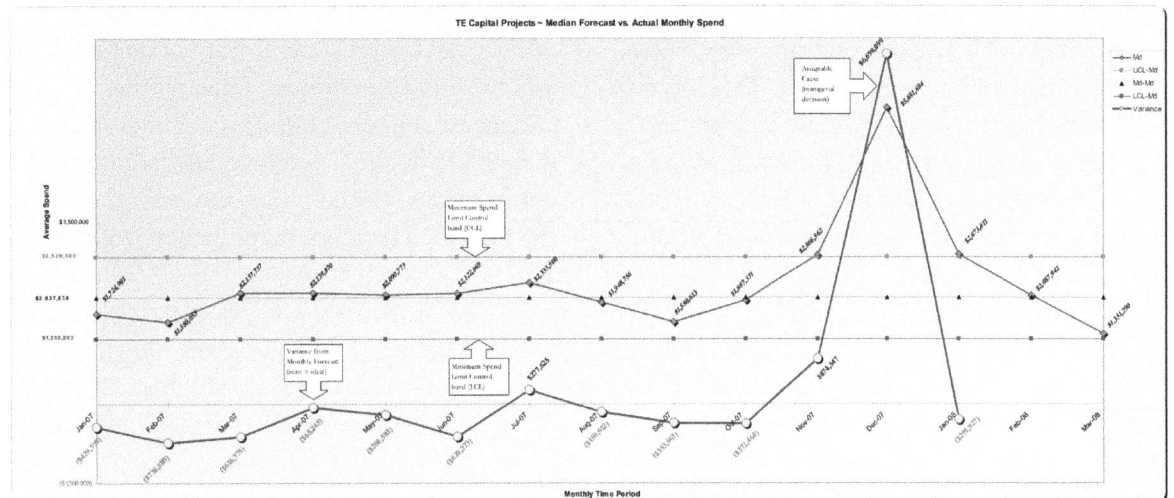

circuit breakers & switchers) experienced in the field. An example of the specific cost elements are in the above-grade construction cost estimates are depicted in Figure 14.

The element details are adjusted in accordance with the relative and prevailing cost structure. The astute planning engineer, project manager, and/ or supply-chain expeditor will investigate the material cost to ensure the expenditures are acceptable. The total project calculations (Above and Below-Grade construction costs) are summarized to include the minimum/maximum, and expected costs. The entire financial process is designed to provide an accurate benchmark for the work effort based on the fundamental historical data of similar endeavors. When implemented appropriately, the process will eliminate cost overruns, unnecessary delays, and communication issues generally experienced with labor-intensive capital projects. A specific electrical company improved its financial forecasting with this technique and developed a viable capital budget for the following year (Figure 15).

The scheduling data is simulated in various software programs as well and occasionally implemented in the project management process. The decision-makers rarely review the *data range* of

a project status report and usually focus on the outliers/anomalies. Nevertheless, the correct performance of the entire process (indicators, data interpretation) is an important tenet for the success of the complete project. The schedule data for the above-grade construction horizon is depicted in Figure 16.

Detailed Regional Cost Breakdowns and Funding Oversight

In this regional study, the 69-kV area encompasses the most projects (30.94%), yet only consumes $513M (Figure 14). This particular distribution segment falls within the "Condition Projects" and require immediate attention before an overload situation is experienced. Conversely, the proposed projects with the highest costs ($24B) have one of the lowest quantities (4.17%) in the overall scheme.

The 765-kV line transmission projects are also required to alleviate the bottlenecks in the electrical grid at these higher cost estimates. The tradeoff between operations & maintenance dollars vs. the allocated capital expenditures becomes indistinct during the project prioritization process (Figure 18). Both entities are competing for the same dollars based on varying need and indistinctive requirements. As the kilovolt value increases, the cost increases proportionally. The conflict lien the lower kilovolt line (distribution) projects where most of the work effort is essential. The upward trend of the project quantity in the chart is clearly in conflict with the estimated costs (decreasing trend). A thorough assessment of each proposed project sector is required to balance the portfolio for venture prioritization.

Fiscal management is a critical component espoused within the engineering and project management spheres. The development of resource plans and budgets for capital projects must be scrutinized for the prearranged expenditures to ensure spending is within the appropriate allotments. Moreover, the administrative control over the funds, contracts, and procurements to is also an essential tenet of the paradigm to prevent mis-management. An independent regional task force assigned must be assigned to investigate problematic capital projects within the electric utility company, and develop penetrating questionnaires for specific managers before scheduled discussions. The scope of one particular business case encompassed the North American Electric Reliability Corporation (NERC) Blackout Report Section 8A mandating minimum relay setting loadability criteria. The project included upgrades to the relay and communications equipment on a

500-kV terminal. There were also control-wiring changes required on a 345-kV terminal. The project was on schedule to exceed its budget by 138.89% and the driving force for the account variance entailed the hiring of contract engineers dictated by internal resource allocation and potential delays. The compliance meetings revealed the following results and mitigation strategies:

- Contractor over-charges was the underlying theme of the meetings.
  - Project Managers automatically identified potential contractors when assigned a venture without investigating alternatives (*any* available contractor satisfied the criteria).
  - The contractor's engineering schedule *generally* dictated the entire project's timetable.
- The solution-based initiatives included the following:
  - The Project Managers must be able to apply and thoroughly understand the three (3) major contract types during negotiations:
    - Fixed-Priced or Lump Sum where the buyer has the least amount of risk.

Figure 13. Below-grade construction cost details

	Engineering	Contracted Construction									
	Cost $	Material Cost $	Labor $	Labor Contingencies	Material Contingencies	Engineering Contingencies	Total Element Cost $	Total Element Cost (Values Only) $	Minimum	Maximum	Simulated
BELOW GRADE CONSTRUCTION				10%	10%	10%					
Clearing of trees/brush	$0	$0	$0	$0	$0	$0	$0	$0	$0	$0	$0
Concrete/foundations	$2,269	$10,992	$21,960	$2,196	$1,099	$227	$38,743	$38,743	$32,877	$44,550	$38,713
Conduit	$812	$2,096	$26,136	$2,614	$210	$81	$31,947	$31,947	$27,107	$36,759	$31,933
Drainage	$0	$0	$0	$0	$0	$0	$0	$0	$0	$0	$0
Fence	$0	$0	$0	$0	$0	$0	$0	$0	$0	$0	$0
Grading	$0	$0	$0	$0	$0	$0	$0	$0	$0	$0	$0
Grounding	$1,369	$2,406	$4,440	$444	$241	$137	$9,037	$9,037	$7,692	$10,394	$9,043
Landscaping	$0	$0	$0	$0	$0	$0	$0	$0	$0	$0	$0
Oil Containment	$1,800	$17,490	$7,680	$768	$1,749	$180	$29,667	$29,667	$25,168	$34,165	$29,667
Roadway	$0	$0	$0	$0	$0	$0	$0	$0	$0	$0	$0
Soil Resistively Testing	$410	$0	$1,140	$114	$0	$41	$1,705	$1,705	$1,453	$1,959	$1,706
Soil Borings	$0	$0	$0	$0	$0	$0	$0	$0	$0	$0	$0
Stone	$51	$836	$960	$96	$84	$5	$2,032	$2,032	$1,730	$2,340	$2,035
	0	$0	$0	$0	$0	$0	$0	$0	$0	$0	$0
Sub-Totals:	$6,711	$33,820	$62,316	$6,232	$3,382	$671	*$113,132*		$96,027	$130,167	$113,097
Cost+Contingencies	*$7,382*	*$37,202*	*$68,548*								
Other Direct Construction Costs	$23,992					Total Calculated Cost:	*$113,132*				

	Total Below Grade Construction Cost Estimate:	*$113,097*

Table 3. Detail of identified major assets

Circuit Breaker/ #Switch	ID#	Wiring Diagram#	Name	Model	Serial #	Manufacturer	Year	Usage	
8W8	1391	151	Deli	FK145	139A66252	GE	1970	Breaker	
9W8	1746	1262	Tart	145PM40		ABB	2002	Breaker	
45W8	173	1117	Layn	AHE54	19272	McGraw	1971	Breaker	
45B7	526	1117	Layn	AHE54	19374	McGraw	1971	Breaker	
199	582	32	Mio	GM4	34Y8V22	Westinghouse	1948	Breaker	
199	436	495	Gay	GM4	14Y8V20	Westinghouse	1948	Breaker	
199	733	680	White	FK439	K656147G20	GE	1956	Breaker	
177	1364	1090	Scott	145PM	10134S02	ABB	1995	Breaker	
188	124	1090	Scott	145PM	10085S01	ABB	1994	Breaker	
35W8	1439	980	Spade	145PM40	138Y2D11	ABB	1968	Breaker	
199	686	276	Twin	GM4A	138Y2Y28	Westinghouse	1968	Breaker	
8B7	368	151	Deli	FK145	139A6625U07	GE	1970	Breaker	
177	862	1305	W. Road	EV	10087Y01	S. States	1993	Switch	
177	901	2023	Project #1	MK40A		HK Porter	2001	Switch	
188	1190	1305	W. Road	EV	B00078T02	S. States	1998	Switch	
188	902	2023	Project #1	MK40A	10273T02	HK Porter	1997	Switch	
199	1499	1117	Layn			Pennsylvania		Switch	

- Cost reimbursable contracts where the buyer has the most cost risk. It may include incentives.
- Time and Material contracts where the buyer has medium amount of cost risk and based on per hour or per item. This is a combination of the above two.
- The contract must include clarification of the responsibilities/authorities and the technical business management approaches.
  - The Project Manager must be engaged in some capacity during contract negotiations. The dialog involves clarification and mutual agreement on the structure and requirements of the contract prior to its signing.
  - Increase competition by placing the contractors in the same room for bidding and/or bundled work.
  - Resource tracking & activity based costing (monitor change orders, testing results) to verify ongoing cost effectiveness.
  - Key performance indicators for project process evaluation strategy:
    - Contingency delay cost/consequences and delay factors.
    - Assessment of construction/material installation.
  - Cost and schedule scenario analysis
    - Resource leveling and coordination.

- Increased resources (crashing techniques evaluation).
- Cost boundary evaluation
- Wait vs. Idle cost resource analysis.

These sessions also revealed the inevitable contract disputes based on the general boilerplate(s) developed by the legal department. One approach of dispute avoidance is to evaluate (proactively) unanticipated impacts of the project's cost and schedule overruns (Godlewski, E., et al., 2012). The establishment of ground rules and the utilization of a collaborative process to involve every team member is intended to facilitate a satisfactory resolution of the conflict (Levasseur, R.E., 2011).

The mitigation strategies included an increased managerial presence (the personnel most familiar with the work effort) during the contract negotiations. The dispute resolution consisted of various levels of mediation, to avoid the high cost of litigation. The general dispute resolution mediation levels are outline below:

1. Mediation (Non-Binding)
2. Mediation (Binding)
3. Mediation (Pre-Mediation Binding Mediation or Arbitration where the parties share expenses)
4. Binding Mediation – Arbitration (Graduated Process)

Figure 14. Above-grade construction cost details

		Engineering	Contracted Construction						Total Element Cost $	Total Element Cost (Values Only) $		Minimum	Maximum	Simulated
		Cost $	Material Cost $	Labor $	Labor Contingencies 10%	Material Contingencies 10%	Engineering Contingencies 10%							
ABOVE GRADE CONSTRUCTION														
1 Lot	Animal Protection	$1,336	$2,610	$8,640	$864	$261	$134	$13,845	$13,845		$11,767	$15,926	$13,847
1 Lot	Battery & Charger	$720	$2,063	$1,800	$180	$206	$72	$5,042	$5,042		$4,286	$5,797	$5,042
	Buildings	$0	$0	$0	$0	$0	$0	$0	$0		$0	$0	$0
	Bus Duct	$0	$0	$0	$0	$0	$0	$0	$0		$0	$0	$0
1 Lot	Cable, Control, 600V	$600	$757	$5,820	$582	$76	$60	$7,894	$7,894		$6,710	$9,088	$7,899
	Capacitors	$0	$0	$0	$0	$0	$0	$0	$0		$0	$0	$0
	Carrier Equip.	$0	$0	$0	$0	$0	$0	$0	$0		$0	$0	$0
1 Lot	Circuit Breakers/Circuit Switchers	$10,000	$38,453	$6,000	$600	$3,845	$1,000	$59,898	$59,898		$50,910	$68,897	$59,904
1 Lot	Conductors & Fittings	$1,710	$19,263	$40,380	$4,038	$1,926	$171	$67,488	$67,488		$57,427	$77,542	$67,485
1 Lot	Conduit	$73	$1,105	$2,400	$240	$110	$7	$3,935	$3,935		$3,339	$4,521	$3,930
	Distribution Cabinets, AC & DC	$0	$0	$0	$0	$0	$0	$0	$0		$0	$0	$0
	Fused Disconnects	$0	$0	$0	$0	$0	$0	$0	$0		$0	$0	$0
1 Lot	Insulators	$202	$726	$540	$54	$73	$20	$1,615	$1,615		$1,371	$1,855	$1,613
	PT & CT's	$0	$0	$0	$0	$0	$0	$0	$0		$0	$0	$0
	Reactors	$0	$0	$0	$0	$0	$0	$0	$0		$0	$0	$0
	Regulators	$0	$0	$0	$0	$0	$0	$0	$0		$0	$0	$0
1 Lot	Relays, Metering, Telephone & RTUs	$7,355	$14,300	$3,360	$336	$1,430	$736	$27,517	$27,517		$23,416	$31,694	$27,535
	Station Service Transformers	$0	$0	$0	$0	$0	$0	$0	$0		$0	$0	$0
1 Lot	Steel Structures	$1,400	$9,562	$4,440	$444	$956	$140	$16,942	$16,942		$14,401	$19,456	$16,928
	Surge Arresters	$0	$0	$0	$0	$0	$0	$0	$0		$0	$0	$0
1 Lot	Switches - Group Operated	$1,800	$6,050	$7,680	$768	$605	$180	$17,083	$17,083		$14,525	$19,659	$17,092
	Switches - Load Break	$0	$0	$0	$0	$0	$0	$0	$0		$0	$0	$0
	Switches - Regulator Bypass	$0	$0	$0	$0	$0	$0	$0	$0		$0	$0	$0
	Switches - High Speed Grounding	$0	$0	$0	$0	$0	$0	$0	$0		$0	$0	$0
	Switches - SPST Disconnect	$0	$0	$0	$0	$0	$0	$0	$0		$0	$0	$0
1 Lot	Switchgear	$11,000	$152,206	$8,640	$864	$15,221	$1,100	$189,031	$189,031		$160,650	$217,460	$189,085
	Transformer - Mobile (Installation)	$0	$0	$0	$0	$0	$0	$0	$0		$0	$0	$0
1 Lot	Transformer - Power	$9,400	$683,000	$7,800	$780	$68,300	$940	$770,220	$770,220		$665,714	$884,492	$770,103
	Contracted Engineering Supervision (10%)	$0			$0	$0	$0	$0	$0		$0	$0	$0
1 Lot	Field Testing / Commissioning			$78,900	$7,890	$0	$0	$86,790	$86,790		$73,554	$99,853	$86,704
	Adjusted engineering	$0			$0	$0	$0	$0	$0		$0	$0	$0
	Sub-Totals:	$45,596	$930,094	$176,400	$17,640	$93,009	$4,560	$1,267,299			$1,088,070	$1,456,240	$1,267,187
	Cost+Contingencies	*$50,156*	*$1,023,104*	*$194,040*									
	Other Direct Above Grade Construction Costs	$61,839											

Total Calculated Cost: **$1,267,299**

Total Above Grade Construction Cost Estimate:	**$1,267,187**

Total Construction Cost Estimate: Cash Required	**$1,380,284**

Maximum Cost Estimate:	**$1,586,407**

Minimum Cost Estimate:	**$1,184,097**

Figure 15. Electric company financial forecast tornado chart

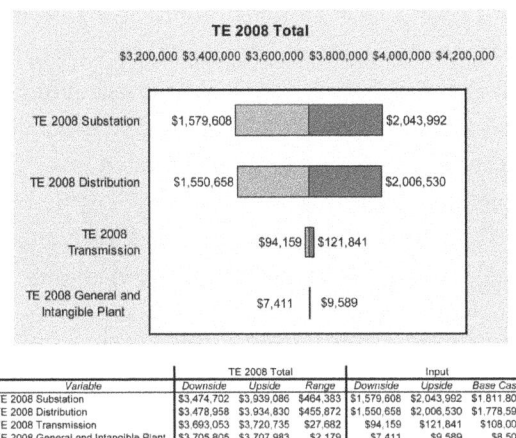

Variable	TE 2008 Total			Input		
	Downside	Upside	Range	Downside	Upside	Base Case
TE 2008 Substation	$3,474,702	$3,939,086	$464,383	$1,579,608	$2,043,992	$1,811,800
TE 2008 Distribution	$3,478,958	$3,934,830	$455,872	$1,550,658	$2,006,530	$1,778,594
TE 2008 Transmission	$3,693,053	$3,720,735	$27,682	$94,159	$121,841	$108,000
TE 2008 General and Intangible Plant	$3,705,805	$3,707,983	$2,179	$7,411	$9,589	$8,500

The dispute resolution clause language encompassed in the contract was similar to the conditions outlined below:

If a dispute develops between the parties to the contract, the parties will first look to the current edition of the "Construction Performance Guidelines." Both the company and contractor agree that the guidelines shall be used as the minimum acceptable level of construction that will be acceptable by the parties

Once implemented, the techniques empowered the regional project managers to mitigate cost overruns and associated schedule delays. In fact, one particular region's capital project cost performance was well within the financial tolerance after an annual review of the process. In addition, the team was instrumental in the development of compliance checklists for specific maintenance and commissioning contracts. It included a synopsis of the basic service requirements the contractor must provide to the authorized user under the contract. The checklist encompassed "Yes", "No" and "Date" boxes where, under the contractual terms, the non-compliance items (No check box) must be brought to the contractor's attention in writing. The contractor was allowed thirty-days (30) to correct the deficiencies. If the contractor did not remedy the non-compliant item, the authorized user notified the company of the non-performance issue. A Stop Work Order was issued immediately for any serious infractions. Sample questions of the compliance checklist are depicted below:

1. Does fixed pricing for full service testing and maintenance work on the electrical equipment include all per diem expenses, testing, adjustments, repairs, lubrication, etc.?
2. Was maintenance work performed with equipment or device de-energized from the normal power system?
3. Did Contractor provide a report within thirty (30) calendar days after the appraisal and testing work including but not limited to:
   a. Executive summary of principal findings and corrective work.
   b. Recommendations on repairs, overhauls and other maintenance.
   c. Complete documentation of all tests and inspection results on each significant item: power circuit breaker, protective relay, power transformer, and all motor controls, etc.

Once the funding is prioritized and approved for specific capital projects, the regional accountants, supply-chain personnel, and project managers provide the necessary oversight to guide the venture to its orderly end. As part of the contract negotiations, these key personnel must be directly and/ or indirectly (via lessons-learned packages, responsibility checklist routing) involved in order to facilitate concurrence and avoid cost overruns. The compliance checklist is an effective, defensible tool for contract agreements. The implementation of this viable mechanism "keeps the service provider" in line and maintains the overall objectives of the company.

The "Color of Money" is a key conception in finance relating to fund allocation (Mann, D., and Domb, E. 1997). Change the allocation of funds to include reallocation from one purpose, account, or budget to another. The concept also encompasses:

- When structuring a contract or agreement, allow (or disallow) the possibility of full or partial reallocation of funds among all or some of the accounts.
- Renegotiate a contract to permit such reallocations.

Based on the regional financial data and project quantities, the prioritization process must be implemented within *each* kilovolt range. A thorough need assessment must be performed for each voltage class category (discussed in the ordinal ranking chapter) for proper fund allocation. Cost estimate business rules must apply as outlined in this section. Decision-makers must be involved in the front-end analysis of the capital projects to not only ensure its success, but to defend its funding as well. A detailed breakdown of the 230-kV capital projects (17 proposed) and associated costs are depicted in Figure 19.

The other capital project requests for this region in descending voltage order are listed below:

- 800-kV Total for two regional projects: $3,525,500,000
- 500-kV Total for four regional projects: $860,600,000
- 345-kV Total for 101 regional projects: $6,485,640,629

Figure 16. Above-grade construction schedule details

Above Grade Schedule		Total Element Schedule (Values Only ~ days)	Minimum ~ Optimistic	Maximum ~ Pessimitic	Most Likely
ABOVE GRADE CONSTRUCTION					
1 Lot	Animal Protection	5.00	4.00	6.00	5.00
1 Lot	Battery & Charger	25.00	19.99	29.97	24.98
	Buildings				
	Bus Duct				
1 Lot	Cable, Control, 600V	10.00	8.00	12.01	10.00
	Capacitors				
	Carrier Equip.				
1 Lot	Circuit Breakers/Circuit Switchers	20.00	16.00	24.02	20.01
1 Lot	Conductors & Fittings	18.00	14.41	21.60	18.01
1 Lot	Conduit	18.00	14.42	21.61	18.02
	Distribution Cabinets, AC & DC				
	Fused Disconnects				
1 Lot	Insulators	20.00	15.98	24.02	20.00
	PT & CT's				
	Reactors				
	Regulators				
1 Lot	Relays, Metering, Telephone & RTUs	28.00	22.43	33.59	28.01
	Station Service Transformers				
1 Lot	Steel Structures	16.00	12.83	19.17	16.00
	Surge Arresters				
1 Lot	Switches - Group Operated	28.00	22.38	33.62	28.00
	Switches - Load Break				
	Switches - Regulator Bypass				
	Switches - High Speed Grounding				
	Switches - SPST Disconnect				
1 Lot	Switchgear	30.00	24.05	35.97	30.01
	Transformer - Mobile (Installation)				
1 Lot	Transformer - Power	25.00	19.99	29.97	24.98
	Contracted Engineering Supervision (10%)				
1 Lot	Field Testing / Commissioning	50.00	39.96	59.99	49.98
	Adjusted engineering				
		293.00	234.44	351.54	293.00

Total Construction Schedule Estimate: Most Likely	*445.98*

Pessimistic Schedule Estimate:	*534.91*

Optimistic Schedule Estimate:	*357.02*

- 138-kV Total for 169 regional projects: $768,304,434
- 120-kV Total for 10 regional projects: $15,260,000
- 115-kV Total for 98 regional projects: $297,526,743
- 69-kV Total for 215 regional projects: $513,776,029
- < 50-kV Total for 13 regional projects: $32,449,100
- Other kV Total for 37 regional projects: $17,460,001

A detailed breakdown of the 765-kV regional capital projects (29 proposed) are shown in Figure 20.

The project pre-planning aspect of the capital endeavors is an essential component for the preliminary action. The importance of *virtual prototyping and numerical simulation techniques* are indispensable tools to smooth the progress of the analysis. The above data clearly proves that a *change in the degree of flexibility is* necessary for project priority funding (Besterfield, D.H., et al. 1999). Once the capital endeavors are vetted at the regional phase, the decision-makers can monitor the respective budgets by making the system self-serving with the techniques discussed in this section.

SAMPLE TRANSFORMER FINANCIAL ANALYSIS

The transformer implementation financial cost mechanism was designed to control spending within the allocated budget. Several situations created the cost overruns. Nevertheless, the simple financial mechanism alleviated more than half of the budgetary problems in the transformer sector alone. As previously discussed, the Substation work estimates and contractor Spending were problems discovered by the Design of Experiment analysis. Part of the Contractor Spend was attributed to the competitive bid process (the vendors were not in the same room as one example). The process alone attributed to 25% of the overall transformer processing costs. Once the procedure was established to improve the process, it was possible to obtain the 60% goal of minimizing the overall transformer processing costs. The area is an input to minimize the transformer installation costs (50%) on the hierarchy chart. Consequently, the financial mechanism assisted with the overall goal to "Minimize Transformer Installation Turn-Around." An overlay of the improved data as compared with the original transformer data is depicted in Figure 21. When the Financial Mechanism was employed (dotted data), the average spend was noticeably reduced.

Figure 17. Regional financial data

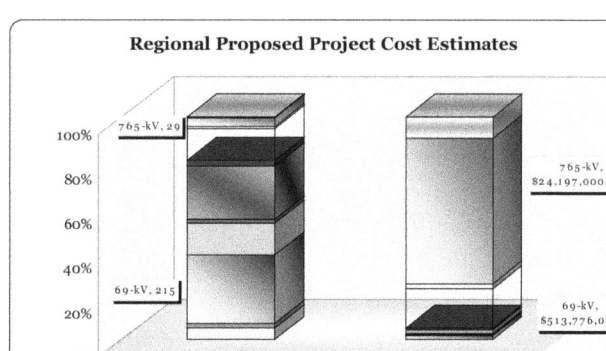

249

Figure 18. Regional financial trend data

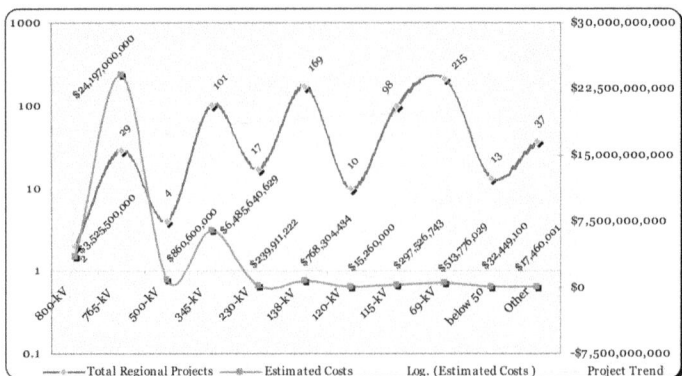

Further analysis of the range chart revealed the data clustered around an acceptable spend range (Figure 22).

There is a significant improvement when the financial mechanisms are utilized. When the decision-makers and project managers are acutely aware of potential cost overruns, the mechanisms are in place to identify and track the data outliers. It is a tool used to augment the processes and methodologies currently employed by project managers. Moreover, it is used as a verification (sanity check) method for the Budget, Schedule, and Forecasting for project management of the construction processes. The financial mechanism embraces the following characteristics:

- The Financial Model allows the Project Manager to control his/her monthly spending.
- The Model also provide key performance indicators over the project's life cycle.
- It also supports the scheduling and forecasting issues encountered by project managers.

When the resource closest to the work effort is allowed to control the assigned budget allocation, a greater understanding of the project parameters is encountered. If a certain financial "bucket" experiences excessive cost overruns (i.e. contractor "other" category), the project manager has the ability to ascertain and mitigate these particular overheads. The capital costs in the majority of renewable energy project implementations are typically "lumped" together of classified in a "loaded" (benefited) category. Nonetheless, the decision-makers must utilize, and comprehend, a financial control mechanism *before* excessive costs are encountered. If the model was developed and utilized in previous years (before the cost overruns), the transformer spending would experience a dramatic cost-savings effort. The historical spend is viewed in four distinct forms:

1. Heuristic
2. Statistical
3. Model
4. 95% Level

It is then compared with the actual data against a "what-if" analysis of the model in a post-mortem analysis. The data is shown in Figure 23.

- The budget experienced an *enormous* increase in accuracy (2202%)
- When the model was compared with the statistical data, a 386% improvement was observed

The importance of budget formulation must include detailed cost estimates of the proposed work effort. That encompasses the general re-

Figure 19. 230-kV Total for 17 regional projects: $239,911,222

230-kV Proposed Project Costs

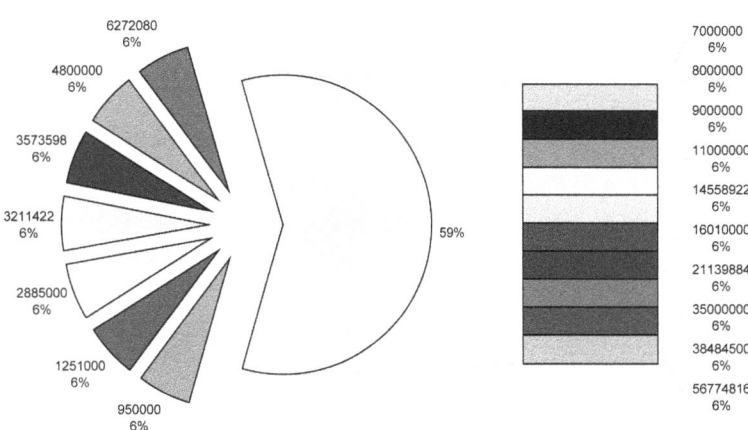

source arena of man, machine, methods, material, methods, measurements, and the environment. Unfortunately, most of the renewable energy efforts only include the time and material costs and do not take into account the imperative "white space" between organization silos. Several "subject matter experts" in the planning, engineering, and finance departments typically formulate a budget with "guesses" and the application of the "ten percent rule" (whatever the previous cost of a project ~ just add 10% for the upcoming fiscal year). A few companies rely solely on one financial "machine" to guide and direct the capital budget. Other organizations consult best practices and in-depth feasibility evaluations to assist with the budget preparation. Nevertheless, budget analysis must encapsulate not only the need base requirements of a capital project, but a total work effort for the plant-in-service implementation cost as well. The data comparison of the typical transformer implementation for the substation is depicted in Figure 24.

- Overall, "best" equates to zero and "worst" is any value greater than 100% or more.

- There is *only* one data point (triangular) from the Heuristic Budget, as the other points failed to fit within the graph.
- The Transformer Budget was not plotted for comparison purposes, as it failed to fit within the graph.
- The closest budget data lie within the statistical data (blue diagonal line).
- The model data (Brown dot/Brown line) forms a perfect circle within the center of the graph.

The simple results: The transformer budget would experience at least a 386% improvement and the project managers would not chase *"a moving target"* with more accurate monthly forecasts from the previous period. The lean methodologies (cost/waste reduction) can be – and should be utilized in the financial arena as well. When the capital project funding decisions and budgetary allotments are discussed in an organization for resource-intensive endeavors, the financial arm of the company, planning, the project/asset management organization, engineering, environmental, legal as well as maintenance must be included at the table. A thorough comprehension of the work effort is essential to further the progress of the project implementation. An examination of the actual spend vs. the financial model is depicted in Figure 25.

- There are more green data points than red (improvement in accuracy).
- If the model was utilized during this time period by the engineers and project managers, an overall improvement in the forecasting is experienced.
- The project managers would have total financial control of his/her project.

The financial forecasting mechanism for any project is an important parameter for the disbursement of expenditures. Several organizations use the "change order" or "cost justification" form for unanticipated alterations in a project. Some changes, unfortunately, modifies the scope of a project and create additional delays as well as cost overruns. The forecasting accuracy of the financial model assists the decision-maker with a thorough comprehension of the project from one

reporting period to another. Moreover, forecasting methodology (as with the budget formulation), can be used throughout the organization. The data comparisons of the actual spend financial data for the transformer implementation is depicted in Figure 26.

- The Heuristic Actual Spend was much better during this time period in spite of the project managers were operating with a flawed budget.
- "Brute Force" methodology was obviously utilized in during the period as several Asset Directors confirmed this suspicion with written policies.
- The model data was equivalent (or exceeded in some cases) with the Heuristic Actual data.
- The Transformer Actual Spend would experience at least a 34% improvement vs. the Heuristic Data.
- The Project Managers would spend more time managing the Transformer Installation Efforts.

Figure 20. 765-kV Total for 29 regional projects: $24,197,000,000

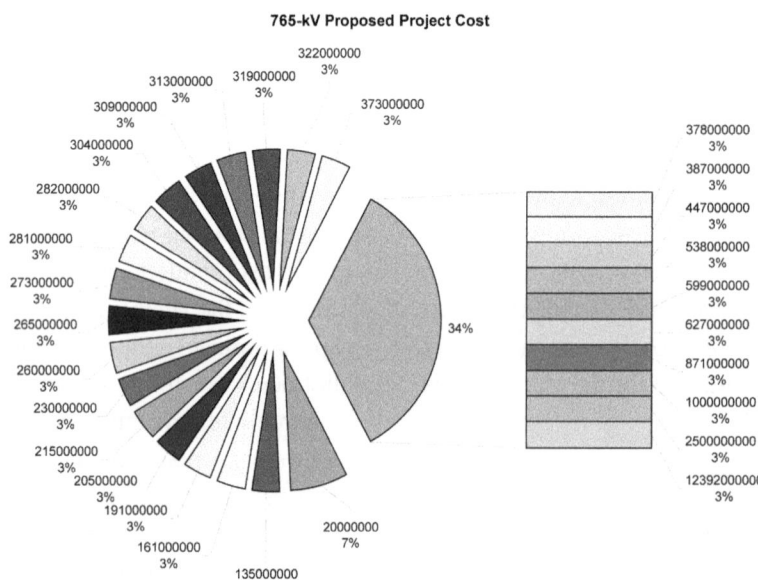

Figure 21. Financial mechanism overlay: median budget vs. actual spend

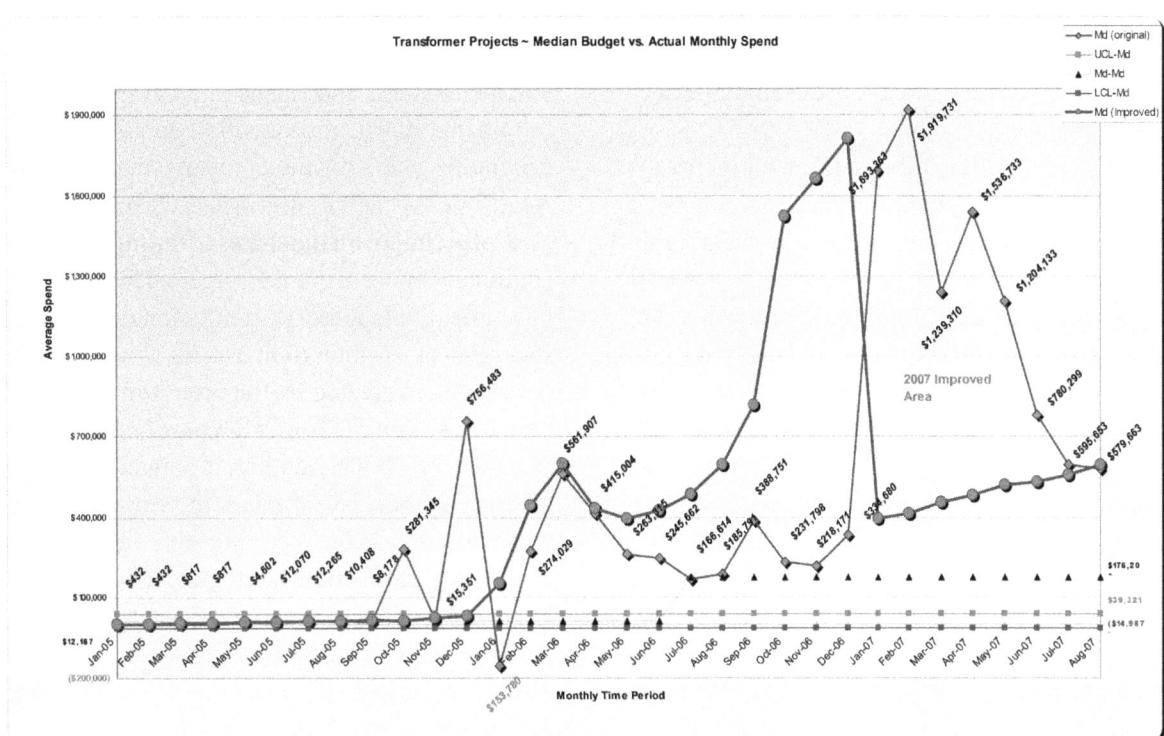

Figure 22. Financial mechanism overlay: range chart budget vs. actual spend

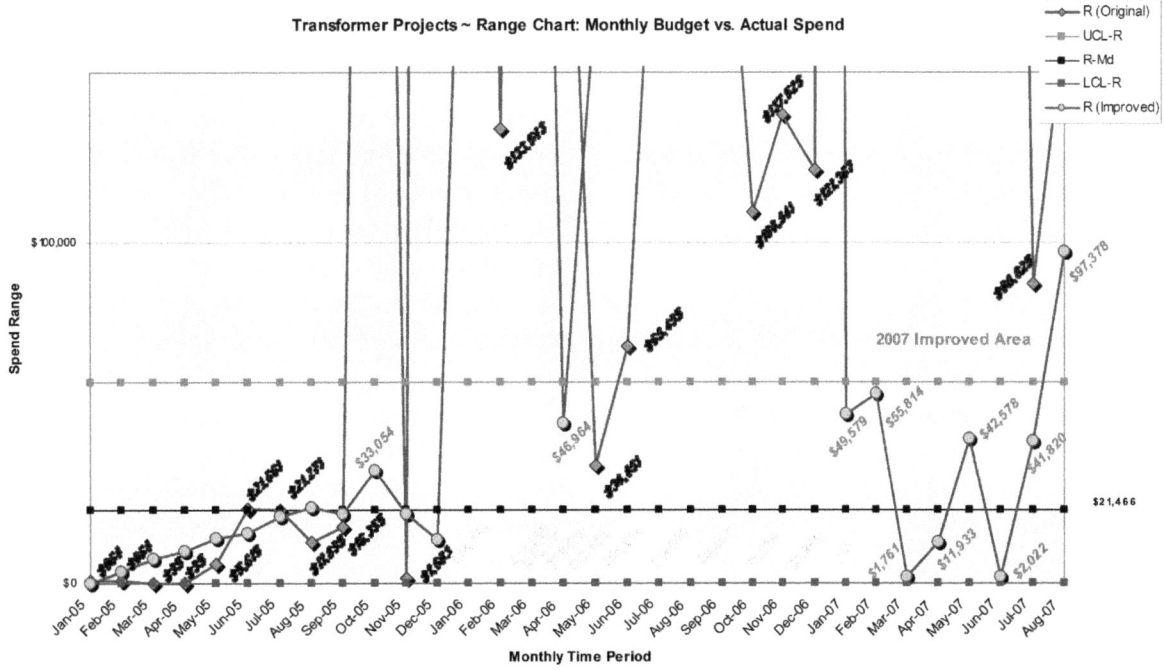

Further investigation of the "what if" scenario revealed the actual spend data protracted to a future year (Figure 27). The Model data excelled in practically every month of the study (Brown dot/Brown Line).

The single computing mechanism generally relies on a single output for the given inputs. In a multi-criteria decision analysis financial problem, every input must be given the appropriate weight for successful project implementation. The project manager must be astute in budgeting, cost estimation, as well as engineering economics. He/she must "manage-by-walking around" and communicate potential problems to the appropriate organizations. The staff accountant must comprehend the "cost of removal" and "retirement" codes associated with the capital project. Moreover, the team member must review and appraise the actual material and/or equipment to make informed assessments of the project as a whole. Members in the supply-chain must expedite the required materials once the endeavor is approved and funded because of the long lead-times. A proactive approach is required for capital-intensive projects as depicted in principle 9 (preliminary anti-action) within the TRIZ model. Also, use mistake-proofing (Poka-Yoke) techniques - designed for foreseeable unintended use.

If it will be necessary to do an action with both harmful and useful effects, this action should be replaced with anti-actions to control harmful effects (Retseptor, G. 2003).

The simple techniques and methodologies will assist the "man, material, methods, machine, and measurements" portions of the assigned resources with renewable energy implementation.

Equipment Cost Acquisition Strategies

The implementation of the renewable energy resources may require existing high-voltage equipment acquisitions from various entities. The goal of the investigation, as a renewable energy implementation example, is to recapitulate the strategic and financial analysis essential in making informed acquisition decisions while generating value for the company shareholders. Acquisitions generally necessitate the payment of a control premium for the shares of the high-voltage equipment being acquired (target). The acquirers must effectively preserve stand-alone values and synergies in addition to managing the new business problem created by the premium (averaged between 40 and 50% over the past two decades) (Sirower, 1997). The acquirer is obliged to be able to identify where performance improvements can be achieved because of acquiring a target when evaluating acquisition strategies. The Position Bargain Model, where management can improve the target's performance by further distinguishing its offering (enhancing existing attributes and/or adding new ones), further improving its relative efficiency or both, is the recommended approach. The current plans to integrate a specific target with a company may or may not be an obvious option, and are typically confidential. Nevertheless, opportunities exist for better management, strategy, and cost position of the target once it is acquired.

A Midwestern electric utility company (company "A") classified its high-voltage equipment strategy in various layers of acquirement. There were fourteen (14) major target areas (and 28 electrical devices) identified with the potential of full company ownership. Sixteen (16) major targeted devices were housed in the company's substations shared with another utility's (company "B") distribution system. Additionally, there were ninety-five (95) major targeted devices housed in company B's substations, and one hundred & sixteen (116) major targeted devices with the company A's high-voltage electrical lines passing through the asset external to the substations within the bulk transmission electric system notwithstanding the associated equipment and joint ownership issues. In total, there were two

Figure 23. Transformer budget statistical data

Date	Data	Fitted	Forecast (prior year accuracy % where zero to 24.9% = excellent, 25% to 49.9% = Fair, greater than 50% = Poor)	Fitted Forecast (prior year accuracy % where zero to 24.9% = excellent, 25% to 49.9% = Fair, greater than 50% = Poor)	Fitted Forecast Comparison Upper: 95%	Forecast Model Comparison vs. Fitted Heuristic Budget Data	Forecast Model Comparison vs.Statistical Budget & Lower: 5%
						2006 Xfmr Budget Statistical Data (Financial Model ~ What-if)	
Jan-06	$278,683		2.74%	100.00%	Worse	97.26%	584%
Feb-06	$340,800	$305,784	7.05%	3.95%	Better	7285.30%	670%
Mar-06	$280,532	$268,788	21.67%	24.95%	Worse	3507.79%	474%
Apr-06	$279,750	$285,833	31.65%	30.17%	Better	2546.77%	423%
May-06	$280,560	$303,527	41.25%	36.44%	Better	2037.02%	381%
Jun-06	$280,409	$305,572	51.07%	46.68%	Better	1625.48%	341%
Jul-06	$280,035	$314,746	60.91%	56.06%	Better	1376.06%	304%
Aug-06	$273,576	$312,139	71.36%	67.32%	Better	1142.73%	260%
Sep-06	$504,078	$301,936	64.81%	78.92%	Worse	952.70%	590%
Oct-06	$286,268	$307,525	90.01%	89.27%	Better	825.59%	219%
Nov-06	$286,935	$309,149	90.91%	90.20%	Better	716.30%	200%
Dec-06	$273,997	$294,843	92.12%	91.52%	Better	597.70%	166%
Jan-07			$294,843		$418,214		$171,471
Feb-07			$305,118		$440,827		$169,409
Mar-07			$305,118		$455,906		$154,331
Apr-07			$305,118		$474,755		$135,482
May-07			$305,118		$498,988		$111,248
Jun-07			$305,118		$531,300		$78,937
Jul-07			$305,118		$576,536		$33,700
Aug-07			$305,118		$644,391		($34,154)
Sep-07			$305,118		$757,482		($147,245)
Oct-07			$305,118		$983,664		($373,427)
Nov-07			$305,118		$1,662,209		($1,051,972)
Dec-07			$305,118		Improvement:	2202%	386%

hundred & fifty-four (254) assets identified for potential ownership. Several line switches and circuit breakers were located within the Midwestern utility's substation/transmission line jurisdiction, however were owned by the company "B" and/or other municipalities. Conversely, several of company B's electrical distribution substations and related equipment were supplied by company "A" transmission lines. The identified equipment (albeit not exclusive and/or an exhaustive listing of every potential ownership target) falls within the sphere and criteria for company "A" to assume full ownership and responsibility of the high-voltage devices. The 138-kV electric system was thoroughly explored and scrutinized for potential asset ownership from the Midwestern electric utility perspective. The general locations of assets in the major target area were acknowledged first. The depiction of the company acquisition strategy is outlined below:

- (A)=Company "A"
- (B)=Company "B"
- (3rd)=Third Party

The operational premium recapture model where the target's financial performance is characterized as an amalgamation of required improvements in expected profitability. Increasing the spread between the returns on capital and the cost of capital of an industrial business is central to the objective of value-creating diversification and operational strategy. The mathematical concept of this paradigm is depicted below:

$$M_1 = \Sigma \, B_t * R_t (1-r_t)/(1-k_e)^t$$

M_1 = Pre-acquisition equity ~ Market Value of the Target (present value of the expected equity cash flows)

B_t = Book Equity at the beginning of the year (t)

R_t = Rate of Return on B_t in year (t)

r_t = Fraction of Earnings Reinvested at the end of the year (t)

k_e = Cost of Equity Capital

The equation is simplified:

$$M_1 = B_1 + B_1 (\theta/k_e - g), \theta = R - k_e$$

Where g (which is less than k_e) designated the target's average rate of growth of equity cash flows forecasted by investors.

If θ = zero, then M_1 will equal B_1;

If θ is positive, then M_1 will be greater than B_1;

If θ is negative, then M_1 will be less than $B_{1...}$

If the acquisition is to create value through full premium recapture, the acquirer will want the target's post-acquisition equity value to be *"no less"* than the magnitude denoted by M*:

$$M^* = M_{1*}(1+p)(1+v)$$

If the target is acquired at a price so that:

$$M = M_1 (1+\rho)$$

Where:

ρ= Control premium expresses as a fraction

v= Acquirer's minimum acceptable value increment

Thus, the acquirer's challenge would be to improve the target's return growth performance sufficiently to transform it from a unit worth M_1 to a unit worth M*.

The Major Identified Assets

Details of the major assets identified in conjunction with the associated costs are shown in Table 3. The generated cost structure was derived from the asset compilation list housed in the local financial software system. The equivalent costs encompassed the comparable actual cost of a circuit breaker and/or switch in company B's substation in certain cases. The fifteen major assets identified for the investigation and the component listing were arranged by substations. The costs were arranged in descending order where the most expensive component displayed first. The actual equipment costs with the associated depreciation

Figure 24. Transformer budget comparison

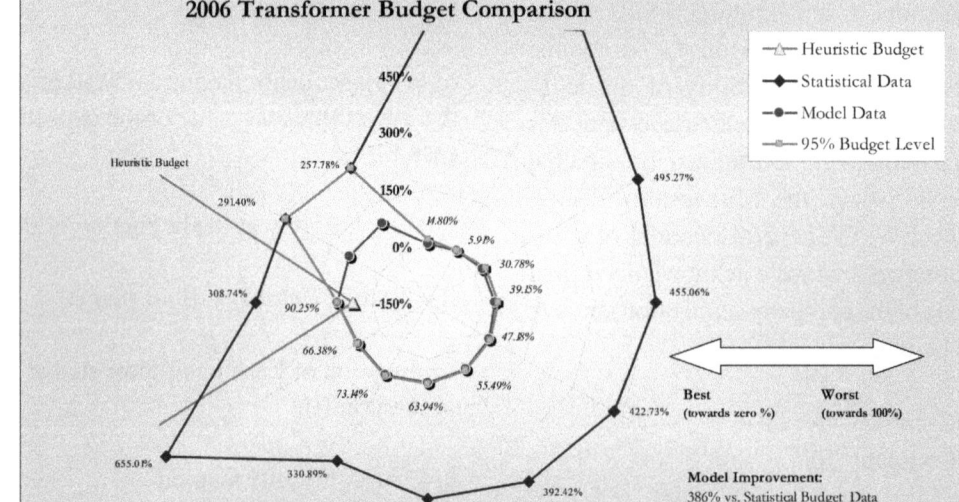

Figure 25. Transformer actual spend data

			2006 Xfmr Actuals Statistical Data (Financial Model ~ What-if)				
Date	Data	Fitted	Forecast (prior month accuracy % where zero to 24.9% = excellent, 25% to 49.9% = Fair, greater than 50% = Poor)	Fitted Forecast (prior month accuracy % where zero to 24.9% = excellent, 25% to 49.9% = Fair, greater than 50% = Poor)	Fitted Forecast Comparison & Upper: 95%	Forecast Model Comparison vs. Heuristic Actual Data	Forecast Model Comparison vs.Statistical Actual & Lower: 5%
Jan-06	($586,242)		3931.55%	100.00% Better		-3621.19%	-3809.64%
Feb-06	$207,257	$238,345	63.34%	57.85% Better		-24.16%	31.48%
Mar-06	$843,281	$969,773	0.68%	15.79% Worse		199.92%	83.51%
Apr-06	$550,257	$632,795	20.60%	38.69% Worse		76.10%	71.14%
May-06	$246,069	$282,979	20.15%	8.17% Better		-7.86%	76.77%
Jun-06	$210,914	$252,124	24.58%	9.85% Better		0.20%	73.16%
Jul-06	$53,192	$411,556	79.33%	59.91% Better		1.67%	20.17%
Aug-06	$98,006	$380,743	58.37%	61.72% Worse		5.80%	40.81%
Sep-06	$273,424	$231,688	34.85%	14.27% Better		10.91%	63.09%
Oct-06	$177,327	$176,321	0.03%	0.54% Worse		38.02%	98.76%
Nov-06	$149,406	$162,573	12.62%	4.92% Better		35.31%	86.44%
Dec-06	$395,364	$150,271	152.69%	3.96% Better		-108.40%	-54.98%
Jan-07			$218,705		$572,482		($135,071)
Feb-07			$218,705		$607,860		($170,449)
Mar-07			$218,705		$651,099		($213,689)
Apr-07			$218,705		$705,148		($267,738)
May-07			$218,705		$774,640		($337,230)
Jun-07			$218,705		$867,296		($429,885)
Jul-07			$218,705		$997,014		($559,604)
Aug-07			$218,705		$1,191,592		($754,181)
Sep-07			$218,705		$1,515,887		($1,078,476)
Oct-07			$218,705		$2,164,478		($1,727,067)
Nov-07			$218,705		$4,110,250		($3,672,840)
Dec-07			$218,705		Improvement:	34%	65%

values were also displayed. Only the assets of interest were utilized for the analysis.

Company A's substations that included Company B's distribution devices were also identified. The criteria was based on Company A's wholly owned substations with Company B's circuit breakers/switchers on the property and modeled after a premier substation where the company exclusively owns the circuit breakers and other high side electrical equipment with company B's distribution system. The equivalent costs of the identified sixteen (16) devices are the actual realized value of the asset and its associated age and reflected in each corresponding drawing. The targeted asset was highlighted on the wiring diagram for visibility and decision-maker review. Company B's substations that included company A's devices were identified as well. The associated premium costs (18%) and applicable state taxes (6%) were calculated as the total value of the asset. The criteria was based on company B's wholly owned substations with company A's circuit breakers/switchers on the property and modeled after a premier substation where the Company B exclusively owns the Substation, however the

circuit breakers and other high side equipment is owned (or potentially owned) by Company A. The equivalent cost of the ninety-five (95) devices is the actual realized value of the asset and its associated age and reflected in Table 4. The through transmission lines external to Company A's substation assets along with the associated targeted devices were identified as well.

There are several managerial consequences to consider when creating value acquisitions and coordination of activities between two companies. Moreover, there are some precautions ~ such as domain definition (turf identification) ~ that are compulsory to avoid redundancy and internecine rivalries that could develop where joint ventures/ownerships evolve into competitors of their owners, which may include the following:

- Exclusive use of their respective High Voltage System (or not).
- Freedom to sell its services to third parties (or not).
- Obligations to use their respective patents, trade secrets, "expertise", and other knowledge the company would consider

Figure 26. Transformer actual spend comparison

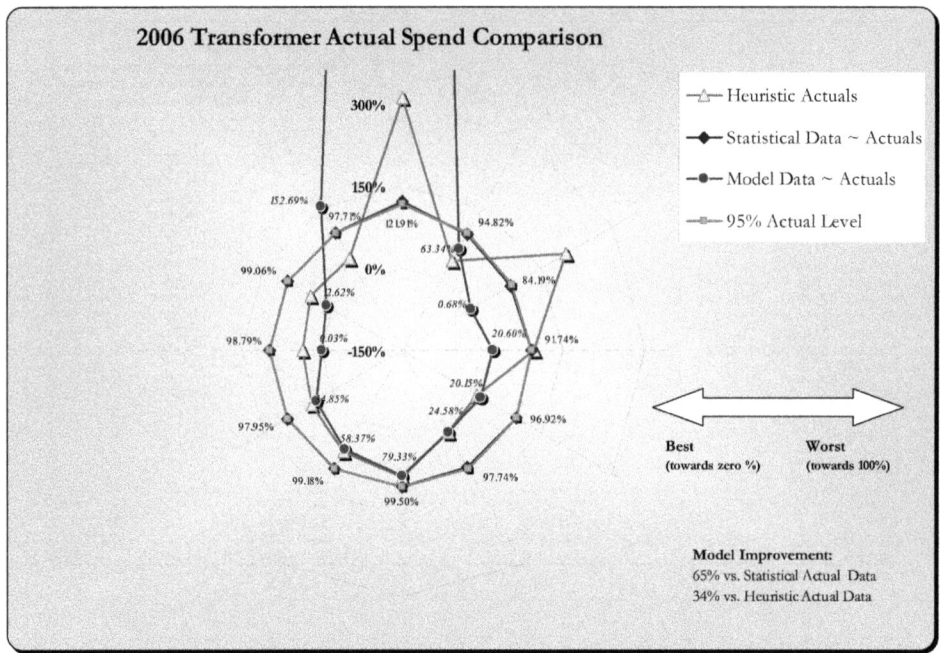

Figure 27. Transformer actual spend: protracted comparison

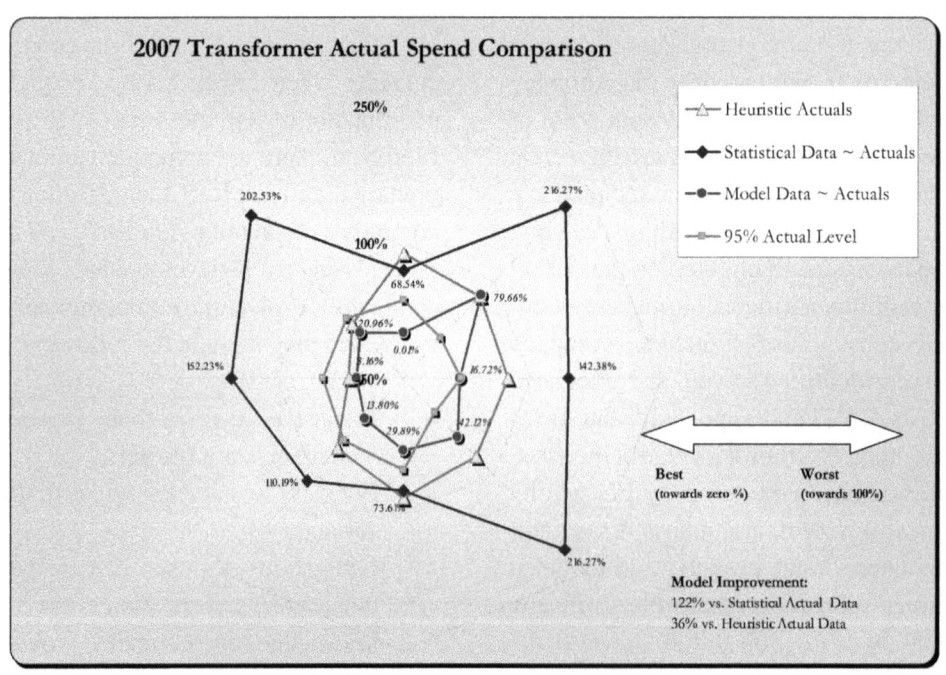

Table 4. Sample of asset total costs

Location Code (Subs)	Name	Component ID	Description	Manufacturer	Model	Year	Equivalent Cost ($)	Premium Cost (18%)	Taxes (6%)	Asset Total ($)
605	Wezad	500	Breaker	Westinghouse	GM5C	1968	$19,658.00	$3,538.44	$212.31	$23,408.75
	Wezad	377	Switch	Delta Star	MK40	2000	$19,689.09	$3,544.04	$212.64	$23,445.77
	Wezad	777	Switch	Delta Star	MK40	2000	$19,689.09	$3,544.04	$212.64	$23,445.77
585	Bleson	188	Breaker	Fed-Pacific	RHE-64	1956	$23,469.35	$4,224.48	$253.47	$27,947.30
	Bleson	277	Switch	Delta Star	MK40	2000	$19,689.09	$3,544.04	$212.64	$23,445.77
684	Iatso	500	Breaker	GE	FK138	1964	$15,477.06	$2,785.87	$167.15	$18,430.08
	Iatso	288	Switch	SWA		1965	$10,849.56	$1,952.92	$117.18	$12,919.66
	Iatso	477	Switch	Delta Star	MK40	2000	$19,689.09	$3,544.04	$212.64	$23,445.77
1263	2 Road	500	Switcher	S&C	Mark II	1975	$13,023.06	$2,344.15	$140.65	$15,507.86
721	Cecite	177	Breaker	GE	FK145	1973	$34,423.37	$6,196.21	$371.77	$40,991.35
	Cecite	188	Breaker	GE	FK439	1948	$18,609.32	$3,349.68	$200.98	$22,159.98
1298	Hasbilt	500	Switcher	Westinghouse	V2-LB	1979	$14,471.22	$2,604.82	$156.29	$17,232.33
713	Emett	35B7	Breaker	Westinghouse	GM4A	1950	$22,311.99	$4,016.16	$240.97	$26,569.12
	Emett	37W8	Breaker	Westinghouse	GM4A	1950	$22,311.99	$4,016.16	$240.97	$26,569.12

proprietary (as opposed to using outsider's technology).

- Close coordination of other germane activities with those of the target company.

In order to avoid these pitfalls during the acquisition process, Company A must be empowered to employ the following strategies:

1. Develop its own high-voltage channels distinct from Company B.
2. Purchase resources/services from outside vendors.
3. Use technical standards, designs, technologies, patents, expertise, and other intangibles supplied by outside firms.
4. Invest in other assets for unique purposes (including acquisitions).
5. Create its own market image.
6. Raise capital from outside sources.
7. Take other actions autonomously.

In addition to the above, Company A must be thoroughly prepared to acquire the targeted assets in a timely fashion. There were numerous sophis-

ticated circuit breaker schemes exclusively utilized by Company B for substation control. Many of the targets were located within these schemes. The acquisition of the identified electrical targets would create severe disputes between the two companies because of the control methodologies. Specific processes and procedures, in conjunction with a modified service agreement, were highly recommended to accomplish the asset acquisition for full ownership.

Planning is essential for the financial monitoring of the capital project. It is designed to aid the astute decision-maker, weather s/he resides in the financial planning office or within the technical resource organization, with tools to study, plan, do, and act upon periodic encroachments of the budget. The responsible manager in any business anticipates cost overruns and is equipped with the visualization as well as the computational mechanisms to handle the predicament. Accurate financial forecasting models in conjunction with periodic spend reports are aimed to keep the capital project within its natural budgetary constraints. When utilized properly and in a timely manner, the efficiency of the financial mechanisms will maintain resource accountability across the organization.

REFERENCES

Besterfield, D. H. (1999). *Total quality management* (2nd ed.). Upper Saddle River, NJ: Prentice Hall.

Godlewski, E., Lee, G., & Cooper, K. (2012). Systems dynamics transforms fluor project and change management. *Interfaces, 42*(1), 17–32. doi:10.1287/inte.1110.0595

Grubaum, R., & Pernot, J. (1999). *Thyristor-controlled series compensation: A state of the art approach for optimization of transmission over power links*. Retrieved 5 July, 2010, from http://www.abb.com/FACTS

Levasseur, R. E. (2011). People skills: Optimizing team development and performance. *Interfaces, 41*(2), 204–208. doi:10.1287/inte.1100.0519

Mann, D., & Domb, E. (1997). *40 inventive (business) principles with examples*. Retrieved 8 November, 2008, from http://www.triz-journal.com

Retseptor, G. (2003). *40 inventive principles in quality management*. Retrieved 24 March, 2009, from http://www.triz-journal.com

Sirower, M. L. (1997). *The synergy trap: How companies lose the acquisition game*. New York, NY: Free Press.

Thomas, K. (1975). *The handbook of industrial and organizational psychology* (Dunnette, M., Ed.). Chicago, IL: Rand McNally.

Wisler, D. (2003). Engineering – What you don't necessarily learn in school. *Mechanical Engineering Magazine Online*. American Society of Mechanical Engineers. Retrieved 15 February, 2009, from http://www.memagazine.org/contents/current/webonly/

APPENDIX

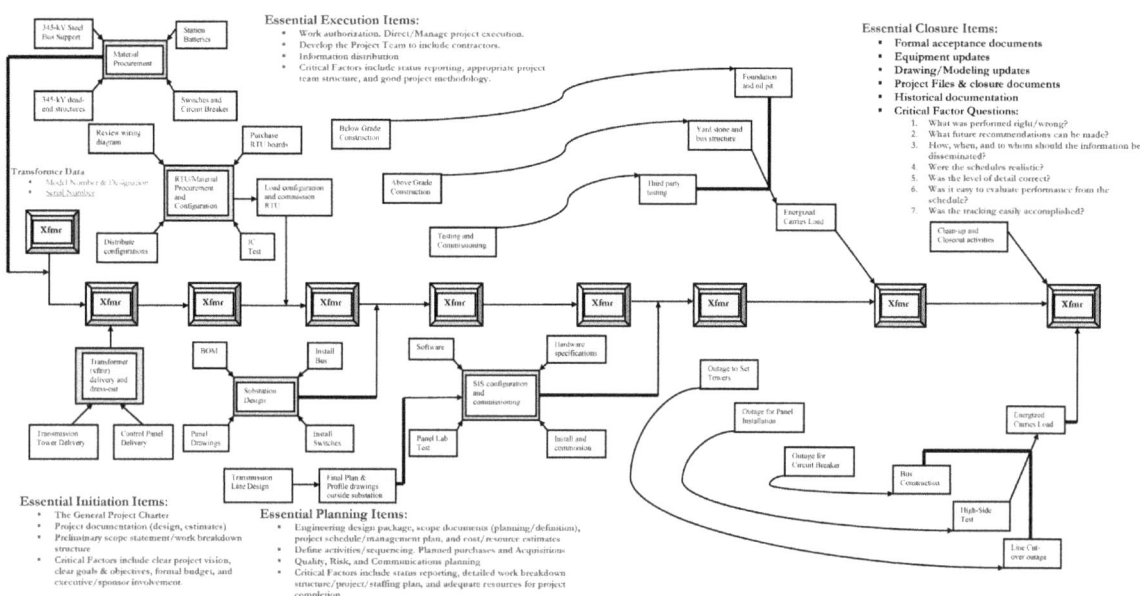

Surveillance & Witness Assessment of a 345-kV Transformer Installation

The objective is to minimize the delay time during system integration. The analysis is based on three control factors, 2-levels, and eight total trials.

Table A
Scheduling Factors

Trial #	Factor "A" Material Integration Time	Factor "B" Processes	Factor "C" Resources	Response	Average Transformer Delay Output (days)
1. Material Integration	+	+	+	Y_1	84
2. Procurement	+	+	-	Y_2	107
3. Engineering Design	+	-	+	Y_3	141
4. Department Coordination	+	-	-	Y_4	21
5. Material Preparation	-	+	+	Y_5	73
6. License, Permits, and Regulation	-	+	-	Y_6	47
7. Crew Experience/Training	-	-	+	Y_7	37
8. Outage Approvals	-	-	-	Y_8	10
Keys:	(+) = High Material integration (-) = Low Material integration	(+) = Satisfactory Process (-) = Unsatisfactory Process	(+) = Experienced Crew (-) = Inexperienced Crew		Average: 520/8 = 65 days late

Conduct the Test:

Table B
Trial and the Delay Results

Trial #	"A"	"B"	"C"
1	84	84	84
2	107	107	-107
3	141	-141	141
4	21	-21	-21
5	-73	73	73
6	-47	47	-47
7	-37	-37	37
8	-10	-10	-10
Δ	186	102	150
Effect: Δ/4	46.5	25.5	37.5

Graph A
Delay by days/Trial

Transformer Construction Delays

Scheduling Delays of the Substation Transformer Installation (continued)

Example Calculation

Material Integration Effect:

Low = 10+37+47+73/4 = 41.75 days

High = 21+141+107+84/4 = 88.25 days

Main Effect = High – Low
= 88.25 – 41.75 = 46.5 days

Translate the test results:

- Factor "A" (Material Integration) will improve the schedule by 46.5 days in

> Material Integration
> Procurement
> Engineering Design
> Department Coordination

- Factor "B" (Processes) will improve by 25.5 days in

> Material Integration
> Procurement
> Material Preparation
> License, Permits, and Regulations

- Factor "C" (Resources) will improve by 37.5 days in

> Material Integration
> Engineering Design
> Material Preparation
> Crew Experience/Training

Resolution & Recommendations:

- Material Integration is common in all three (3) effects:

  1. Resource Management and Performance
  2. Based on material logistics, delivery times, and system integration into the substation
  3. The Project Manager can reduce delays with proper communications (progress reports, department coordination) and via monitoring/controlling techniques

- Engineering design problems **must** be resolved during parallel operations:

  1. Perform Quality Checks for accuracy (i.e. phasing, general scope, alignments, Bill of Materials, etc). Utilize "Checklists" where applicable
  2. System Configurations

- Material Preparation and Staging

  1. The Project manger **must** "Manage-By-Walking-Around" to ensure adequate part compatibility, free of defects, etc.
  2. Provide frequent project site updates/vendor performance reports
  3. Provide documented quality inspections, corrective & preventative actions. Investigate previous "Lessons Learned" packages.

The Optimal Performance:

A   B   C
46.5 +   25.5 +   37.5 =   109.5

Improvement = 109.5/2 = 54.75

Optimum = Average - Improvement
= 65 -   54.75
= 10.25 days

General Analysis:

- Two (2) primary common areas are targeted for immediate improvement:

  1. Material Integration/Delivery
  2. Procurement Processing (Lead times)

- Two (2) secondary common areas are targeted for improvement:

  1. Substation Engineering Design
  2. Material Preparation/Testing

High material Integration (+)   Satisfactory processes (+)   Experienced Crew (+)

MUST FOCUS EFFORTS IN:
- The Supply-Chain Department
- Engineering Design (with an emphasis on quality)
- The Construction Crews

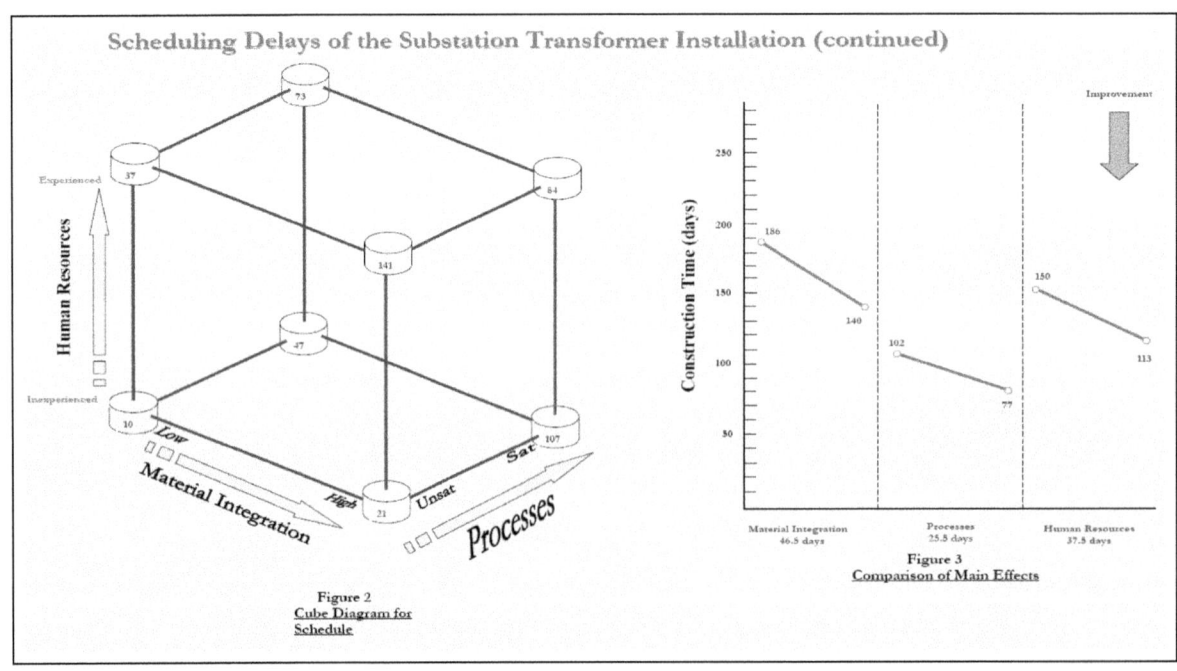

Figure 2
Cube Diagram for Schedule

Figure 3
Comparison of Main Effects

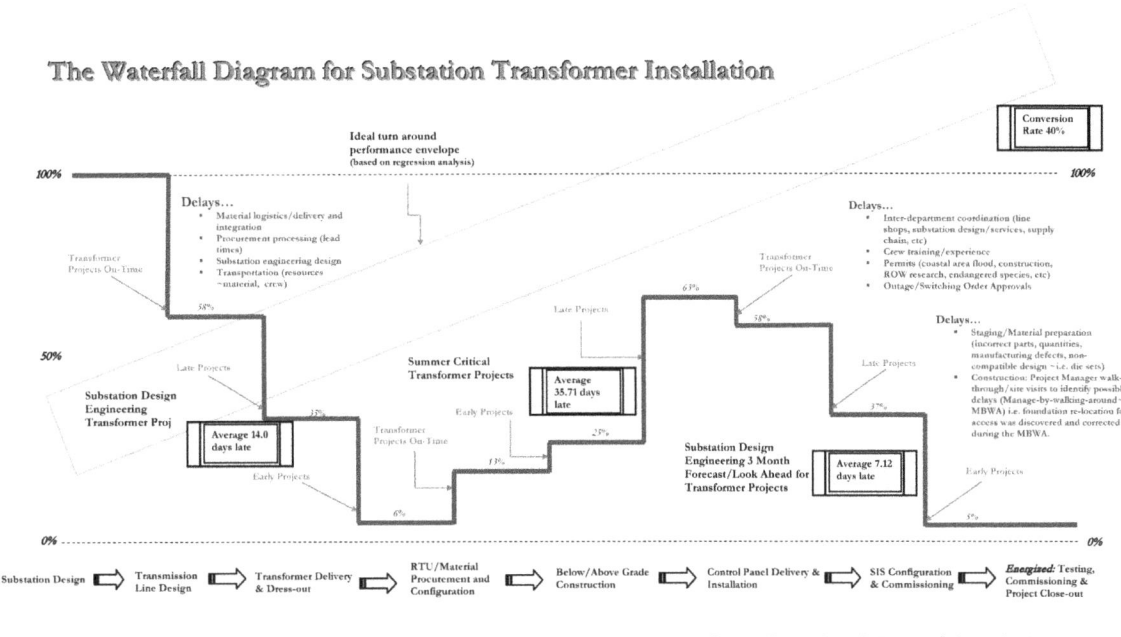

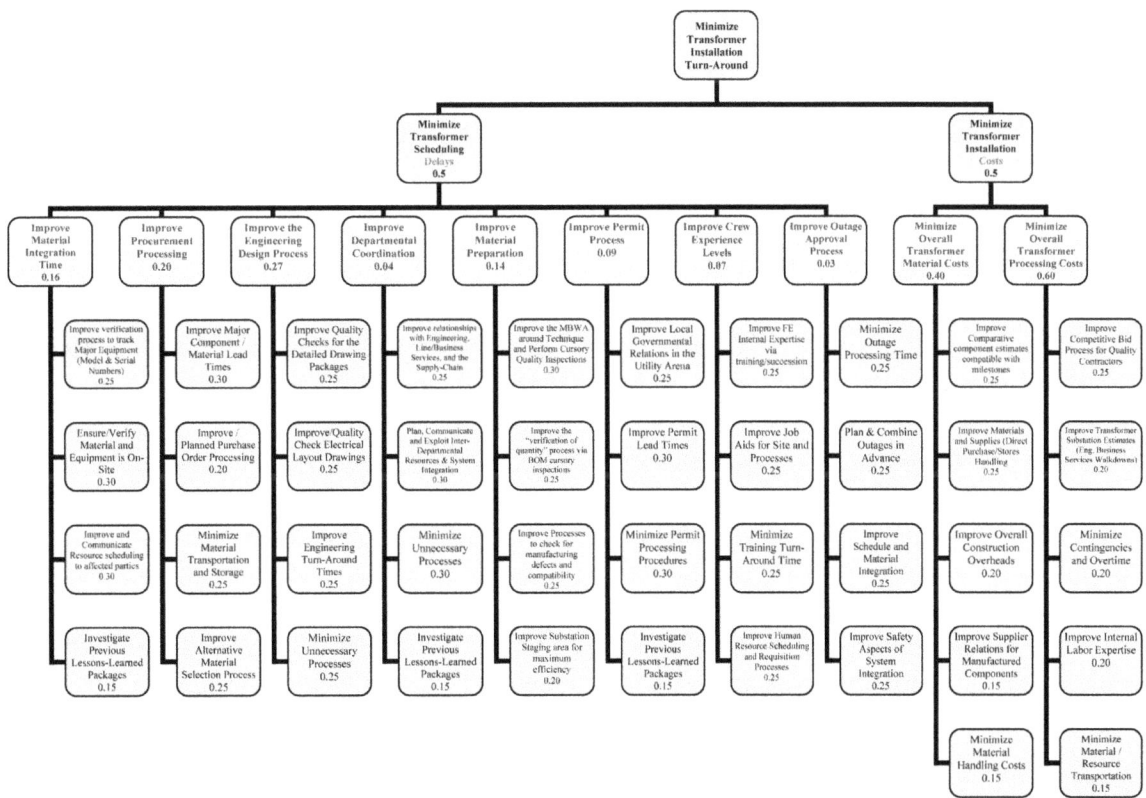

Chapter 10
Project Management Substation Guidelines

ABSTRACT

Well-organized and astute project management methods are indispensable in the equipment installation development. The strategy team evaluates the correct electrical equipment installation process from a flawless good start standpoint. The transition from the conceptual renewable energy incorporation activities to the detailed requirements of process design as well as the resource commitment is imperative for the implementation of the capital-intense project. The model must have exceptional management commitment to execute the technology within the current infrastructure. One of the key elements for delivery of the desired results within the scheme is the specification or scope. Strict adherence to the plan is expected throughout the project lifecycle. The manager is responsible for control of the project by monitoring its progression, identifying risks and resolving issues, and providing status of completed milestones.

PROJECT MANAGEMENT OVERVIEW

Efficient and judicious project management techniques are essential in the equipment installation processes. The various electrical components housed within a substation, for example, requires several degrees of equipment integration, complex methodologies that involve many levels of quality, specifications, resources, and time. The management of the various aspects of the installation, specifically for electrical equipment upgrades, is a business requirement that encroaches on practi-

cally every responsible department as well as the suppliers. Moreover, the planning, organizing, implementing, and control of capital-intensive ventures are fundamental elements in project management (Jones, R., 2007). All of these processes are highly-integrated and inter-related with tentacles in the financial, engineering, quality, and construction arena. Additionally, effective project management necessitates some form of business, system, and technical responsibilities that defines the field, structure, technology, techniques and applications of the equipment installation process. The coordination is in essence, a delicate balanc-

DOI: 10.4018/978-1-4666-2839-7.ch010

ing act between man and machine to achieve a desired goal when properly implemented. Some electrical utility companies excel in the project management field whereas others consistently produce cost/schedule overruns in practically every single venture. The equipment may arrive on time at the wrong substation with the supporting resources. Conversely, excessive delays are incurred when the specialized resources are diverted to other projects (demand maintenance, fix-it-now activities, etc.) with little or no notice. The electrical equipment integration problems are compounded by other forces affecting the venture – land, permit, engineering issues – to mention a few, that will cease all activity in the assign project. The political will to shepherd a project to a successful completion is encompassed within every responsible department. Key individuals in these organizations must be accountable for every aspect of the electrical equipment installation process – not just specific departmental elements of the implementation. Project status meetings must entail some form socio-technical system (goal setting, job enrichment, participative approaches, etc.) to increase the effectiveness of the work effort to date (Feigenbaum, A., 1991). Effective leaders are motivators of men and women. The utilization of a project visibility program, for example, is strategically displayed for all to see of the progress for each venture. The apparent psychological effect imposed on the employees of the organization is designed to create a spirited competition for a successful project implementation.

Fall down seven times. Stand up eight. (Japanese proverb)

A report from a regional manager at an update meeting stated that the turbine blade shipments was late should prompt several questions as to why. In fact, with incisive project management techniques in conjunction with logistical contingency schedules from high-performance vendors, the turbine blades should have been installed on time, on budget, and most importantly, *on strategy*. The astute project manager is expected to perform due diligence on the supplier that include the correct utilization of vendor surveillance and quality conformity within the strategic bandwidth. Additionally, vendor audits, past performance results, and supplier conferences are expected project management criteria to minimize any corrective action and material discrepancies. Therefore, the key communication channel is between the project manager and the vendor (as opposed to the capital project status meeting) to ensure that the causes of the discrepancy are eliminated and corrective action is permanent. The plan and resolution development is implemented well before project update sessions occur.

The project risk assessment and management is a systematic process of categorizing and identifying potential threats and seeking ways to mitigate these dangers. Risk is generally classified as two types – certain and uncertain. The decision-makers must choose an approach and plan the risk management activities for a project. Potential risks are active within the astute project manager's arena at all times. The identification of these risks assists with the determination of the specific characteristics of the potential threat and how it can affect the project (Kerzner, 1992). Ideally, a qualitative analysis is performed and prioritized to evaluate the conditions and effects of the project outcome. Subsequently, a prioritized list of quantified risks are developed and measured against the venture's time, budget, and performance constraints. Modeling tools utilized in this process are analytical, mathematical, statistical, and graphical for the quantitative evaluations. Managerial, anecdotal, empirical, and to some extent management games, are employed for the qualitative area. The project modeling tools may utilize some form of simulation, graphs, project software, or statistical analysis as part of the expansion process (Taguchi, 2001). Procedures are developed as part of the risk response plan, to enhance the opportunities while reducing the threats from the risk, aligned with the scheme objectives. The project manager is expected to monitor and control any residual and secondary risks as a consequence of incorporating the risk response plan. Moreover, the risk management includes documenting corrective actions, change requests, work-around, and evaluating the effectiveness of the risk reduction plans throughout the project lifecycle. Alternative approaches may be included in a simulation for high-risk areas to evaluate decisions even further for complex systems (Clymer, 2009). The risk reduction is enhanced with the utilization of an earned value metrics. The system permits the project management team to mitigate impacts by facilitating early adjustments as well as providing accurate estimates of schedule completion and projected final cost.

The complex phases involved with renewable energy system integration includes various tenets of efficiency, quality, flexibility, environmental, globalism, and nanolity. The management of the project equates to allocating and timing specific resources to achieve the goal as efficiently and expeditiously as possible while managing the time, cost, and performance objectives. In order to optimize the entire paradigm, the focus is directed toward the resources, processes, and technology areas as well as their associated systems. Nevertheless, the project constraints of time (schedule requirements), cost (budget limitations), performance (quality specifications), and strategy are continually balanced throughout the project lifecycle to improve electric service in the region. The larger, complex projects in today's system integration environment often entail shorter timescales with more variability. The quality realities must be congruent with the project objectives. Consequently, more than half of the complex projects fail because of requirements related to problems (missing user specifications, limited or mis-aligned) and /or incomplete, unrealistic, dynamic requisites. The majority of failures can be traced back to communication problems (a qualitative issue). The mitigation strategy includes

communication, cooperation, and coordination. The actual linkage between these three important areas are connected by the simple questions of *who, what, why, when, where, and how* within the project management sphere. The resources are expected to *communicate* the project scope, organization, direct/indirect benefits, potential adverse impacts, anticipated cost, required personnel contribution, and the pros/ cons of the project justification. *Explicit cooperation* entails components of the resource role expectations, the time frame, the organizational impact, the importance of cooperation as well as its rewards. The emphasis of *coordination* is within the internal /external interfaces of the project organization. The successful project system integration efforts encompass effective team management, schedule control, human resources utilization, contract administration, information management, quantitative modeling, and quality/cost management. All require unique administration approaches to optimize soft-side project management areas in the system integration process. A general project lifecycle overview is depicted in Figure 1.

The entire model is based on project control, quality, risks, and the fractal aspect of a capital project process. The Project Lifecycle Model simply illustrates the *progress philosophy* of ventures to promote a better understanding and a better communication process within the process. The Technical Project Management Practices includes:

- Preliminary Design Reviews
- Critical Design Reviews
- Production Readiness Reviews
- Progress Reviews
- Close-out Reviews

As a consequence of a review, an activity that was supposedly completed may be re-worked (an additional activity neither planned nor scheduled), may be carried out, or simply to go ahead. Project managers must quantify these possible additional

Figure 1. Project lifecycle overview

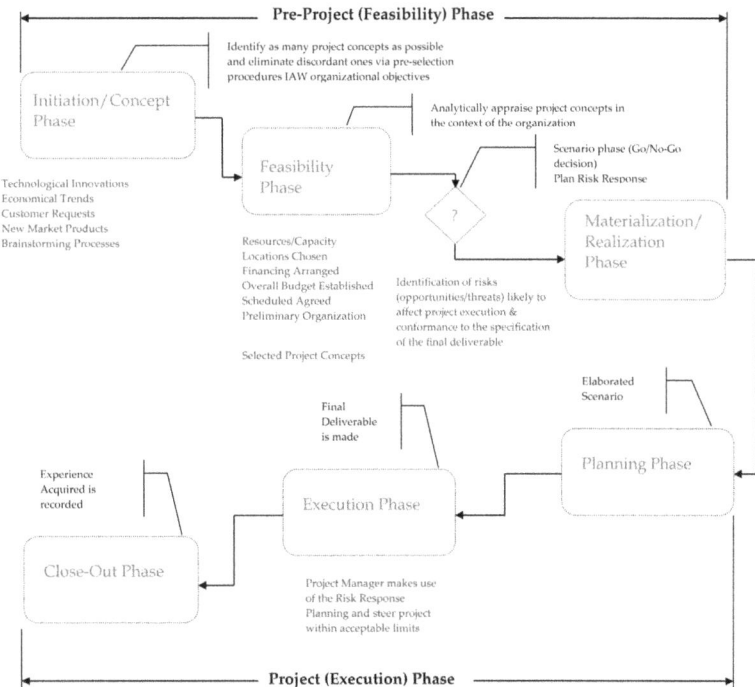

activities and spread them throughout the length of the control span as contingencies along the project execution phase. Strategic Issues are typically found in the first level of a Planning and Scheduling System. The Master Schedule is designed to reflect the strategy of the project. Coordination schedules between the installation and commissioning phases of a newly launched project is comprised of several overlapping schedules as shown in Figure 2.

The design phase of the project implementation is crucial for its success. The decision-makers set policy and parameters of the entire scheme for downstream activities. The project management requirements for the major activities are depicted in Table 1.

The Project Life-Span Model

This model is based on project managerial thoroughness with emphasis on "Front-End" work, especially in capital-intensive projects. The time spent up-front includes:

- Defining Needs
- Exploring Options
- Modeling
- Testing
- Reviewing various business benefits

The "Front-End" technique is central to producing a successful project. Decisions made at the early definition stages set the strategic framework within which the project will develop. Defining the problem is half of the solution; ninety percent of the outcome is defined in the first ten percent of the project. Thus a well thought-out project life span plays a key function in the control strategy for the evolution of a project. The phases represent significant changes as the project progresses through succeeding levels of maturity to include:

- Progressive levels of detail in management decision-making
- Required management style

Concept Phase: Explorer type project manager

Development Phase: Coordinator type project manager

Execution Phase: Driver (assertive) type project manager

Closing Phase: Administrator (clean-up) type project manager

The overall scheme is illustrated in Figure 3.

Suggestions for Effective Project Evaluation

Effectual valuation and selection of capital project prospects is vital to the overall scheme's success. The majority of the circumstances are too multifaceted to use simple economic models as the sole basis for decision making with the ever-increasing dynamics of the business environment. The capital project evaluation procedures must include an extensive spectrum of variables for defining the venture's value to the organization to attain maximum effectiveness. Structure, discipline, and manageability must be designed into the selection process by grouping the evaluation variables into four major categories:

1. Reliability and effectiveness of the project with the business mission, strategy, and plan;
2. Multifunctional Ability to produce project results, including cost, technical, and time factors;
3. Achievement in the end-user environment;
4. Economics, to include profitability.

Effective project evaluation and selection requires a broad-scanning progression that can deal with the risks, uncertainties, ambiguities, and imperfections of data available at the commencement of the development cycle. It also requires managerial leadership and proficiency in planning, organizing, and communicating. Moreover, assessment team leaders must be social architects in unifying the multifunctional process and its people. These leaders must share risks and promote an environment that is professionally stimulating and strongly linked with the sup-

port organizations eventually needed for project implementation. There are alternative project management processes that exploit the resources while adhering to the standards espoused in the project management community. These processes are based on the *project activities* as appropriate to the project manager. Several methods encompass and empower interdisciplinary team associates to support the objectives and business strategies of the company for project implementation. The results provide structure and standardization in planning, scheduling, and control in project management and are exceptionally beneficial for the electric utility company. The project evaluation methodology is outlined below:

- Search for the significant information. Evocative project evaluations require relevant quality information.
- Acquire a *top down* view as details approach in the later stages. Features are less important than information relevancy and assessor capability. Avoid problems on lack of detail during the early phases of the project evaluation. Incorporate an iterative process. Significant effort is wasted assembling perfect data to justify a "no-go" decision.
- Choose the right resources. Whether the project evaluation consists of a straightforward economic analysis or a intricate multifunctional assessment, competent people from those functions critical to the overall success of the project must be implicated.

Figure 2. Project schedule overlap

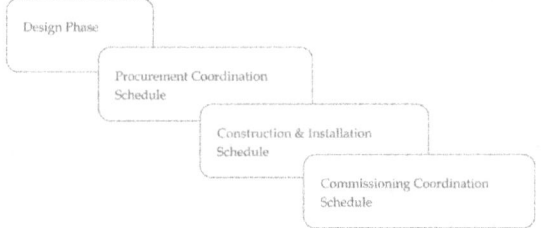

Table 1. Project management parameters

Project Management	Assessment	Initiate Define/Plan	Design Specify	Deliver Construct/Integrate	Close Deploy/Transition
Key Deliverables	Feasibility Report	Project Charter Business Requirements Technical Requirements	Detailed Business rqmts Systems Analysis	System construction System Integration/ pilot test Implementation Plan	Project Deliverables Evaluation
Approval	Feasibility Report Review Assessment Approval	Project Approval	Design Approval	Construct Integrate Approval Deployment Approval	Project Audit Completion Approval
Scope	Scope Boundaries	Scope/deliverables Benefit/value Assumptions and Alternatives Strategic & Tactical: Impact/priority/alignment	Change request procedures Issue management procedures	Change management Issue management	Manage delivered value
Human Resources	Resource identification	Roles & responsibilities General resource capacity Training rqmts. Business sponsors	Resource impact and assignment Team training	Resource management Resource performance Knowledge transfer End customer training	Resource performance evaluation
Time	"Window of Opportunity"	Preliminary project schedule Time reporting database	Work breakdown structure Project plan	Execute and monitor plan	Verify activity/ Completion Close time buckets
Cost	Cost projections	Capital budget Operating cash flow Return on investment	Cash flow details	Execute and monitor budget	Close out cost centers
Procurement	Alternatives evaluation	Hardware/software Consulting services Vendor RFPs	Vendor selection Contract finalization	Purchase hardware/ software Vendor performance report	Ongoing maintenance agreements Vendor performance evaluation
Quality	Contractor assessment Quality rqmts	Quality plan Previous lessons learned	Test approach Review lessons learned Walk throughs / reviews	Test plans Test (unit, system integration, acceptance)	Process review Post-implementation review Capture to lessons learned
Risk	"Opportunity Costs"	Risk assessment	Risk management plan	Risk mitigation	Capture to lessons learned
Communication	Program coordination	Communication rqmt project site	Progress reports Meetings schedule Project site update	Project site update	Administrative closure

Figure 3. Project life span model

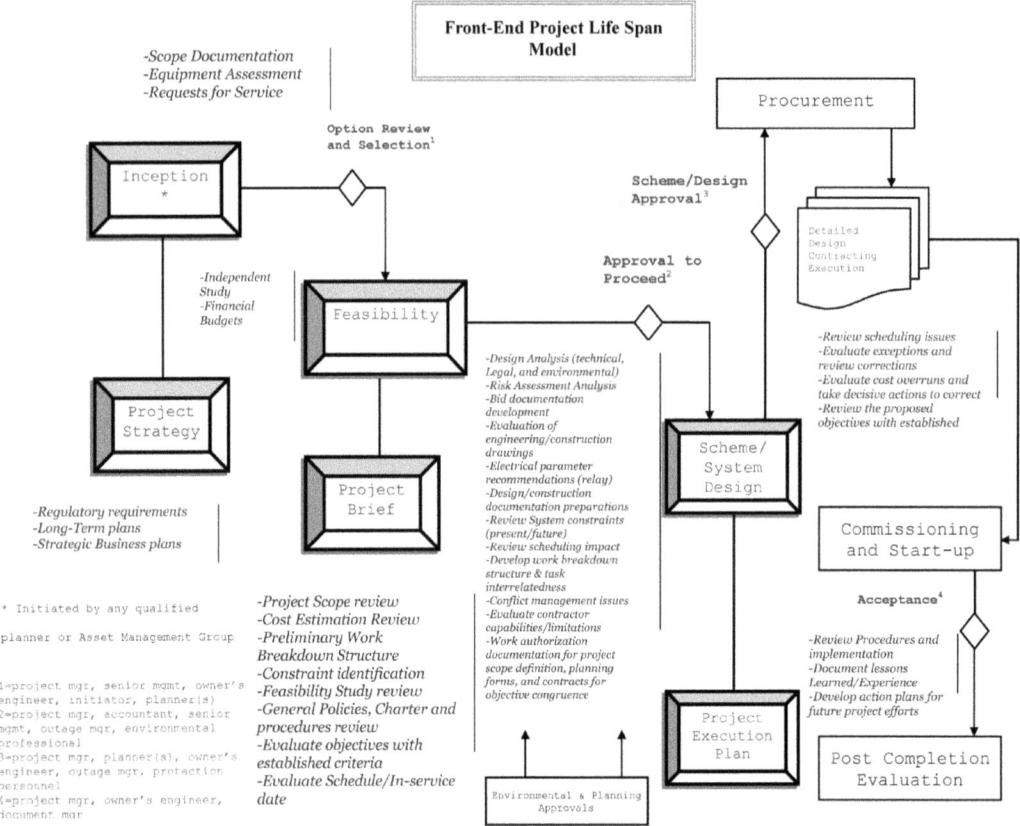

- Define the criterion for success. The evaluation criteria can be quantitative (i.e. return on investment) or qualitative (probability of winning a contract). The criteria must cover the accurate range of factors affecting the accomplishment or stoppage of the project.
- Evaluate exacting quantitative criteria; it is capable of containing misleading data. Be attentive of evaluation procedures based only on quantitative criteria (cost, return on investment, etc.). The input data used to determine these criteria are typically based on rough estimates and are often unreliable.
- Compact the criteria list. Unite the evaluation criteria, particularly among the judgmental categories, to maintain a manage-

able list (as a goal, remain within twelve criteria).
- Communicate. Assist with communications among evaluators and functional support groups. Characterize the development for arranging the project team, conduct the evaluation, as well as the selection process.
- Make certain of cross-functional team cooperation. Resources on the evaluation team must have a strategic vision across organizational lines. The purpose, goals, and objectives of the project should be apparent, along with the relationship to the business mission.
- Focus on the *big picture*. As discussions evolve into detail throughout the evaluation, the team must maintain a broad perspective.

- Research between iterations. As project evaluations are most likely performed progressively, action items for more information, amplification, and further analysis arise. These action items must be properly allocated (with follow up) thus enhancing the quality of the valuation with each consecutive assessment.

- Encourage originality and improvements. Senior management must cultivate an innovative ambience for the evaluation team. Evaluating multifarious project situations for prospective success or failure engages elaborate sets of variables, associated among organization, technology, and the business environment (also involves dealing with risks and uncertainty). Innovative approaches are necessary to evaluate the accurate potential of success for these projects. Risk sharing by senior management, acknowledgment, visibility, and a constructive image in terms of high priority, interesting work, and importance of the project to the organization has been found as strong drivers toward attracting and holding quality resources to the evaluation team and gaining their active and innovative contribution in the process.

- Manage and lead. A decision-maker who has the trust, respect, and leadership credibility with the team members must chair the evaluation team. Additionally, senior management must be able to confidently influence the work environment and the process by providing some procedural guidelines, charters, visibility, resources, and active support to the project evaluation team.

THE STAGE-GATE PROCESS

The strategy team evaluates the correct electrical equipment installation process from a flawless *good start* standpoint. The transition from the conceptual renewable energy integration activities to the detailed requirements of process design, the resource commitment is imperative for the implementation of the capital-intense project. The chief components that comprise the *good start* paradigm are *business readiness*, where the completed business plans are financially viable, and *technical readiness*, where the project feasibility satisfies the stakeholder's requirements on the strength of the demonstrated technology (Cooper, R., 1993). The model must have exceptional management commitment to implement the technology within the current infrastructure. In the phase-gate (stage-gate) model, forward-looking preparation for entry into a subsequent phase creates a more positive environment than backing-looking inquisitional evaluations against exit criterion. The intent of a "help/go" style phase-gate is to accelerate progress by resolving issues. The performance dimension consists of metric sets (typically determined by senior management), metrics, and the overall performance measurement system (Melnyk, S., et al., 2004). Therefore, the metric must be a measurable item, qualitative or quantitative in nature, financial or non-financial, described with respect to a target of reference point to align with the organization's strategy. The deliverables at each gate consists of the following documentation:

- Policies
- Forms
- Procedures
- Checklist

These documents are in place to ensure critial steps of the process are not omitted. The typical questions asked at each gate is shown below:

- Where are we today? (time, cost, strategy)
- Where will we end up?
- What are the present and future risks?
- What assistance is needed from management?

The gatekeepers major responsibilities and associated decisions are outlined below:

- Authorized to evaluate performance to date issues against predetermined criteria.
- Provide the team with additional Business and Technical information.
- Must make four common decisions:
  - Proceed to the next gate based upon original objectives.
  - Proceed to the next gate based upon revised objectives.
  - Delay the gate decision until further notice.
  - Cancel the project.

The following diagrams (Figures 4 and 5) depict the stage-gate process as it pertains to Project Management in the electrical utility business.

As a special note, the project may be terminated at any gate if prescribed conditions are not met.

Gatekeeper A: Team appointed by senior management to include a planner, a scheduler (outage manager), owner's engineer, accountant, environmental professional.

Gatekeeper B: Team consists of: planner(s), engineering manager, environmental professional, owner's engineer, scheduler (outage manager).

Gatekeeper C: Same team as "B".

Gatekeeper D: Same team as "A".

Gatekeeper E: Same as "B" with the addition of the Document manager (save scheduler, environmental professional).

Comprehensive Substation Procedures and Specifications

One of the key elements for delivery of the desired results within the scheme is the specification or scope. Strict adherence to the plan is expected throughout the project lifecycle. The manager is responsible for control of the project by monitoring its progression, identifying risks and resolving issues, and providing status of completed milestones. Nonetheless, the intricate balance of maintaining the inter-related variables of the required resources, time, quality, and the specification is essential in project management. If the resources are increased, for example, the allotted time for the venture's completion is reduced to sustain specifications and quality. If the specifications are reduced in conjunction with a diminished resource pool, the project is completed on time without compromising quality. The delicate balance between these factors can drastically affect the outcome of the scheme. However, the project manager and the decision-team may elect to seek the most effective version of the plan, which must be understood by all stakeholders.

The specification and/or drawing utilized in electrical installation are vital components for system implementation. The drawings are part of the assets. It must be written as clearly as possible for total comprehension by field personnel. The plan illustrates the ideal equipment condition as it is installed within the confines of a substation. Nevertheless, the requirements for interpretation must not be utilized as a crutch for inferior drawings or engineering packages. The obligatory additional information must be accompanied by the project change order request and included in the drawing to enhance comprehension. These vital drawings and specifications must be interpreted to the field personnel by utilizing various communication methods (i.e. project strategy sessions, amplified visual aids, lead project engineer breakout sessions, historical database, etc.). The project management team must have the attitude and motivation to maintain the *get it right the first time* principle espoused in quality and performance of the project implementation.

Development of thorough, absolute plans and specifications is very important to ensure substation construction and operation at the lowest attainable cost proportionate with the quality of

Figure 4. The stage-gate process

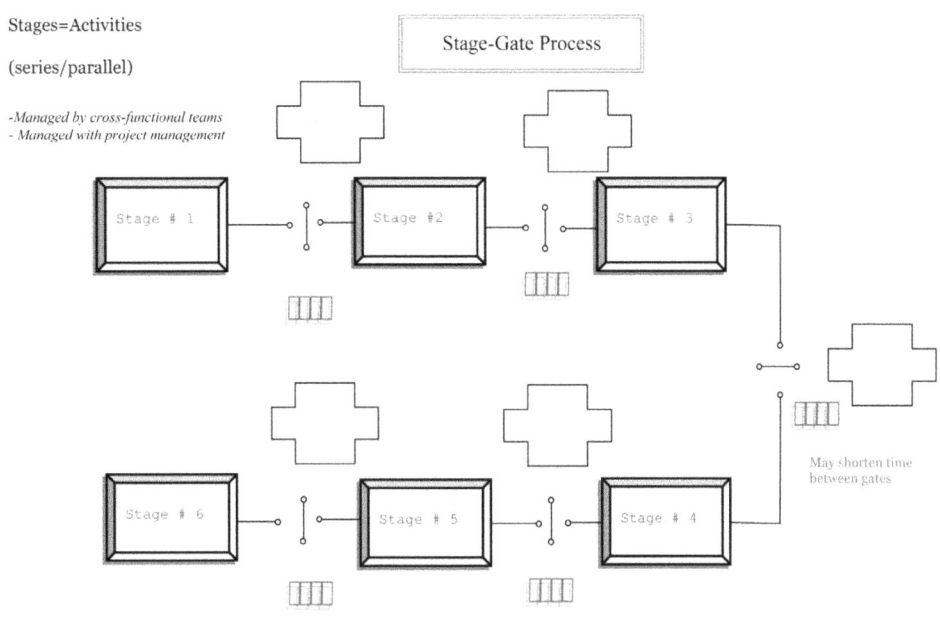

Stages=Activities

(series/parallel)

-Managed by cross-functional teams
- Managed with project management

Stage-Gate Process

Stage # 1   Stage #2   Stage # 3

Stage # 6   Stage # 5   Stage # 4

May shorten time between gates

Gates=Structured decision points... The purpose is not only to obtain authorization to proceed, but to identify failure early enough so that resources would not be wasted, but would be assigned to more promising activities. (No more than six gates – avoids from too much focus on gate review preparation instead of actual project management).

Gate Keepers=Individuals (i.e. sponsors) or groups, not project managers, designated by Senior Management.

service desired. The drawings accompanying the specifications must be of adequate detail and precision to avoid any possible construction delay and/or errors/misinterpretations during the itinerary of the project. The following partial list of item requirements must be considered when preparing substation drawings:

- **The One-Line Diagram:** This representation must be stringently reviewed since the general arrangement and ratings of all major electrical equipment are depicted on this drawing.
- **The Plot Plan:** The point of reference of the substation structures on the plot plan must be coordinated with the direction of the incoming and outgoing lines with all distances between adjacent structures and clearances properly dimensioned.
- **Elevations:** The elevation depiction must detail all electrical equipment with special

emphasis given to identifying significant clearances. These electrical clearances must be in accord with specific industrial procedures, with complementary allowances made for heavy snow, high altitude, high contamination, and special problem areas.

- **Foundations:** The illustration must include the type and design of the foundation and/or footings for the various substation structures. The drawings must be comprehensive enough to define the required construction as 1) Steel reinforcement bar size, spacing, and location, 2) Top of concrete elevations, 3) Anchor bolt mark number, projection, and number required, 4) Number of foundations required for each type, 5) Depth of foundation in relation to frost line, and 6) Cable trench.
- **Grading:** The grading depiction must exhibit the plan and elevation of the finished

sub grade and existing contours of the substation and surrounding area. These drawings must also present cross-sectional views indicating slopes for cut and fill areas, berms, access roads, and graveled surfaces. Orientation to a north arrow and a horizontal tie must be shown on the plan. It is beneficial for record purposes to display the location and log of the soil borings.

- **Grounding and Fence Details:** The grounding calculations are used to determine whether the overall substation ground grid will be of sufficient size, length, and impedance to provide for proper operation of protective equipment as well as personnel safety. Inappropriate grounding may produce such adverse effects as inappro-

priate relay operation, transformer insulation failure, and possible serious injury to substation personnel.

- **Structural:** Structural illustrations must display the design loading requirements, manufacturing details, and erection of structure components, including embedded materials such as stub angles and anchor bolts. Major structures used for transmission line take-off (deadend) and bus (strain) tower must be accompanied by a drawing indicating design conditions. This drawing serves as the criteria for the designer and a check to insure that the actual installation does not exceed design limits. Construction drawings for structure erection and stringing are required to control

Figure 5. Project management stage-gate process

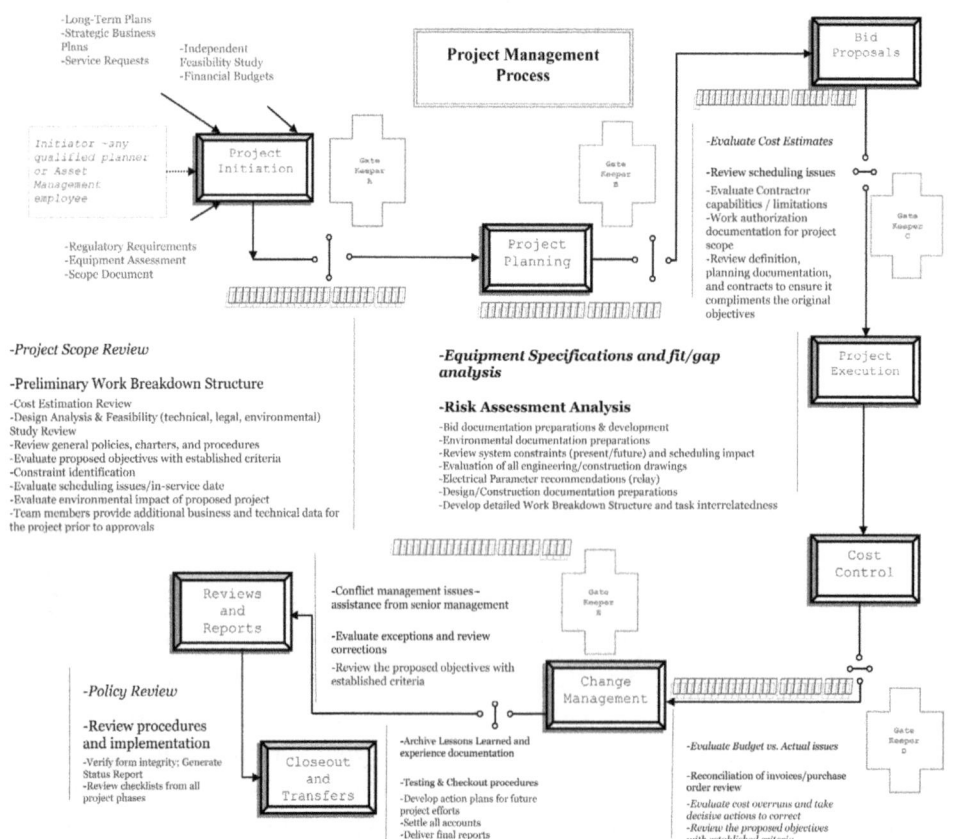

and expedite construction activities at the site.

- **Station Service (AC):** Site and utility depictions must include all details of auxiliary substation equipment that will be installed. It must include details of AC and DC power panels, yard lighting, and wire sizing.
- **Station Service (DC):** The DC system has the most critical loads. The construction specifications must ensure that the materials and construction practices meet and comply with industry standards, and/or local requirements.

The Typical Capital Project Outline Background

An electrical transmission request for 110 MW of long-term, firm, Point-to-Point service for generation in the West Maple control area to a point of delivery in the East Ridge control area beginning 1/01/2013 and ending 1/01/2019, the Independent Service Operator performed system analyses to determine the influence of such transmission service on the electric power system. The System Impact Study determined that a capacity constraint developed at the Maple Cabin 345/115-kV transformer during the 110 MW power flow evaluation. The proposed project requires the installation of a second 345/115-kV transformer at Maple Cabin to mitigate the system constraint.

Typical Project Assumptions

The preeminent estimates of cost and lead-time for the project are discussed and outlined in the following sections. There are many factors that influence the actual cost and project duration estimates that includes the construction requirements of permitting authorities to secure approvals, unexpected increases or changes in labor expenditures, permitting authorities and required siting approvals, unexpected increases in material costs,

equipment delivery, company and non-company labor scheduling and availability, ability to schedule outages on the electric systems of the local utility company and other electric companies in the footprint, inclement weather and other Acts of God, emergencies occurring on the systems of the local utility company or other electric companies in the footprint, and other factors not specifically identified. For the purposes of the Facilities Study Estimate the following assumptions have been made:

- Environmental permitting is not required.
- Sub-surface soil conditions are normal.
- Grading is not required.
- Fifteen month lead-time from issuance of purchase order has been allowed in the Project Schedule for delivery of the 345/115-kV transformer.
- The control building has adequate accommodations for a future secondary 320 A/hr battery system, i.e., batteries, charger, DC cabinets, etc. The secondary DC battery system will not be installed as part of this study.

Future Considerations

The additions being constructed as part of this project will allow for future expansion of both the 115-kV and 345-kV yards with minimal cost and associated outages.

Proposed Construction Cycle and Outage Requests

Construction Progression

1. The proposed work effort is designed to install the 345-kV circuit breaker 9M27 and connect it to switches 9M27B1 and 9M27B2. *Note: The switches 9M27B1 and 9M27B2 are to remain open during this phase.*

2. The proposed work effort will close switches 9M27B1 and 9M27B2 as well as breaker 9M27.

3. The proposed work effort will open breaker 9M26, switch 9M26B and switch 9M33A. The installation of breaker 9M34, switches 9M26A and 9M35A will follow. The extension of a five inch bus is required as well as the installation of switch TR 15.

4. The proposed work effort is designed to install transformer TR 15.

5. The proposed work effort is designed to install an electrical bus to TR15. *Note: Switch TR15 is maintained in the open position during this phase.*

6. The proposed work effort is designed to install the TR 15 115-kV strain-bus dead-end structure.

7. The proposed work effort is designed to install new circuit breakers 3M579 and 3M628. The units will connect to new panels 12E and 12W.

8. The proposed work effort is designed to remove the Jackson line relays and install new TR 15 relays into panels 15N and 15S.

9. The proposed work effort is designed to install new 115-kV switches 3M579A, 3M579B, 3M628B1 and 3M628B2. Also, the installation of an interconnecting five inch bus is required.

Outage Request

- A one week outage is requested of the 115-kV Main Bus #1 to connect breaker 3M579 into the differential relaying and to connect switch 3M579A to 115kV Main Bus #1.
- A 2 day outage is requested of the Jackson Line to remove the slack span to row 6 deadend structure and re-tap to row 3 bus.
- A one week outage is requested of the 115-kV Main Bus #2 to connect breaker 3M628 into the differential relaying and to connect switch 2A to 115-kV Main Bus #2.

- A one day outage is requested of the 115-kV Main Bus #2 to connect 115-kV double 1210 strain bus.

Risk Check List

Check all elements that are risk factors.

- Are there supplementary probable alternatives under consideration? *No.*
- Is significant labor cost modifications expected? *No.*
- Are material and resource lead-times expected to be longer than anticipated? *Transformer lead time of 15 months should be adequate.*
- Is significant material cost modifications expected? *Transformer price increase due to copper escalation is possible.*
- Are the presented drawings inclusive and exact? *Yes.*
- Is the survey information accessible? *See Contour and Grading Layout Drawing*
- Is the soil boring outcomes accessible? *Yes, as discussed in the design guide.*
- Are there existing erosion tribulations? *No.*
- Are there extraordinary soils or ecological conditions? *No.*
- Are there key materials or items that necessitate judgments or authorizations? *Yes, the 345/115-kV auto-transformer needs to be ordered ASAP.*
- Are there impending permit delays or abnormal requirements? *No.*
- Are there intricate or cyclic electrical outage requirements? *No.*
- Are there differing outage requirements (conflicts with other projects, etc.)?
- Is the project at risk for the reason of the construction project resources and the associated availability?
- Are there risks related with policies to recycle existing material? *No.*

- Are uncharacteristic construction methods compulsory? *No.*
- Are the presented oil containment elements sufficient? *Yes.*
- Is the presented lightning protection adequate? *Yes, additional mast and overhead static wires are being installed.*
- Is the existing bus and equipment ampacity sufficient? *Yes.*

Amplifying comments concerning the project outcomes: _____
_____
_____
_____
_____
_____
_____
_____
_____
_____

Engineers, project managers, and specialty labor consume an inordinate amount time completing project documentation usually in the incorrect phases of the endeavor. The majority of these types of questions must be reviewed with anticipated, incisive responses by the strategic leaders and stakeholders involved with the work effort. The scenario analysis and in-depth evaluations are expected to occur in an ideal design condition to minimize the effects of the external control factors. The leaders subsequently compile the solicitation of the general feedback from the subject matter experts as well as lessons learned historical documents and are compelled to take action in specific phases of the strategic planning process to facilitate a seamless transition of the electrical equipment installation.

Problem Formulation

Although the utility industry is closely tied to the environmental and economic issues for which it serves, only fifty-five percent of surveyed U.S. companies have sustainability initiatives (Savitz, A. et al., 2002). Further analysis revealed several major issues facing the sector today to include generation capacity, lack of green power progression, and an aging workforce. The concerns were subsequently categorized in critical areas in accordance with existing federal regulations and national interests. The five most critical issues facing the energy industry today entail the environment, improved infrastructure, labor force administration, technology, and additional regulation. Nearly seventy percent of the total business output is represented in the service industry (Woods, R., 2009). Forward-thinking organizations are investing in the human capital supply-chain to mitigate the skill shortage problem in the utility industry. The establishment of necessary structures and processes are obligatory to plan and manage the orderly implementation of transformation. The energy policy change issues of nationwide concern are outlined below:

- **The environment:** Comprises of such issues as global warming/climate change and emissions/carbon requirements.
- **Industry infrastructure:** Comprises of recuperating costs for constructing additional generation and transmission facilities.

Specific renewable energy market-based rate tariff proposals are ascertained and evaluated to determine if the requesting entities meet the generation and/or transmission market power standards with respect to barriers to entry. The analysis of test energy of new units (particularly unauthorized power sales) warrants a customer refund of the revenues collected, and calculated in accordance with the Federal Electric Regulatory Commission (FERC) regulations 18 CFR § 35.19(a) for the entire period that the rate was collected without authorization. The assessment of the generator special feature agreements must include the cost differentials between the applicant and the company in accordance the regulation, 180 days of the completion of the work.

- **Labor Force Administration:** Comprises of substitution of an aging workforce and amplified competition for talent.

Knowledge transfer is a major challenge facing the utility industry. Many company workers possess several decades of experience with little or no effort to capture/transfer the vital data and making it available for the next generation. The traditional baby boomer holds industrial information close to the vest (Christopher, R., 2012). The development of a viable methodology to evaluate retirement dates in conjunction with regularly scheduled meetings in an effort to record information before an individual (engineer, business analyst, project manager, control operator) departs the utility is necessary. In addition to exit interviews that covered processes and technical expertise, the establishment of a 90-day post-retirement window to continue the discussion

with the retirees will augment the process and increase the knowledge database. The additional knowledge the organization gathers about an asset, employee experiences, and specific user details, the better the informed decision-making by management (Meehan, B., 2011). The results are expected to produce informative notations for future drawings, interconnection agreements, and crucial data utilized for planning purposes.

The assessment of a plethora of interconnection agreements between the transmission providers and the local distribution companies are also required to ensure annual compliance with the filing deadline date. Specifically, the review of the projected voltage levels under normal system conditions in accordance with the anticipated annual peak load and 80% of the anticipated annual peak load for each interconnection point with planned additions for the next five-year horizon. The policy document of FERC 715 planning standard and the adopted fundamental planning principles as the foundation for specific reliability criteria is employed to ensure that the appropriately

Figure 6. One-line diagram of proposed maple cabin project

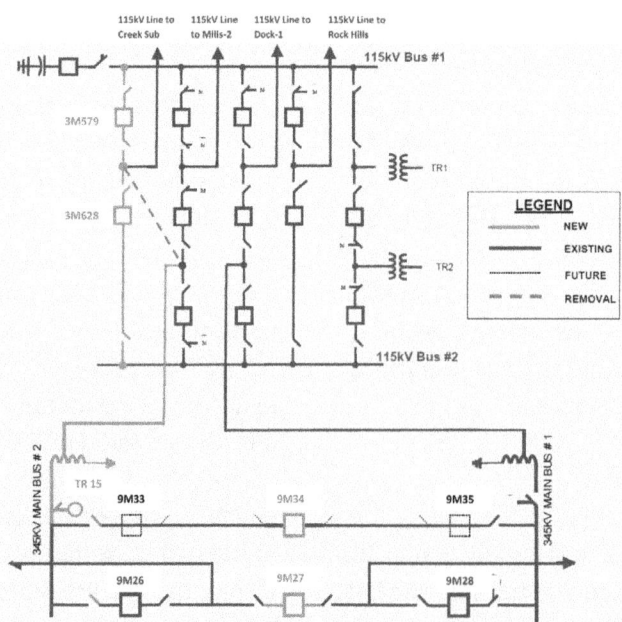

designed transmission system provides a reliable distribution of power flows by avoiding excessive geographical concentrations of generating sources or transmission paths. The current utilization of simulation techniques for targeted congested areas and the development of "what-if" analysis of increased load scenarios, as well as parallel power flows must be included in all planning activities.

- **Technology:** Comprises of smart meters, automated metering and clean-coal technology, which was found to be critical to meeting future challenges.

The evaluation and monitoring of the progress/implementation of the smart grid evolution currently experienced in New England and other sections of the country is important for the future expansion of the electric infrastructure. Particularly, the demand-response resource component of the endeavor in the pilot program that included the ancillary services (reserves) and the guidance, standards, and protocols the North American Electric Reliability Corporation, the Federal Electric Regulatory Commission, and the National Institute of Standards & Technology (NIST) provided in its implementation. The protocols are required by these "smart devices" to enable the communication capabilities for the automated real-time control of the electric supply and demand. The Alternative Technology Regulation (ATR) pilot program is monitored as well. An 18-month program in response to FERC Order 890 will assess the impact of non-generating technologies on the Regulation Market and allow owners of ATR resources to evaluate the technical and economic suitability of their technologies as possible regulation-service sources.

- **Additional regulation:** Comprises of environmental regulations on new and existing plants as well as costs associated with incorporating renewable resources into the energy portfolio.

Because of these national policy issues, electric utilities are concerned with new and innovative system integration processes not previously experience – especially in the area of the environment. The progressive company will utilize an assortment of financial models and strategies to account for the required system upgrades and renewable energy integration. The "Triple Bottom Line" index system, for example, provides a set of three dimensions of sustainability – economic prosperity, environmental quality, and social justice (Wang, L., and Lin, L., 2007). The intersection of the main components are further reduced and classified as eco-social, social-environmental, eco-environmental, and eco-social-environmental. The electrical utility must understand and comprehend the effects of these dimensions for equipment installations as well as its affect on the fiscal bottom line. Moreover, the paradigm is best incorporated in every facet of capital-intense ventures, the project/asset management organization, and system planning (both demand and supply-side power management programs).

Nevertheless, the major substation equipment installation delays and project failures were evaluated and subsequently categorized in specific levels of problems. The majority of the failures are attributed to people issues – commonly referred to as management causal factors (Levasseur, R.E., 2010). A sample of capital-intense electrical utility projects was derived from a larger pool of ventures from a regional Midwestern and west coast utilities. The resolutions to all of the problems encountered are essentially eliminated with the effective project management techniques and profound strategies originating from the executive arm of the organization. The perilous sample ventures were analyzed and arranged into the following categories:

- Land Issues
- Permit Issues
- Logistical & Scheduling Issues
- Labor Division Issues
- Financial & Cost Estimation Errors

These foremost categories were frequent with a recurring theme for project delays and cancellations. In fact, if the proper communication channels were utilized during the early phases of one particular land issues case, where the farmer refused to sell specific acreage to the utility company for renewable energy development, it would have been a *non-issue*. The company was forced to develop a resolution plan to appease all affected parties at additional cost, or resort to an alternative right-of-way at a much higher outlay of cash. It is here where knowledge is power and the effectiveness of a positive, renewable energy community outreach program is beneficial. This one scenario was obviously excluded from the strategy and planning sessions *before* the execution phase of the project. There were absolutely no contingency plans developed for this single farmer as the utility assumed, incorrectly, that every stakeholder was aligned with the master plan. Consequently, the farmer created a work stoppage for a critical system integration project. The mitigation strategy of cooperation, coordination, and the vital communication element was never developed, or simply ignored. The problem was detected during a quarterly project status session when the essential probing questions (*how, what, who, when, why, and where*) were asked. Projects fail because of lack of communication in an organization. The spirit of departmental cooperation as well as coordination with all assigned project resources must breakdown the barriers of the "silos" within the business structure. The matrix organization is specifically designed to eliminate these barriers allocated roles and responsibilities for a successful project implementation[1]. When resources interact in a more frequent basis, potential problems are identified and resolved in the early stages of the process.

In the permit issuance category, a specific project was destined to fail without the required documents. Moreover, the capital venture was developed without a contingency plan for this particular area. The utility scheduled a hearing with the local municipality to identify the constrained areas of the permit in order to expedite the process. Consequently, the project managers redistributed the resources to other higher priority endeavors, abandoned the materials on site, and requested the planning engineers to re-evaluate the resource constrictions in the electrical footprint. Problems such as these are easily mitigated with the incorporation of the proper communication channels, coordination, and cooperation in the strategy area of the process. It is designed to minimize re-work efforts and the associated costs as well as providing a path to develop viable scenarios to potential problems, which oftentimes are ignored.

The root cause of the majority of the evaluated capital project problems were fundamentally caused by inferior communication and lack of coordination with the project management, planning, and strategy organizations. The multi-million dollar ventures experience excessive scheduling delays and cost overruns as a result of such failed efforts. The case of the wandering transformer on the high seas for an eastern utility became a serious logistical and scheduling delay issue. The transformer was purchased and tested, monitored in the manufacturer's country of origin and subsequently placed on a freighter that sailed to another country. Because of the pending holiday season (or for some unknown reasons), the ship returned to dock with the undelivered transformer presuming still onboard. The after-the-fact expeditor was dispatched from the utility company to the second country in order to facilitate the process (horse-out-of-the-barn scenario). The logistical questions should have been asked at the point of sale and thoroughly vetted during the strategy sessions. Moreover, the *expensive* unit – once purchased – should have possessed *inexpensive* tracking devices (data logger, global positioning system, etc.) to monitor the status throughout the supply-chain. Contractual obligations that highlighted the penalties for such a long delay should have been clearly outlined as well as departmental cooperation, as both items were obviously non-existent in this case. In fact, procurement of the most expensive component in the substation was treated as if one would purchase a postal stamp from another country with the expectation of receiving the goods in due time. Consequently, additional installation time was requested at a project status update session to include a second outage. There was absolutely no float time on the schedule to unload and install the transformer with only a limited outage window. This activity was obviously scheduled on the venture's critical path. As a result, the entire project was delayed.

The scheduling of free float time would have allowed an event delay without affecting the subsequent activities and the project outcome. The *no float time,* tight schedule scenario creates excessive resource expenditures as well as safety issues under such conditions. The performance and quality of work suffers in the reduced time/increase resource scenario with abnormal circumstance and intermediate work effort. Although penalties are imposed on vendors for product delays, the situation should have never occurred with incisive planning and logistical integrations. High-performance vendors with the quality component embedded in the process must be placed at the top of the "preferred supplier list" with the associated delivery case histories, and thoroughly vetted by the project organization for these critical components. The effort will ensure the scope, schedule, and budget adherence of the multi-million dollar project implementation.

The financial estimation errors are another major problem area in the project management sphere. It is often cited for the cost overruns during the "finger pointing" breakout sessions of project update sessions in some electrical regions. It is true that the material expenditures (i.e. the unit cost of copper quadrupled in price from project-to-project in one region alone) increased for certain substation elements; yet it is not the sole reason for the cost overruns. The incisive leader has included this element in the bandwidth of the metrics during the strategy sessions. This same leader is well aware of the optimal levels to achieve the project's goals, especially from a bottom line perspective. The model of commu-

nication, coordination, and cooperation – when utilized effectively – empowers every individual in the project organization with the cost containment strategy. If the material cost was an outlier beyond the limits of the bandwidth, the element is subsequently competitively-bid with the high-performance vendors for preferred or volume discounts. The high-priced item (problem) is essentially eliminated.

One particular utility project cited high-cost estimates of the contractor performing rock and soil testing on a proposed site as the sole reason for the entire venture's cost overruns without asking the critical questions. This external entity is a vital member of the project management team where data collection and conducting the experiments are included in the planning process. The detailed-oriented work is highly competitive and essential for proper electrical equipment installation. The engineers and technicians performing this work typically has access to specialized equipment/material, possess a wealth of knowledge about the instrument capabilities, test procedures, past design failures, and other institutional expertise not readily available to the project office from other sources. It will behoove the astute project manager to promote an inclusive relationship with the testing facility to comprehend the steps and procedures involved in the process. Moreover, an assessment of the site complexities and resource requirements are also in order before a formal bid is developed. This effort improves the quality of the overall test results and incorporates another beneficial resource in the project organization. Consequently, costs are driven down when the soil testing procedures are thoroughly evaluated, compared, and understood.

The cost control strategy of "re-balancing" the capital project portfolio to reduce labor expenditures was a general theme in several electric utility regions. This technique was liberally employed for problematic ventures continually over-budget. The division of labor between the internal resources and the contractual workers was a contentious area for the equipment installation ventures. One particular Midwestern region was eleven million dollars over-budget with the assigned capital projects and immediately requested to re-evaluate the portfolio as a cost containment strategy, midstream in the construction cycle. The proposal included the transfer of specialty labor to internal personnel and outsources the rudimentary tasks to the higher-cost contractors within the substation work effort. The plan was designed to free up labor for capital work to balance the portfolio while simultaneously meet the operation and maintenance commitments in the region. As a short-term solution to an over-budget situation, the resolution will certainly decrease the expenditures – temporarily. The region will experience moderate cost results for a couple of project status update sessions before the expenditures spiral out of control. The simple reason again is based on the communication, project coordination, and project cooperation efforts. In order to essentially transfer the labor from contractor-to-internal personnel, a specific level of communication is required to accomplish the task. This is akin to "crashing" or increasing the resources at the construction site and must be carefully planned. The augmented resources will subsequently increase the communication channels and coordination efforts. The traditional bathtub model illustrates the number of resources assigned to specific tasks along the x-axis and the task duration on the y-axis. There is an optimal point on the curve that depicts an ideal relationship between the maximum number of resources assigned and the minimum duration of the completion for the tasks involved. However, as the resources increase, the curve in the bathtub model will increase proportionally which will lead to an amplified duration of task achievement. Consequently, the communication element to update the team on the immediate tasks/specifications and the newer resources with the project goals increased the timescale of the entire project. The communication channels are derived by the following relationship:

$$N(N-1/2)$$

Where

N= number of project stakeholders

If a substation project, for example, has twelve stakeholders assigned with an expected increase of seven in the following month, the communication channels will increase from 66 to 171. Therefore, the seven additional resources effectively increase the communication channels to 105 in a one month period on time. The work effort on the substation project will require more time to efficiently transfer critical information via additional communication channels.

The evaluation of sample capital projects in this particular region produced substation ventures under-budget for a short period of time, yet behind schedule. Further analysis revealed, after a significant amount of time had past, projects were both over-budget (again), behind schedule (failed to meet key milestones), or simply re-assigned to another budget year. The communicative element is a key component to complex capital-intensive projects. How the labor division is optimized, in this case, requires the robustness of a system to identify the sources of noise (number of resources, delays, etc.) and minimize its affects at the strategic level. It is desired to seek an optimal mix of personnel to reduce process cycle time and minimize cost as the solution to the labor division problem. The situation is similar to the individual strengths of a batting line-up in baseball, with high on-base percentage players bat ahead of the powerful, clean-up hitter. Once the critical resources are identified and secured in the strategy sessions of a matrix organization, the overall project costs are drastically diminished.

THE ANALYSIS BACKGROUND

The mean is utilized to compare results of experimental data analysis with traditional statistical methods. The standard deviation metric is employed to ascertain the significant differences between two groups of data. The Signal-to-Noise Ratio – a different statistic to compare experimental results – is utilized in the Taguchi methodology. The technique exploits the variables under study by dividing the data into two general groups:

- Factors that are controlled (a process parameter whose value can be selected and controlled by the strategic design engineer) or;
- Factors that either cannot be controlled or too expensive to control (called *noise factor* – an uncontrollable factor that causes the process quality to vary).

There are generally three types of noise classifications: 1) Noise due to external causes (i.e. operator, weather, equipment downtime, etc.; 2) Noise due to internal causes (i.e. wear and tear, deterioration, etc.); and 3) Noise due to process-to-process variation (Taguchi, G., 2005). The primary goal in the quality approach to engineering is to optimize the process design for the minimal sensitivity to noise. The greater the noise effect, the greater the inconsistency. The Taguchi method elects control factors not only to produce the desired result (such as stronger) but also to direct a process that is less sensitive to noise. Although the noise cannot be eliminated, its effect can be minimized. The results are therefore stronger with minimum variability. The SN Ration calculation encompasses the mean as well as the variation from one result to the next. The analysis is considered as a two-dimensional component as opposed to the typical one-dimensional evaluation (Peace, G., 1993).

The objective of the "Design of Experiment" is to balance or separate the data with the utilization of orthogonal arrays. The process provides the necessary flexibility and capability to assign numerous variables for evaluation. The orthogonal array in quality engineering is used exclusively as a design mechanism to evaluate a process with respect to *robustness* against *noise* and its related

cost. A process is considered robust when it experience restricted or reduced functional variation, even in the presence of noise (Phadke, M., 1989). It is a strategic inspection device employed by incisive project management organizations to prevent an inferior design from traversing to downstream activities.

The Parameter Design

The parameter design philosophy focuses on *success* instead of *failure* as espoused in quality control. The goal of the paradigm is to define the normal value that will determine the design. Parameter design is the second of the three major evaluation steps that follows the system design phase, where engineering and managerial knowledge is applied to produce a viable process characteristic in the initial settings of the implementation.

In the division of labor issues for the substation work effort, the design will therefore seek the smallest variability and highest efficiency of the operation. A subsequent goal on parameter design is to eliminate the effect of noise *before* it occurs as the process is essentially immune to variation sources (Fukahori, M., 1995). For example, resources are assigned immediately after the detailed list of tasks, or work breakdown structure, are compiled – not during the project execution phase. The role assignment and responsibilities are developed in accordance with the applicable skill sets of the labor pool. The associated matrix is created and disseminated to the parties involved with the project[2]. Nevertheless, evaluation of the company internal and contracted labor pool is in order to compare the effects utilizing the SN Ratios. The investigative tasks involve the installation of conductors and fittings within two sample substations and differing labor hours/costs. The preliminary experiment was conducted with only noise factors to ascertain the tendencies and direction of these conditions. The noise factors in this case derived from the sampled engineering estimated completion hours of the installation on similar substation project as compared with the average expected values of the venture acquired from the historical (lessons-learned) database. Internal (A_1), Contracted (A_2), and Blended (A_3) labor were evaluated to determine the magnitude of the output per input as illustrated by the slope of the input/output relationship, classified as sensitivity. A sample of these data were calculated and captured as illustrated in Table 2.

The decomposition of the data and variation is calculated from the squares of kr_o pieces of data ($f = kr_o$ degrees of freedom):

$$S_t = y^2{}_{11} + y^2{}_{12} + \dots + y^2 k_o$$

- Resultant
  - **Contractor:** 207,396
  - **Internal:** 193,683
  - **Blended:** 16,252

The error variation S_e, included the deviation from linearity, V_e:

$$S_e = S_t - S_\beta \text{ and } f = kr_o - 1$$

- Resultant
  - **Contractor:** 519.47
  - **Internal:** 72.66
  - **Blended:** 2.04

The error variance V_e is the error variation divided by the degrees of freedom:

$$V_e = S_e/kr_o - 1$$

- Resultant
  - **Contractor:** 173.16
  - **Internal:** 24.22
  - **Blended:** 0.68

The variation caused by the linear effect:

$$S_\beta = 1/r_o r (M_1 y_1 + M_2 y_2 + \dots M_k y_k)^2 \text{ and } f = 4$$

- Resultant
  - **Contractor:** 206,876.53
  - **Internal:** 193,610.34
  - **Blended:** 16,249.96

The SN Ratio is[3]:

$$\eta \rightarrow 10 \log \beta^2/\sigma^2$$

The SN Ratio in decibel (dB) units:

$$\eta = 10 \log (1/r_o r) (S_\beta - V_e)/ V_e \text{ dB}$$

- Resultant
  - **Contractor:** 0.0194
  - **Internal:** 0.1301
  - **Blended:** 0.3887

Transformed into decimal value, for example, the contractor calculation:

$$\eta = 10 \log (0.0194)$$

$$= -17.1162 \text{ dB}$$

- **Internal:** -8.858 dB
- **Blended:** -4.1034 dB

The sensitivity in decibel units:

$$S = 10 \log (1/r_o r) (S_\beta - V_e)$$

Estimated value of the sensitivity coefficient for the contractor calculation:

$$\bar{\beta} = (112,750)/61,450$$

$$= 1.834$$

- **Internal:** 1.775
- **Blended:** 0.514

The relationship(s) between M and y:

- Resultant
  - **Contractor:** y=1.834M
  - **Internal:** y=1.775M
  - **Blended:** y=0.514M

Therefore:

- -17.162 dB-(-8.858 dB) = -8.304 dB difference between A_2 and A_1 where internal labor appears to be the best alternative over contracted labor.
- ((-8.858 dB-(-4.1034 dB)) = -4.755 dB difference between A_3 and A_1 where blended labor is the best alternative over internal labor.
- ((-17.1162 dB-(-4.1034 dB)) = -13.013 dB difference between A_3 and A_2 where blended labor again is the best alternative over the contracted labor.

Nevertheless, the augmented labor resources will increase the time spent in the substation. The goal is to have high-quality technical personnel involved in the work effort. The sensitivity analysis between internal and contractor labor is relatively small when compared with the blended resource option. The comparative data is illustrated in Figure 7. The analysis of the impacts of the variances caused by the factors (ANOVA) is depicted in Tables 3, 4, and 5.

The qualitative assessment was not encompassed within the calculations. The characteristics of the system were developed to find only the SN Ratio and related sensitivities. Although the internal labor appears to be the best alternative for the short-term, the enlarged communication efforts (previously discussed) will subsequently increase the resource time in the substation and ultimately the cost, from turnover activities alone. The system produced the best alternative with the utilization of a *blended labor pool* initially developed at the strategic level to reduce noise. By selecting the premium labor talent from both internal and the contractor pool, the highest SN

Ratio is achieved. However, a more detailed assessment is required to determine the absolute system gain based on the available resources, site conditions, skill levels, engineering plan, the depth of resource experience, and most importantly, dedicated project time to implement electrical system upgrades and/or renewable generation integration. The team resource utilization approach is espoused in the TRIZ principle #15 of "Dynamics" as outlined below (Mann & Domb, 1997):

- Allow (or design) the characteristics of an object external environment, or process to change to be optimal or to find an optimal operating condition.
  - Continuous Process Improvement,
  - Optimum Quality Cost Model,
  - Identify team members with the ability to adapt to changing situations.
- Divide an object into parts capable of movement relative to each other.
  - Work teams oriented to achieve the same goal, but at different rates on different objectives.
- If an object or system is rigid or inflexible, make it movable or adaptive.
  - **Flexible staff:** Use of temporary workers and overtime,
  - Control charts, run charts, trend charts for illustrating dynamic display of the process behavior,
  - Changing the supervisor's role (avoid the "firefighting" syndrome).

The robust design philosophy is insensitive to variation in the strategic planning processes instead of the actual, real-time equipment implementation phase, where the attempts to control the many variations oftentimes fail. The emphasis is on the routine optimization of the process prior to its implementation rather than quality through inspection. The quality and reliability are forced back to the design phase where it really belongs. The goal is to identify the optimal settings to

Table 2. Resource-man hour data for contracted labor

M ~ Resource	Contracted Labor
	Y in man-hrs
2	3.669650122
5	9.174125305
10	18.34825061
15	27.52237592
20	36.69650122
25	45.87062653
30	55.04475183
35	64.21887714
40	73.39300244

produce a robust process that can survive project after project, piece after piece, while meeting the requirements.

The energy transformation experienced within the equipment installation of a substation involves eight major stages (scope, ideal function, design factors, noise reduction strategy, simulated test, data analysis, optimum design performance, and confirmation trial) in experimental design. The entire system is captured in the P-diagram illustrated in Figure 8.

The signal factors, noise, and control factors act as inputs to the system – the electrical equipment installation process. The desired output is the amount of time spent in the substation based on the optimal resources. The *scope* of the entire system is to minimize the substation system cost during the equipment installation processes. In order to accomplish such a goal, the ideal function is to evaluate the input demanded from an optimal resource perspective. The selected strategic questions that must be asked are:

- What company manufactures the preeminent, cost-effective components?
- What are the equipment lead times and probable delays?

- What specialized equipment, and the related costs, is required to complete the installation?
- What are the labor requirements and skill levels?
- What is the composition and complexity of the site?

The *ideal function* is expected to encompass the least variability and highest efficiency. The acquisition of the best, talented resources to install the equipment in the least amount of time is a desired condition as part of the strategy. Perhaps an unallocated spare piece of equipment may be obtained from inventory or another substation to reduce costs and minimize installation time. The advantageous scheduled outage of the spring construction cycle must coincide with the available labor resources to facilitate equipment installation *before* the summer critical period. These are a few of the *design parameters* and alternatives that must be determined within the arena of the P-diagram. It is here where the bandwidths are established in the robust design philosophy. If a major piece of equipment is damaged in transit to the construction site (i.e. transformer damage as it was unloaded, handled, and transferred from the freight train to the truck) an alternative plan (contingency) is quickly enacted to include a spare unit located earlier in the logistical analysis process. The project manager will avert a work stoppage of the downstream activities as the replacement unit is secured, dressed, and prepared for installation. These "what-if" scenarios originate in the parameter design stages at the strategic level, especially for high-visibility, capital-intensive ventures.

Once the parameters are identified, the establishment of a *noise reduction* strategy follows. The noise is a component inherent within a process where minimization must be achieved. The number of delays, for example, is an area of chief concern in the electrical equipment installation process. The utilization of a second or third lab, for example, to perform soil testing is designed to minimize delays and reduce noise. The alignment of several consultant specialty engineers on an "as-needed" basis to play a back-up role for the prime technical engineering talent is designed to avoid absenteeism and minimize delays. Again, this type of contingency planning is performed upstream at the strategic level to avoid the downstream "firefighting" syndrome most project managers face on a daily basis. Noise will not be eliminated, only minimized. The incisive decision-maker can identify its source and perform the correct steps to lessen the noise factor. Once this is accomplished, the *experimental test* is conducted to obtain the data (Taguchi et al., 1999). The simulation information is arranged in arrays for efficient data analysis with the typical naming convention as described below:

$L_a (b^c)$

Where

a = number of experimental runs

b = number of levels of each factor

c = number of columns in the array

An L_8 array, for example, can handle seven factors at two levels each under eight experimental conditions. Once the data is obtained from the array, it is analyzed using the SN Ratio, a method to test for robustness. The L_8 array will produce eight SN Ratio resulting outputs as well as the mean. The performance of the *optimal design* is subsequently compiled and predicted from these results. Finally, the *confirmation trial* is conducted using the optimum design to evaluate the gain estimates.

The L_{18} Model

The control factors defined in the P-diagram are based on the efficient methodologies and resources to implement electrical equipment within the substation environment. Each category is comprised of three levels (save the specialized equipment category (C) with only two – available and unavailable). The levels within the *Technical Experience* category (A), for example, are classified as novice (apprentice), intermediate, or expert (journeymen, senior engineer) to accomplish the assigned tasks. Some equipment installation capital projects were lacking in the experience, skill, and knowledge levels. Consequently, these particular ventures were delayed based on the increased communication channels via the project management

Figure 7. SN ratio comparative results

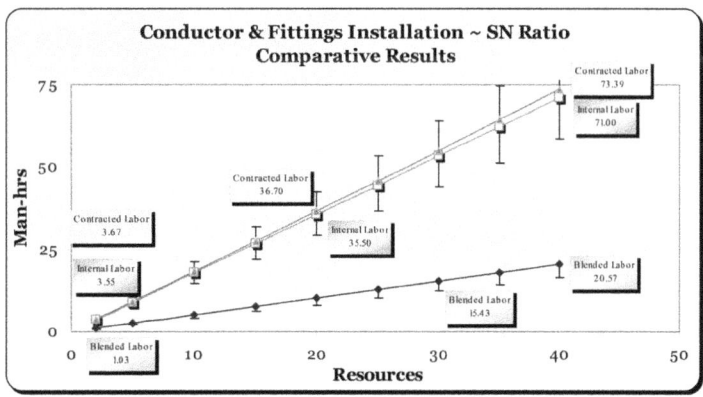

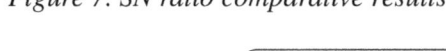

Table 3. Contracted Labor ANOVA

ANOVA Table of Contractor (A2):				
Factor:	f	S	V	E(V)
β	1	206,876.53	206,876.53	$\sigma^2 + 61,450\beta^2$
e	3	519.47	173.16	σ^2
Total:	4	207,396.00		

office, engineering design, line shops, and the supply-chain. It was difficult to administer the "soft-side" area of the project management realm because of the insufficient knowledge factors and processes involved with an inexperienced crew mired in a complex situation. *The Contractor* (G) *and Internal Labor* (H) categories are similarly classified based on the skill level (low, medium, and high). The contractual labor will typically tap into its *best* resources for a construction venture whereas the electric utility capital project team will utilize the most *available* resource. There are occasions where a mixture is desired to achieve the optimal results. The *Site Complexity* (B) category includes the conditions of the construction area and classified as degraded, neutral, or superior. The ease of equipment transportation for installation, mobility, constraints, etc., are vital attributes for consideration in the robust design process. The *Material Cost* (F) category is a major area of consternation for electrical equipment upgrades or new installation. The decision-makers establish the bandwidth for premium-select vendors and applicable volume discounts based on historical supplier relationships. The most cost efficient component may (or may not) come from the same

vendor. The levels are classified as low (cost), median, and High, within the spectrum of the array. As previously discussed, the *Engineering Plans/Details* (E) category is an important asset for the equipment installation process with varying degrees of progression. It is classified as incomprehensible, somewhat comprehensible, and fully comprehensible. Oftentimes, resources are idle because of incomprehensible (incorrect) drawings, misinterpretation of the plans, or communication errors in the design process. How well these plans are organized and communicated to the resources is indicative of the inherent quality in the final product. Nevertheless, a highly-experienced vetted crew, thoroughly knowledgeable of past installation endeavors can succeed with somewhat comprehensible plans. The corrections are consequently transferred back to engineering design as re-work – an inefficient process. The final category of the control factors is *Dedicated Project Time* (D), or a measure that involves the length of time a selected resource remains with the venture. This important factor is classified as low, medium, and high. An internal resource, for example, may begin a wiring procedure within a substation and diverted to another project with-

Table 4. Internal labor ANOVA

ANOVA Table of Internal (A1):				
Factor:	f	S	V	E(V)
β	1	193,610.34	193,610.34	$\sigma^2 + 61,450\beta^2$
e	3	72.66	24.22	σ^2
Total:	4	193,683.00		

Table 5. Blended labor ANOVA

ANOVA Table of Blended (A3):			
Factor:	f	S	V	E(V)
β	1	16249.96	16249.96	$\sigma^2 + 61{,}450\beta^2$
e	3	2.04	0.68	σ^2
Total:	4	16,252.00		

out replacement. The unattended wiring is left incomplete until the resource returns or another resource is immediately available (same skill set, minimal knowledge transfer, and minimal delay time). The dedicated project time is designed to allow a team to start and finish assigned tasks with minimal interruptions. Strategic elements of this paradigm encompasses the construction cycle schedule (electric utility storm season when demand maintenance is high), the customer demand (summer critical season when all available generation units are required and outage requests are low), or when the electrical demands is low and outages are abundant. The optimal combination of the design must include some form of the control factors to produce high-quality equipment installation and minimal substation time. The L_{18} model array is illustrated in Table 6.

SN data analysis (sample calculation):

$$S_m = (5+10+10+10+5+50+5+5)^2/8 = 1250$$

$$S_t = 5^2+10^2+10^2+10^2+5^2+50^2+5^2+5^2 = 2900$$

$$s_e = S_t - S_m$$

$$= 2900 - 1250 = 1650$$

$$V_e = s_e/8\text{-}1 = 1650/7 = 235.714$$

$$\eta = 10 \log (1/\eta)\, (S_m - V_e)/ V_e = 8(1250\text{-}235.714)/235.714$$

$$= 10 \log (0.5378)$$

$$= \text{-}2.693 \text{ dB}$$

Now, the average SB Ratio: for A_1 = -2.693+7.601+…+1.210/9 = 2.601

for A_2 = 4.173+0.695+…+3.701/9 = 4.316

The resulting SN Ratio data is compiled and ranked for each simulation as illustrated in the response Table 7.

The goal of parameter design is to reduce variability and increase the SN ratio. The optimum combination from the SN ratio response table is: A2B2C2D2E2F1G2H2. Factors D, E, G, and H have a very strong impact on the variability whereas B, C, and F are relatively weak. Therefore, C2D2E2 G2H2 is selected to optimize based on the strong impact.

$$\eta_{opt} = {}^-T + ({}^-C_2 + {}^-T) + ({}^-D_2 + {}^-T) + ({}^-E_2 + {}^-T) + ({}^-G_2 + {}^-T) + ({}^-H_2 - {}^-T)$$

$$= {}^-C_2 + {}^-D_2 + {}^-E_2 + {}^-G_2 + {}^-H_2 - 4\,{}^-T$$

$$= 4.32+4.60+4.53+4.45+4.37-(4) \qquad (3.45)$$

$$= 8.47 \text{ dB}$$

$$\eta_{baseline} = T + ({}^-C_1 + {}^-T) + ({}^-D_1 + {}^-T) + ({}^-E_1 + {}^-T) + ({}^-G_1 + {}^-T) + ({}^-H_1 - {}^-T)$$

$$= {}^-C_1 + {}^-D_1 + {}^-E_1 + {}^-G_1 + {}^-H_1 - 4\,{}^-T$$

$$= 2.60+1.92+1.87+1.82+2.41- (13.80)$$

$$= \text{-}3.18 \text{ dB}$$

The gain between the optimal and the baseline:

$$= 8.47 \text{ dB} + (\text{-}3.18 \text{ dB})$$

Figure 8. Substation P-diagram

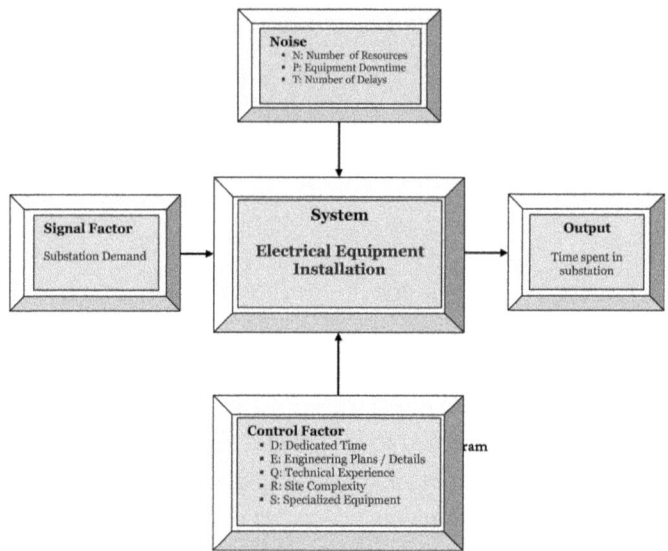

Table 6. The L$_{18}$ model array

	C	A	B	D	E	F	G	H	Mean	SN
1	5	10	10	10	5	50	5	5	15.000	-2.693
2	5	10	25	30	25	25	25	25	20.000	7.601
3	5	10	50	50	50	5	50	50	28.333	3.282
4	5	30	10	10	25	25	50	50	17.500	3.099
5	5	30	25	30	50	5	5	5	24.167	0.699
6	5	30	50	50	5	50	25	25	31.667	3.793
7	5	50	25	30	5	5	25	50	20.000	1.924
8	5	50	50	50	25	50	50	5	38.333	4.495
9	5	50	10	10	50	25	5	25	25.000	1.210
10	25	10	50	50	50	25	25	5	35.000	4.173
11	25	10	10	10	5	5	50	25	10.833	0.698
12	25	10	25	30	25	50	5	50	27.500	4.372
13	25	30	25	30	50	50	50	25	35.000	9.329
14	25	30	50	50	5	25	5	50	30.833	3.793
15	25	30	10	10	25	5	25	5	17.500	4.055
16	25	50	50	50	25	5	5	25	34.167	3.561
17	25	50	10	10	50	50	25	50	32.500	5.167
18	25	50	25	30	5	25	50	5	26.667	3.701

Table 7. The average SN ratio response table

	Average SN Ratio	Response Table							
	C	A	B	D	E	F	G	H	Average
Level 1	2.60	2.38	3.23	1.92	1.87	4.08	1.82	2.41	20.31
Level 2	4.32	4.13	3.74	4.60	4.53	3.93	4.45	4.37	34.07
Level 3		3.34	3.40	3.85	3.98	2.37	4.10	3.61	24.65
delta:	1.72	1.75	0.51	2.68	2.66	1.71	2.63	1.96	79.03
Ranking	6	5	8	1	2	7	3	4	3.4359

$= 5.29\ dB$

The goal of beta is to seek a control factor with a large impact on the mean and a minimum impact on the variability. The beta response table and calculations are similar and summarized in the associated tables with the SN Ratio. A confirmatory experiment based on the factors C, D, E, G, and H was conducted to estimate gain and test for reproducibility. The data is illustrated in Table 8.

The Result Analysis

The confirmatory experiment proved good reproducibility and a significant improvement of the SN Ratio results. It is a state where the conclusions inferred from the small-scale experiment is valid under actual usage conditions. The gain estimates was based on factors C, D, E, G and H with the largest effects. The reduction of variability in the *Dedicated Project Time* is the chief concern of the entire experiment. The response graphs (Figure 9) for the electrical equipment installation process clearly illustrates the large factors of variability in these areas.

Ideally, the high dedicated project time is required from either the internal or contracted resource for project completion. Nevertheless, the data in both panels (A and B) of the response tables rank this category as the number one problem in the analysis. Although the skill levels of the resource from either pool is not a major effect in the system (factors G and H), it is expected the labor must have a moderate comprehension of the tasks involved. Moreover, the technical experience level of the resources is optimized within the intermediate range, which minimizes the cost of highly-skilled personnel. However, exceedingly successful consequences occur with experienced personnel and expert judgment by management (Iward & Mwasha, 2012). The sensitivity response graph (panel B of Figure 9) illustrates absolutely no variation in the resource category, and slightly higher gains in the SN Ratio graph in the contractor category. The response (G2) depicts the contractor work effort at these higher levels

Table 8. The confirmation test results

Configuration:	SN Ratio (dB)		Sensitivity	
	Estimation:	Confirmation:	Estimation:	Confirmation:
Baseline:	-3.18	-3.12	14.508	15.52
Optimal:	8.47	8.53	27.01	27.02
Predicted Gain:	5.35	5.40	12.50	12.50

over the preferred internal labor category, contrary to the initial cost assessments. It is *not* a viable method to balance a regional capital portfolio. Ideally, the contractor is assigned specific projects within its area of expertise to maximize the work effort in a minimal amount of time. In the labor distribution arena within the capital project sphere, this activity is obviously ignored as attested by the $11M cost overruns in the evaluated region. The material cost is significant in the analysis as the price of copper quadrupled in particular capital projects. Moreover, other material costs (1000 kcmil ~ $18.75 to $34.28 unit cost) practically doubled from project-to-project and region-to-region. Consequently, the capital project esti-

mates were erroneous and required standardization from premium quality vendors to support the large ventures. The supply-chain re-engineering is compulsory in order to minimize cost and improve incoming material quality. The ideal condition (F1) illustrated in the response graphs where low material costs are desired can be obtained from the premium vendors, volume discounts, etc., to reduce the variability. In order to design a viable robust system, it is imperative to reduce the noise. All of the components are expected to perform flawlessly. A portion of the elements reside within the *Engineering Plans/Details* category where an inordinate amount of time is consumed with communication/data interpretation. The

Figure 9. SN ratio and beta response graphs

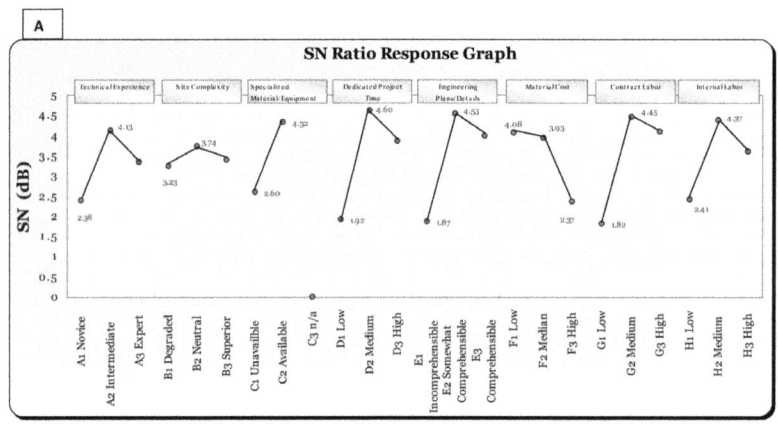

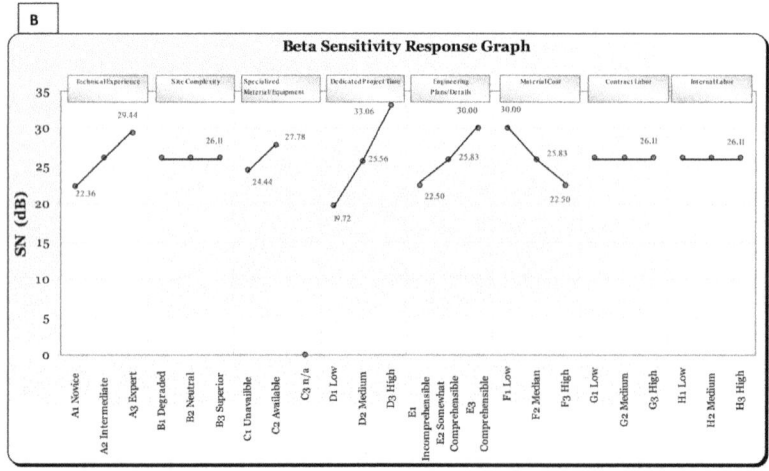

optimal resource (E2) in the SN Ration response graph (panel A of Figure 9) depicts the "somewhat comprehensible" engineering package will produce positive work efforts with the electrical equipment installation process. Fully comprehensible engineering plans does not necessarily equate to faster and complete equipment installation as the skill sets and experience of the assigned resources generally override the common sequences and procedures. The importance of this attribute is the foundation of a potentially successful project or one that is destined for failure. A correct and ideally comprehensible engineering package will minimize delays and reduce overall system costs. The confirmatory experiment successfully estimated gains between the baseline and optimal levels – specifically in the *Dedicated Project Time* areas. Again, the ideal situation of the technical experience at the expert level, with available specialized equipment and fully comprehensible work plans along with a high-level of dedicated project time in conjunction with low material cost is the over-arching desired results of the robust experiment. It requires an experienced, aging workforce to acquire new technological skill sets or, as the experiment demonstrated, a transitional knowledge transfer from the experienced worker to the technologically advanced younger worker. The formulation of a risk-based capacity planning optimization model is required to avert human talent loses and retain existing skill levels. Additionally, the investment decision to manage the shortages in particular is necessary over the planning horizon (Cao et al., 2011). The results also proved that once these workers are on the site, the amount of concerted effort has a strong influence on the project success as opposed to individuals drawn from either labor pool (whereas contracted labor is slightly more dedicated in accordance with the experimental gain). The larger variability in the engineering work plans are certainly a major concern for electrical equipment installation. This area must be incisive and direct to avoid downstream problems (idle time, material logistics, resource utilization, etc.) and excessive project delays or failures. The material, schedule, and cost estimates must be exact. In the parameter design phase of robust engineering, the optimization of the nominal values is defined by the materials, configurations, and dimensions required. Subsequent analysis involves tolerances for maximum effect of quality efficiency and thrift of the design. The quality loss function is then utilized to assist with the evaluation of the effectiveness of manipulating of the dimensions or tolerances (fine tuning). The technique is designed to assist the management team with developing a process with six sigma quality while maintaining a low-cost threshold. The "soft-side" project management realm is expected to be incorporated within the quality performance plans during the strategy sessions for a successful renewable energy system integration effort.

REFERENCES

Cao, H., Hu, C., & Jiang, C. (2011). Integrated stochastic resource planning of human capital supply chains. *Interfaces*, *41*(5), 414–435. doi:10.1287/inte.1110.0596

Christopher, R. (May 2012). Survive and thrive: Adapting to the multigenerational workplace. *Quality Progress*, pp. 48-49.

Clymer, J. R. (2009). *Simulation-based engineering of complex systems* (2nd ed.). New York, NY: John Wiley.

Cooper, R. G. (1993). *Winning at new product development: Accelerating the process from idea to launch*. Reading, MA: Addison-Wesley.

Feigenbaum, A. V. (1991). *Total quality control* (3rd ed.). New York, NY: McGraw-Hill.

Fukahori, M. (1995). Optimization of an electrical encapsulate. *Journal of Quality Engineering Forum*, *3*(2).

Iward, J., & Mwasha, A. (2012). The effects of ISO certification on organization workmanship performance. *Quality Management Journal*, *19*(1), 53–67.

Jones, R. (2007). *Project management survival: A practical guide for leading, managing, and delivering challenging projects*. Philadelphia, PA: Kogan Page.

Kerzner, H. (1992). *Project management: A systems approach to planning, scheduling, and controlling* (4th ed.). New York, NY: Van Reinhold.

Levasseur, R. E. (2010). People skills: Ensuring project success- A change management perspective. *Interfaces*, *40*(2), 159–162. doi:10.1287/inte.1090.0473

Mann, D., & Domb, E. (1997). *40 inventive (business) principles with examples*. Retrieved 8 November, 2008, from http://www.triz-journal.com

Meehan, B. (July 2011). As season utility staff retire, will they take wisdom with them? *Utility Automation Engineering T&D Magazine*. Retrieved 15 May, 2012, from http://www.elp.com/elp/en-us/index

Melnyk, S. A., Stewart, D. M., & Swink, M. (2004). Metrics and performance measurements in operations management: Dealing with the metrics maze. *Journal of Operations Management*, *22*(9), 209. doi:10.1016/j.jom.2004.01.004

Peace, G. S. (1993). *Taguchi methods: A hands-on approach to quality engineering*, (pp. 1-236; 292-312). Reading, MA: Addison-Wesley.

Phadke, M. (1989). *Quality engineering using robust design*. Englewood Cliffs, NJ: Prentice Hall.

Savitz, A. W., Besly, M., & Booth, K. (2002). *2002 sustainability survey report* PriceWaterhouseCoopers LLP. Retrieved 15 November, 2010, from http://www.pwc.com/fas/pdf/sustainability%survey%20report.pdf

Taguchi, G. (2001). Robust design by simulation: Standard SN ratio. *Quality Engineering*, *9*(2).

Taguchi, G., Chowdhury, S., & Wu, Y. (1999). *Robust engineering*. New York, NY: McGraw-Hill.

Taguchi, G., Chowdhury, S., & Wu, Y. (2005). *Taguchi's quality engineering handbook*. Hoboken, NJ: John Wiley & Sons, Inc.

Wang, L., & Lin, L. (2007). A methodological framework for the triple bottom line accounting and management of industrial enterprises. *International Journal of Production Research*, *45*(5), 1063–1088. doi:10.1080/00207540600635136

Woods, R. A. (2009). Industry output and employment projections to 2018. [Washington, DC: Bureau of Labor Statistics, U.S. Department of Labor.]. *Monthly Labor Review*, *132*(11), 53–81.

ENDNOTES

1. See appendix III for detailed roles and responsibility matrices and project management templates.
2. See Appendix III.
3. Note: V_e is used to estimate σ^2.

Chapter 11
Electrical Contractor Work Comparisons

ABSTRACT

The decision theory approach determines the optimal configuration by minimizing its cost for a given risk environment. Prioritization is necessary, as is some quantity of information. Supplementary information gathering may add value, but this depends on how much data the Project Manager already has and how much new information is to be gathered with analytic resources available. The payoff from spending time on project decision analysis is already high, even with the brute force approach of analyzing all projects. The payoff could be higher if project decision analyses plans are focused by taking advantage of knowledge about the general characteristics of a project even before individual projects are assessed. The objectives of the study are to identify the cost structures, utilizing simplistic management science mechanisms, in the contractual work arena. A comparison between The Premium Select's actual contractual costs against competitively-bid similar projects by other contractors forms the basis of this investigation.

OVERVIEW

Contractors are invariably assigned to electrical utility reliability, capacity, and/or condition projects as either the exclusive sole resource or as an additional reserve for the implementation. The renewable energy project execution consists of a general mix of expertise from a variety of resources. Six (6) major projects and their associated historical costs were investigated in this study. The chapter examines the cost structures of competitively-bid utility contractors as compared

with the cost of service provided by a Premium Select Company. Several scientific management methods are employed to analyze the data in order to select the most cost-effective decision for future projects. The selection of the "smallest of the worst regret that could have happened" (minimize the maximum possible) with budgeted capitalized and maintenance dollars was the overall goal of this investigation. However, this ideal situation was not always the case. It was discovered that most of the circuit breaker retirement/replacement program costs were absorbed in overhead expenses

DOI: 10.4018/978-1-4666-2839-7.ch011

when The Premium Select Company performed the work. Conversely, when the same corporation provided services for capacity improvement projects, it excelled in several areas of the work. Third party contractors were very competitive with substation assignments and displayed a slight weakness in transmission line projects. *The overall preliminary study revealed an average cost savings of forty (40%) when external contractors performed substation work, and (0.2%) savings when The Premium Select performed transmission line work.* It is important to first comprehend the framework (i.e. the parameter values) and simply select a project management policy for decision analysis strategies. The decision-maker must exploit a disciplined project selection process for funding at the very minimum, especially with renewable energy generator grid integration. It may be sufficient to simply apply a threshold where projects with high Benefit-to-Cost Ratios are funded and ventures with lower ratios are discarded. In such cases, setting the appropriate threshold might be a reasonable objective for analysis in support of the Senior Management Team. The decision-maker(s) are evaluating data from numerous sources (economic, load growth, financial, performance, environmental) for the successful implementation of the power resource. For these reasons, it may be complicated to choose such a threshold. In such cases, unequivocally ranking schemes adds substantial significance. Judgment matrices and priority vectors are utilized in the typical least cost objective function. In decision-analytic practice, the ranking of select projects is often preceded by the formation of detailed estimates of endeavor's worth. Such improved value estimates are also beneficial, but unless uncertainty is high, enhanced estimates are not nearly as imperative as the basic use of a regimented process. For the astute decision-maker, it is essential to distinguish between these two sources of value. Sensitivity analysis of the data revealed that even when uncertainty seems high and intricate to resolve, there is considerable value to adopting a decision process that incorporates a disciplined

approach to project prioritization. Outside the use of thresholds, two other time savings approaches of the Lean Quality paradigm were considered. The decision-maker would surrender a lot of value by ranking projects in terms of Expected Net Present Value instead of explicitly considering the Benefit-to-Cost Ratio (the preferred method of the electric utility industry), and this approach is not recommended. Executive leadership who already employ a disciplined methodology may be able to save effort by analyzing in detail only those ventures that materialize as close calls. In such cases, a quantitative priority methodology (ranking of the level of importance based on experience and judgment) is utilized to differentiate the essential need of the project. It is realistic to exploit this technique in extreme power supply cases, but the efficiency gains are small enough that if it is worth making the effort to analyze the marginal plan, it is likely that most ventures must be analyzed.

INTRODUCTION

Decision analysis is a chief component for capital project selection of an electrical utility executive leadership team in today's society. The methodology is utilized as a minister to the allocation of resources across sectors of capital investment opportunities (reliability, capacity, and condition projects). The Integrated Resource Plan is a vital input for the project selection process. The primary object of the resource plan is to develop a least-cost arrangement that can meet customer-energy service requirements and environmental improvements. Once the ventures are vetted at this stage, the project decision analysis paradigm predominates. In its fundamental form, the approach consist of applying decision-analytic techniques to the candidate projects in a sector one at a time to estimate the cost and value (or expected value) for each scheme. The opportunities are then categorized or *prioritized* in declining order of one of several essentially equivalent measures. The

Benefit-to-Cost Ratio (defined as the expected value of the proceeds resulting from a project divide by its expenditures). The existing financial plan is then utilized to fund the higher-priority projects until it is exhausted. The situation creates two provisional scenarios for project funding:

- The benefit of decision analysis due to the improved value estimates of the discipline, or
- How much is simply a result of the objective categorization (prioritization) to decide which project will receive the funding.

Enhancements of the value estimates is time intensive and cost prohibitive in some electrical utility organizations, and necessitate expert interviews and complicated financial models. Conversely, prioritization alone can be accomplished with models that must be able to perform a technically effortless task as sorting. As the renewable energy system integration becomes more complex, the models are expected to alleviate the mundane procedures for prompt and effective decision-making. Several managerial studies comprise of estimated values augmented by the decision-analytical approach in particular by project analysis efforts. These investigations characteristically compare the value of the new sector of funded projects with the expected net present value of the sector of ventures that would have been funded without doing the analysis. Enrichment in value due to decision analysis estimated in these studies have ranged from 17% to well over 100% of the value of the original sector that would have been selected.

The expansive series of decision analysis is outlined below (Decision Process, 2005):

1. The problem is designed (matrices and decision trees) by classifying sufficient actions, one that must be decided upon; potential events, one of which occurs thereafter, and outcomes, each which results from a combination of decisions and events.

   a. Decision under certainty issues are those in which each alternative results in one and only one product and that conclusion is sure to occur.

   b. Decisions under probabilistic uncertainty issues are those in which several conclusions can result from a given action depending on the state of nature with known probabilities.

   c. Decisions under probabilistic imprecision issues are those in which several conclusions can result from a given action depending on the state of nature with unknown and imprecise probabilities.

   d. Decision under information imperfection issues are those in which one of several conclusions can results in a given action depending on the state of nature and occur with imperfectly specified probabilities.

   e. Decision under conflict and cooperation issues are those in which there is more than a single decision-maker and where the objectives and activities of one decision-maker is not necessarily known to all decision-makers.

2. At the time of the decision-making, the incident that will actually occur cannot be predicted absolutely.

3. Personal Preferences: The subjective value the decision-maker appends to a conclusion is quantified and termed as the utility of a result.

4. Probabilities of incidents and utilities of outcomes must be measured independently of one another.

5. The probability predetermines the current state of information. Additional information can be acquired in the expectation of reducing uncertainty.

6. The information of the source (field tests, subject matter experts, simulated model) is described in terms of a probabilistic relation between the information and the incident.

7. Given the additional information, prior probabilities can be replaced by subsequent probabilities, and the analysis repeats.

Multi-criteria decision making and decision analysis under uncertainty are two challenging issues in electric utility resource planning, particularly with renewable energy. Strategic planning engages the exploitation of extensive analytical tools, such as investment optimization and production costing simulation modes, and requirements of the support of advanced decision analysis techniques. Instead of determining a single optimal solution, the strategic planning approach will yield a range of satisfactory plans or proficient decisions (Yoon & Kim, 1989). The final resource strategy can then be chosen from among the acceptable alternatives or assessments based on certain well-defined decision criterion. A robust, aging transformer, for example, requires a well-planned strategy conjoined with condition assessments and non-intrusive testing to properly determine to equipment's health and reliability of the electrical system (Ledel & Baker, 2012). Capital investment, system impact, and long-term data management planning are prioritized as fundamental strategies for the decision-makers and preferred over shirt-term planning and cash flow considerations. In an integrated Multi-Attribute Decision Making framework for multi-attribute planning purposes under uncertainty, the strategy provides a structured decision analysis platform compliant to both the probabilistic assessment approach and the risk evaluation methodology and allows the application of hybrid decision methodology. A key procedure in strategic planning is to identify the acceptable alternatives with respect to the given attributes and future conditions. The most often used decision analysis techniques in the electric utility industry are multiple utility function method and tradeoff/risk analysis method.

The multiple utility function method ranks alternative planning strategies based on the measure of projected performance (i.e., probabilistic evaluation approach) while tradeoff/risk analysis method ranks alternative planning strategies based on the measure of robustness (i.e., risk evaluation approach). The robust methodology is particularly suited in situations where the parameters are unknown when the optimization model is solved. Moreover, it is utilized to assess measurement errors in the parameters as well as errors created by numerical instability (McLay et al., 2012). The multi-attribute decision making methods have been widely used in strategic planning of electric utilities which provides an efficient decision analysis framework to assist the decision maker in selecting the best resource strategy with regard to the chosen attributes. The information available to the decision-maker, on the other hand, is often imprecise owing to inaccurate attribute values and inconsistent expert judgments. Consequently, it may not be totally satisfactory to use point estimation as the sole criterion to make a decision, and it appears necessary to incorporate the measure of variance into the decision analysis. Study results using detailed models associate each scenario with a set of quantified attributes, yielding the decision database with which alternative resource strategies can be compared with advanced decision analysis techniques and the best solution or the most desired resource development strategy can then be determined. A practical multi-attribute decision-maker model should be able to display tradeoffs among different attributes, quantitative and qualitative, economic as well as non-economic, and quantify the preferences held by different interests.

The utility industry has used a variety of decision analysis methods in their strategic planning to deal with multiple objectives and concerns for risk due to uncertainties. Essentially, the following questions must be addressed in the decision process:

• Which planning strategy or plan materializes to be superior to others?

- What is the rationale of ranking pertaining to a range of acceptable alternatives?
- What is the relative importance of one attribute with respect to other attributes?
- How the performance of each alternative can be properly measured in the presence of uncertainties?

The utility function model and tradeoff/risk analysis are two most commonly used decision analysis methods in the electric power industry for multi-attribute planning in the existence of uncertainties. The concept of the utility function model is to optimize a formulated function which transforms a multi-attribute decision problem into a scale performance measurement, i.e., the measure of composite utility or the measure of composite distance. Utilizing this method, the plans are ranked based on the rule of probability such that the best solution would be the plan for which the expected value is optimal. The non-technical factors (stakeholder support, team member commitment, department conflicts, etc.) involved in project ranking are direct causes of their failure (Levasseur, 2010). The concept of tradeoff/risk analysis is to treat all attributes individually and to identify the plans that significantly dominate other plans conditional on each specified future. This method also identified non-dominated plans that are ranked based on the measure of robustness, i.e., the number of supporting futures. As for decision analysis under uncertainty, the utility function method is often used in conjunction with the decision tree modeling approach to provide a graphical interpretation of alternative planning strategies, decision variables and uncertainty factors. This decision process is typically termed as probabilistic valuation approach by which the best solution is determined based on the expected performance of respective planning strategy under various future conditions, i.e., the expected value of composite utility (Nour & Bhevaraju, 1989). It is realized that multi-stage decision analysis is required to address the influences of various improbability factors and flexibility enhancement options. Nevertheless, the application of multi-stage decision analysis is restricted by its computational complexity which increases exponentially as the number of uncertainties, decision options and time periods represented in the decision model are increased. Therefore, a practical solution is to develop a two-stage decision analysis model, a compromise between the simplified one-period modeling approach and the complicated multistage modeling approach. The embedded Nash Bargaining Mode is axiomatic with its approach in the decision-making process (Lippman & McCardle, 2012). A set of propositions are anticipated and analyzed where the solution provides a compromise between two extreme positions. Only one decision-maker grants all of the *take-it-or-leave-it* offers.

The intent of the study is to investigate a set of projects currently under consideration by decision-makers advising project managers. Because it is impractical to evaluate the value of all different information acquisition strategies in the renewable energy sector, project simulation techniques are employed using parameters based on empirical data. The cost and expected values of each project are recorded, as managed under several analytical strategies. These strategies are modeled as diverse levels of information that may be available at the time funding decisions were made (a critical tenet considering the long lead times of electrical equipment installation processes). The strategies range from a minimum-value subterfuge in which resources are allocated to projects without making use of relevant information, to a maximum-value model of excellence stratagem in which projects are prioritized based on perfect information (Logan, D. M., 1990). In between are several strategies that utilize partial information about some or all of the projects. These results are used to identify circumstances where the simplest strategies, especially threshold-based rules, perform almost as well as the more time intensive and costly strategies.

STRATEGIES FOR ANALYSIS AND FUNDING

There are four basic strategies with variations on each item. All of the strategies consist of a rule for information-acquisition decisions (i.e. determining which projects will undergo a thorough analysis) and a rule for making funding decisions based on the information available. The strategies are outlined below:

PQ1: Random Financial Support

Projects are selected and funded in random order until the financial plan constraint is achieved. The expected value amplifies at a linear rate with the expenditure at the rate of the average benefit-to-cost ratio of all projects. The rational for assuming decision-makers employ random process even when they have some information is that, without a regimented value-maximizing process, the decision-making sequence may preclude the restricted use of that information. Political and organizational pressures, conflict avoidance, or a historical pattern of haphazard "first-come-first-served" funding could all lead to this situation. Problems of this type are commonly cited as reasons for using a decision-analytic process. This strategy corresponds to a state of no information or, consistently, no prioritization.

PQ2: Apply the Threshold Rule without Resolving Improbability

Projects are considered one at a time in random order and project "A" is financially supported if a specific parameter meets or exceeds a predetermined threshold level, until the financial plan is exhausted. If there is additional funds in the financial plan remaining after all of the projects are indexed in the order they are considered, the decision rule can be expressed in terms of a first pass.

This strategy corresponds to a state of partial information and partial prioritization.

PQ3: Prioritize without Resolving Improbability

Projects are ranked in order of a specific parameter and funded in descending order until the budget constraint is reached.

This stratagem corresponds to a state of limited information and complete prioritization. The funding decision for a single project also includes information about the value of other projects to set the proper threshold level for the financial plan.

PQ4: Resolving Uncertainty then Prioritize

Projects are ranked in order and funded until the budget constraint is achieved. This strategy corresponds to complete prioritization with ideal information about project values.

The normative decision theory formalizes and rationalizes the decision-making process (Decision Process, 2005). The theory depends on the following assumptions:

1. Historical preferences are legitimate indicators of present and future predilections.
2. Decision-makers correctly perceive the values of the uncertainties that are associated with the results of decision alternatives.
3. Decision-makers are able to assess decision situations correctly, and the resulting decision situation structural model is well formed and complete.
4. Decision-makers execute decisions that actually reflect their true preferences over the alternative course of action, which may have uncertain results.
5. Decision-makers are able to process decision information appropriately.

6. Authentic decision circumstances provide the group with decision alternatives that permit them to articulate their true preferences.

7. Decision-makers accept the axioms that are assumed to extend the various normative theories.

8. Decision-makers execute decisions without being so besieged by the complexity of actual decision situations that they would necessarily use suboptimal decision strategies.

While the post-mortem study of the substation and transmission line projects assume the disciplined analytical decision processes for funding, irregularities consistently appeared in the data (i.e. cost estimates, project bundling, excessive overheads, adjustments, miscellaneous expenses, etc). The decisions rendered were partially based on imperfect cost estimates, personal preferences, and implicit contractual obligations. Moreover, the lack of a disciplined approach in the contractor selection process was evident in other areas (planning, asset management, environmental, maintenance) as well. In fact, a subjective vendor rating system (Table 1) was utilized on occasion throughout the organization as the preferred selection method. The financial indicators of the regulatory organization were the impetus of a thorough investigation for specific budgeted expenditures.

Data normalization was required to assess the actual cost comparisons between third party contractual work and similar tasks performed by the Premium Select Company. If the strategies were thoroughly utilized during this process based on the Expected Net Present Value alone, the Premium Select Company would retain the majority of the workload. Nonetheless, the high uncertainty of project cost estimates by this company in particular produced favorable results for the contractors performing similar tasks at a lower rate.

BACKGROUND AND GENERAL PROJECT OVERVIEW

The objectives of the study were to identify the cost structures, utilizing simplistic management science mechanisms, in the contractual work arena. A comparison between The Premium Select's actual contractual costs against competitively-bid similar projects by other contractors formed the basis of this investigation. The substation projects examined for the study are the ARGENTA problematic circuit breaker retirement/replacement program performed by the Premium Select Company and the competitively-bid THETFORD substation master circuit breaker replacement plan within the condition project scheme. The overall scope of the work for both projects involved the short circuit requirement and upgrade of substation capacity by replacing faulty and defective circuit breakers with newer units to meet the future demands (renewable energy interconnection) of the system.

Table 1. Organization current contractor selection criteria

Rank	Attribute	Details
5	**Apparent Strength**	Supplier proposal specifically addressed this item. Ability to perform activity appears to be strength.
4	**Apparent Capability**	Supplier proposal specifically addresses this item, and it APPEARS to be within supplier's capabilities.
3	**Assumed Capability**	Supplier proposal did not specifically address this item, although it IS ASSUMED to be within supplier's capabilities.
1	**Minimal/No Basis**	Supplier proposal did not specifically address this item, and there is either no basis for judgment or activity appears to be beyond supplier's capabilities.

The expenditure estimates and the actual costs for the circuit breaker replacement varied between The Premium Select and the Third Party Contractor in this program. A volume-to-volume circuit breaker analysis and individual unit replacement was utilized to compare these cost structures on specific work orders.

Circuit Breaker Decision Analysis

The basic question posed to the maintenance department is to replace or overhaul (planned fixed operating time) a circuit breaker at failure (repair).

The objective:
Minimize Cost $_{(total)}$
Minimize Downtime $_{(total)}$
Maximize Availability
Cost Model:

$$E\left(C\left(t,t_s\right)\right) = C_f E\left(N_f\left(t\right)\right) + C_s E\left(N_s\left(t\right)\right)$$

where:

$C\left(t,t_s\right) = $ Total cost of repair/maintenance when planned maintenance interval is t_s

$C_f = $ Cost of unscheduled system maintenance

$C_s = $ Cost of scheduled maintenance (assumption of $C_s < C_f$)... Failure replacement more expensive than scheduled replacement

$N_f\left(t\right) = $ Number of system failures in (0,t) which causes maintenance

$N_s\left(t\right) = $ Number of scheduled system maintenances in (0,t)

$T_s = $ System planned maintenance interval (equipment reaching its planned maintenance age)

Preventive Replacement Policy

The basic equipment replacement model (equipment will be replaced when it reaches a specified age t_s) is shown below and in Box 1.

$$C_f(h(t_s)\int_{-\infty}^{t_s} R(t)dt + R(t_s) + T_s h(t_s) - 1) - C_f\{h(t_s)\int_{-\infty}^{t_s} R(t)dt + R(t_s) + h(t_s)T_f\} = 0$$

Where:

$$T_f > T_s$$

Hazard function:

$$h\left(t\right) = f\left(t\right) / 1 - F\left(t\right) = F\left(t\right) / R\left(t\right)$$

Constant Maintenance Interval (Common)

Preventive replacements are performed at fixed intervals. The model is shown below.

Box 1.

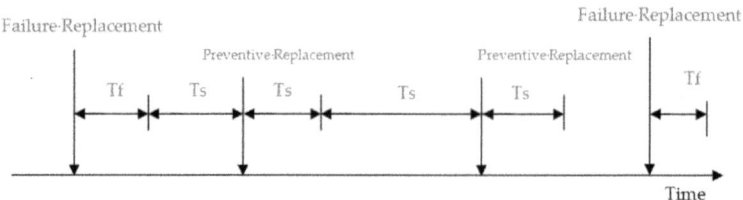

302

$$C\left(t_s\right) = C_s + C_f E\left(N_f\left(t_s\right)\right) / t_s + T_s$$

Renewal Function:

$$H(t_s) = \sum_{R=1}^{\infty} F_x(t_s)$$

Where:

$F_r\left(t\right)$ = Cumulative distribution function of the time up the rth failure.

Two major policies are derived from this model.

Block Replacement Policy

1. The replacement of equipment is carried out at fixed intervals and at failure.
2. Periodic replacement with minimal repair at failure model:

$$h\left(t_s\right)\left(t_s + T_s\right) - \int_0^{t_s} h\left(t\right) dt = C_s C_f$$

Renewal function:

$$H\left(t_s\right) = \int_0^{t_s} h\left(t\right) dt$$

Both policies determine optimal replacement and cost intervals.

Lifecycle Costing

Lifecycle cost analysis is the general measurement utilized to verify and compare the asset performance. It is developed in calculations of the total economic worth of an asset by analyzing initial expenses and discounted future costs, such as maintenance, operating efficiency, repair, rehabilitation, and replacement costs, over a func-

tional life of the asset. The expenditure is typically segmented into four major separate components and outlined below:

- **Initial capital expenditure:** This is the initial capital cost of the asset including the cost of capital.
- **Maintenance costs:** This includes all costs related to predictive, preventative, or breakdown repair maintenance including parts and labor.
- **Operating Efficiency:** This refers to changes in operating costs associated with the degradation in the asset's performance because of equipment aging or poor maintenance.
- **Replacement Costs:** Most assets will need to be replaced eventually. By including the cost and timing of replacing the asset in the lifecycle cost analysis, the benefits of increasing the useful life of the asset maintains accountability.

Asset Optimization

Based on this model, it is clear that, once the original expenditure has been prepared by purchasing the equipment, asset performance, as measured by lifecycle cost, can be improved by reducing maintenance costs (especially breakdown repair maintenance, which is more costly than predictive or preventative maintenance), maintaining operating efficiency (directly proportional to maintenance more so than age), while increasing an asset's useful life.

Knowledge Management

The fundamental challenge of asset optimization is obtaining the required knowledge to determine the best economic equilibrium among an asset's financial performance, operational performance, and risk exposure. A key to meeting this challenge is knowledge management, which can be defined as an amalgamation of related technologies and expert analytical services that transform data into information, and information into knowledge. The effectual knowledge management tools are essential to control the three (3) variables of an asset's lifecycle cost after the initial purchase (maintenance cost, operating efficiency, and length of useful life). The paradigm links decision-making and action with real-time information such that decisions are driven by actual condition and performance of assets both individually or system wide.

The Transmission Line Projects examined for the study consisted of the following: VEVAY-DELHI/TOMPKINS-DELHI capacity improvement and the GAINES-BRADLEY capacity improvement line work. Both projects were performed by the Premium Select Company. The competitively-bid PERE-MARQUETTE-STRONACH transmission line was utilized as a cost comparison connecting the projects. The overall scope of the work of the transmission line reconductoring project encompassed rebuilding to improve the import capacity of the selected regions. The wood pole replacement, as stipulated in the project's scope, was also included in the general re-work of the transmission line. The per-unit cost (wood poles, conductors, static wire, etc.) are also analyzed as an offshoot of the actual costs. The estimated cost structures include engineering services, project management, and direct construction costs.

The Regulatory Body

As an independent dogmatic entity, the Federal Electric Regulatory Commission (FERC) primarily regulates the hydroelectric, electric, and natural gas industries. The electric power regulation portion (sales of electric energy for resale, power transmission) of the FERC relates to the public utility sector in accounting cost functions

as well. The organization provides oversight of all regional transmission organization (RTO) and/or independent system operator (ISO) products and services under approved tariffs. Furthermore, the Federal Power Act affords FERC plenary and exclusive authority over the organized energy markets and transactions, including financial contracts that inhabit through RTO and ISO systems, both within and between organizations subject to their regulation. In addition, FERC is involved in the renewable energy integration initiatives, transmission cost allocation/siting issues, smart grid protocols /standards, and demand response areas. Future areas of FERC involvement includes electric storage, electric vehicles, operational efficiency for utilities/gas pipelines, and greenhouse gas issues. The industrial greenhouse gas emission reduction improvement processes are problems identified within the substation sphere. One of the components of high voltage equipment in an electrical substation is Sulfur Hexafluoride gas (SF_6). Because of its unique chemical properties, the gas is used to provide high voltage insulant /arc quenching medium and is frequently found in high voltage electrical equipment utilized by utilities and their major industrial customers. Unfortunately, the same properties that compose SF_6 so effectively for utilization in electrical equipment also make it a potent greenhouse gas (GHG) with significant global warming potential if it is released to the atmosphere. Consequently, the gas is now targeted for emission reduction and, as an environmental priority, warrants replacement technology in the electrical industry. There are currently independent alternative analysis and evaluation strategies being performed of the replacement equipment for the gas circuit breakers of targeted utilities and process (primarily gas) industries. The exploration and evaluation of numerous potential solutions of each circuit breaker is required before accepting a preferred methodology. A thorough investigation of the high voltage (≥ 765-kV) air blast breaker as well as the low voltage (≤ 38-

kV) vacuum type breaker must be included in the in-depth alternative analysis. Moreover, evaluations of the service leak frequency, disposal, and maintenance best practices are also necessary to facilitate the program. Additionally, the FERC is heavily engaged in the development of the smart grid, efficient competitive electric power markets, and barrier removal to demand response issues.

The General FERC Codes

The general FERC codes for the projects investigated are described below:

- **Account 352:** Structures and Improvements: This account shall include the cost in place of structures and improvements used in connection with transmission operations.
- **Account 353:** Station Equipment: This account shall include the cost installed of transforming, conversion, and switching equipment used for the purpose of changing the characteristics of electricity in connection with its transmission or for controlling transmission circuits. The item details are shown in Table 2.

The specification for construction covers right-of-way (ROW), clearing and access, wood poles, pole top assemblies, structure assembly and structure erection, guys and anchors, grounding and bonding, insulators and hardware, and phase conductors and overhead ground wires.

- **Account 354:** Towers and Fixtures: This account shall include the cost installed of towers and appurtenant fixtures used for supporting overhead transmission conductors. The item details are shown in Table 3.
- **Account 355:** Poles and Fixtures: This account shall include the cost installed of transmission line poles, wood, steel, con-

crete, or other material, together with appurtenant fixtures used for supporting overhead transmission conductors. The item details are shown in Table 4.

The installation processes are intricate and detailed. A description of FERC account 355 section 1 (anchors, head arm and other guys, including guy guards, guy clamps, strain insulators, pole plates, etc.) and section 4 (excavation and backfill, including disposal of excess excavation material) is outlined in general terms.

When constructing power lines, the electrical designer will need a means of strengthening poles and keeping them in position (2). To accomplish this, the use of guys, anchors, and braces are required. Anchors are buried in the ground, and guy wires are connected to the anchors and attached to the pole, or a push brace may be used. The guys and braces are used to counter the horizontal strain on the pole caused by (conductors, pole-line components, and abnormal loads, such as snow, sleet, or wind). Anchors are designed to meet specific soil conditions. The electrical designer must know the type of soil before selection of a certain type of anchor. The devices come in many forms and have different methods of installation. The Expanding Anchor, Screw Type Anchor, and Never Creep Anchor are commonly used because of ease of installation (Figure 1)[12].

Anchors placed in a hole must be approved by the engineer in writing before the anchor hole is backfilled. The holes must be backfilled and tamped in the same manner as is required for wood pole backfilling. Only suitable native soil or approved imported granular material must be used for anchor backfill. Where required by the engineer, anchors must be tested to 50 percent of their designated ultimate rated capacity. Power installed screw anchors must be installed with the appropriate size and type of equipment in accordance with the engineer's requirements and manufacturer's recommendations.

Table 2. FERC account 353 details

	353 Items:
1	Bus compartment, concrete, brick, and sectional steel, including items permanently attached
2	Conduit, including concrete and iron duct runs not a part of a building
3	Control equipment, including batteries, battery charging equipment, transformers, remote relay boards, and connections.
4	Conversion equipment, including transformers, indoor and outdoor frequency chargers, and motor generator sets, rectifiers, synchronous converters, motors, cooling equipment, and associated connections.
5	Fences
6	Fixed and synchronous condensers, including transformers, switching equipment blowers, motors and connections.
7	Foundations and settings specially constructed for and not expected to outlast the apparatus for which provided.
8	General station equipment, including air compressors, motor hoists, cranes, test equipment, ventilating equipment, etc.
9	Platforms, railings, steps, gratings, etc. appurtenant to apparatus listed.
10	Primary and secondary voltage connections, including us and supports, insulators, potheads, lightning arresters, cable and wire runs from and to outdoor connections or to manholes and the associated regulators, reactors, resistors, surge arresters, and accessory equipment.
11	Switchboards, including meters, relays, control wiring, etc.
12	Switching equipment, indoor and outdoor, including oil circuit breakers and operating mechanisms, truck switches, and disconnect switched.
13	Tools and appliances.

Screw anchors (Figure 2) must not be reversed to meet the requirements of project of the rods above the ground. The anchor rod serves as the connecting link between the anchor and the guy cable. Anchor rods must be installed in line with the guy wire and installed so that not more than 8 inches of rod (including eye) remain out of the ground after guy tension is applied.

A guy is a brace or cable fastened to the pole to strengthen it and keep it in position (Figure 3). Guys are used whenever the wires tend to pull the pole out of its normal position and to sustain the line during the abnormal loads caused by sleet, wind, and cold. Guys counteract the unbalanced force imposed on the poles by dead- ending conductors, by changing conductor size, types, and tensions; or by angles in the transmission or distribution line. The guy should be considered as counteracting the horizontal component of the force with the pole or supporting structure as a strut resisting the vertical component of the forces. Guys must be installed and attached to the structures as shown on the transmission line structure drawings before conductors or overhead ground wires are strung. Each guy must be pretensioned to remove any slack in the guy. Guys must be re-tensioned after the conductors and overhead ground wires are installed to plumb the poles and to equalize tensions in the guys. If slack guys are found, they must be readjusted so that all guys in any structure have approximately equal tension. The final tension in the guys and the plumb of the poles must meet the approval of the engineer.

There were safety instances where guys had been damaged, leaning poles had dropped cables, or conductors below required clearances. Likewise, improper guying or tensioning of conductors or cables created either lower cables to sag below the required ground clearance or upper conductors

Table 3. FERC account 354 details

	354 Items:
1	Anchors, guys, braces.
2	Brackets
3	Crossarms, including braces.
4	Excavation, backfill, and disposal of excess excavated material.
5	Foundations
6	Guards
7	Insulator pins and suspension bolts.
8	Ladders and steps.
9	Railings, etc.
10	Towers

to sag downward too close to lower ones at mid-span (or had pulled a lower cable so tight that it approached an upper conductor too close under design conditions). If the lowest conductor or cable in an overhead span is immovable by a truck or other vehicle and lines or structures are broken, all occupiers of a joint-use structure can be unfavorably affected.

- **Account 356:** Overhead Conductors and Devices: This account shall include the cost installed of overhead conductors and devices used for transmission purposes. The item details are shown in Table 5.

A description of FERC account 356 section 2 (conductors, including insulated and bare wires and cables) and section 4 (insulators, including pin, suspension, and other types) is outlined in general terms.

Transmission lines carry power from generating plants to the distribution systems that feed electricity to domestic, commercial and industrial users. Electricity is normally generated away from load centers because of environmental and safety reasons. Transmission lines can be either overhead or underground cables. Electricity is usually sent over long distances through overhead power transmission lines. Underground power transmission is used only in densely populated areas (such as large cities) because of the high costs and losses. Most power is transmitted as 60 Hertz (cycles per second) alternating current (AC) power (Clapp, A. 2006) Transmission lines vary from a few miles/kilometers long in an urban environment to over 621.31 miles (1000 km) for lines carrying power from remote hydroelectric plants. An overhead transmission line is made of a conductor, insulators, and a tower. The three phase conductors carry the electric current. Insulators provide support and electrically isolate the conductors. Tower support the insulators and conductors. It is firmly grounded with special foundation and optional shield and ground conductors protect against lightning. Transmission line conductors are normally made from Aluminum with certain reinforcements. Copper is not usually at high voltage because of its costs even though it has a very low resistance. The conductors are made of aluminum strands which are reinforced

Table 4. FERC account 355 details

	355 Items:
1	Anchors, head arm and other guys, including guy guards, guy clamps, strain insulators, pole plates, etc.
2	Brackets
3	Crossarms and braces
4	Excavation and backfill, including disposal of excess excavation material.
5	Extension arms
6	Gaining, roofing stenciling, and tagging
7	Insulator pins and suspension bolts
8	Paving
9	Pole steps
10	Poles, wood, steel, concrete, or other material.
11	Racks complete with insulators.
12	Reinforcing and stubbing.
13	Settings
14	Shaving and painting.

Figure 1. Typical anchor

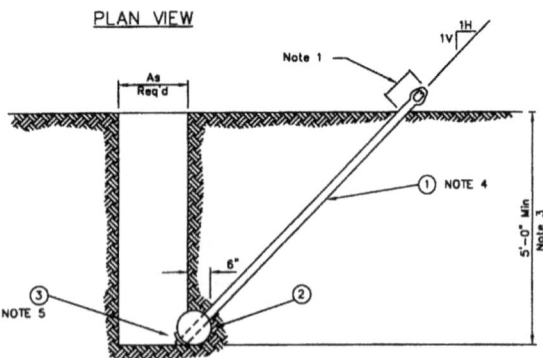

by another material. Aluminum conductors are classified as follows:

- Aluminum Conductor Steel Reinforced (ACSR)
- All Aluminum Conductor (AAC); and
- All Aluminum Alloy Conductor (AAAC)

The conductor is made of 30 Aluminum strands and 7 steel strands. The electric current is carried by the aluminum strands while the steel strands provide mechanical support. Overhead line conductors are usually bare without any insulation. Bare conductors have excellent heat dissipation characteristics. Also insulating high voltage conductors will be economically prohibitive.

Reels of wire must be stored off the ground and adequately supported so as to avoid damage to the reel, protective covering, and wire. Wire and reels must be kept free of standing water, excessive dust, and mud, and stored no closer than 50 feet from an energized portion of a substation or transmission line. Protective covering must be removed at the job site and the outside layer of each reel must be examined by the contractor and the engineer for the reasons listed below:

- To be sure that the wire is in good condition,
- And that no nails, staples, or other sharp objects, which could damage the wire dur-

ing unreeling, protrude on the inside of the reel heads.

Identification tags and markers must be retained on the reels. Conductor reels should not be rolled. The reels must be lifted or transported by a reel dolly. The special equipment and handling instructions are included in several procedural field manuals and engineering data sets. Nevertheless, several contractors diminish the use of specific measures based on inaccurate assumptions/past experiences of the processes.

Insulators are used to support, anchor and insulate conductors from ground. They are made of porcelain, glass and several synthetic materials. There are two types of insulators:

- Pin
- Suspension types

Suspension types are usually used for high voltage lines. A number of insulators usually form a string between the conductors and the tower cross arms. The number of insulators is dictated by the voltage level of the line. Insulator and hardware assemblies must be fully assembled and installed as shown on the drawings. Items of hardware and

Figure 2. Typical screw type anchor installation

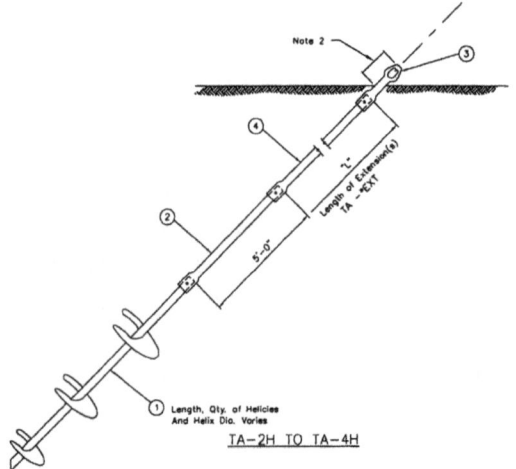

Figure 3. Typical guy installation

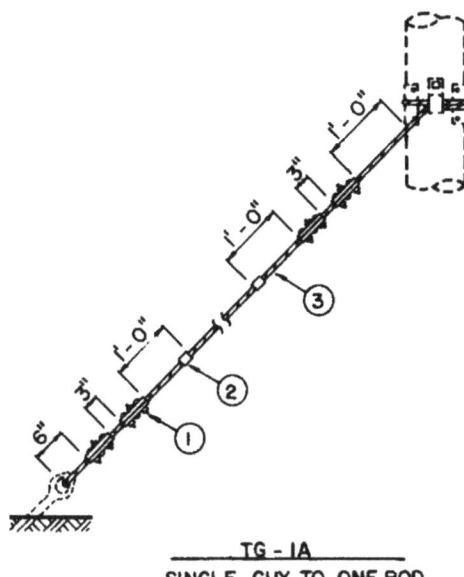

TG - IA
SINGLE GUY TO ONE ROD

Table 5. FERC account 356 details

	356 Items:
1	Circuit Breakers
2	Conductors, including insulated and bare wires and cables.
3	Ground wires and grounding clamps.
4	Insulators, including pin, suspension, and other types.
5	Lightning arresters.
6	Switches
7	Other line devices.

insulators must be inspected for missing parts, defects, and proper fit before installation. Defective or missing pieces must be replaced. Insulators and hardware must be stored in their appropriate shipping containers until installation. The insulator must be properly supported and stacked so as not to damage the individual items. The insulator must be blocked up off the ground so that they cannot come in contact with the ground or standing water. Insulators must be carefully handled to prevent damage to the porcelain skirts, pins, galvanizing, and cotter keys. A cradle or other suitable device must be used to hoist all insulator strings whenever the quantity exceeds 6 units per string. Insulators that are cracked, chipped, or damaged in any way must be replaced with units that are not defective. All insulators must be wiped clean with a clean, soft, nonabrasive cloth.

All connections must be made in accordance with the drawings. Bolts must be torqued to the manufacturer's specifications. Cotter keys, where required, must be fully inserted. Cotter key eyes on insulators and hardware items must be oriented toward the structure, or in such a way as to facilitate easy removal during hot line maintenance. Pins and bolts to insulator string assemblies must be

oriented with the head upright wherever possible. Pin-type insulators must be tight on the pins. On tangent structures, the top groove must be in line with the conductor after tying in. A typical insulation diagram is shown in Figure 4.

For transmission lines, tree contact and insulator failure are the two most common failure modes. For transmission lines, maintenance histories on tree-trimming and insulator cleaning would be essential; in addition, underlying foliage inspection histories including underlying foliage characterization would be important. The level of risk resulting from high power flows depends on routine inspections to establish the physical state of the line and the presence of any changes in surrounding objects and overhead conductors. Clearly, the addition of a new distribution line beneath an existing line, if not noted, may alter the probability of an electrical flashover at maximum load. The undetected presence of an improperly made or severely corroded compression splice or worn or fatigued insulator string hardware can greatly affect the risk of high temperature operation.

PROJECT COST COMPARISONS

Substations

A thorough examination of the historical cost structures for substation circuit breaker replacement as performed by the Premium Select Company revealed fairly consistent overhead costs in the general work orders reviewed. The estimated cost of the ARGENTA substation breaker work 10 units, 5 Charge Capacitor Voltage Transformers (CCVT), 84 cubic yards of concrete removed) totaled $1,933,000.00 less the cost of removal (estimated at $24,500.00). A detailed analysis of the overall cost for the selected major accounts of the project is summarized in Table 6.

The actual costs are compiled as of a summation of the work order invoices obtained from the *Financial Software System*. Also included in this table are the project's overhead costs which encompass the dollars utilized for engineering services, project management, and general oversight. The actual cost of this project was below the estimate by approximately thirty (30%) percent. The inaccurate assessment of the capitalized dollars created budget inefficiencies as the expenditures could have been utilized more effectively in other projects (previously discussed). Further examination of the data (Table 7) revealed the cost per unit for the major selected material and its associated overhead cost allocation (material + labor) is spread across the largest asset, the circuit breaker in this case, then the station material, then the Charge Capacitor Voltage Transformers, as a typical accounting practice in this utility region. The actual total cost remains well below the estimate (inefficiencies) detailed in this breakdown structure and only depicts a single circuit breaker replacement/retirement on average. Nevertheless, the cost estimates versus the actual costs on work performed by the Premium Select Company look as if it was obtained on a case-by-case, substation-by-substation basis (inconsistencies).

Single Circuit Breaker Work

The cost structure examination of a single circuit breaker retirement and replacement (performed by the Premium Select Company) at the ARGENTA substation was in order to verify the inconsis-

tencies and determine the standard cost of the service as performed by this particular company. The estimated cost for the proposed project was $177,900.00-$1500.00 (removal) = $176,400.00 (labor cost included). The selected major cost comparison utilize the same logic discussed in the previous section and is depicted in Table 8 (save the labor cost). However, the final cost was well over the estimated value by approximately twenty (20%) percent (inefficiencies).

A detailed breakdown of the cost allocation and per-unit analysis is depicted in Table 9 (every line item is not shown and does not total 100%).

The total estimated material cost of $81,900.00 includes the circuit breaker, relays and switchboard equipment, tubing, cable & clamps, etc. whereas the final cost was exceeded by almost threefold. Although the workforce failed to clear quantity cost discounts for circuit breaker service installation, the higher cost item clearly is depicted in the single circuit breaker project. Further examination of the data revealed the station material cost more than *doubled* from the estimates to the actual rate in the single circuit breaker project and practically remained unchanged in the multiple circuit breaker installation projects. Moreover, the work was performed at the *same* substation by the *same* company within a twelve month period. The inconsistent estimated expenditures can and will lead to inefficient project funding in capital budgetary decisions. The allocated resources in under-funded projects especially could employ the decision analysis strategy where uncertainty is resolved before the projects are prioritized. Accurate estimates and thorough examination of historical construction cost data is required to achieve this goal.

The same philosophy was employed in the examination of competitively-bid cost structures, where only selected major items were analyzed. The main focus was on the estimated material overhead cost versus the actual cost of the project account. The study is a "snap-shop" of a project, THETFORD substation breaker retirement-replacement program (7 units, 20 CCVT without carrier, 1 CCVT with carrier) in progress awarded to the lowest bidder. The project is based on comparable substation work performed by the Premium Select Company. The detailed analysis of the overall costs (Table 10) is based on the cost of removal Contractual Cost Structures.

Although the contractor predicted the project cost within a specific degree of confidence (9.16%), the probability of the actual cost to exceed the estimate is extremely high at this early stage.

A selected per-unit detailed cost analysis of the highest expenditures is shown in Table 11. The substation breaker retirement/replacement program is in the infancy stages of development. However, a profound improvement with the cost structures and process are expected as economies of scale are achieved.

Transmission Lines

The actual cost vs. estimated project costs in Transmission Lines as performed by the Premium Select Company is examined in the next section. The overall cost of the capacity improvement project on the VEVAY-DELHI/TOMPKINS-DELHI transmission line is considered and referenced to a general work order. The total estimated cash required is ($2,207,906.00-$115,360.00 removal=$2,092,600.00) for the project, with an actual cost of $2,201,760.00. Again, the cost estimates are an integral part of the project funding decisions. The accuracy of these estimates aid in the decision analysis model and the associated strategies for the funding of capital projects. The data for this particular line is fairly accurate for a project of this scope. Detailed analysis of the overall cost (shown in Table 12) is based on 147 wood poles and approximately 53.03 miles of conductor/station wire.

The per-unit cost detailed analysis (shown in Table 13) includes the costs of the wood poles (a 80.29% change between the estimated unit costs vs. the actual costs), towers and fixtures (a 70.30%

Figure 4. Typical insulator installation

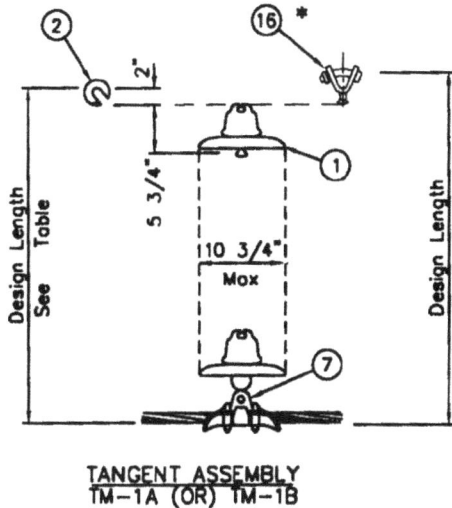

change between the estimated unit costs vs. the actual costs), conductors (a 43.02% change between the estimated unit costs vs. the actual costs), and the static wire (an 86.55% change between the estimated unit costs vs. the actual costs). Many of the components exceeded the estimates when the data was thoroughly scrutinized. Although these costs missed the mark by wide margins, the overall estimated project cost was within the expected range for the entire project.

Nevertheless, if the decision analysis strategies were utilized during the project evaluation phase of the process, budgeted capitalized dollars may have been diverted for other causes (as previously discussed).

Single Wood Pole Structure Work

The reconstruction and replacement of a single wood pole structures and the rebuild (12.3 miles of 138-kV on existing centerline using 795 ACSS) associated conductors for capacity improvements were examined on the GAINES –BRADLEY transmission line as performed by the Premium Select Company. The total estimated cash required ($2,177,011.00-$164,147.00 removal=$2,012,864.00). The actual cost of the

Table 6. Overall cost analysis

Account #	Estimated Cost	Actual Cost	
352	($24,500)	$22,835	
353	$1,401,700	$1,515,325	
Overhead (engineering services, project management)	$531,300	------ (overhead cost allocated across the asset)	
Totals:	$1,933,000	$1,538,160	$394,840

Under estimated cost
30.56% Positive Change

project totaled $1,779,072.00 and produced an 11.61% positive difference between the estimated cost and the actual cost. Both of the line projects conducted by the Premium Select Company were fairly close (and within a reasonable margin of error) to the estimated values however it lacked precision as depicted by the moderate swings in either direction. The uncertainty involved in this case is minimal when this company performs the required tasks. According to the figures, the company has a solid historical base for this type of project. A detailed analysis of the overall cost is shown in Table 14.

A per-unit detailed analysis (Table 15) displays the estimated cost of the material (wood poles, conductors, static wire, etc.) as compared with the actual costs of this material. The emphasis remains on the accuracy of cost estimation and allocation from a Premium Select perspective for capitalized budgetary decisions. Although the experience level The Premium Select has with this type of project exceeds its competitors, the

data revealed the cost estimation was slightly inaccurate for the scope of the work performed.

The actual cost versus the estimated costs of a competitively bid project was examined and compared with The Premium Select transmission line work. The overall estimated cost of the PARE MARQUETTE-STRONACH line work was awarded to the lowest bidder at $3,459,145.00. The projected actual amount is $4,940,756.00. The detailed analysis of the overall cost is depicted in Table 16. The total estimated cost structures are well above the prescribed limits (previously discussed) for capitalized budgetary decisions. The imprecise estimate created an error over the actual cost well over thirty-percent.

Per-unit cost detailed analysis (Table 17) based on 105,600 feet @ 3 lines=316,800 feet (60 miles) and 400 structures broken down to the unit level. The FERC codes 355 (Poles & Fixtures) and 356 (Overhead Conductors & Devices) include the following parameters ~Surveying Cost ($105,000.00), Clearing & Access ($410,972.00),

Table 7. Selected per-unit detailed cost analysis

	Estimated Cost/Unit	Actual Cost/Unit (Material + Labor Overhead)	Actual Total Cost	Overhead Cost (% Allocation)
Circuit Breakers (10)	$61,630	$117,953	$1,179,535	76.68%
CCVT (18)	$4,440	$8,498	$152,958	9.94%
Station Material	$181,465	$182,829	$182,829	11.88%
Totals:	$247,535	$309,280	$1,515,322	98.51%

Table 8. Overall cost analysis

Account #	Estimated Cost	Actual Cost	
352	($1,500)	$7,392	
353	$81,900	$223,603	
Overhead (engineering services, project management)	$78,500	------ (overhead cost allocated across the asset)	
Totals:	$160,400	$230,995	($70,595)

Over estimated cost
20.42% Negative Change

Grounding & Clean-up ($150,321.00) within the work order. The conductor cost/unit increased from The Premium Select cost by (10.47%) along with the static wire (31.36%). The projected overall cost will exceed the most liberal estimates for a project of this scope if performed by the Premium Select Company. It is imperative for a manager to evaluate the entire process and the associated parts, suppliers, etc. to produce the "maximum value" in estimating the project's cost for sound decision making. The project selection strategies employed by senior management (previously discussed) may allocate the correct funding for the correct project at the correct time and efficiently save capitalized dollars where the estimates are imprecise.

RESULTS

Decision Analysis - Substations

The "Overall Cost for Service" decision is divided into several plausible choices and shown in Table 18. It is based on the premise that The Premium Select performs all work, partial work, a combination of work with a designated contractor, or none of the work. The decision theory approach determines the optimal configuration by minimizing its cost for a given risk environment (Anderson, 1999).

The "Loss/Regret Analysis" (Table 19) is formulated based on the "Mini-max Criterion" where the minimum of the maximum values are selected as the recommended decision. It is based on the strategy concerning the smallest of the worst regret that could have happened and utilizes the parameters in the "Overall Cost for Service" (table #13) as a foundation for the arguments.

The basic formulation:

Table 9. Selected per-unit detailed cost analysis

	Estimated Cost/ Unit	Actual Cost/Unit	Actual Total Cost	Overhead Cost (% Allocation)
Circuit Breaker	$70,204	$197,666	$197,666	85.57%
Station Material	$11,696	$25,937	$25,937	11.22%
Totals:	$81,900	$223,603	$223,603	96.79%

Table 10. Overall cost analysis

Account #	Estimated Cost	Actual Cost*	
352	($3,550)	$3,905	
353	$942,675	$1,172,273	
(engineering services, contingencies, project management)	$235,585	$120,990	
Totals:	$1,178,259	$1,297,168	($118,909)

Over estimated cost
9.16% Negative Change

$$Vj^*_{(max)}$$

$S1$	$S2$	$S3$
$\$1,933,000$	$\$1,957,500$	$\$1,538,160$

and:

$Rij = |Vj^* - Vij|$

Minimax Criterion

See Table 19.

Conservative Criterion

Select the Maximum Gains of the Minimums:
D1 → D4

D1 = $1,538,160 → *Maximum of the 4: select D1 as the recommended decision*

D2 = $1,006,860

D3 = $1,276,910

D4 = $1,129,897

Optimistic Criterion

Select the Maximum Gains: D1 → D4

D1 = $1,957,500 → *Maximum of the 4: select D1 as the recommended decision*

D2 = $1,426,200

D3 = $1,696,250

D4 = $1,297,168

Table 11. Selected per-unit detailed cost analysis

	Estimated Cost/ Unit	Actual Cost /Unit*	Actual Total Cost*
Circuit Breaker (7)	$64,872	$81,090	$567,630
CCVT (20)	$4,781	$5,976	$119,525
Station Material	$388,094	$485,118	$485,118
Totals:	$457,747	$572,184	$1,172,273

*The project was in progress. The figures are based on average completion cost structures of historical and comparable un-unitized data, interpolation, and extensive document reviews with key statistics from (project managers, senior tax accountant, owner's engineer, contractor, etc.) involved with the project.

Table 12. Overall cost analysis

Account #	Estimated Cost	Actual Cost	
354 (Towers & Fixtures)	$140,600	$117,315	
355 (Wood Poles)	$782,500	$1,324,067	
356 (Conductors)	$741,800	$760,378	
Overhead (engineering services, project management)	$427,700	------ (overhead cost allocated across the asset)	
Totals:	$2,092,600	$2,201,760	($109,160)

Over estimated cost
4.95% Negative Change

Middle-of-the-Road Criterion

Averages of the Maximum/Minimum Gains: D1 → D4

D1 = $1,957,500 +1,538,160=$3,495,660/2= $1,747,830 → *Maximum of the 4: select D1 as the recommended decision*

D2 = $1,426,200+1,006,860=$2,438,060/2= $1,216,530

D3 = $1,696,250+1,276,910=$2,973,160/2= $1,486,580

D4 = $1,297,168+1,129,837=$2,427,065/2= $1,213,533

Expected Value Criterion

Probability of Low estimated cost P(S1)=0.10

Probability of High estimated cost P(S2)=0.60

Probability of exceeding High estimated cost P(S3)=0.30

D1= E(G1)=(1,933,000)(0.10)+(1,957,500)(0.60)+(1,538,160)(0.30)= $1,829,248 → *Maximum of the 4: select D1 as the recommended decision*

D2= E(G2)=(1,401,700)(0.10)+(1,426,200)(0.60)+(1,006,860)(0.30)= $1,297,948

D3= E(G3)=(1,671,750)(0.10)+(1,996,250)(0.60)+(1,276,910)(0.30)= $1,567,998

Table 13. Selected per-unit detailed cost analysis

	Estimated Cost/Unit	Actual Cost/Unit	Actual Total Cost	Overhead Cost (% Allocation)
Towers & Fixtures	$17,417	$58,657	$117,315	5.32%
Wood Poles	$1,775	$9,007	$1,324,067	60.13%
Conductors (795 kcmil)	$1.43	$2.51	$704,110	31.97%
Static Wire	$0.16	$1.19	$56,268	2.55%
Totals:	$19,194	$67,668	$2,201,760	100.00%

Table 14. Overall cost analysis

Account #	Estimated Cost	Actual Cost	
355 (Wood Poles)	$764,200	$898,810	
356 (Conductors)	$726,300	$880,261	
Overhead (engineering services, project management)	$522,364	------ (overhead cost allocated across the asset)	
Totals:	$2,012,864	$1,779,071	$233,793

Under estimated cost
11.61% Positive Change

D4= E(G4)=(1,129,897)(0.10)+(1,178,259) (0.60)+(1,297,168)(0.30)= $1,209,096

In Figure 5, a visual depiction of the expected value decision is shown in the B-C-H portion of the diagram (decision tree). The heavy line determines the highest values with the given parameters of the proposed path for the decision. Conversely, the dashed line is also shown on the same diagram to reveal the low cost option as the basis for the decision.

In this case, D1 (The Premium Select performs all work) is the recommended decision for the ARGENTA substation breaker retirement/replacement for all of the criterion parameters. Twelve (12) different outcomes with four (4) decision alternatives were employed to determine the most logical approach for the breaker retirement/replacement program. The decision analysis is modeled and referenced to the expected value parameters for ease of understanding. The Pre-

mium Select decision in this case (with a 60% probability of a high estimate for the project) was the correct choice from a *posteriori* criterion (maximizes the expected gain with respect to the updated probability distribution). However, maximum capitalized dollars were utilized. By contrast, the most cost effective approach is the contractor performing the work with a 10% probability of a low estimate for the project. The path is depicted by the red dashed line to display minimal cost for equivalent work performed by the Premium Select Company.

Minimized Cost - Substations

A capital budget situation where the selection of a project is driven by cost is depicted in the next scenario. Table 20 displays the services available based on the average industry cost of selected substation work.

If Sps=1…

Table 15. Selected per-unit detailed cost analysis

	Estimated Cost/Unit	Actual Cost/Unit	Actual Total Cost	Overhead Cost (% Allocation)
Wood Poles	$1,692	$4,685	$955,677	53.71%
Conductors (795 kcmil)	$1.46	$3.42	$726,015	40.80%
Static Wire	$0.24	$1.47	$97,379	5.47%
Totals:	$1,694	$4,686	$1,779,072	100.00%

Table 16. Overall cost analysis

Account #	Estimated Cost	Actual Cost*	
355 (Structures)	$1,236,261	$1,286,261	
356 (Conductors)	$2,092,684	$3,509,145	
Overhead (engineering services, project management)	$130,200	$145,350	
Totals:	$3,459,145	$4,940,756	($1,481,611)

Over estimated cost
30.05% Negative Change

$$Cost=157,809(11)+64,872(0)+104,383(1)+388,094(0)+8498(1)+4781(0)$$

$$= \$1,848,780$$

If Soc=1...

$$Cost=157,809(0)+64,872(11)+104,383(0)+388,094+8498(0)+4781(1) \quad (1)$$

$$=\$1,106,467$$

This specific case depicts the minimum cost for the equivalent work performed by the Premium Select Company, where the total cost of the project was under the estimate by 30.56% and juxtaposed against the competitively-bid contractors (where the total cost is projected to be 9.16% over the estimate). The actual quantities (11) were normalized for comparison purposes. Although the actual costs incurred by The Premium Select Company were under the estimated cost of the project, the data revealed the overall cost exceeded the third party contractor's costs for similar work. A savings of *$742,313 (40.15%)* is realized with the selection of the outside contractor.

Decision Analysis - Transmission Lines

The "Overall Cost for Service" decision is again divided into several plausible choices and shown in Table 21. It is based on the premise that The Premium Select performs all work, partial work, a combination of work with a designated contractor, or none of the work. The Loss/Regret Analysis is depicted in Table 22 where the minimum of the maximum values is selected as the recommended decision.

Table 17. Selected per-unit detailed cost analysis

	Estimated Cost/Unit	Actual Cost /Unit*	Actual Total Cost*
Structures	$2,953	$2,953	$1,181,200
Conductors (795 kcmil)	$3.82	$3.82	$1,210,176
Static Wire	$1.009	$1.009	$319,930
Totals:	$2,958	$2,958	$2,711,306

*The project was in progress. The figures are based on average completion cost structures of historical and comparable un-unitized data, interpolation, and extensive reviews of key statistical information from (project manager, senior accountants, owner's engineer, contractors, and surveyor) involved with the project.

Table 18. Overall cost for service - ARGENTA circuit breakers

Service Decision	S1: Low Estimated Cost	S2: High Estimated Cost	S3: Actual Total Cost
D1: PS work with Overhead	$1,933,000	$1,957,500	$1,538,160
D2: PS work w/o Overhead	$1,401,700	$1,426,200	$1,006,860
D3: PS work (contracted labor @ 50% estimated)	$1,671,750	$1,696,250	$1,276,910
D4: Contractor performs Service	$1,129,897	$1,178,259	$1,297,168

Minimax Criterion

See Table 22.

The basic formulation:

$Vj^*_{(max)}$

S1	S2	S3
$2,092,600	$2,207,906	$2,486,953

and:

Rij = |Vj* - Vij|

Conservative Criterion

Select the Maximum Gains of the Minimums: D1 → D4

D1 = $2,092,906 → *Maximum of the 4: select D1 as the recommended decision*

D2 = $1,664,900

D3 = $1,961,550

D4 = $1,530,432

Optimistic Criterion

Select the Maximum Gains: D1 → D4

D1 = $2,207,906

D2 = $1,780,206

D3 = $2,076,856

D4 = $2,486,953 → *Maximum of the 4: select D4 as the recommended decision*

Table 19. Loss/regret analysis - ARGENTA circuit breakers

Loss/Regret	S1:	S2:	S3:	Maximum
D1:	$0	$0	$0	$6423
D2:	$531,300	$531,300	$531,300	$531,300
D3:	$261,250	$261,250	$261,250	$261,250
D4:	$803,103	$779,241	$0	$803,103

Select Minimum: D1 is the recommended decision

Figure 5. Decision tree for substation work

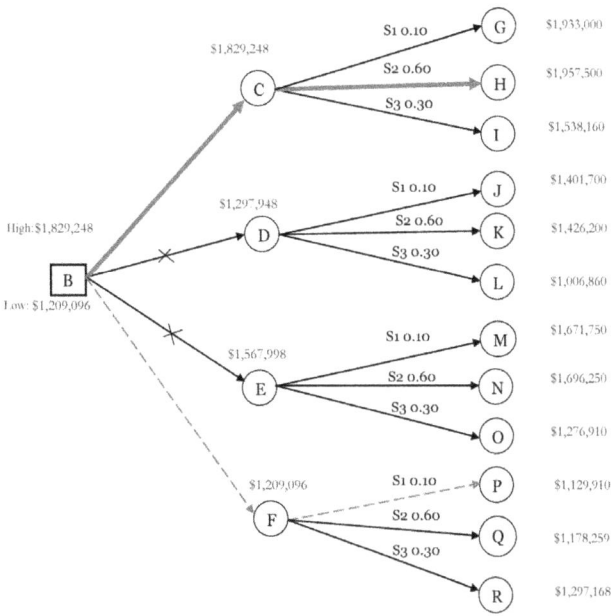

Figure 6. Decision tree for transmission line work

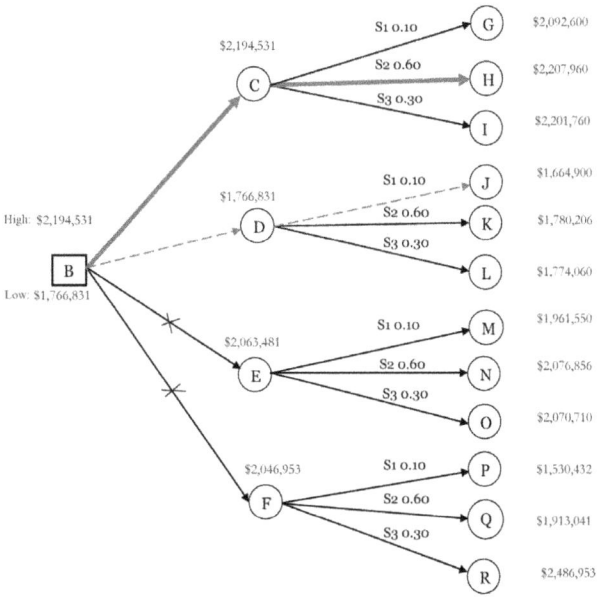

Table 20. Service contractors - substation work

Which contractor to select?	Actual Avg. Circuit Breaker Cost (w)	Actual Avg. Station Material Cost (x)	CCVT Avg. Actual Cost/unit (y)	Qty. Units (z)
The Premium Select (PS)	$157,809	$104,383	$8498	11
Other External Contractor (OC)	$64,872	$388,094	$4781	7

Minimize:

Z=157809(Wps) +64,872(Woc)+104,383(Xps)+388,094(Xoc)+8498(Yps)+4781(Yoc)

Subject to:

Wps=1,788,925(Sps) ....Wps, Woc>0

Woc=967,246(Soc) ...Sps, Soc = 1,0

Sps+Soc=1 ... Forces set-up for either Sps or Soc

Based on per-unit circuit breaker quantities

Where:

Wps= Number of circuit breakers installed by Premium Select

Woc=Number of circuit breakers installed by Other External Contractors

Xps=Station Material installed by Premium Select

Xoc=Station Material installed by Other Contractors

Yps=Number of CCVT installed by Premium Select

Yoc=Number of CCVT installed by Other Contractors

Sps=1 for PS Installation

0 if otherwise

Soc=1 for Other Contractor Installation

0 if otherwise

Middle-of-the-Road Criterion

Averages of the Maximum/Minimum Gains: D1 → D4

D1 = $2,207,906 +2,092,600=$4,300,506/2 =$2,150,253 → *Maximum averages of the 4: select D1 as the recommended decision*

D2 = $1,780,206+1,664,900=$3,445,106/2= $1,722,553

D3 = $2,076,856+1,961,550=$4,038,406/2= $2,019,203

D4 = $1,530,432+2,486,953=$4,017,385/2= $4,008,692

Expected Value Criterion

Probability of Low estimated cost P(S1)=0.10

Probability of High estimated cost P(S2)=0.60

Table 21. Overall cost for service - Vevay-Delhi/Tompkins-Delhi transmission line

Service Decision	S1: Low Estimated Cost	S2: High Estimated Cost	S3: Actual Total Cost
D1: PS work with Overhead	$2,092,600	$2,207,906	$2,201,760
D2: PS work w/o Overhead	$1,664,900	$1,780,206	$1,774,060
D3: PS work (contracted labor @ 50% estimated)	$1,961,550	$2,076,856	$2,070,710
D4: Contractor performs Service	$1,530,432	$1,913,041	$2,486,953

Table 22. Loss/regret analysis - Vevay-Delhi transmission lines

Loss/Regret	S1:	S2:	S3:	Maximum
D1:	$0	$0	$285,193	$285,193
D2:	$427,700	$427,700	$712,893	$712,893
D3:	$131,050	$131,050	$416,243	$416,243
D4:	$562,168	$294,865	$0	$562,168

Select Minimum: D1 is the recommended decision

Probability of exceeding High estimated cost P(S3)=0.30

D1= E(G1)=(2,092,600)(0.10)+(2,207,906)(0.60)+(2,201,760)(0.30)= $2,194,531 → *Maximum of the 4: select D1 as the recommended decision*

D2= E(G2)=(1,664,900)(0.10)+(1,780,206)(0.60)+(1,774,060)(0.30)= $1,766,831

D3= E(G3)=(1,961,550)(0.10)+(2,076,856)(0.60)+(2,070,710)(0.30)= $2,063,481

D4= E(G4)=(1,530,432)(0.10)+(1,913,041)(0.60)+(2,486,953)(0.30)= $2,046,953

In this case, under the maximum gain criterion, D1 is the recommended decision where The Premium Select Company performs the work with a 60% probability of a high estimated cost. The contractor performing the same work is *approximately 11.5% higher than The Premium Select for this type of project*. Although The Premium Select Company exceeded the original estimate by approximately 5%, the entire project was based on a competitive cost structure. Moreover, the low

Table 23. Service contractors - transmission line work

Which contractor to select?	Actual Avg. Structure Material Cost (w)	Actual Avg. Conductor Cost (x)	Actual Avg. Static Wire Cost (y)	Qty. Units (z)
The Premium Select (PS)	$58,657	$59,588	$6,402	12
Other External Contractor (OC)	$59,060	$60,508	$5,332	20

Minimize:
Z=58,657(Wps) +59,060(Woc)+59,588(Xps)+60,508(Xoc)+6402(Yps)+5,332(Yoc)
Where:
Wps= Number of structures installed by Premium Select
Woc=Number of structures installed by Other External Contractors
Xps=Cost of Conductors installed by Premium Select
Xoc=Cost of Conductors installed by Other Contractors
Yps=Cost of Static Wire installed by Premium Select
Yoc=Cost of Static installed by Other Contractors
Sps=1 for PS Installation
0 if otherwise
Soc=1 for Other Contractor Installation
0 if otherwise

estimate for the project is in The Premium Select's favor (red dashed line with a 10% probability of a low estimate). The logic revealed the company's overhead (engineering, project management, adjustments, miscellaneous) was extremely high for the scope of work. The decision to select the contractor, D4, for the work is recommended only in the "Optimistic Criterion," however, the maximum strength of the remaining categories clearly depicts The Premium Select as the favorite in this competitive field.

Minimized Cost - Transmission Lines

A capital budget situation where the selection of a project is solely driven by cost is depicted in the next scenario. Table 23 displays the services available based on the average industry cost of selected transmission line work.

If Sps=1…

Cost=58,657(12)+59,060(0)+59,588(12)+60,508(0)+6402(12)+5,332(0) = *$1,495,764*

If Soc=1…

Cost=58,657(0)+59,060+59,588(0)+60,508(12)+6402(0)+5,332(12)(12) =*$1,498,800*

This case depicts a 12 mile normalized line project compared against The Premium Select and external contractors. The criteria were based solely on the major components of the project (conductor/static wire installation, structure material). A savings of *$3,036 (0.02%)* is realized if The Premium Select is selected to perform the work.

In summary, the simple uncertainty resolution in project decision analysis is clearly not always the primary source of economic value. Prioritization is necessary, as is some quantity of information. Additional information gathering may add value, but this depends on how much information the Project Manager already has and how much

new information is to be gathered with analytic resources available. The payoff from spending time on project decision analysis is already high, even with the brute force approach of analyzing all projects. The payoff could be higher if project decision analyses plans are focused by taking advantage of knowledge about the general characteristics of a project even before individual projects are assessed. The value-added decision analysis process for several capital improvement projects is worth the additional effort for improved cost estimates and financial planning in the equipment installation endeavors.

REFERENCES

Anderson, D. (1999). *An introduction to management science: Quantitative approaches to decision making* (9th ed.). South-Western College Publication.

Clapp, A. (2006). *NESC handbook: A discussion of the national electrical safety code* (6th ed.). New York, NY: Institute of Electrical and Electronics Engineers, Inc.

Decision Process. (2005). *McGraw-hill concise encyclopedia of science and technology* (5th ed., pp. 697–698). New York, NY: McGraw-Hill.

Ledel, D., & Baker, C. (May 2012). Modern transformer maintenance, life extension strategies. *Utility Automation Engineering T&D Magazine*. Retrieved 15 May, 2012, from http://www.elp.com/elp/en-us/index

Levasseur, R. E. (2010). People skills: Ensuring project success- A change management perspective. *Interfaces, 40*(2), 159–162. doi:10.1287/inte.1090.0473

Lippman, S. A., & McCardle, K. F. (2012). Risk aversion and impatience. *Decision Analysis, 9*(1), 31–40. doi:10.1287/deca.1110.0224

Logan, D. M. (1990). Decision analysis in engineering-economic modeling. *Energy*, *15*(7/8), 677–696. doi:10.1016/0360-5442(90)90014-S

McLay, L., Rothschild, C., & Guikema, S. (2012). Robust adversarial risk analysis: A level-k approach. *Decision Analysis*, *9*(1), 41–54. doi:10.1287/deca.1110.0221

Nour, N. E., & Bhevaraju, M. P. (1989). *A comparison between two decision analysis methods*. EPRI Report, RP2537-2

Yoon, K., & Kim, G. (1989). Multiple attribute decision analysis with imprecise information. *IIE Transactions*, *21*(1), 21–25. doi:10.1080/07408178908966203

ENDNOTES

1. From: The United States Department of Agriculture *"Rural Utilities Service"* Summary of Items of Engineering Interest - 2004 Figures 1-4.

APPENDIX

Exhibits

	CB Cost	Material	CCVT	Totals	Project Mgmt
Premium Select Service Provider	$157,809	$104,383	$8,498	$1,848,780	$168,071
Other Contractor	$64,872	$388,094	$4,781	$1,106,467	$100,588

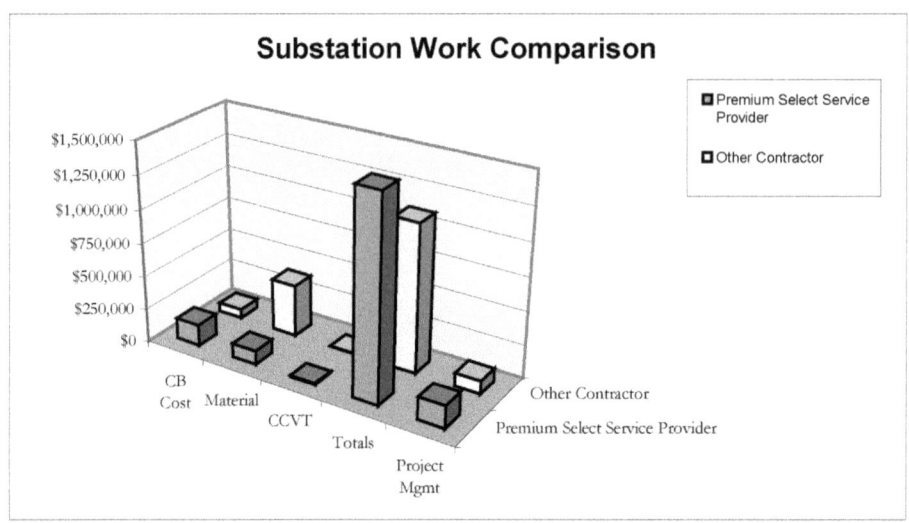

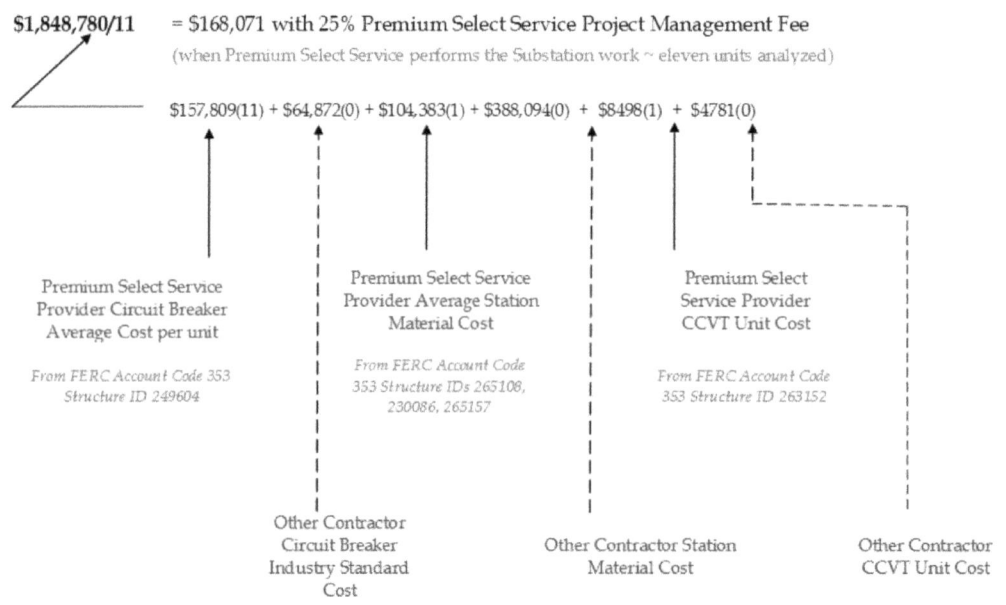

$1,848,780/11   = $168,071 with 25% Premium Select Service Project Management Fee

(when Premium Select Service performs the Substation work ~ eleven units analyzed)

$157,809(11) + $64,872(0) + $104,383(1) + $388,094(0) + $8498(1) + $4781(0)

Premium Select Service Provider Circuit Breaker Average Cost per unit

From FERC Account Code 353 Structure ID 249604

Premium Select Service Provider Average Station Material Cost

From FERC Account Code 353 Structure IDs 265108, 230086, 265157

Premium Select Service Provider CCVT Unit Cost

From FERC Account Code 353 Structure ID 263152

Other Contractor Circuit Breaker Industry Standard Cost

Other Contractor Station Material Cost

Other Contractor CCVT Unit Cost

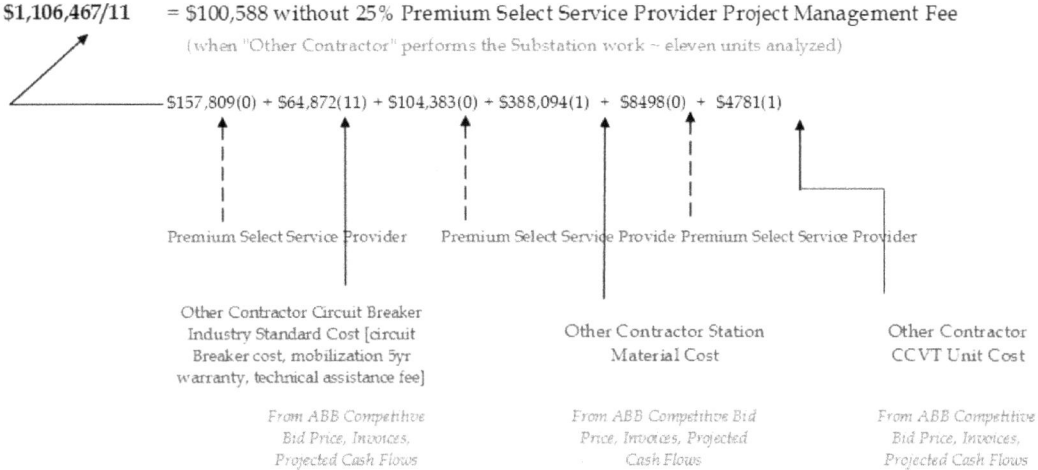

$1,106,467/11 = $100,588 without 25% Premium Select Service Provider Project Management Fee

(when "Other Contractor" performs the Substation work ~ eleven units analyzed)

$157,809(0) + $64,872(11) + $104,383(0) + $388,094(1) + $8498(0) + $4781(1)

Premium Select Service Provider   Premium Select Service Provide Premium Select Service Provider

Other Contractor Circuit Breaker
Industry Standard Cost [circuit
Breaker cost, mobilization 5yr
warranty, technical assistance fee]

Other Contractor Station
Material Cost

Other Contractor
CCVT Unit Cost

*From ABB Competitive
Bid Price, Invoices,
Projected Cash Flows*

*From ABB Competitive Bid
Price, Invoices, Projected
Cash Flows*

*From ABB Competitive
Bid Price, Invoices,
Projected Cash Flows*

Variance: $168,071(CE) - $100,588 (Other Contractor) = $67,483

Total Savings: $1,848,780(CE) - $1,106,467 (Other Contractor) = $742,313 [40% Company Cost Savings]

Chapter 12
Above Grade Analysis

ABSTRACT

Generally, in order to build the initial substation for the wind turbines or any other renewable supply source, specific tasks are evaluated and subsequently allocated to the skilled personnel. The equipment installation varies tremendously as determined by the complexity of the work effort. Some electric utilities utilize a standard boilerplate for labor and cost whereas other entities rely on the most experienced planners and engineers to provide the most plausible estimates. However, the accuracy of the entire capital project cost estimates depends upon the process analysis data as an established baseline parameter. Once the major preliminary components (meteorology, public relations, electrical engineering studies, land issues, and civil engineering development) and pre-construction activities (negotiations with the wind turbine manufactures, contractor mobilization, and development of the environmental construction plan) are completed in a renewable energy capital project, the venture enters the execution phase. There are variations of the theme based on the grade conditions, remote locations, and the availability of assets. The resources at some locations may perform construction activities in sensitive areas where site-specific environmental procedures and/or best practices are strictly followed.

TYPES OF RENEWABLE GENERATION AND INITIATIVES

Introduction

There are several power supply resource options available for the electric utilities. The renewable energy alternatives comprise of system trade-offs and environmental regulatory issues. The trend towards new power supply resources include wind, tidal, and solar arrays (among others) in an effort to primarily decrease the exploitation of fossil fuels. The implementation strategy of these relatively newer technologies must encompass the electrical region's economic situation, resource availability, and load growth. Strategic planning in the solar industry, for example, incorporates several decision scenarios to optimize the investments (Thomas & Bollapragada, 2012). Smaller investments are made by risk-adverse executives which restricts market capacity for growth opportunities. Conversely, other executives may miss opportunities

DOI: 10.4018/978-1-4666-2839-7.ch012

with a *wait-and-see* approach. System planning and careful scrutiny of the resource alternatives are essential tenets for renewable energy penetration. Several legislative state bills across the U.S. were introduced in recent years to include renewable energy in the electrical footprint within a specified period of time. The State of Michigan, for example, launched such legislation[1] to provide accommodation of 600 megawatts of innovative renewable capacity in a seven year span. The assessment of potential interconnection points, siting, and optimal structure heights for wind turbine generators are components of successful implementation of the technology. Moreover, detailed generation /congestion feasibility studies are required to determine the most effective utilization of the medium. The Renewable Energy Resource is a supply that naturally replenishes over a human, not a geological, time frame and that is ultimately derived from solar, water, or wind power. The resource does not consist of petroleum, nuclear, natural gas, or coal. It is expected to originate from the sun or thermal inertial of the earth and minimizes the output of toxic material in the energy conversion process. Although these initiatives are forthcoming, various disadvantages to renewable energy expansion exist. Solar thermal generation, for example, necessitates large swathes of land, and typically impinges on the natural habitat of the indigenous wildlife. The surroundings are also affected when structures, electrical transmission lines, access roads, and transformers are constructed. Additionally, the solar or photovoltaic cells are manufactured by means of the same technologies as those used to produce silicon wafer chips for computer processors, a practice that may employ toxic chemicals. While the renewable power resource fails to discharge air pollution or utilize fossil fuels, it still has a significant consequence on the environment.

Under-developed hydroelectric power potential is a vital factor for the Integrated Resource Plan as well. Generally, the plan for the new generation resource must include the following:

- Financial/Performance Criteria
- Energy Security Criteria
- Environmental Criteria

The Financial Plan, for example, must include the overall cost (capital, fuel, and other costs), utility net income, the interest coverage ratio, return on equity, and the internal generation of funds. The basis of the Performance Plan encompasses the system reliability criteria (amount of customers served, loss of load probability, reserve margin, and the efficiency of energy use). Deeper analysis of the plan reveals the requirements within the Energy Security arena. The use of the renewable resources, diversity of the supply (fraction of each fuel utilized), and the use of domestic resources are the major tenets of the plan. The Environmental Criteria consists of the amount of CO_2 produced over the life of the plan, the liquid/solid waste production of the renewable energy resource, and the amount of land use for the energy facility, and the impact on the wildlife/biodiversity. Other factors affecting the Integrated Resource Plan are the aesthetic issues (recreation/tourism), the impact on the economic sectors (positive and negative), political acceptability/feasibility of the plan, social implications, cultural impacts, and employment impacts of the plan.

Transmission line constraints are cited as one of the main issues that impede the implementation progress of the renewable energy technology. The constrictions of these transmission lines promote price differences among network generators and profit margins. However, the higher unit availability does not equate to increase revenues based on more intense competition (Shanbhag, et al., 2011). The hurdle is one of many impediments involved with the successful performance of the additional resource. Several upgrades of the high-voltage transmission system are required to facilitate any type of renewable energy initiatives. These expansion efforts are required to meet a portion of the escalating renewable energy requests in specific regions while providing a cost effective method to access/deliver economic energy. The wind power expansion efforts are the primary reason to implement renewable portfolio standards (Wiser et al., 2010). The philosophy of the transmission line upgrade is designed to expand the renewable energy markets as well as improve the reliability of the transmission grid. The efficiency of the transmission and distribution lines equates to a percentage of the power/energy lost during conduction. An example of the renewable energy efforts where a transmission expansion need was recognized and evaluated occurred with the High Plains Express initiative in the Rocky Mountain and Desert Southwest districts of the Western Electricity Coordinating Council. The effort was deliberated to accommodate the renewable energy expanding markets, strengthen the regions' transmission system, and provide economic benefits for the states of Arizona, Colorado, New Mexico, and Wyoming. The evaluated alternative encompassed two 500-kV lines with a combined capacity of 3500 MW that would substantially expand the transmission connections between the four states. The intent of the effort removes the flow-based restrictions of the under-sized / non-existent transmission lines to accommodate renewable resources.

The renewable energy interconnection requests within the queues of the regional operating authorities have escalated over the last few years. Some of the project benefits cited espouses the general requirements outlined above (reduce transmission congestion, reduce system losses, enhance regional reliability, provide relief of the transmission constraints, and support the renewable generation development) and includes a sense of urgency. In fact, based on the high wind potential in one particular region, the interconnection generation requests indicated over 40,000 MW of wind resources were proposed for the area at one time. Realistically, the lack of available transmission provides a huge challenge to accommodate most requests.

An additional proposed renewable energy project focused exclusively on the transmission line constraints in the Upper Midwest region, by proposing four additional line upgrades. The goal of on particular 345-kV line was to provide access to wind generation resources in Southwestern Minnesota and eastern South Dakota to increase generation delivery of the supply rich area by 700 MW. The other transmission lines were primarily proposed to support infrastructure reliability and future wind development. Another transmission project to accommodate potential renewable generation resources in Colorado was proposed in support of Senate Bill SB07-100. The endeavor consisted of constructing ninety-five miles of new, double-circuit 230-kV transmission line between two major substations (one proposed to house 230-kV and 345-kV equipment and autotransformers) with additional right-of-way[2].

Generally, substations are required at the point-of-interconnection. Some of these substations exist in the electrical footprint while others are planned in the renewable energy resource. The additional land is usually a point of contention for the environment and the community as a whole. An interconnection request from an independent power producer in the Pacific Northwest prompted the regional operating authority to re-evaluate the initiatives. The new substation requires approximately 3 acres to operate. The general concerns for large capital-intensive projects is outlined below:

- **Transmission Line Concerns:** How will it affect farming (irrigation and aerial spraying disruptions) and effects on crops during construction.
- **Human Health Risks Concerns:** Associated with the electromagnetic fields produced from the high-voltage transmission line.
- **Property Value Concerns:** The impact on land sales and the inherent value of family homesteads.

- **Visual Impact Concerns:** Views in the Columbia River Gorge National Scenic Area, other recreation areas, and homes.
- **Natural Preserve Area Concerns:** Wildlife habitat and protected animal and plant species.

The concerns are vetted at the regional meetings to promote a better understanding of the project. Some people vocalized their concerns while others prefer to utilize written statements. Nevertheless, the alternatives for major projects are meticulously detailed and presented to the stakeholders with the associated cost-benefit analysis. The project refinements are typically reviewed, evaluated, and implemented once the commenting period closes. An example of a concerned voice from the community is depicted below:

We have concerns about the safety of the high voltage lines being so close to our home because we have two young children. We also currently use satellite connections for phone, television and computer. In the event that the powerlines interfere with reception of the phone and an emergency ensued, then possible dire consequences could happen. We currently have cell phones but at best they only work occasionally. (Hanson public comment: 2009)

Other organizations are most concerned about the environment and the construction process of a renewable energy resource. A typical public comment from one such organization is shown below:

The Washington Natural Heritage Program has records of two high-quality native plant communities in the canyon to the west of East Road, located in the northwest corner of section 33 and east side of section 32 of township 5N, range 21E. These have been identified as Big Sagebrush/ Idaho Fescue Shrub Herbaceous Vegetation and Rock Buckwheat/Sandberg's Bluegrass Dwarf-shrub Herbaceous Vegetation. These canyon shrub steppe and grassland communities should

be protected by avoiding or minimizing side-casting materials from above the canyon. Lithosol communities and scabland ecosystems adjacent to the canyon provide buffers to the canyon and additionally have biodiversity value of their own. (McMillen/Washington Natural Heritage Program public comment: 2009)

The lean quality initiatives introduced within the electrical equipment installation construction processes are quite powerful and effective. It is designed to allow the company to utilize fewer resources while achieving more of an output. The method is proven to reduce waste, labor, costs, as well as cycle times while maintaining and/or increasing throughput and higher quality. The below grade and transmission line evaluations espoused the use of these lean tools for a continued improvement effort by the capital project personnel. The profound improvement measures with the above grade analysis provide similar gains. Planning is imperative within the renewable energy generation sphere. It entails the evaluation of the optimal timing of the proposed generator, the size, and location in a region over future years. Planning also identifies priority investments within the major transmission interconnection between sub-regions to optimize the power generation requirements over a specified period. The planning also encroaches into the construction cycle for renewable energy resources. As part of the Preliminary Action of a process (*perform, before it is needed, the required change of a system ~ either fully or partially*) planning includes strategic business /quality planning, project pre-planning, concurrent engineering, capability studies, early supplier involvement, and timely supply of information (Mann & Domb, 1997). A profound statement from a former president summarized the circumstances skillfully:

If I had eight hours to chop down a tree, I'd spend six hours sharpening my ax. (Abraham Lincoln 1861)

The equipment installation process is certainly beneficial for complex system integration of renewable energy resources.

Detailed and oftentimes independent analyses of specific regulations for utility planners are essential inputs for regional generator assessments. The Federal Energy Regulatory Commission's (FERC) Final Rule on preventing undue discrimination and preference in transmission service requires all transmission providers, including independent system operators, to implement and document a coordinated, open, and transparent transmission planning process that complies with the planning principles and other requirements articulated in Order No. 890. The compliance with FERC Order 890 and an Enhanced Transmission Planning Process is a problematic proposition for many independent system providers. Each organization must initiate a compliance filing that clarifies the ISO's obligation and authority to plan for, and promote, the enhancement and expansion of transmission capability within its footprint. The planning function is responsible for developing a forward-looking, coordinated transmission plan that provides for full compliance obligations as well as proactive infrastructure planning initiatives that facilitate a robust market, supporting resource adequacy program(s) and the implementation of generator interconnection studies, renewable integration analysis, and other responsibilities. The planning must encompass the feasibility and assessment of all long-term requests, performing annual congestion studies as required by FERC Order 890, seasonal operating studies, maintaining operating procedure, perform required analysis for transmission and generation outage requests, seasonal local area operating assessments, supporting real time operations, providing planning and operational engineering/technical analysis, and coordinating with surrounding control area operators on engineering issues. Several tenets of the order necessitate amplification within the regional planning entities to close feedback loops once the equipment is in the Plant-in-Service (PIS) phase. The information, analysis, and the relative recommended position must be presented in a clear and convincing manner to the regional board for review and implementation.

An independent analysis revealed several additional problems once a wind turbine generator was implemented in a specific region. Feasibility studies occasionally neglect vital maintenance and outage issues (downstream and/or in parallel) associated with a major grid predicament. A loop flow problem existed in the region created by the addition of the new wind farm connection on the 115-kV line. The dilemma was further exasperated during winter peak load conditions when little or no wind generation was produced. The predicament was discovered *after* the connection of the wind farm. The Independent Service Operator studies from the planning engineers ineffectively captured and modeled the condition when the new expansion connection was executed. The response loop, particularly in renewable energy integration technologies, must include the re-examination of critical issues at each phase of the endeavor before the Plant-In-Service completion stage.

Clarification and interpretation of specific regulations for regional energy attorneys, independent power producers, and community planners are required as well. The FERC Order 693, for instance, makes the electric reliability standards mandatory and applicable to all users, owners and operators of the bulk power system within the contingent United States. The Order 693 also directs the North American Electric Reliability Corporation (NERC) to further improve 62 of the 83 approved standards and reserve an additional 24 proposed standards for further review. The NERC rules contain provisions for reporting and disclosure, including self-certification, compliance audits, spot checking, self-reporting, compliance investigations by both the NERC (and its regional reliability representatives) and FERC, periodic data submittals, exception reporting, and complaints. If a notice of a violation is issued by FERC, NERC or a regional reliability representative, FERC and NERC rules provide protection, as well as procedures that ensure due process of law, including settlement procedures, evidentiary standards, *ex parte* communication rules, disposition without a hearing (involving no issues of material fact), shortened and full contested hearings, and appeals. FERC, NERC and its delegated regional authorities can assess a penalty for violation of the 83 electric reliability standards. Depending on the severity of confirmed violations, the penalty can reach as high as $1 million per violation, per day.

The General Renewable Energy Types

Less than one percent of renewable energy capacity is sanctioned in a particular Independent Service Operator's territory. However, the overall renewable energy consumption in the electric power sector is trending upward (for the periods 2004 to 2008). Wind, Solar Thermal/PV, and Biomass depicts a slight uptick during this period[3]. The electric utility companies are attempting to follow the demand curve, nonetheless the power supply system and associated protocols (reliability and condition prioritized capital projects) tends to encumber the proposed growth of the capacity expansion ventures. Moreover, the current infrastructure must combat the overload conditions first before entertaining the new expansion initiatives. Nevertheless, some regions have found compatible methods to permit the renewable energy

expansion to co-exist with the current demands of maintaining the bulk transmission backbone that comprises the electric grid. The general trend is illustrated in Figure 1.

The number of planned wind turbine generators for a small portion of the period (2007 to 2008) equated to forty[4]. The number of wind turbine units ranged from as low as 4 to as high as 274. The net summer capacity of the planned units ranged from 3.2 MW to 300 MW for the period as well. Conversely, there were only five planned geothermal units (maximum net summer capacity of 20.46 MW) for the same period and qualifies as an upward trend for the energy source. The data also revealed a relatively stagnate trend for the solar (photovoltaic /thermal) power with three planned units and a maximum 14 MW of net summer capacity. There are positive planning initiatives for the renewable energy technologies with successful penetrations in specific regions of the country. There is also a competing requirement to reinforce the existing infrastructure and increase the bulk transmission grid to accommodate the additional generating capacity. The challenging interests of the environmental issues, socioeconomic concerns, as well as the cost-benefits of the renewable technologies are major impediments of wind turbine generator implementation.

Figure 1. Renewable energy consumption

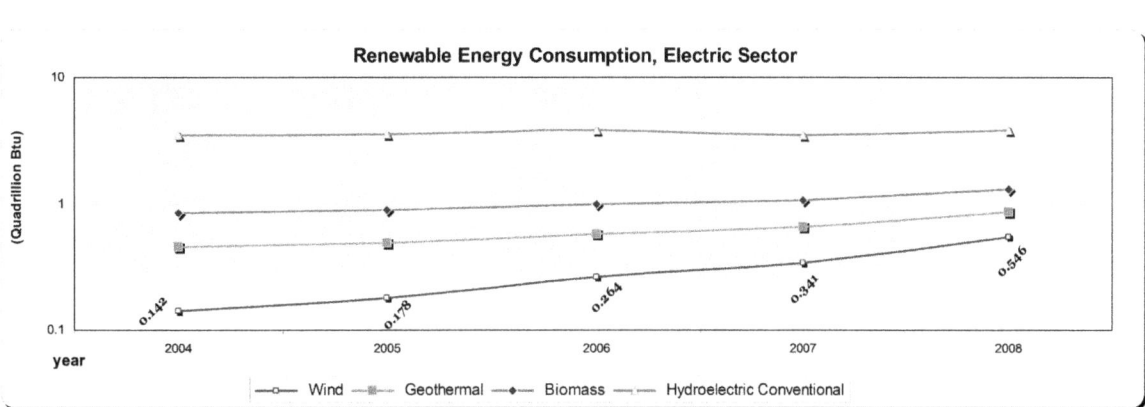

The various types of resource options are shown in Table 1.

The comparisons of the technologies illustrate the general distinctiveness of each. The tidal wave energy is expanding in Europe as well as some parts of the United States. The technology requires additional transmission cables to transport the energy to shore and augment the infrastructure. The system comprises of an underwater substation to accumulate the power, and a submarine cable to deliver the renewable energy to the electric grid. The most recognizable and presumably prudent renewable energy source are wind turbine generators. The predictability characteristics (synchronization of the wind with the load, dispatch ability, and voltage regulation) are major concerns for this source of energy. Wind is a capricious power resource that is difficult to predict. It is a considerable challenge to match the available generation with demand on a second-by-second basis. If these objectives are unmet, transmission system will destabilize and the system becomes unbalanced, resulting in blackouts. In order to preserve system balance in the high-voltage grid, electric utilities exploit balancing reserves, or generation held available to control fluctuations between the load and power generation. In most climates, the winds are unsteady. The energy source tend to ramp up or down relatively quickly and often unexpectedly. The strain on the system operators to invent novel techniques to balance the load and generation is the ever increasing challenge of wind turbine generator penetration. The solar (photovoltaic/thermal) technology is the costliest for the associated load factor. The traditional fossil-fuel plants are predictable for the load factor. However, the units are typically differentiated as an environmental hazard based on the annual amount of carbon emissions produced. The renewable energy initiatives necessitate penetration in specific areas where the technology can eventually assume the load of the baseload (fossil-fuel) unit. The additional generation must co-exist with these baseload units until the renewable technology is self-sustaining. Moreover, the fate of the fossil-fuel units (sunset over a period of time, complete shutdown) are politically as well as locally deter-

Table 1. Resource technology general comparisons[5]

	Tidal	Solar	Wind	Biomass	Natural Gas and Coal
Energy Density	High.	Low – Moderate	Low	Moderate	Very High
	Approx. 1000 x denser than wind				
Predictability	High.	Moderate	Low except in some sites	Dispatchable, subject to fuel supply	Dispatchable
	Accurate forecasts days in advance				
Load Factor	30% - 45%	10% - 20%	25% - 35%	50% - 90%	50% - 90%
Potential Sites	Extensive	Limited for large capacity sites	Moderate	Widespread; permitting process can be lengthy	Wide-ranging; permitting process can be lengthy
Visual Impact	Minimal. The equipment, in general, is not visible from shore	Inconspicuous	Moderate	High	Very High
Cost Per Kilowatt Hour – Utility Power	~ 15¢	24 - 34¢	8 -16¢	14 - 20¢	4 - 7¢

mined by means of these state initiatives and mandates. The effort necessitates high-level strategic planning, deep consternation, and proven renewable technologies to harness the required energy to accomplish these multi-objectives.

OVER-GENERATION, FISH AND WILDLIFE, AND POTENTIAL PITFALLS

The Balancing Act

The overarching goal of any regional system operator is to balance the generation with the load. The prevailing objective is particularly complicated with the introduction of additional, non-dispatchable generators. Operational violations may occur when non-controllable (and volatile energy sources such as the wind) are unable to reduce generation within a specific time period. In one western region, there were unscheduled wind generation fluctuations in excess of 1000 MW in less than an hour[6]. The load is characteristically more predictable in one to two hour time spans than wind generation. If the load is comparatively flat and the wind suddenly plummets within a ten to twenty minute period, the system operators are obligated to simultaneously ramp up additional generation at the same speed the wind turbine generator is ramping down in order to maintain the load-generation balance, as well as the frequency. Similarly, if the wind increases without warning, the operators must ramp down other generators to sustain the critical load/power-supply balance. Moreover, since the wind resource is difficult to predict, the energy scheduling aspect of the operations component is equally complicated to forecast. Consequently, the reserve amount essential by an entity to hold must increase accordingly in order to balance the load with the generation. A supplementary combination to the power supply quandary is the hydroelectric system capacity and flexibility to assist with the load balancing act.

The system planners and operators must assess the annual energy requirements, evaluate the installed generating capacity, and analyze the flexibility of each unit's output to meet the rapid demand to balance the load and generation. The generating unit's flexibility is characterized by its ramp rate (measured in MW/minute), its minimum generation level (physical and economic limits), and its rated capacity. This flexibility is required to follow the load, especially during high-demand periods. Additionally, the regulatory authority-based on realistic utility practices- requires the regional entities to hold a specific amount of operating reserves (contingency reserve) to maintain load and resource balance in case of an outage (transmission line/generator) and to meet the inevitable instantaneous variations in the load. The constraint also incorporates the fluctuations in the production of the wind turbine generators.

The load balance of the available power resources encompasses the imports and exports of hourly transactions as well. The regional entity assess the operators by the use of the area control error (ACE), which is a measurement (calculated every four second) of the load imbalance with the available generation. The ACE encompasses the interconnection frequency as well as the planned transaction imports/exports, and is maintained by means of a combination of operator interaction and automatic generation control. The solutions essential to alleviate the renewable energy integration predicaments embraces the rapid response natural gas-fired generators, pumped storage hydroelectric plants, and electric storage resources. Secondly, the utility demand response programs and "smart grid" applications are required to assist with the wind farm system integration. Thirdly, the utilization of increased accuracy of short-term wind forecasting designed to reduce the amount of balancing reserve capacity essential to cover an error as well as inter-hour scheduling practices of renewable generation. The high cost to integrate the variable generation necessitates additional operating procedures, standby emission-based (CO_2,

SO_2, NO_x) generation units, and viable business practices to provide the essential regulation and load-following services.

Over-Generation

Traditionally, the wind is strongest during the low-load nighttime hours. The wind farm is primed to ramp up generation in this time span when the energy is not desired. As a result, the system operators typically ramp down other generation to maintain the crucial load-generator balance. There are significant minimum generator requirements (physically and economically) for the baseload units during this time period. There are times in specific regions where undesirable high winds as well as high waters simultaneously occur within the electrical footprint[7]. The broad consequences of this phenomenon can wreak havoc on the electric grid and the environment (endangered species) as a whole. The regional entity experienced the potential divergence amid environmental goals, renewable energy development, and endangered species protection during the crisis. The fundamental problem was producing too much electricity for the regional load. In fact, the entity was faced with generation reduction in all facets in their portfolio because of the excessive spill at the hydroelectric units, which produced twice as much power to meet the loads. The surplus power was provided to other utilities ($0/MW) at the time in order to maintain the crucial generation-load balance. Moreover, the regional thermal generation declined considerably as utilities consumed very low or no cost hydropower instead with moderately flat and low loads because of the cool weather. The wind energy fluctuated between "0" to practically full output as storms prevailed throughout the region, creating very large ramps (up and down) that were difficult to precisely predict. The hydroelectric units were operated consistently at approximately 1000 MW higher than the minimum generation as a standby initiative to instantly reduce hydro generation in case

the wind increased above its schedule. Conversely, approximately 850 MW of hydroelectric units were consistently operated below the maximum generation as a standby initiative to instantly increase the hydro generation in the wind decreased and the wind farm fell below the schedule within the hour. Deviation from the schedule of the wind turbine generators are closely monitored by the system control operators to inform the wind farm to reduce their output or the entity will curtail the transmission to limit additional hydro generation. The high flow condition occurred at night, when the demand for electricity was minimal. The hydroelectric generation was summarily reduced and the interconnection capacity curtailed tremendously. The smaller "pipe" on the transmission line restricted the exporting capacity of the units and subsequently, increased the spill. The requirements for increased bulk transmission lines to shore up the infrastructure as well as reconductoring of the existing lines (condition project) are prominently illustrated in this case. In response to the rapid increase of the wind within the regional footprint, several interconnecting states developed policies to protect the reliability of the balancing authority by limiting the renewable generation throughout severe ramp periods. Moreover, rate methodologies were implemented to assure that acquired costs were recovered appropriately. The over-generation conditions experienced during periods of light loads is a major dilemma in the electric utility industry. The utilization of energy storage systems to shift off-peak production to delivery electricity during peak-load conditions is a viable alternative to alleviate the predicament (Ahlstrom et al. 2003). An example of this system is currently utilized as a commercial unit in Huntorf, Germany (Energy Storage Systems, 2007). The off-peak electric energy is converted to mechanical energy by propelling air into a suitable cavern where it is stored at appropriate pressures up to 80atm. The expanding air must be heated by the combustion of either natural gas or a bed of pebbles before it could be used to drive a generator/turbine. The system is illustrated in Figure 2.

Battery storage lasting more than 2000 cycles or greater is another viable alternative. The technology may possibly be exploited in installations with capacities of several hundred thousand kWh in various on the electric power grid. Potential candidates for the medium are outlined below:

- Nickel-Zinc
- Nickel-Iron
- Zinc-Chlorine

More wind energy demands for additional flexibility of the existing generation units, especially during high-wind, low-load circumstances. Additionally, more reserves are required to accommodate the renewable energy technology as the existing units must reduce their outputs. The nuclear power unit is categorized as a full-load, must run electric unit and consequently, unaffected by the wind energy integration. However, the wind turbine generator will diminish the full-load hour equivalents of the fossil-fueled units. The decrease in operation and profits of these units are directly attributed to the innovative renewable technology, as wind energy will always replace the most expensive generation plant on the system. The wind turbine generator therefore decreases the total system operating costs by saving fuel and the associated emission expenditures. The technology also diminishes electric transaction imports while increasing exports in the integrated region. The net generation of selected hydroelectric and wind power generation for the top ten states are illustrated in Table 2.

The data depicts the net generation of hydroelectric and wind power in a given year where a disproportion of the sources exists in some states. More hydroelectric net generation was produced in the State of Washington, for example (78,829 kWh), than net wind generation (2,438 kWh). Conversely, the State of Texas produced far more kilowatt-hours in wind generation (9,006 kWh) than hydroelectric generation (1,644 kWh) for the same period. The surplus capacity has the potential to create over-generation within a highly interconnected electric grid in specific circumstances – especially when intense environmental issues and extreme weather conditions are introduced. The regional planners are obliged to recognize and act upon potential power flow restrictions and congestion during the generation system interconnection impact evaluations. The simulations must be conducted and assessed frequently with real-time data as new generation and/or bulk transmission lines are implemented in the system.

Figure 2. Compressed air storage system

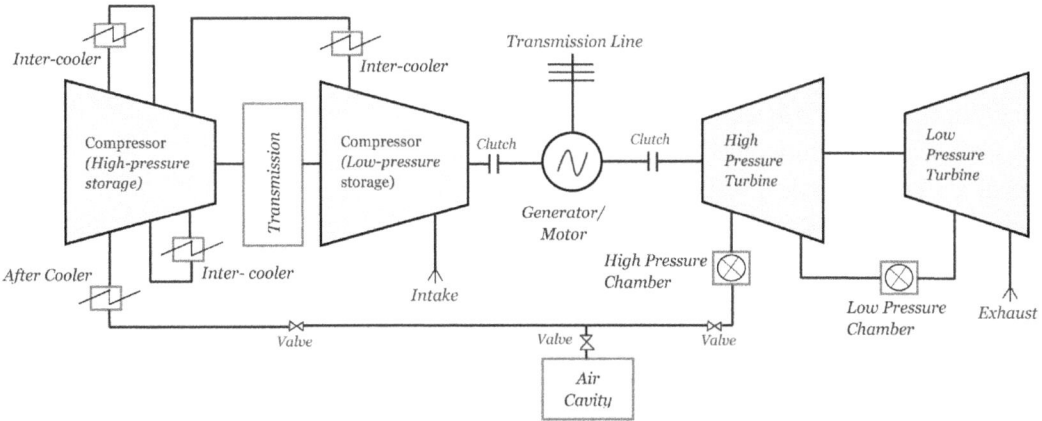

Fish and Wildlife Issues

The goal of the electric utility integrated resource planning is to strategically integrate supply-side and demand-side options to meet customer energy-service requirements and environmental improvements in a least-cost approach. The integration of renewable energy in the power supply portfolio creates divergent objectives for the environment while simultaneously adding value for the electric utility. Substantial efforts have been made to reduce environmental tribulations related with hydropower operation, such as providing safe fish passage and improved water quality in the past at both Federal facilities and non-Federal facilities licensed by the Federal Energy Regulatory Commission. Nevertheless, numerous unrequited questions remain concerning how best to maintain the economic viability of hydropower with increased demands to protect fish and other environmental resources. The balancing act is exasperated with widespread curtailments of wind energy over-production during non-peak hours. Cost-effective demand side management programs and portfolios can be selected and compete with supply-side options in the least-cost resource planning process. Screening mechanisms are required to determine an optimal generation mix over the planning horizon or a reference resource plan from which appropriate project configurations can be scheduled, i.e., the number of units for each candidate generation type that may be selected each year or during a certain time interval. Generation flexibility is also necessary for renewable energy optimization to include increased short-term wind forecasting accuracy, dispatchable generators, and dynamic scheduling. The least-cost scenario priority methodology for new resources encompasses the cost-effective conservation initiatives as well as the regional load obligation. The electrical system is operated in such a manner to protect fish, including salmon, steelhead, and trout listed as threatened or endangered species. The fish are protected by numerous initiatives to include flood control, the

Table 2. Total renewable energy net generation of electricity by top ten hydroelectric and wind source combinations and state: 2007[8]

State	Hydroelectric	Wind
WA	78,829	2,438
OR	33,587	1,247
CA	27,328	5,585
NY	25,253	833
MT	9,354	496
ID	9,022	172
OK	3,066	1,849
TX	1,644	9,006
IA	962	2,757
MN	654	2,639

(In millions of kilowatt-hours (78,829 represents 78,829,000,000))

Endangered Species Act, and compliance with the Clean Water Act all of which takes precedence over power production. For example, certain amounts of water are required for spillage under these acts at specific projects on the Columbia River rather than passed through the turbines for fish passage each spring and summer. On the other hand, excessive spill can result in high levels of total dissolved gas in the water creating bubble trauma (bends) in fish. Voluntary spill for fish passage is practiced as part of the compliance to prevent these conditions. Another scenario experienced by the electric power industry occurs during high stream flows when the turbines at a dam are operating at capacity or when an insufficient market for the power reduces the use of some turbines and there is scarce reservoir storage. In this condition, spill is added incrementally across the projects to avert exceptionally elevated total dissolved gas levels as part of the Gas Abatement Plan for federal dams. The reduction of the balancing reserves for wind power units in high water events aids the fish protection requirements as well.

The fish and wildlife conservation efforts must be combined with the run-off pattern of the hydroelectric facility in an effort to mitigate additional problems. Moreover, existing power plants can be uprated and new power plants added at current sites without a considerable effect on the environment. The new facilities can be constructed with concern of the environment. For instance, dams can be built at remote locations, power plants can be placed underground, and selective withdrawal systems can be used to control the water temperature released from the dam. Hydroelectric facilities can incorporate features that aid fish and wildlife, such as salmon runs or resting places for migratory birds. In unification of our natural and built environments there will be tradeoffs and compromises. Analyses of social problems, environmental preservation and its related contacts are tenets to identify the lowest-performing, highest impacts of opportunity for the organization (Nejati & Ghasemi, 2012). As we learn to live in harmony as part of the environment, we must search for the best alternatives amongst all ecologic, economic, technological, and social perspectives.

THE ABOVE-GRADE PROCESS

The Anatomy of the Processes

Once the major preliminary components (Meteorology, Public Relations, Electrical Engineering Studies, Land Issues, and Civil Engineering Development) and pre-construction activities (Negotiations with the wind turbine manufactures, contractor mobilization, and development of the environmental construction plan) are completed in a renewable energy capital project, the venture enters the execution phase. Generally, in order to build the initial substation for the wind turbines, specific tasks are evaluated and subsequently allocated to the skilled personnel. The wind turbines consist of a small number of basic apparatuses: a rotor, an electrical generator, a speed control system, and a tower. Once the foundation concrete is cured, the tower section, rotor components,

and nacelle (The structure at the top of the wind turbine tower just behind - or, in some cases, in front of - the wind turbine blades that houses the key components of the wind turbine, including the rotor shaft, gearbox, and generator) for each turbine is transported to each site and unload by crane. A large erection crane is designed to set the tower segments on the foundation, place the nacelle on top of the tower, and place the rotor onto the nacelle, following the ground assembly. The towers are expected to be delivered in three to four major sections where the first section will be hoisted into place on top of the foundation pedestal by cranes and bolted securely. The remaining tower sections will then be mounted by crane and subsequently bolted together. Succeeding steps include hoisting the nacelle into place on top of the tower by crane, and attaching the turbine blades to the rotor hub. The crane crew is expected to erect the turbines after all of the components arrive to minimize the amount of time the equipment is on the ground. The generated energy of the turbines is promptly delivered to the substation via an underground collection system (typically 34.5-kV) where a transformer will increase the voltage to the transmission level of 230-kV. Additional equipment (specifically capacitor banks) is installed to provide the obligatory voltage support to meet the interconnection requirements. The substation will also comprise of a control building to house the electrical metering equipment with a concrete slab to form the floor. The above-grade work effort generally consists of the following activities:

- Install the substation steel structures and control enclosures.
- Install the above-grade ground stringer.
- Install the substation electrical equipment (circuit breakers, transformers, switches, and electrical protection).
- Install the bus conductors and jumpers.
- Install the control and relay communication equipment.

- Install the secondary control/power cable terminations.
- Install the final layer of crush rock surfacing.
- Perform testing and commissioning activities.
- Energize the substation.

There are variations of the theme based on the grade conditions, remote locations, and the availability of assets. The resources at some locations may perform construction activities in sensitive areas where site-specific environmental procedures and/or best practices are strictly followed. The environmental compliance processes are outlined in the capital project permits stipulating the site-specific conditions and contingency plans.

The equipment installation varies tremendously as determined by the complexity of the work effort. Some electric utilities utilize a standard boilerplate for labor and cost whereas other entities rely on the most experienced planners and engineers to provide the most plausible estimates. A well-balanced mix of technical and managerial skills is required for the successful deployment of the analytics and system integration organization. Hiring the best individuals to perform the assigned tasks is the key to success in equipment implementation (Davenport et al., 2010). However, the accuracy of the entire capital project cost estimates depends upon the process analysis data as an established baseline parameter. The methodology accounts for the resources assigned to a particular process performing highly-specific tasks within a matrix organization where the time and labor costs are accurately computed. An example of the technique is espoused in the value-added circuit breaker and electrical switch installation process. The two resources designated to perform the work effort are the design engineering group ($125/hour @ 75 hours maximum) and the construction labor ($95/hour @ 50 hours maximum). The maximum (upper control limit) time is calculated from historical similar projects with the *same degree* of installation difficulty and simulated using the most recurrent frequency within the statistical analysis. All responsible electrical installation entities do not evaluate their capital projects in such detail. The material cost ($35,117) is included in the process as the scheduled resources design and install the circuit breaker/electrical switch in place. Obviously, the engineers evaluate (from prior studies) and determine the most feasible location for the devices (electrically and physically) as a separate component in the joint process. The construction labor interprets the engineering drawings - and barring any contingencies – install the circuit breakers/switches within the allotted time. The construction management service element comprises of the engineering group performing material and equipment approval tasks and field audits. Depending on the scope detail and the complexity of the equipment installation, the resource will provide a concerted, part-time effort parallel to the critical path activities. These duties and the major deliverables for a typical large-scale capital project work effort within the substation are illustrated in Table 3.

The insulators, switch gear, group switches, bus installation, and steel structures, employ similar processes with varying material costs for each. Any contingencies with the labor or material will subsequently increase the time and cost for the entire project, as most expansion ventures generally utilize the optimistic completion paradigm. The foundation of the Lean Quality techniques is to reduce cycle time and cost by exploiting the available resource utilization. The techniques espouse a business process characteristic of performing more work with fewer resources. Innovation, reliability, and feedback are the key principles for the quality continuing improvement efforts and Lean methodologies (Senthilmaran, 2010). The essential outage coordination element is scheduled within the project parameters as well to set towers, circuit breakers, and for the panel installation. These processes are equally important as interconnected relationships in the entire scheme for the renewable generation implementation.

There are numerous project meetings, pre-bid sessions, document reviews performed by various resources (technology manufacturer, engineering, construction oversight, business services, procurement, project management, legal, and finance) are obligated to have a seat at the table to facilitate any inter-related activities and associated approvals. The decision-makers are required to scrutinize the cost estimates based on the historical trend data and lessons-learned details of similar projects, realistic schedules, as well as accurate resource utilization forecasts. The front-end analysis of a capital project oftentimes only receives a cursory inspection of detailed results. The time to decide if blasting is required to level the grading for transporting the turbine blade, for example, is an expected outcome from these sessions – not as a contingency or scope change. The decision to order the correct terminations, compatible insulators, or the proper cable length is determined during these sessions – not as a costly re-work activity. The evaluation of specific mitigating circumstances and the effects on other parallel projects in the Independent Service Operator's queue are obligatory planning action items as well. The decision to re-evaluate the peak seasonal power flows and adjacent capital project clearance requirements (i.e. cross-arms, transmission lines) are segregated and parsed during these crucial front-end implementation meetings. The upper control limit of each scheduled process and the associated contingencies are compulsory elements established as a front-end metric of the project implementation. These data are required to further the discussions where the time, resource utilization, and financial characteristics of the project are of a great concern. Breakout sessions are expected to produce feasible outcomes of the design's compliance, civil/structural review, and the material procurement lead-time analysis. The information is expected as a feed-in component for the general session with accountable – and meaningful – action items

assigned to the responsible entities. Frequent scheduled project progress follow-up meetings are designed to ensure the overall accountability of the assigned entities – specifically at major milestones. Quality is built into the process and is essential. Re-work reduction and cost mitigation strategies are basic tenets of the lean techniques. The analysis of these items as a business process is imperative for each renewable energy project implementation.

The Above-Grade Model

The over-arching, multi-criteria objectives of the electrical equipment installation for renewable energy projects are to minimize cost, labor (engineering, construction, testing), material, and contingencies. Additionally, the minimization of the time component (labor) is a direct reflection of the cost objective. The Above-Grade work effort usually follows the Below-Grade segment with minor parallel operations (animal protection, fence, crush rock, etc.) bridging a small time gap. Material delivery and engineering design efforts are other continuing work labors as the below-grade processes conclude. Nevertheless, the above-grade process analysis comprises of equally detailed tasks performed by highly-skilled resources. Several of the tasks were analyzed from the raw data processes to determine the constrictions and vulnerabilities within the system. As such, many of these tasks were combined - from a simulated modeling perspective – while maintaining system integrity. For example, the pessimistic value of 5688 hours (critical path) consumed by the engineering design resource encompasses the essential telephone line installation, remote terminal unit in-service processes, punch list completion, final point list, and transmission line equipment protection preparation. These resources are included in the tally to represent the high-level conditions of the process (baseline – brute force model) compared with an optimized (realistic – scheduled) version of the model. Moreover, as

Table 3. Project deliverables

Task	Description
1100	Field Survey Reports
2210	Meeting Minutes
2215	General Cost Estimates
2220	Cost Estimates with Monthly Cash Flows
2225	Design & Construction Schedule
2230	Scope Single Line with Equipment List
2235	Typical Layout Sketch for the Substation
2240	Typical Layout Sketch for the Control Building
2255	Typical Layout Sketch for the High-Voltage Line Work
2260	Pre-Approved Prime Contractor List
2265	Pre-Approved Sub-Contractor List
2270	Request for Proposal Documentation
6605	Material & Equipment Approval
6610	Field Audit Reports

the contractor selection and job assignments are vetted and allotted respectively, the construction site is expected to experience between 25 to 75 personnel performing various task levels of the installation work effort. The model accurately captures the essential elements of the above-grade analysis for a modular substation installation and is illustrated in Figure 3.

The major labor resources are depicted on the left of the diagram, the processes are displayed as boxes in the middle, and the transmission/distribution asset (T&D) in service completes the drawing on the right hand side. The baseline – brute force model employs infinite resources constantly in the system to complete the assigned tasks. The regional dispatch operation resource develops the switching clearances as well as the plan to energize the equipment within a 44 day

limit and an added-value cost of $75/hour (non-value added cost $38/hour). More importantly, the waiting ($25/hour) and holding costs ($23/hour) are calculated in the system queues assigned to the entity. Since the regional dispatch function is near the latter stages (downstream) of the entire equipment installation process, a possible device delay or logistical problem will have a tremendous effect on the system accumulated cost. The potential problems (contingencies) are routed and classified with the engineering design group delay based on the entity wait time decision in excess of 150 hours. Similarly, the relay protection group dedicated purpose is to prepare the relay tap settings and review/issue changes to said items. The work effort encompasses relay site visits, revenue meter installation, and the designing of the control function to augment the engineering services resource allocation. Unique contingencies captured in this category are germane to the engineering service, substation construction, and relay protection delays. If the relay protection group exceeds the 68 days of its allotted time, the associated holding ($25/hour) and waiting ($18/hour) are assessed accordingly. Even a minute delay to deliver the electrical materials (4 days allowed) will increase the time and cost in the baseline – brute force model. The decision parameters are shown in Table 4.

Because of the high-visibility of a renewable energy capital project integration, a delay in excess of 35 hours in the engineering service/relay protection categories will greatly affect the venture's outcomes and idle several resources. It is imperative that these entities (as well as the engineering design resource) are consuming productive time to take the project to an orderly end. The coordination of material delivery and resource utilization at the construction site are critical communication components within the project management sphere. There were several occasions where the critical electrical devices were hurried through the supply-chain only to receive an incorrect part at the correct site and vice-versa. Correspondingly, resource availability (construction labor, engineering oversight, etc.) must be prioritized for critical electrical equipment installation to avoid delays. The utilization of scheduled personnel for demand maintenance work (storms) creates further project work stoppages. The contingency plans must embrace the resource substitution paradigm to continue the uninterrupted work effort on the capital project. The resources are scheduled based on the percentage of involvement in a specific process. The entity utilization is depicted in Table 5.

The processes for the above-grade equipment installation embrace the Lean Quality Tracking methodologies for the minimum and maximum control limits for the time in the system. As previously discussed, the decision-makers may elect the contract parameters for resource utilization (lower control limit equates to an incentive if the completion time is met or a penalty if the time exceeds the upper control limit). These processes are inter-related with associated waiting queues to determine system vulnerability. Moreover, several outputs are dependent on the inputs of the other processes (similar to the matrix organization). Consequently, an upstream contingency will create problems downstream and within other processes. The successful implementation of the renewable energy system integration involves astute, fundamentally sound business processes.

The Results

The average idle cost is illustrated in Figure 4.

The Average Idle Cost: The ideal baseline (brute force) model depicts the average idle costs near the expected minimum for sixty percent of the resources. Although the Engineering Service Labor and the Relay Protection Labor are exceedingly high in the model (and creates waste), it is usually acceptable in the process industries. This ideal model where sixty percent of the resources are efficiently performing at a 98% efficiency level (without breaks, without time-off, devoid of

major problems), is effectively a problematic system for project implementation. The assumption of a minor contingency is a fallacy as an assortment of situations will occur based on the schedule, cost, and overall project strategy. There will be unforeseen events where a "work-around" effort is required to accomplish the goal while maintaining the schedule and budget. Other project management initiatives must be included in the overall scheme to mitigate the potential contingencies. The idle cost in the optimized model is a direct reflection of "real-world" circumstances. Several of the resources are not truly engaged in the project until the communication of the approvals is received. Idle costs are depicted in several formats for each entity, especially in the engineering area where a concerted effort is warranted usually by cascading events (impact studies, permit approvals, construction feedback reports, contractor availability, etc.). Nevertheless, the capital project in the engineering sphere is only escalated after a sequence of steps from other members in the supply-chain. Consequently, the Relay Protection Labor will incur an idle cost in excess of $13,305 when not actively engaged in the project effort. However, for the 11% of dedicated schedule work for the capital project implementation in the optimized model, the resource will apply the just-in-time quality philosophy to utilize its talents for the assigned relays within the substation. These idle costs are barley noticeable in the project cost overruns, yet are persistent in practically every submitted invoice. In fact, the process analysis are typically lumped together by the electric utility company accountants in various regulatory financial "buckets" for the final plant-in-service reconciliation statements and generally do not reflect the original engineering estimates. Common costs (construction overhead, contract administration, equipment specialist, etc.) may be arbitrarily dispensed in these financial buckets as well. The increased idle costs (in both models) clearly depicts where these expenditures could be

Figure 3. Above-grade equipment installation general process flow

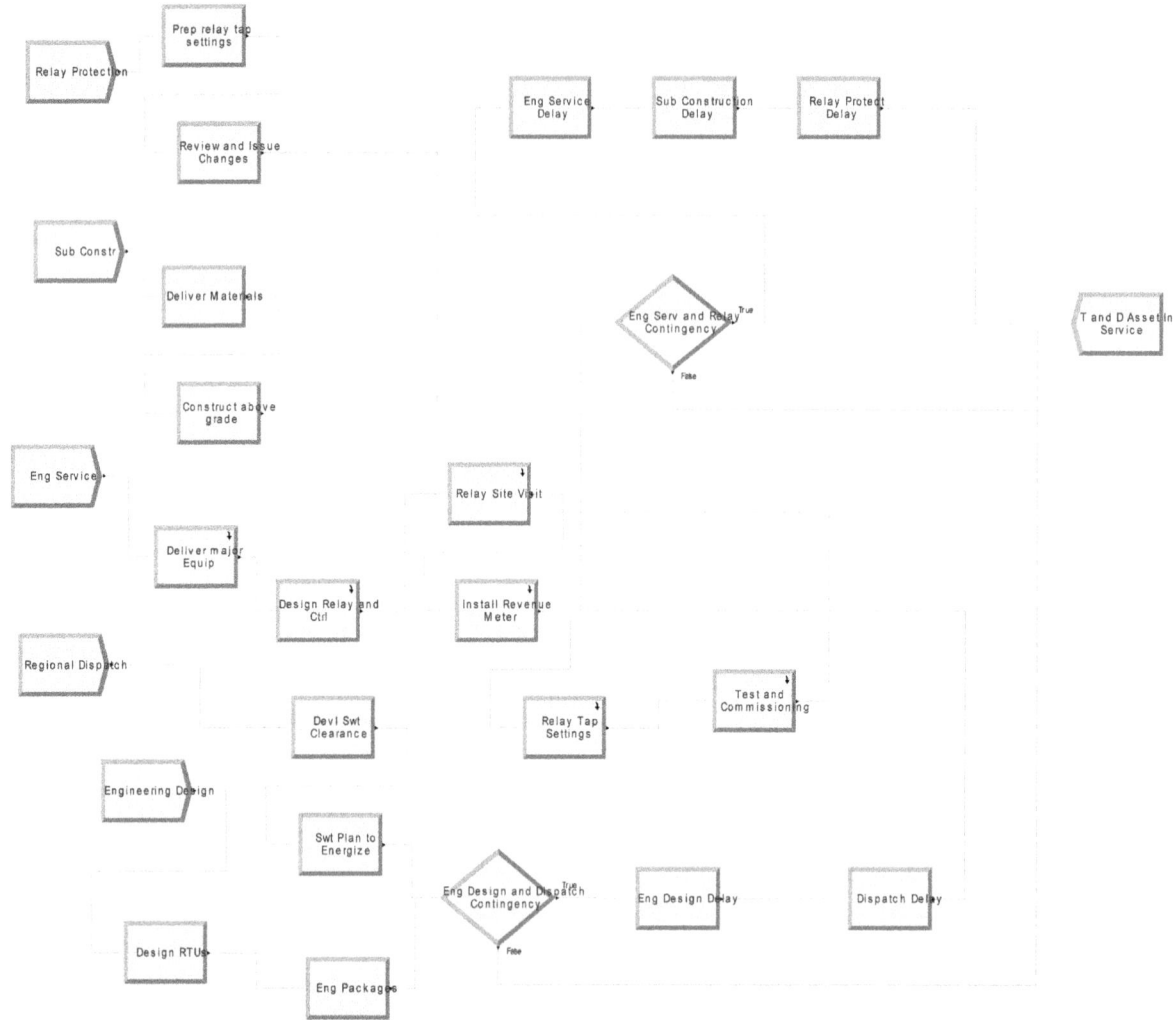

mitigated in the capital project implementation. Principle twenty, germane to the equipment installation for quality management, in the TRIZ model states (Mann, D., and Domb, 1997):

- Continuity of useful action
- Eliminate all idle or intermittent actions or work

Process sequencing, resource leveling (the act of smoothing peaks and valleys, minimizing the effect of conflict in demand for the same resource, and scheduling activities during slack periods),

and the use of multi-skilled bottleneck functions to improve workflow, are in order for the Dispatch Labor ($14,793) as well as the Engineering Design Labor ($10,790) to minimize costs. The parallel efforts of both of these resources will exercise lean quality principles to reduce waste. The maximum average idle cost is illustrated in Figure 5.

Maximum Average Idle Cost: The idle data reflects the cost of all resources based on the average number of inactive units. Additionally, the idle data for the resources is based on the realistic weekly project schedule, the resource inactive cost per hour, and the cost of the resource.

Table 4. Decision parameters

Decision Parameters						If			Time (hrs)
Eng Serve and Relay Contingency	2-way by Condition	50	Attribute	Variable 1		Entity. Wait Time	Entity 1	>	35
Eng Design and Dispatch Contingency	2-way by Condition	50	Attribute	Variable 1		Entity. Wait Time	Entity 1	>	150

The Engineering Service Labor, for example, is utilized 32% per hour in the optimized model. The associated maximum average idle cost for this entity at point #4 on the chart is $6851. A delay in the equipment delivery will obviously add to the cost, as the Engineering Service Labor component (with an initial loaded value-added cost of $105/hr) major assignment is to install the electrical devices. Yet, the same entity employed in the Baseline (brute force) Model reflects an idle cost of $3697, which is relatively high considering the resources involved to perform the work. Moreover, the material implementation is extremely transparent for this resource in both models as a cascade of contingencies (material delay, weather, engineering packages, and demand-side management) will ultimately increase the idle cost. Ideally, these costs are minimized when one process is completed and another immediately begins. Nevertheless, in the electrical equipment installation processes, these costs are oftentimes ignored and included in the final invoice. The Dispatch Labor cost on the same chart is extremely high as well. The resource is in essence aligned to develop and implement the clearance plans with project contingencies. Consequently, the entity's time and efforts are best utilized within the project sphere in anticipation of the more effective use of its talents. The eventualities typically increase the budget throughout the supply-chain. The idle costs are an emblematic reflection of the capital project implementation and act in total opposition of the lean quality principles, especially within the construction (hands-on) arena. The engineering labor is dedicated to several projects within their jurisdiction and may contribute only a fraction amount of time (approximately 37% for the Engineering Design Labor, 11% for the Relay Protection Labor in the optimized model) required to this project for a successful implementation. Consequently, the increased idle cost is a direct manifestation of the outlay as most of time is spent performing re-work activities. Total "cradle-to-grave" dedication to the project is required to minimize cost, waste, and successfully implement above-grade electrical equipment for renewable technologies.

Average Instantaneous Utilization: The actual resource utilization in the models is shown in Figure 6.

Table 5. Entity utilization

Entity	Type				Busy/hour	Idle/hour	
Eng Serve Labor	Fixed Capacity	1	1	Wait	0.32	0.65	0
Relay Prot Labor	Fixed Capacity	1	1	Wait	0.11	0.85	0
Constr Labor	Fixed Capacity	1	1	Wait	0.11	0.85	0
Eng Design Labor	Fixed Capacity	1	1	Wait	0.37	0.62	0
Dispatch Labor	Fixed Capacity	1	1	Wait	0.11	0.85	0

Figure 4. Average idle cost

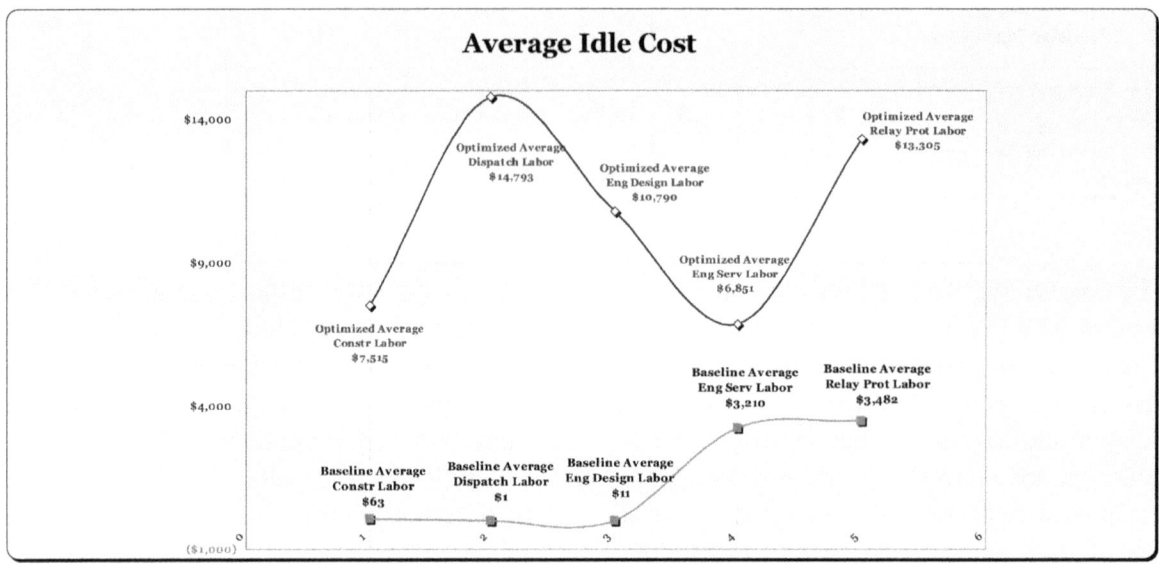

The data is a compilation of the resource's utilization at any instant in time and examines the "number of busy" resources in conjunction with the "number scheduled." The baseline (brute force) model depicts sixty percent of the resources performing at an extremely high efficiency rate (98% and above). Conversely, the optimized model (where the resources are selectively scheduled) depicts the more realistic approach where the Construction Labor, for example, is operating at a 49% efficiency rate. The scheduled resources are effectively utilizing the production perspective of the Lean Quality tools for electrical equipment installation. The productivity measures, performance efficiency, quality rate, lead-time (processing/transportation time, queue and set-up times), and inventory turn over (average inventory of the work-in-process/finished project) are all components of the production perspective in the Lean Quality techniques (Jones, D., and Womack, J., 1996). The above-grade materials (circuit breakers, transformers, switches, control panels, power cables and terminations, conduit, steel structures, etc.) are expected to be on station and implemented as scheduled. Nevertheless, the highly-skilled, selective labor involved in the process is executed at the right time and position for a successful renewable energy placement. The problematic area in the chart (point #4 ~ Engineering Service Labor) increased to approximately 39% in the optimized model (from 13% in the baseline model) based the resource schedule utilization. The resource is effectively performing an additional effort of twenty six percent to achieve the desired results. However, the expertise of the resource is essential and highly-valuable to deliver/install the major equipment, design the relay control panels and tap settings, and the test/commissioning activity involvement.

Management productivity is a more appropriate term than labor productivity. Improved productivity means less human sweat, not more. (Henry Ford 1919-1987)

The optimized model also encompasses reduced cycle time for eighty percent of the resource utilization based on the schedule implementation. The inherit improvement in the quality and safety issues are built-in to each of the processes as an

Figure 5. Maximum average idle cost

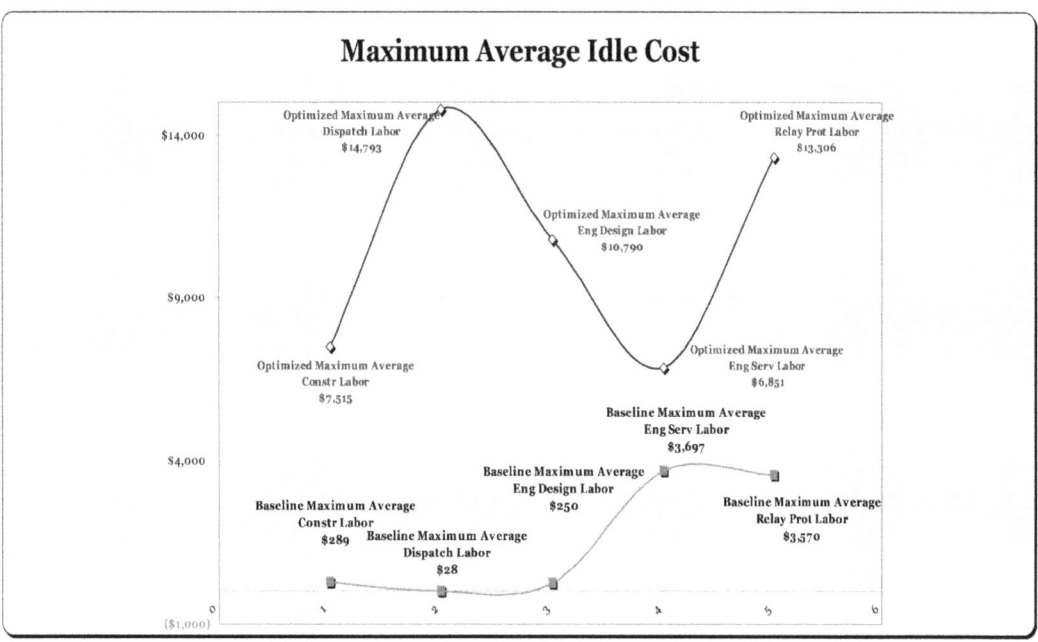

enhancement and cost reduction component in the entire system.

Minimum Average Instantaneous Utilization: This metric is used to determine the minimum control limit within the resource utilization process. It is the smallest average recorded across all replications to assess stability. In the optimized model, the data points are extremely stable when compared with the baseline model. An example of this is observed with the Construction Labor data points (Baseline Model =0.9869-0.9402) versus the Optimized Model = 0.4903-0.4902. The resource is busy enough to sustain the optimal service level while producing the requisite system output. The difference in the bandwidth is designed to support the decision-makers with adequate information to establish the minimum resource utilization involved in a fastidious above-grade construction project. The data is shown in Figure 7.

The small resource utilization difference (0.1592 at point #5) depicted in the Relay Protection Labor data points derive from the Lean Qual-

ity processes achieving the optimal output. Several capital-intensive projects are proposing an equipment reduction covered by a single differential relay protection scheme in order to perform the same related tasks. Pilot relaying, for instance, is employed whenever high-speed protection is vital for all varieties of short circuits and line faults anyplace within the protected segment. For two-terminal lines, and for many multi-terminal lines, all the terminal breakers are tripped with no intentional delay, thereby authorizing high-speed automatic reclosing. The technological advances embedded within the "smart" devices allow the Relay Protection group to include these flexibilities in the overall relay design, thus reducing time and waste. The blend of high-speed tripping devices and high-speed reclosing permits the transmission system to be loaded virtually to its stability limit, thus providing the maximum return on the investment. The equipment will perform at capacity with minimal resource input for the implementation. The optimized model illustrates the benefits of precise scheduling to obtain a cost-effecting renewable energy equipment integration process.

Total Accumulated Average Cost per Entity: The data in Figure 8 below represents the total average cost that was accrued by the entities in the process associated with the activity area.

Overall, the data in this cluster experienced a 76.28% accumulated average cost reduction with the utilization of scheduled resources characterized in the optimized model. In the baseline (brute force) model, the Above-Grade Construction (average cost $136,289) and the Testing and Commissioning activities (average cost $118,975) are prominent as the highest accrued expenditures. Both are value-added activities with associated high priorities. However, with ideal (infinite) resources engaged to accomplish the assigned tasks, the costs are expected to exceed the allotted budget. The Commissioning and Developing of the Switching Clearance activities alone embodies detailed switching and tagging procedures, coordination of skilled personnel (job supervisor, authorized switchman, authorized workman ~ lineman, mechanics, technicians, inspectors, and

engineers, clearance holder, and dispatcher), and extensive/recurrent training. The communication, scheduling, and project coordination between these resources are critical for a successful renewable energy equipment implementation. Many project managers fail to embrace the accrual concept with the assigned capital ventures. Conversely, the optimized model depicts reduced costs in these activities, (Construct Above-Grade to $53,955 and Commissioning activities to $0) especially after careful scrutiny of the entire system. The only caveat within the entire scheme where a cost reduction was not experienced lien the Engineering Service Delay category where it increased from $0 to $68,644. Lean quality techniques (resource leveling, smoothing, etc.) will partially alleviate the excess cost in this category. However, a thorough investigation (and documentation as to why to avoid future accruals) of the engineering delay would eliminate the problem entirely. The realistic schedule embedded within the optimized model emphasizes the use of "just-in-time" resource utilization in the system (Schroer, B., 2004).

Figure 6. Average instantaneous utilization

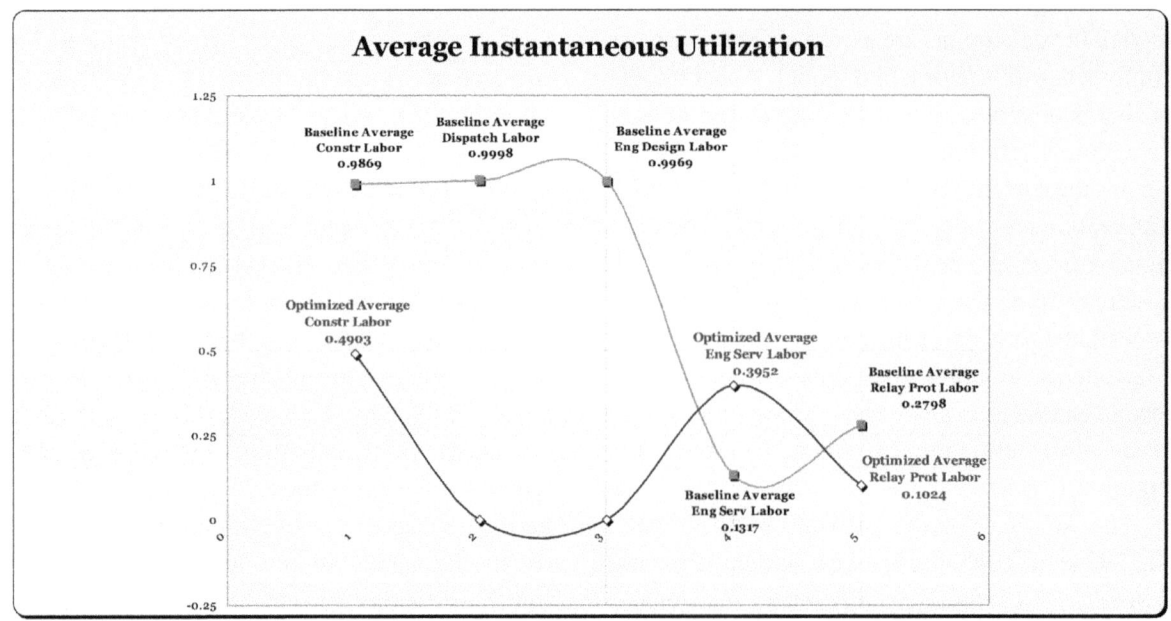

Figure 7. Minimum average instantaneous utilization

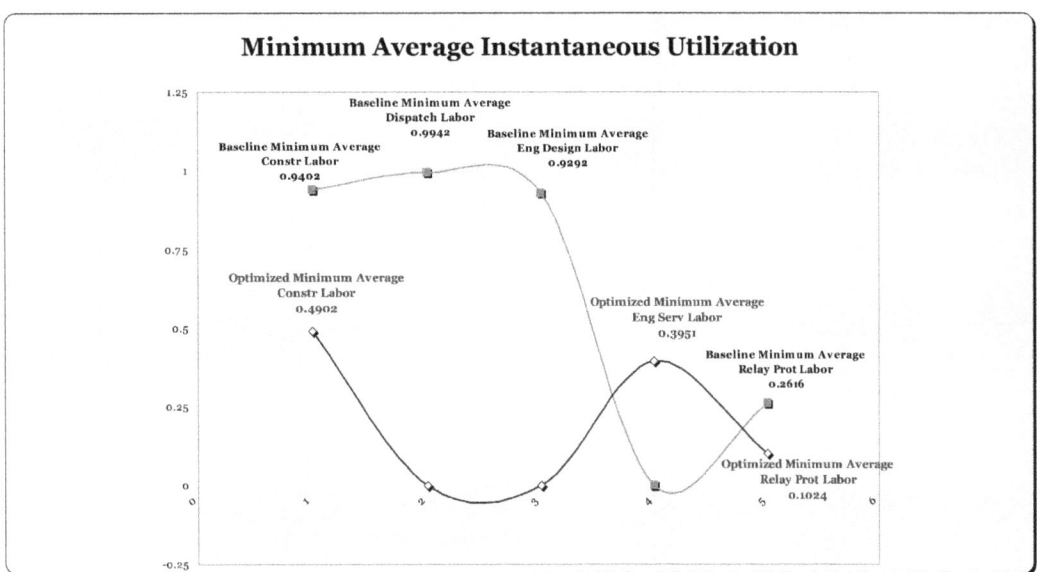

Maximum Accumulated Average Cost per Entity: The maximum average values of the accumulated costs per entity are shown in Figure 9.

Again, the data depicts the highest accrued cost per entity in the system between the baseline and optimized models. Overall, the data in this cluster experienced an 82.46% accumulated maximum average cost reduction embodied within the optimization model. The utilization of the 5S concept from the Lean Quality toolbox is illustrated in the optimization model. Particularly, the "Sort" (organize, separating the needed

Figure 8. Total accumulated average cost per entity

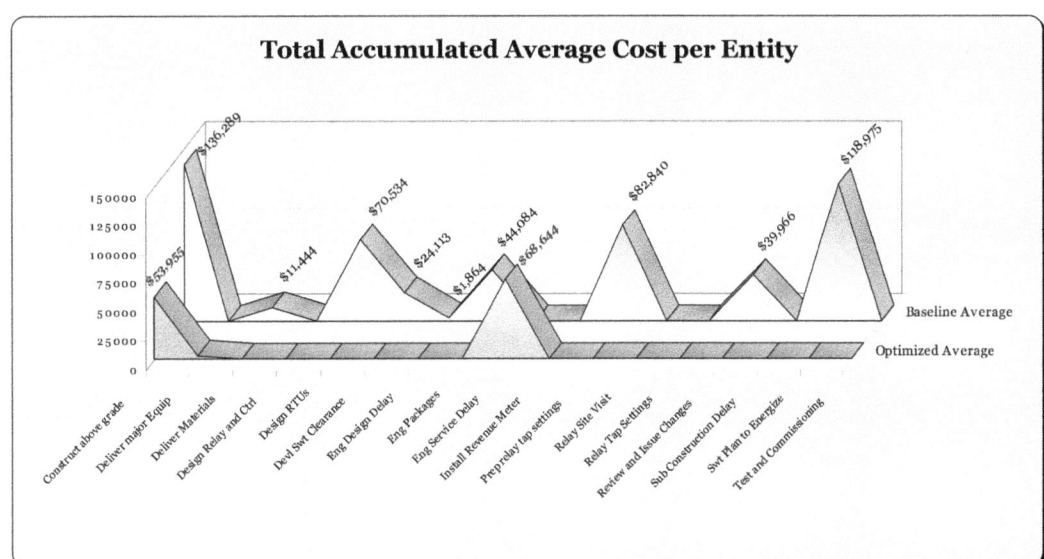

from the unneeded) and "Straighten" (arrange and identify for ease of use) concepts are utilized to not only save time, but cost as well. The unorganized methodologies in the baseline model created excessive queue waiting times as well as substandard resource utilization based on an idealized schedule. The same arrangement is employed in the optimized model with the exception of the "Straighten" tactic for ease of use of the resources in the system. Moreover, the "Sort" concept obviously is implemented as part of the tactic to assist with the elimination of the "unneeded." Several of the processes were reduced to $0 accumulated dollars when the overall strategy was formulated. The additional resources in key areas along with a feasible schedule for critical personnel to lend their expertise on the project were instrumental with the time and cost reductions. The maximum accumulated average cost for a baseline model can be included in the vendor contract to demonstrate project expectations. Moreover, it will determine the cost of unplanned stoppage times and the associated consequences (cost/time) within the supply-chain.

The dramatic accumulated cost increase in the Engineering Design Package category (baseline model), for example, is a resource utilization issue reduced with astute organization and key scheduling (illustrated in the optimized model). These packages consist of the engineering cost estimates (materials, contractor costs, labor analysis ~ design, engineering, line, substation, cable, relay, electronic technicians, etc.) as well as the construction/electrical drawings for the project. The drawings include One-Line Diagrams (to include busses, feeders, transformers, and generators), control schematics of the protective relaying used to isolate faults on the electric system, and the location of the major assets within the substation. Several of the plan and profile drawings are contracted to external electrical engineering "shops" where computer-aided drafting expertise is prevalent. This action necessitates coordination with the renewable energy project milestones as well as the contractor's internal schedule (sponsor need date versus electrical in-service date). The schedule mis-alignment between the two entities alone may cause a potential project delay. The

Figure 9. Maximum accumulated average cost per entity

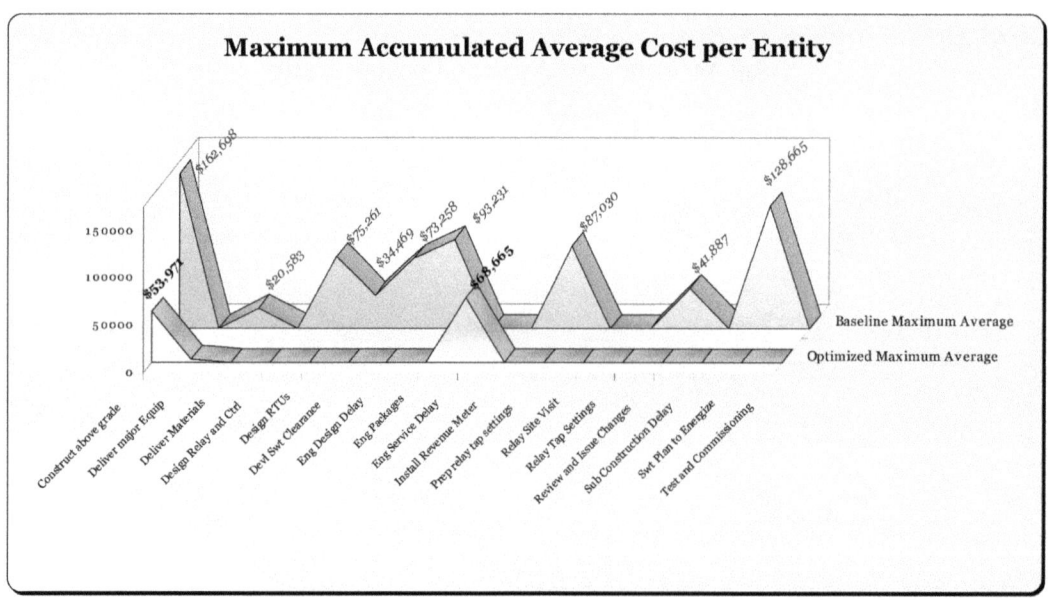

Figure 10. Total average time per entity

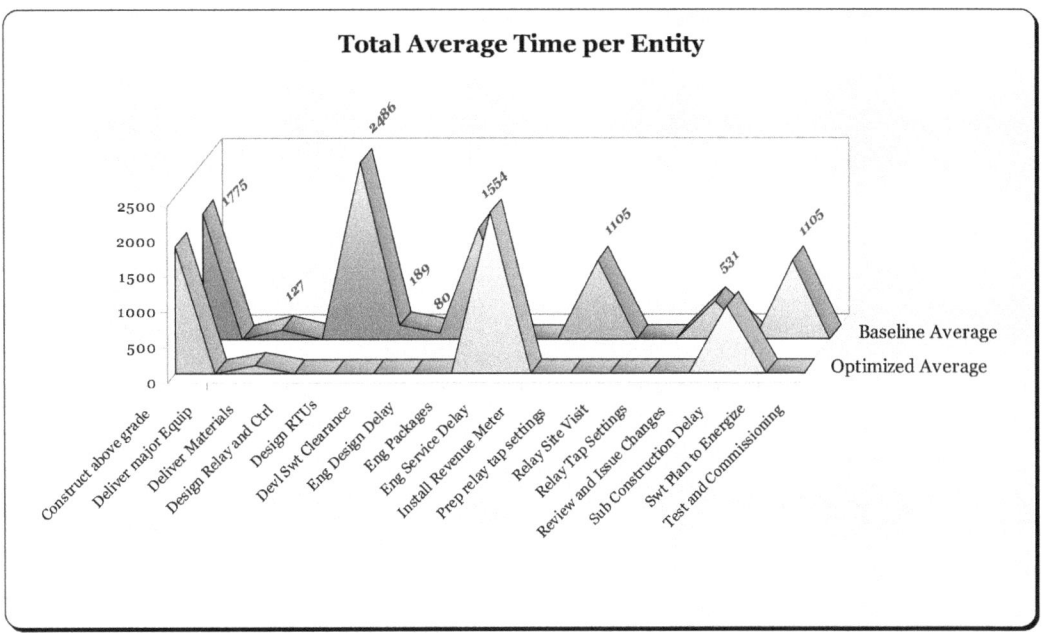

details involved in the drawings typically require additional resources which increases the costs in the Engineering Design Package category. When the data in the maximum accumulated cost per entity is utilized properly, it can not only impose penalties for work delays, but may also create bonus incentives for performing a quality work effort on-time as well as on-strategy.

Total Average Time per Entity: The metric is a compilation of the total average time (busy/idle/wait) an entity spends in a process. It is classified as the cycle time when a resource enters the organization, conducts the assigned process before leaving the system. In the Above-Grade baseline (brute force) and optimization models, the average times are illustrated in Figure 10.

Overall, the optimized model demonstrated a 42.98% improvement in the average time reduction group with the utilization of the Lean Quality tools and strategic resource scheduling techniques. A noticeable shift was experienced in the Engineering Service Delay and the Substation Construction categories. The resource-leveling modus

operandi eliminates the delays. An expeditor and schedule re-alignment with the associated contractors for the capital project will assist with the work effort to reduce the impediment. Nonetheless, astute planning and organization essentially reduced the average time per entity to a more manageable constraint for the above-grade equipment installation. The Design RTUs (remote terminal units) is the largest consumer of time (2486 hrs) in the baseline model. The utilization of constant resources simply increased the process time in this category. A parallel work effort and increased resources are essential to normalize the process. The remote terminal units, in conjunction with programmable logic controllers for wind turbine generators are sophisticated and comprises of a high degree of security protocols. It is part of the broader Supervisory Control and Data Acquisition (SCADA) system remotely implemented to monitor/control the wind technologies and equipment within the substation. Because of the inherit vulnerabilities of the system, the design engineers typically include methodologies to circumvent operation interruptions, asset mis-configuration for circuit protection, and dangerous circumstances (Young, W. et al., 2003). The utilization of wireless bridges to exchange data with the turbine generators includes source/destination details, storage locations, and communication paths. The endeavor is time consuming if the resource expertise is deferred to other capital projects. It involves a concerted effort to accomplish the enormous tasks involved within this group. However, the optimization model clearly reduced the essential time by means of appropriate resource scheduling, reduction of the queuing delays, and the elimination of the average time involved of the assigned processes by 82.35%. The Lean concept is a proven, cost-effective method that can be effectively utilized within the renewable energy equipment installation implementation.

Maximum Average Time per Entity: The maximum average time per entity is illustrated in Figure 11.

Overall, the cluster experienced a 64.41% improvement with the maximum average time allowed in the processes for the assigned entities. There are expected tradeoffs between the upper

control limit (maximum average time) and the average time embedded within the processes, as contingencies invariably will occur. However, when correctly executed, the Lean processes are designed to smooth the peaks and valleys of time and cost in a system. The reduced average maximum time (from 228.35 hours to 104.4 hours) in the Deliver Materials category, for example, depicts a value-added, high priority activity where the quality techniques are proven to be advantageous. In the optimized model, the scheduled construction labor is exploited logistically to perform the task at the right time and place. An accurate amount of skilled laborers is utilized to facilitate the operation. If the materials were delayed with the resources on station, the misalignment of supply would be reflected in the associated final cost of the project. This scenario happens all too often in the construction industry. Conversely, if the supply-chain releases the material too soon without the appropriate construction labor resource, the compensation for the transportation labor would be exceedingly high based on the delay. The maximum average time per entity is utilized by the decision-makers and stakeholders with the development of specific

Figure 11. Maximum average time per entity

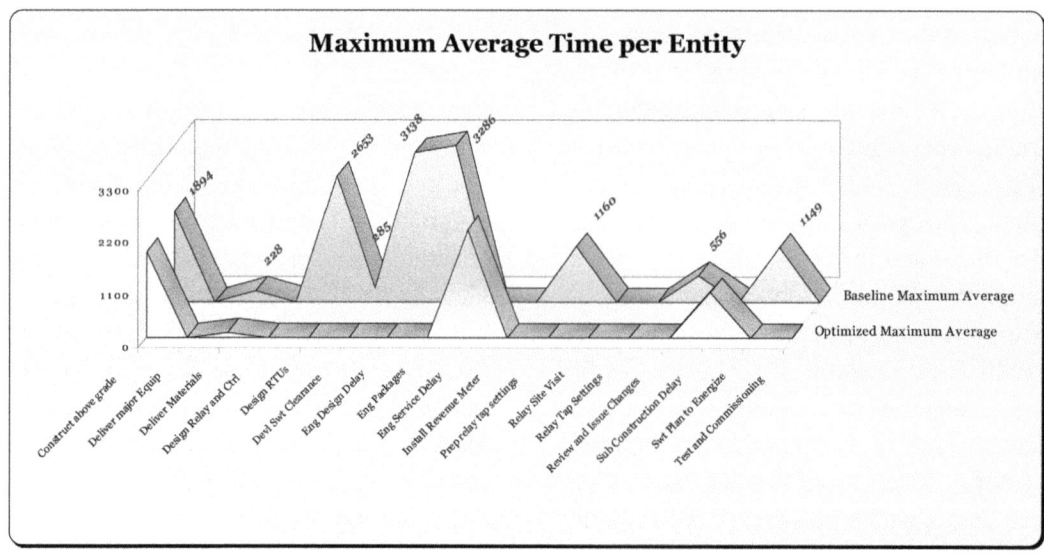

language in a contract to produce the desired results. Based on similar projects and scope, each entity is capable of designing localized parameters for time and cost. A transformer capital project installation will certainly have different costs and time requirements in New Jersey versus the same implementation processes in Oregon. Nevertheless, the maximum average time boundaries of similar performance and tasks can be established to accomplish the goal. The significant analysis of time and cost is imperative for accurate renewable energy equipment installation.

REFERENCES

Ahlstrom, M., Grant, W., Jones, L., & Zavadil, L. (2003, November/December). The future of forecasting and utility operations, planning for improved system operations. *IEEE Power and Energy Magazine*, *3*(6), 57–64. doi:10.1109/MPAE.2005.1524621

Davenport, T., Harris, J., & Morison, R. (2010). *Analytics at work: Smarter decision, better results*. Boston, MA: Harvard Business Press.

Energy Storage Systems. (2007). *Encyclopedia of science and technology* (10th ed.). New York, NY: McGraw-Hill.

Jones, D., & Womack, J. (1996). *Lean thinking: Banish waste and create wealth in your corporation*. New York, NY: Simon & Schuster.

Mann, D., & Domb, E. (1997). *40 inventive (business) principles with examples*. Retrieved 8 November, 2008, from http://www.triz-journal.com

Nejati, M., & Ghasemi, S. (May 2012). On the right course: DMAIC can help organizations navigate tough decisions on social responsibility projects. *Quality Progress*, pp. 29-31.

Schroer, B. (2004). Simulation as a tool in understanding the concepts of lean manufacturing. *Simulation*, *80*(3), 171–175. doi:10.1177/0037549704045049

Senthilmaran, K. (August 2010). *Electric utility deploys powerful approach for continuous improvements*. Making the Case for Quality: American Society for Quality Knowledge Center. Retrieved 1 April, 2011, from http://www.asq.org/knowledge center/index.html

Shanbhag, U., Infanger, G., & Glynn, P. (2011). A complementary framework for forward contracting under uncertainty. *Operations Research*, *59*(4), 810–834. doi:10.1287/opre.1110.0947

Thomas, B. G., & Bollapragada, S. (2010). General Electric uses an integrated framework product costing, demand forecasting, and capacity planning of new photovoltaic technology products. *Interfaces*, *40*(5), 353–367. doi:10.1287/inte.1100.0518

Wiser, R., Barbose, G., & Holt, E. (2010). *Supporting solar power in renewables portfolio standards: Experience from the United States*. Lawrence Berkeley National Laboratory. Retrieved 9 January, 2011, from http://www.eetd.lbl.gov/ea/ems/reports/lbnl-3984e.pdf

Young, W., Stamp, J., & Dillinger, J. (2003). *Communication vulnerabilities and mitigations in wind power SCADA systems*. Paper presented at the American Wind Energy Association WINDPOWER 2003 Conference, Austin, TX.

ENDNOTES

1. Enrolled State Bill #213 Act No. 295 Public Acts of 2008 Effective October 6, 2008.
2. Source: Transmission Projects Supporting Renewable Resources. Edison Electric Institute.
3. Sources: Analysis conducted by U.S. Energy Information Administration, Office of Coal, Nuclear, Electric and Alternate Fuels and specific sources described as follows. Residential: U.S. Energy Information Administration, Form EIA-457A/G, "Residential

Energy Consumption Survey;" Oregon Institute of Technology, Geo-Heat Center; and U.S. Energy Information Administration, Form EIA-63-A, "Annual Solar Thermal Collector Manufacturers Survey" and Form EIA-63B, "Annual Photovoltaic Module/Cell Manufacturers Survey." Commercial: U.S. Energy Information Administration, Form EIA-906, "Power Plant Report," Form EIA-920, "Combined Heat and Power Plant Report," and Form EIA-923, "Power Plant Operations Report;" and Oregon Institute of Technology, Geo-Heat Center. Industrial: U.S. Energy Information Administration, Form EIA-846 (A, B, C) "Manufacturing Energy Consumption Survey," Form EIA-906, "Power Plant Report," Form EIA-920, "Combined Heat and Power Plant Report," and Form EIA-923, "Power Plant Operations Report;" and Oregon Institute of Technology,

Geo-Heat Center; Government Advisory Associates, Resource Recovery Yearbook and Methane Recovery Yearbook.

4. Source: Energy Information Administration, Form EIA-860, "Annual Electric Generator Report".

5. Sources: IEA 2005, "Project Costs of Generating Electricity", Ernst & Young 2007 "Impact of Banding the Renewables Obligation – Costs of electricity production", Pöyry 2007 "Compliance costs for meeting the 20% Renewable Energy Target in 2020".

6. Source: Bonneville Power Administration "Fact Sheet" DOE/BP-4146 March 2010.

7. Source: Bonneville Power Administration DOE/BP-4203 September 2010.

8. Source: Energy Information Administration, *Renewable Energy Trends 2007* (published April 2009).

Appendix 1: Below Grade Scenario Analysis

The decision-maker is typically confronted with resource scheduling issues spanning several key projects. Oftentimes, the unforeseen project delays consume an inordinate amount of time and resources during the endeavor's timeline. When a systematic approach is utilized for the project implementation, the resources are proven to be more effective. The scenario analysis for the below grade work effort, for example, explores two simulated conditions:

1. Constant Resources (theoretical)
2. Scheduled Resources (realistic)

The realistic perspective provides a simulated scenario of the scheduled work effort for the engineering, construction, and test labor. It includes ramp-up times as well as actual process work effort of the individual entities. Moreover, the assumption of the constant material in the supply-chain and a "wait time" that exceeds 2500 hours as a contingency are encompassed in the simulation. Conversely, the theoretical simulation assumes constant resources dedicated to the actual project. The results are compiled and compared in various charts as a resource-leveling paradigm. A sample model is depicted in Figure 1.

Figure 1. Scenario analysis below-grade model

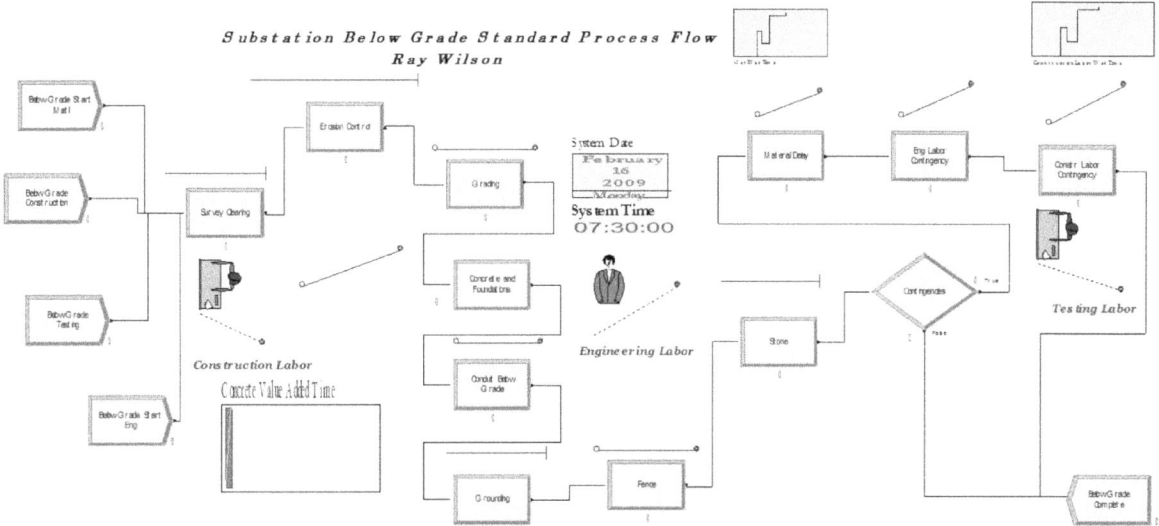

The results of the simulation are in the order of 38.10% ($788,994 - $488,319= $300,675) cost avoidance when the scheduled resource model is properly implemented. A sample schedule is shown in Table 1.

There are scheduled union breaks (15 minutes in the morning/afternoon, one hour lunch break) for the construction labor as well as ramp-up/down periods (25%, 50%, 75% of full capacity) for the professional labor (engineering and testing). Holidays and weekends are integrated in the master schedule as well. The key to the resource utilization is embedded within the matrix organization in the processes. The "Concrete & Foundations" process *only* utilizes 25% of the engineering labor, 50% of the construction labor, and full capacity for the material. The simulation depicts an absent resource during non-scheduled time. This scheme, employed in conjunction with the schedule (engineering labor performing on multiple projects, weather/permitting delays, equipment problems, etc.) emulates real-world activity. Consequently, the cost control and resource tracking parameters are easily maintained. The efficiencies of the resources are displayed in Table 2, where the busy percentage provides an enhanced representation of the project work effort as the material progress in the system.

Table 1. Scenario analysis schedule

Schedule Entity:	Below Grade Work Schedule							
	Monday	Tuesday	Wednesday	Thursday	Friday	Total (hrs)	Time Allotted (hrs/wk)	Number of Weeks
Engineering Labor	---	8	8	8	---	96	20	4.75
Construction Labor	8	8	10	10	4	523	40	15.70
Testing Labor	2	---	---	---	---	8	2	4.00
Material	8	8	8	8	8	1920	40	48.00

Table 2. Busy resource comparison

Number Busy	Standard	Schedule		Minimum	Maximum	Minimum	Maximum
	Average	Average	Half Width	Average	Average	Value	Value
Contstr Labor	0.9176	0.1531	0.01	0.07307689	0.2270	0.00	1.0000
Eng Labor	0.8369	0.1391	0.01	0.01453961	0.2269	0.00	1.0000
Matl	0.8343	0.3677	0.02	0.1965	0.5167	0.00	1.0000
Test Labor	0.06883553	0.04716719	0.00	0.02882285	0.06902786	0.00	0.5000

Another indicator of an efficient system is depicted in the "Work-In-Process" and "Instantaneous Utilization" charts Figures 2 and 3). The regulatory entities are interested in these data especially during the audits and annual reporting periods. It is designed to replace the ad hoc procedures of speculation and inference most companies employ. The resources are performing at an efficient level in the "Schedule Average" scenario.

The resources are completing the assigned tasks in the above scenario (Schedule Average) for this particular project. Notwithstanding, the associated costs with the scheduled utilization is concurrently reduced as the professional resources in particular execute the obligatory responsibilities. The "Average Wait Time" increased from the "Schedule Average" perspective as depicted in Figure 4. The realistic approach exhibits the resources in actual work-effort conditions as an underlying basis for this particular circumstance.

Figure 2. Work-in-process comparison

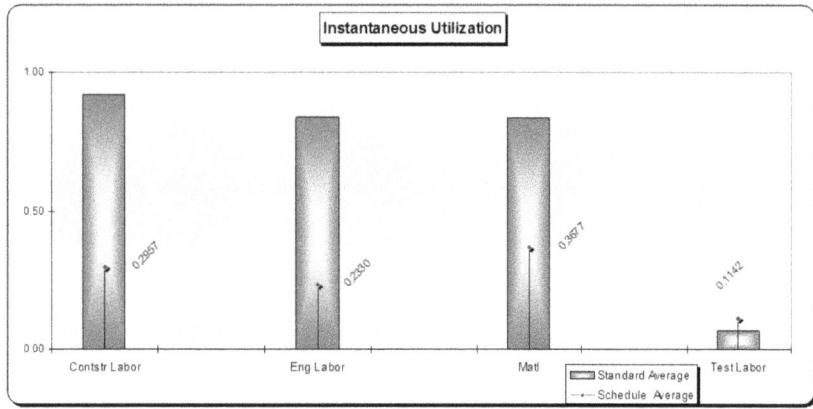

Figure 3. Instantaneous utilization comparison

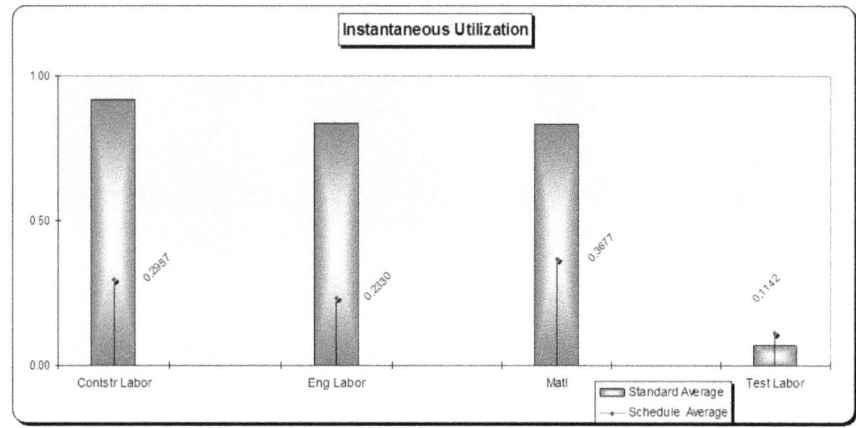

356

The "Survey Clearing Queue" is exceedingly high based on the engineering preparation as well as the test entities (waiting for results, permits, documentation, etc). Further inspection of the waiting queue depicts *only* 1.954 units in the queue when compared to the "Standard Average." This situation is shown in Figure 5 where the resource-leveling scheme is functioning as planned.

The throughput for the other queues is effective. The "Total Average Time per Entity" (Figure 6) and the "Total Accumulated Average Time" (Figure 7) both depicts outliers in the "Concrete & Foundations" sections from the "Standard Average" perspective. Several resources are utilized during this key phase of the endeavor and consequently, a bottleneck is formed in the scenario. The "Schedule Average" scenario performed the same task *only* at the 447.77 hour level. The caveat lies in the "Survey & Clearing" task where the project preparation is predominated (previously discussed).

Figure 4. Average waiting time comparison

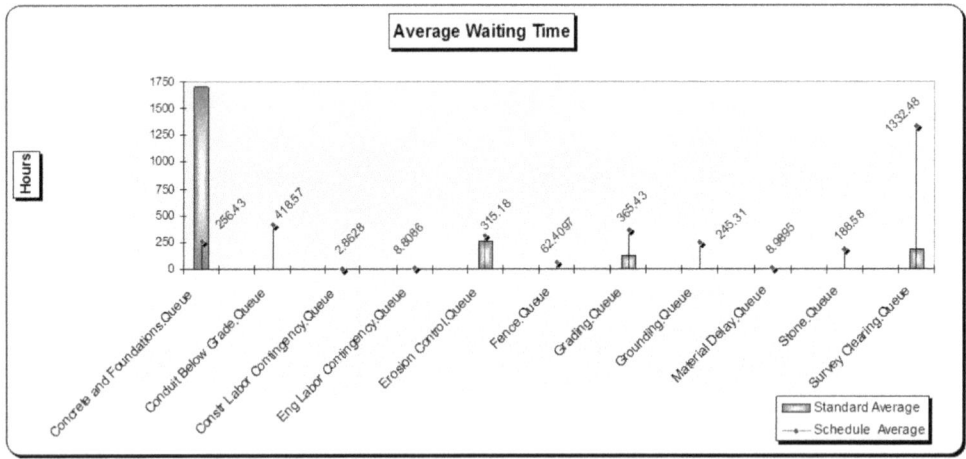

Figure 5. Average number waiting comparison

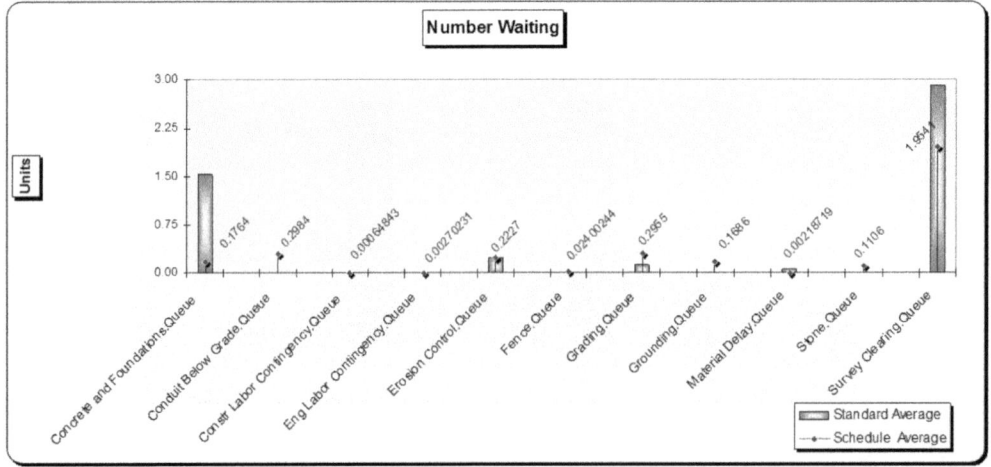

Figure 6. Total average time per entity comparison

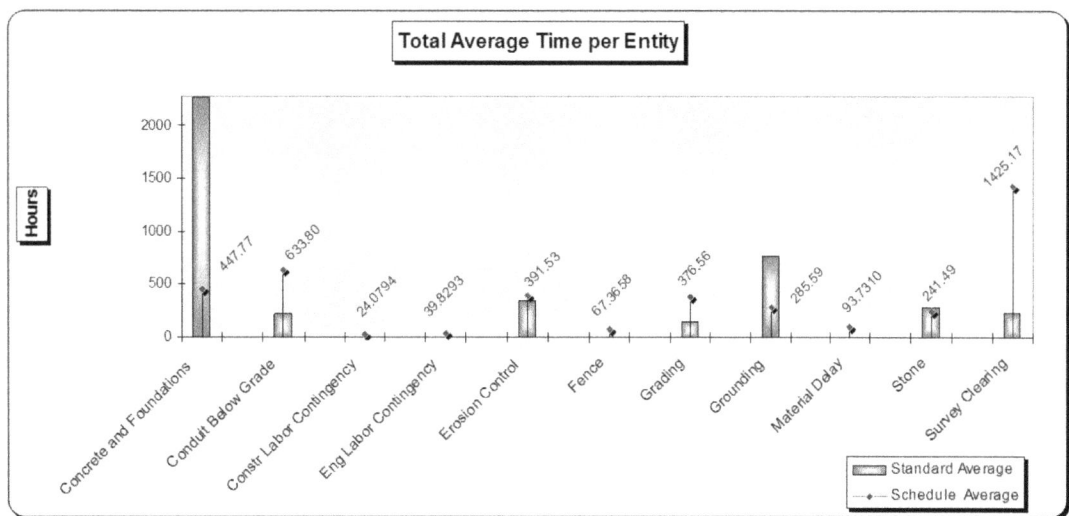

Figure 7. Total accumulated average time comparison

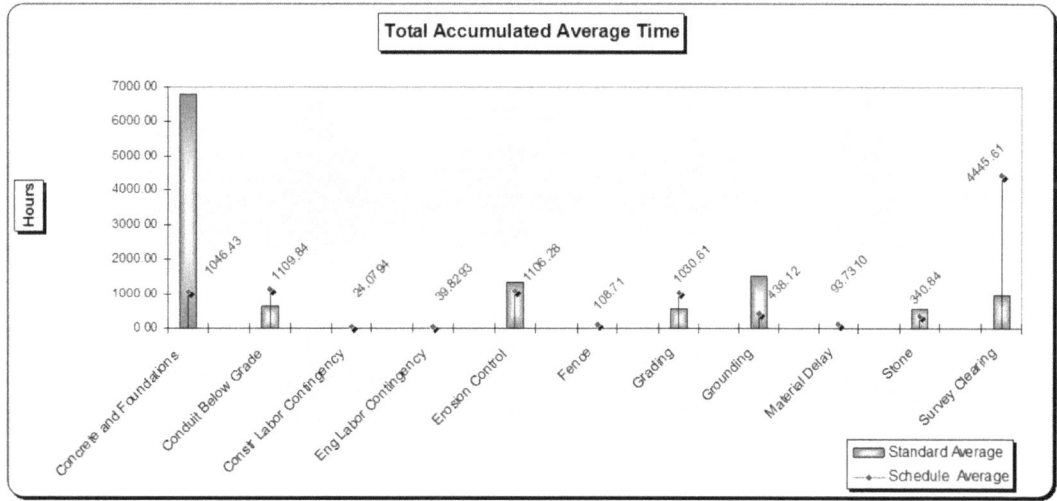

The scenario cost analysis provides an insight into the resource utilization distribution. The resource idle cost alone exhibits improved efficiency (Table 3) in the test labor sphere. When the schedule is properly utilized, a dramatic cost reduction is experienced in the "Schedule Average" scenario (Figure 8). Although the brute force technique is deemed effective in some short-term strategy circumstances, the schedule utilization paradigm is the most cost-effective solution.

The wait cost of the resources is detrimental in the "Schedule Average" scenario (Figure 9). When the resources are available to receive the next process in the queue, these costs will reduce dramatically. The cost is associated as cost accrued by the entity while it is in an added value activity and includes

Table 3. Idle resource comparison

Idle Cost	Standard	Schedule		Minimum	Maximum
	Average	Average	Half Width	Average	Average
Contstr Labor	$29.34	$77.89	1.65	60.2216	95.4616
Eng Labor	$0.00	$89.25	2.00	71.4228	113.49
Matl	$73.26	$73.16	2.37	51.3540	97.8340
Test Labor	$2,983.63	$48.76	0.46	44.6832	53.0345

Figure 8. Total cost per entity comparison

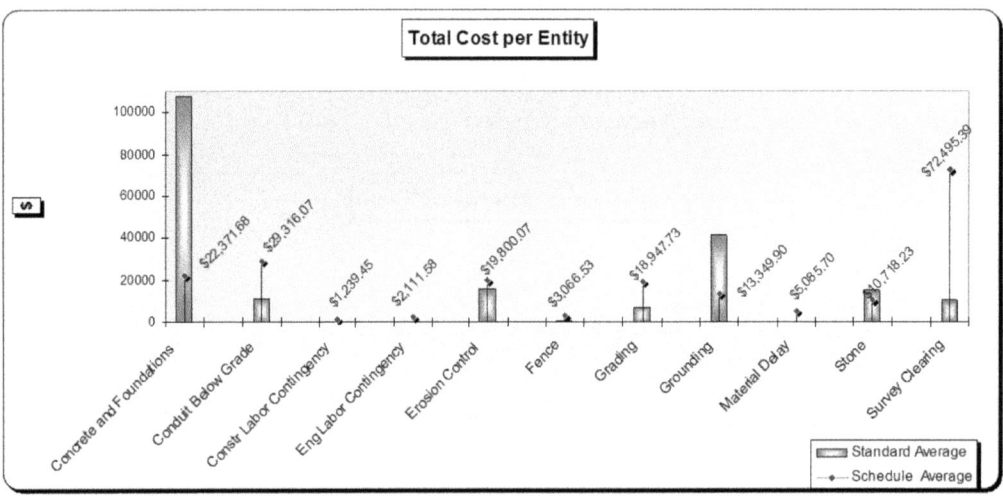

Figure 9. Wait cost per entity comparison

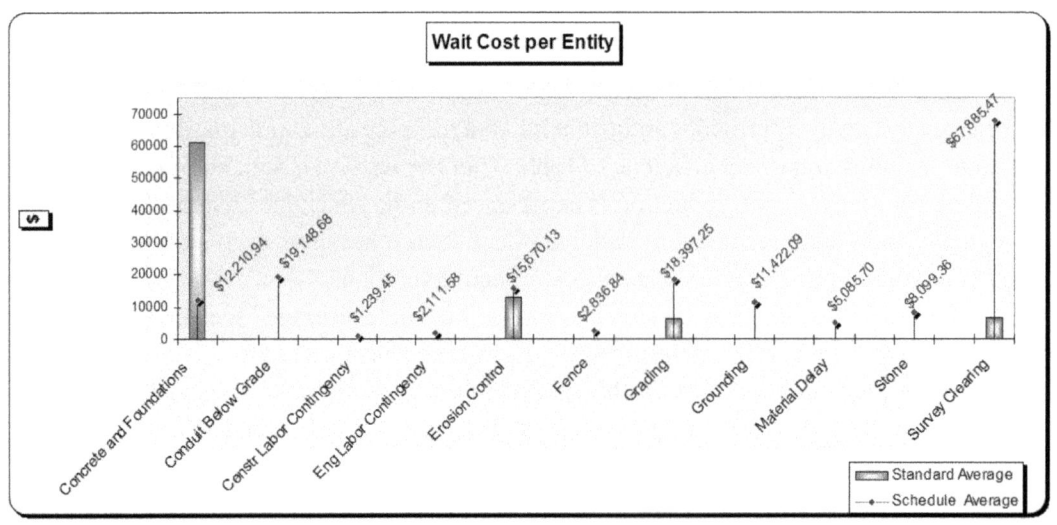

359

the overhead/hold costs. Nevertheless, in the overall system scheme, the "Wait Costs" exhibits a profound improvement when compared to the "Standard Average."

The "Accumulated Wait Costs" represents the total wait cost that was accrued by the entities in the station logic directly associated with the activity area as well as cost "roll up" from related activity areas. When engineering is scheduled to perform its tasks for example, it is imperative that all of the elements are "on station" to minimize this cost. By comparison, the "Schedule Average" exhibits improvement in key areas of the process (Figure 10). Consequently, an efficient operation is achieved in this scenario when implemented to the prescribed specifications.

Figure 10. Total accumulated cost comparison

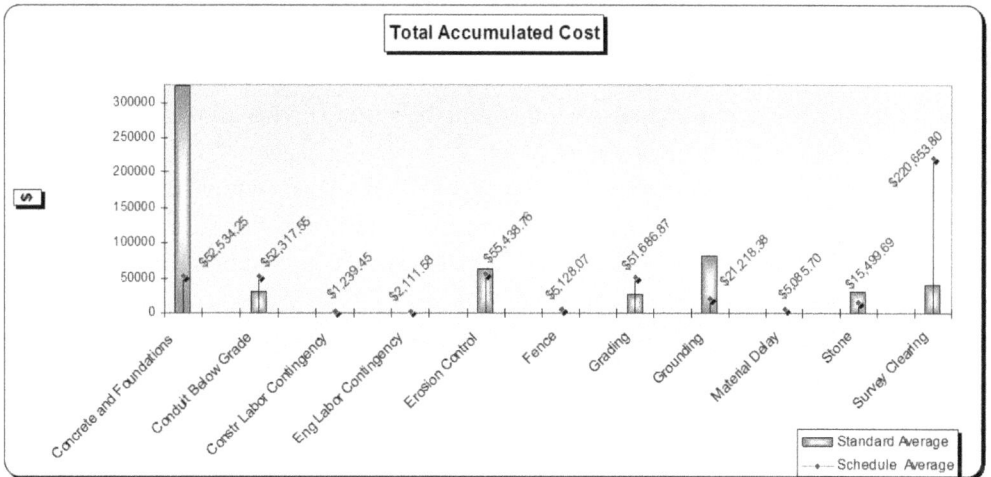

Appendix 2: Selected Regional Transmission Line Design Parameters

Conductors must be selected with sufficient thermal capability to meet continuous and emergency current ratings. The overhead line conductor and static wire should be chosen from those used by the Transmission Owner. This provides the ability to quickly repair a section of line with utility stock material should an emergency arise. Standard transmission conductor types are ACSR, ACSR/AW, ACSS, and ACAR. The ambient temperature range listed covers the system and is used for the electrical ratings of the conductors as well as the structural loads upon the towers or poles.

Wind Pressure: The pressure resulting from the exposure of a surface to wind. The pressure values provided are for wind acting upon objects with circular cross section. Pressure adjustments for other shapes shall be as set forth by the ASCE Guidelines for Electrical Transmission Line Structural Loading (ASCE Publication 74) [ASCE 74].

Radial Ice: Radial ice is an equal thickness of ice applied about the circumference of the conductors and static wires. Ice density is assumed to be 57 lbs per cubic foot. For the purpose of transmission line design, ice is not applied to the surface of the structure, insulators, or line hardware.

Temperature: Used for calculating conductor and static wire sag and tension.

Transverse load: Forces or pressures acting perpendicular to the direction of the line. For angle structures, the transverse direction is parallel to the bisector of the angle of the transmission centerline.

Longitudinal load: Forces or pressures acting parallel to the direction of the line. For angle structures, the longitudinal direction is perpendicular to the bisector of the angle of the transmission centerline.

All wires intact: A condition where all intended spans of conductors and static wires are assumed to be in place. In the case of a Line Termination Structure, conductor and static wire spans are only on one side of the structure.

Broken Conductor or Static Wire: A condition where one or more conductors or static wires are specified as broken. It is assumed that the broken conductor or static wire is in place on one side of the tower, and is removed from the other side. The span length for determination of loads from the conductor or static wire weight, wind pressure, and radial ice shall be not less than 60% of the design span length for the intact condition.

Load Factor: A value by which calculated loads are multiplied in order of provide increased structural reliability. For the purpose of structural design, Overload Capacity Factors as specified by NESC shall be considered Load Factors.

Design Loading Conditions

NESC: The provisions of the NESC Heavy Loading District, Class B Construction shall apply to all structure types. All wires intact. The latest NESC edition in effect at the time of line design shall apply. For informational purposes, the 1997 edition of NESC specifies the following requirements. Wind pressure – 4psf. Radial Ice – 0.5in.

Temperature: 0°F ~ For the purpose of calculating conductor or static wire tensions, a load constant of 0.3lbs shall be added to the resultant of the per linear foot weight, wind, and ice loads on the conductor or static wire. For steel structures, the load factor for wind load is 2.50; the load factor for vertical loads (dead weight and ice) is 1.50; and the load factor for conductor and static wire tension is 1.65. The associated factors for wooden transmission line structures shall be obtained from the Transmission Owner.

Extreme Wind Loading Condition: Applies to all structure types. All wires intact.

Line voltage 230kV and greater: Wind pressure applied to the wires shall be 25psf. The ambient temperature is to be 60°F. The wind pressure applied to the structure shall be 31.25psf. Load factor is 1.00.

Line voltage less than 230kV: The provisions of the NESC Extreme Wind loading shall be applied, subject to a minimum wind pressure of 17psf. The load factor is 1.00. The provision in NESC permitting exclusion of structures less than 60ft in height from extreme wind criteria shall not apply.

Heavy Ice Loading Condition: Applies to all structure types. All wires intact.

Line voltage 230kV and greater: Radial ice thickness on the wires only is to be 1.50in. No wind pressure. Temperature is 32°F. Load factor is 1.00.

Line voltage less than 230kV: Heavy ice loading (if any) shall be as specified by the Transmission Owner. Ice loading will not be more severe than that required for voltages 230kV or greater.

Foundation Loading: The ultimate strength of overturning moment and uplift foundations shall be not less than 1.25 times the design factored load reactions of the structure. The ultimate strength of foundations subjected to primarily to compression load shall be not less than 1.10 times the design factored load reactions of the structure. Overturning moment foundations designed by rotation or pier deflection performance criteria shall use unfactored structure reactions for determination of the foundation performance, but shall use factored reactions for the 1.25 time ultimate strength check.

Personnel Support Loading: Structures shall be designed to support a point load of 350 lb at any point where a construction or maintenance person could stand or otherwise be supported.

ELECTRICAL DESIGN PARAMETERS

Right-of-Way Width

The transmission line is to be designed with adequate right-of-way width to provide access for line maintenance, repair, and vegetation management. These widths are based upon the listed number of circuits on the right-of-way. For additional circuits, a wider right-of-way should be utilized. Vehicle or other means of access to each structure site is required for both construction and maintenance activities.

Wire to Ground Clearance

The minimum allowed clearance between the lowest transmission line conductor(s) shall meet the required NESC minimum plus a safety envelope of 3 feet. (The safety envelope is required to allow for sag and clearance uncertainties due to: actual conductor operating temperature, conductor sagging error, ground topography accuracy, plotting accuracy and other sources of error. The inclusion of a safety envelope is considered to be prudent). The NESC minimum shall be calculated with the conductor at maximum operating voltage and the maximum operating temperature or maximum conductor loading. The minimum clearances should take into account the limitation of a 5 mA shock current as given in NESC Rule 232D3c. All areas beneath the line shall be assumed to allow vehicle access beneath the line. For agricultural areas that may utilize farming equipment, additional clearance will be provided to assure public safety and line reliability during the periods of farming and harvesting activities.

Wire to Signs, Structures, under the Wires

The minimum allowed clearance between the lowest transmission line conductor(s) shall meet the required NESC minimum plus a safety envelope of 3 feet. The NESC minimum shall be calculated with the conductor at maximum operating voltage and the maximum operating temperature.

Wire to Structure Clearances

The minimum clearances between the phase conductors and the supporting tower or pole shall not be less than specified. These clearances are to apply for all anticipated conductor positions from an every day condition to a displaced condition due to a 9-psf wind or ice loading. These clearances do not have any adders provided for birds or other animals, but are based upon the switching surge values.

Wire-to-Wire Clearances

Clearance between the bottom transmission conductor and any lower wire shall meet the required clearance of NESC Rule 233 and 235 as a minimum. When the lower wire is a non transmission wire, then the clearance should be at least 10 feet for voltages less than or equal to 230 kV, and 20 feet for voltages above 230 kV. This will allow safe personnel access to the non-transmission conductors. These clearances should be calculated with the transmission conductor at maximum operating temperatures or heavy ice, whichever provides greater conductor sag, and the non-transmission conductor at 0°F. Clearances between transmission conductors should be either the larger of clearances based upon switching surges, or clearances based on the NESC.

Line Design

For transmission conductors of different circuits, the clearances should be increased so that any wind induced dynamic conductor movement does not result in any breaker operations and subsequent reduction in transmission circuit reliability.

Conductor Operating Temperature and Conductor Sag

The conductor will be assumed to operate at or above the minimum temperature shown below, and at temperatures less than the maximum shown below. While the line conductor may be designed to operate at a lower temperature, the line must be sagged assuming the conductor temperature is at or above the minimum shown. For designed operating temperatures above the minimum shown, and still below the maximum, the line sag and clearances will be calculated for that operating temperature after rounding up to the nearest 10°C.

Insulation Requirements

The insulation system for the transmission line shall have values in excess of the leakage distance, 60 Hz wet, and Critical Impulse flashover. These values shown are minimum conditions and may need to be increased in specific locations such as coastal environments, industrial smokestack sites, or high altitudes.

Lightning Performance and Grounding

All transmission structures will be individually grounded through a dedicated earth driven grounding system composed of ground rods and/or buried counterpoise. This system is to be measured on each individual structure prior to the installation of any overhead conductors or wires. The maximum acceptable resistance measurement of this grounding system for voltages up to and including 230 kV is 25 Ohms, and 15 ohms for voltages 345 kV and greater. The grounding system may include radial counterpoise wires, Equipotentiality rings, or both. The Transmission Owner must approve all grounding methods, and connections to the grounding system that are below grade. These resistance requirements are to assure acceptable lightning performance on the line as well as provide for the safe grounding of the line by construction and maintenance forces. Individual tower grounding measurements will be allowed to exceed the 25 or 15 Ohms required only if the average value for the 5 adjacent structures along the line is less than the 25 or 15-Ohm restriction. To assure acceptable lightning performance, a shield wire is required above each transmission line. Each new structure design is to be analyzed to determine that the line design and actual grounding design provides the required lightning performance. In instances where it is very difficult to provide the required lightning performance, the Transmission Owner may grant permission to utilize a limited application of transmission lines arresters. In no case will chemical ground treatments be allowed to improve structure grounding.

EMF, RFI, TVI, and Audible Noise

The transmission line system is to be designed so that radio and TV interference is just perceptible at the edge of the right-of-way. This is typically the case with radio signal to noise ratios above 20 db, and TV signal to noise ratios above 40 db. The achievement of this level of performance is more of a problem for lines above 230 kV, so a radio frequency survey and investigation should be performed to measure actual radio and TV signal strength and calculate the signal to noise ratio. Audible noise at the edge of the right-of-way should be calculated for the designed transmission line using wet conductor as the design condition. The resultant noise level must not exceed the level limited by the state and local authorities. Typically the limitation is 55 dbA during the daylight hours, and 50 dbA at night. Electric and Magnetic Field (EMF) levels are to be calculated and compared to any state or local limits. Modifications are to be made through phasing, structure height, ground clearance, etc. to assure these limitations are met. If no specific limitations exist, the line should be designed to the level of EMF on and adjacent to the right-of-way. A typical example of such an effort is the appropriate choice of phasing on the right-of-way.

Inductive Interference

A study should be done to determine the inductive impact upon other utilities due to the power flow in the new transmission line. The power flow may induce unusual currents and voltages in magnetic and electrical conductors that run parallel to the transmission line.

When it is determined that the currents or voltages are being induced in nearby utilities or other facilities, the engineer for the new or modified line being constructed must take the appropriate corrective actions to eliminate or lower the currents or voltages to an acceptable level.

Appendix 3: Project Management Substation Guidelines (General Process Flow Template)

Checkpoint Methodology

The schedule is issued prior to the "go" decision. It covers the duration of the project. It is designed as a tool for the project manager to communicate with the external world.w

PMM-1 Project Initiation Responsibility Matrix

Legend

- ○ General Management Responsibilities
- ● Specialized Responsibility
- △ May be consulted
- ▲ Must be consulted
- ☐ Must be notified
- ▣ Must Approve

PMM-2 Project Planning Responsibility Matrix

Legend

- ○ General Management Responsibilities
- ● Specialized Responsibility
- △ May be consulted
- ▲ Must be consulted
- ☐ Must be notified
- ▣ Must Approve

PMM-3 Project Execution Responsibility Matrix

Legend
○ General Management Responsibilities
● Specialized Responsibility
△ May be consulted
▲ Must be consulted
▢ Must be notified
▧ Must Approve

PMM-4 Project Controlling Responsibility Matrix

Legend
○ General Management Responsibilities
● Specialized Responsibility
△ May be consulted
▲ Must be consulted
▢ Must be notified
▧ Must Approve

PMM-5 Project Closing Responsibility Matrix

Legend
○ General Management Responsibilities
● Specialized Responsibility
△ May be consulted
▲ Must be consulted
▢ Must be notified
▧ Must Approve

INITIATION

Background Information

Project managers not only must rely heavily upon the traditional on-time, on-budget, and on-quality performance measures, but also with the added *on-strategy* dimension central to managing project success and to alleviate leadership challenges. A quadruple constraint is created with the insertion of the *on-strategy* dimension. The main thrust of this point is illustrated in Figure 1.

Figure 1. Project dimensions

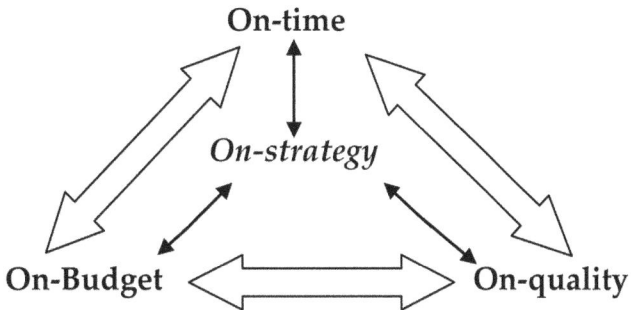

Many leadership tasks relate to developing a vision of the project outcome that is practical, yet capable of mobilizing and motivating team members to accomplish the project's goals and objectives. The leadership vision engages stakeholders who are not actively involved in the project; it also inspires them to maintain their support over the project's timeline. The solution-building negotiation approach to defining the scope of a project, and then clearly communicating this to the project team and other stakeholders defines a strategy for realizing the vision, and translating the strategies into operational plans and results. The core problem is the insufficiency of instability of strategy to properly develop and express project vision connected through measurement to tangible business outcomes. Organizations should link planned outcomes to their corporate strategy using a measurement framework, referred to as *performance management*, and is common within corporations.

Details

The process is designed to facilitate the formal authorization to start a new project. It is often performed external to the project's scope of control by the organization, which may blur the project boundaries for the initial project inputs. Clear descriptions of the project objectives are developed, including the reasons why a specific project is the best alternative solution to satisfy the current requirements (Figure 2).

Essential Initiation Items

General Project Charter **Initial**_____ **Date**_____
Project Document (design, estimates) **Initial**_____ **Date**_____
Preliminary Scope Statement **Initial**_____ **Date**_____
Preliminary Work Breakdown Structure **Initial**_____ **Date**_____

Critical Factors in the Project Initiation Phase

- Clear Project Vision
- Clear Goals and Objectives
- Formal Budget
- Executive/Sponsor Involvement

Figure 2. Project initiation process

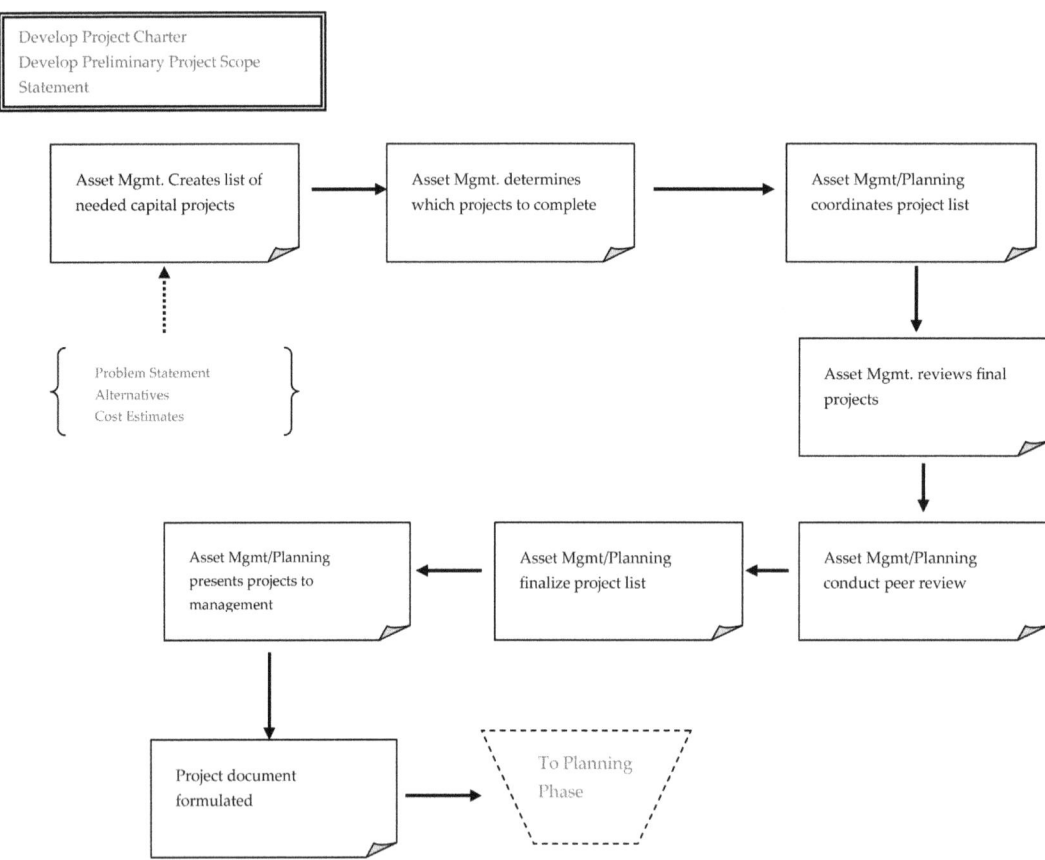

PLANNING

Background Information

The most important responsibilities of a project manager are *planning, integrating, and executing plans.* Project planning is described as a predetermined course of action within a forecasted environment. The project requirements set the milestones on the schedule. The planning must be systematic and flexible enough to handle unique activities. The course of action must be disciplined through reviews and controls. Moreover, the process must be capable of accepting multifunctional inputs in order to achieve the desired results. The nine (9) major components of planning are:

- **Objective:** A goal, target, or quota to be achieved over time.
- **Program:** Strategy followed/major actions taken in order to achieve objectives.
- **Schedule:** A plan showing activities/accomplishments started/completed.
- **Budget:** Planned expenditures required to achieve or exceed objectives.
- **Forecast:** A projection of what will happen by a certain time.
- **Organization:** Positions/duties/responsibilities to achieve objectives.

- **Policy:** Guide for decision-making/individual actions.
- **Procedure:** A detailed method for carrying out a policy.
- **Standard:** A level of performance defined as adequate or acceptable.

Details

The process is designed to gather information from several sources with each having varying levels of completeness and confidence. This process identifies, define, and mature the project scope, cost, and the project activities that occur within the project. As new project information is discovered, additional dependencies, requirements, risks, opportunities, assumptions, and constraints will be identified (Figure 3).

Figure 3. Project planning process

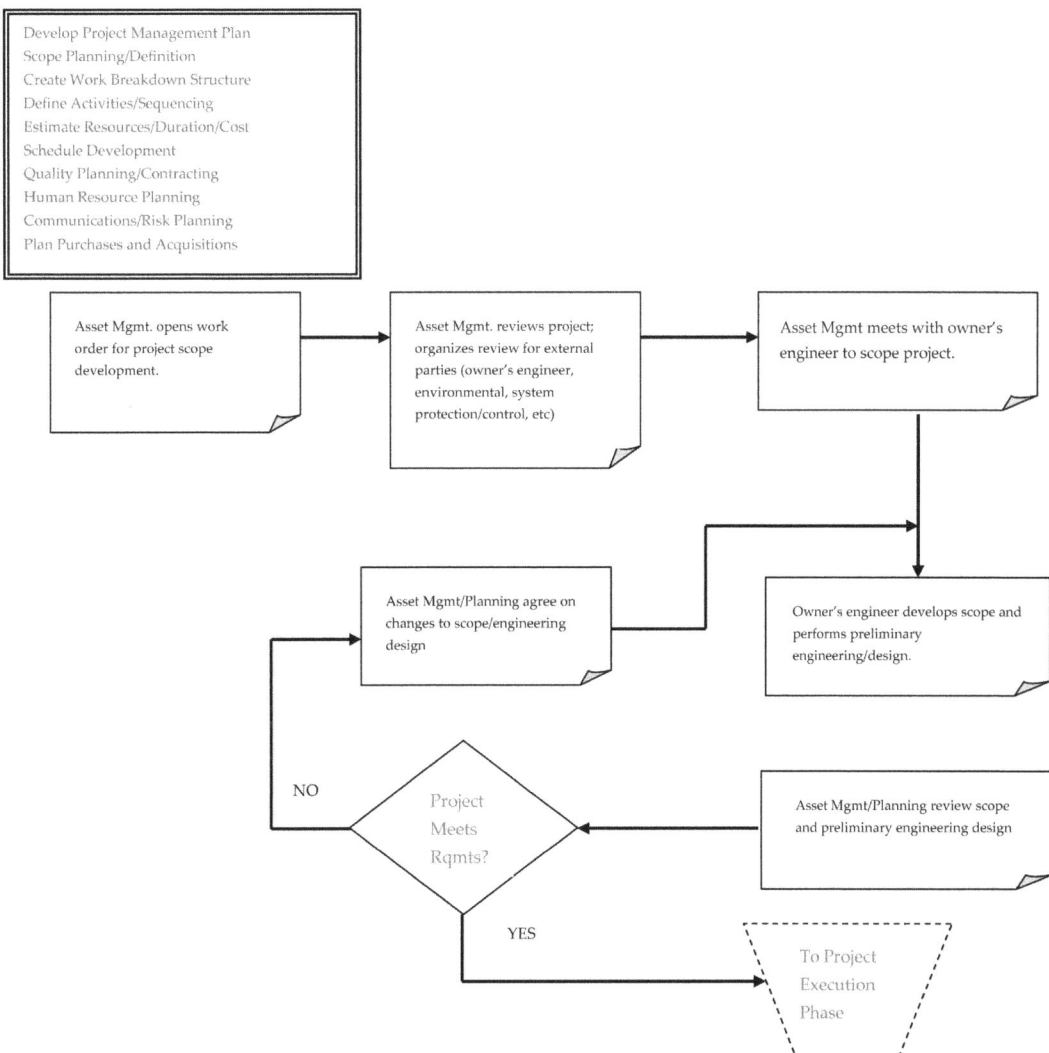

Essential Planning Items

Scope Documents **Initial**_____ **Date**_____

Approved Work Orders **Initial**_____ **Date**_____

Approved Purchase Orders **Initial**_____ **Date**_____

Work Breakdown Structure **Initial**_____ **Date**_____

Engineering Design Package **Initial**_____ **Date**_____

Project Schedule **Initial**_____ **Date**_____

Project Management Plan **Initial**_____ **Date**_____

Cost/Resource Estimates **Initial**_____ **Date**_____

Critical Factors in the Project Planning Phase

- Clearly communicating project status
- Detailed Project Plan
- Detailed Work Breakdown Structure
- Detailed Staffing Plan
- Adequate Resources
- Appropriate Project Team Structure

EXECUTION

Background Information

The execution phase is a key aspect of the project progress philosophy. The project manager must assemble the required resources and management plans soon after the "go" decision is made. By demanding that project team members link their own actions and decisions with the overall intended strategy of the project (an extension of the corporate strategy) can assist with *on-strategy* project execution. It extends a virtual leadership presence, which injects itself into every critical project event and decision. The associated administrative tasks involved in this phase (work orders, bid documents, status reporting, etc.) are extremely important in project management. Many of the key characteristics of the project (scope, cost, resources, procurement, quality, risk, communication) are identified and highlighted during this phase.

Details

The process is designed to coordinate people and resources as well as integrating and performing the activities of the project in accordance with the prescribed objectives/scope documentation (Figure 4).

Essential Execution Items

Bid Package **Initial**_____ **Date**_____

Contractor Evaluation **Initial**_____ **Date**_____

Work Authorization **Initial**_____ **Date**_____

Project Team **Initial**_____ **Date**_____

Figure 4. Project execution process

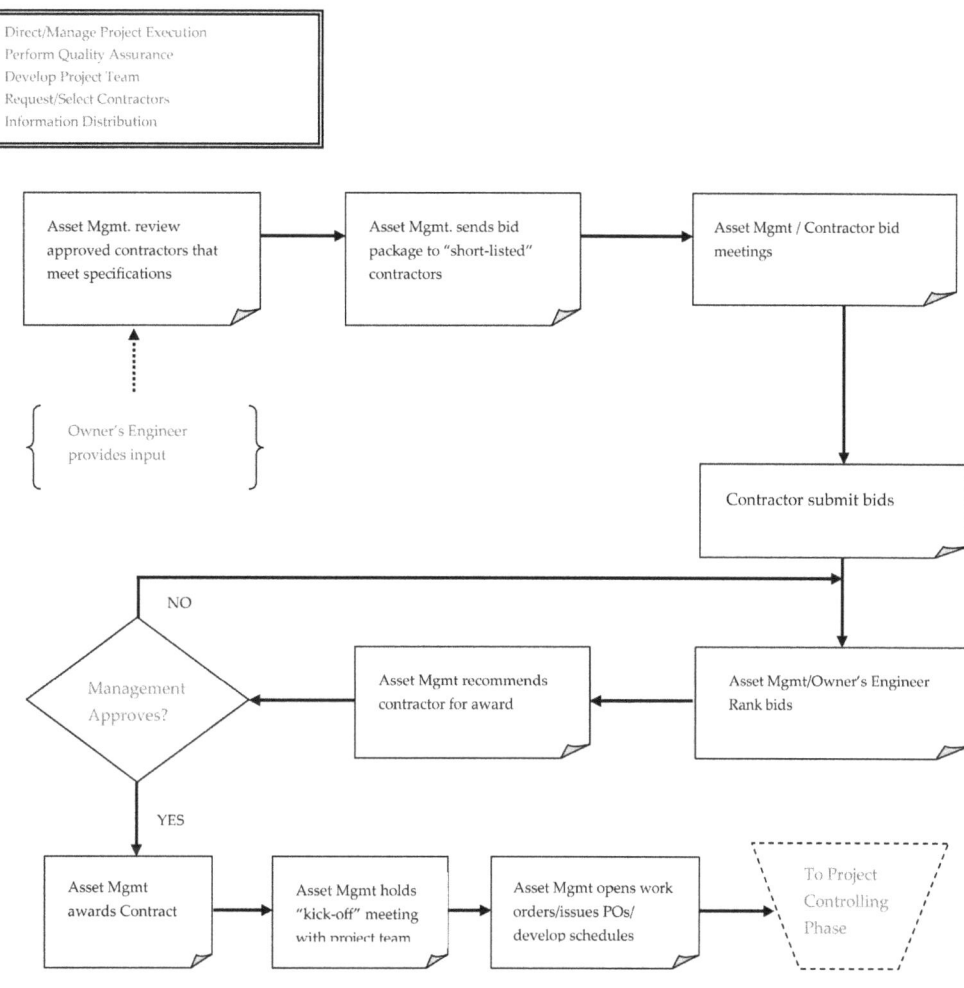

Project Schedule **Initial**_____ **Date**_____

Critical Factors in the Project Execution Phase

- Good Project Methodology
- Status Reporting
- Appropriate Project Team Structure

CONTROLLING

Background Information

Many project managers are not fully schooled – either in formal academic settings or through on-the-job project experience – to necessarily note the subtle but important difference between perceived power and actual power. Learning to distinguish between *influence* and *control* to achieve results often means the difference between temporarily controlling an outcome by forced compliance versus creating a lasting change in people's behavior. Some project managers, however, may not see the two techniques as very different. Consequently, project managers eventually conclude that they really cannot be everywhere at once to vet every decision to ensure that the team appropriately conducts itself in performing its roles and realizing the importance of the project. As a result, most managers revert to some kind of *exception-based or situational leadership* method to address ongoing challenges as recommended by established theory. While somewhat effective, this tactic does not completely address the issue. Project managers therefore are taxed to employ a variety of techniques to achieve the desired outcome to include a practical balance between traditional proactive risk management and reactive project troubleshooting.

The proactive risk management approach minimizes the occurrences of anticipated problems to the extent that it is practical to do so. Project troubleshooting minimizes the impact of all unexpected disruptions, both large and small. The project manager's goal should not focus on avoiding risk; the project manager should strive to manage risk to prevent risk from disrupting the project. The aim is not simply risk mitigation and risk avoidance; the manager's goal is the total management of risk – proactively and reactively.

Details

The process is designed for project management procedures associated with initiating, planning, execution, and closing. Corrective or preventive actions are taken to control the project performance. Monitoring includes collecting, measuring, and disseminating performance information and assessing measurements/ trends to effect process improvements (Figure 5).

Essential Monitoring & Control Items

Documented Corrective Actions **Initial**_____ **Date**_____

Documented Preventive Actions **Initial**_____ **Date**_____

Schedule/Cost Forecasts **Initial**_____ **Date**_____

Documented Quality Inspections **Initial**_____ **Date**_____

Documented Change Requests **Initial**_____ **Date**_____

Critical Factors in the Project Controlling Phase

- Project Tracking & Control
- Status Reporting
- Good Risk Management Practices
- Contingency Funds
- Clearly Communicating Project Status

PROJECT CLOSE-OUT

Background Information

The orderly closure of a project is an essential function in project management. Projects are considered completed or *closed-out* after the sponsor receives and approves all reports as required by the terms and

Figure 5. Project controlling process

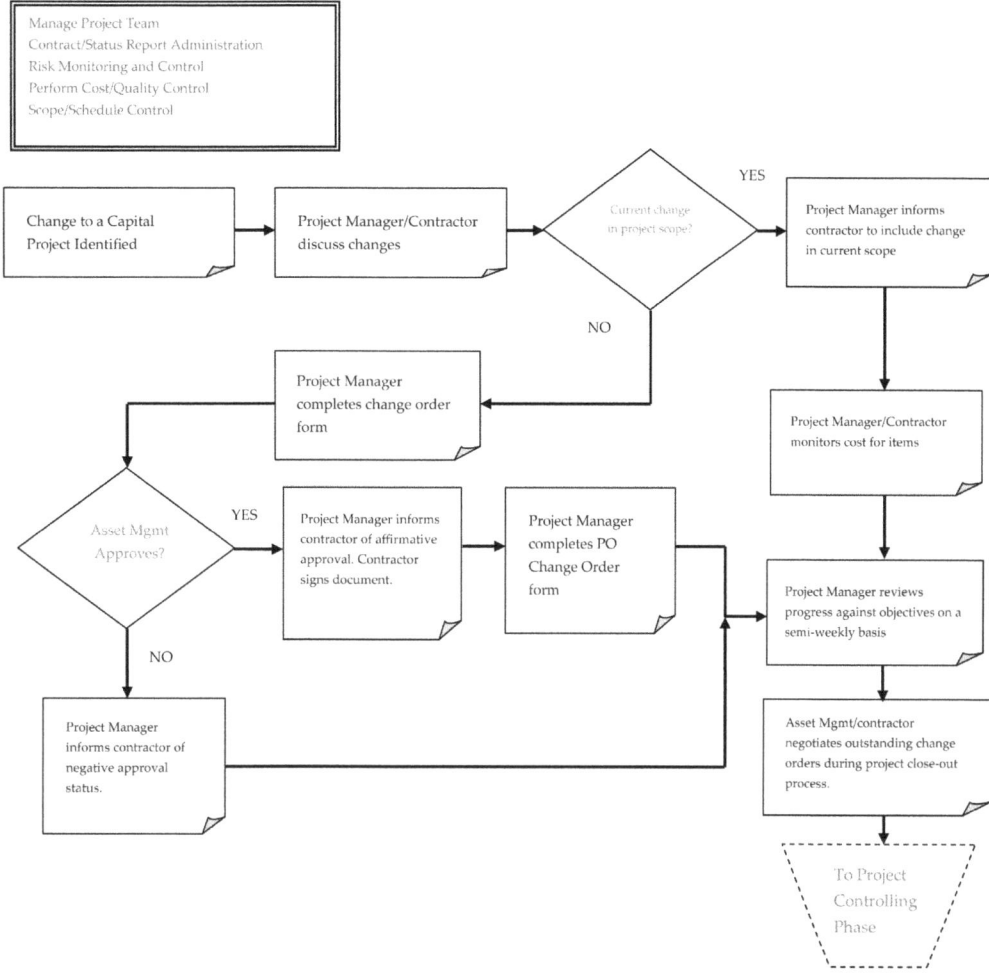

conditions of the award, and notifies the responsible parties of its acceptance and closure of the project. It is the sole responsibility of the project manager to be aware of when reports are due and to submit them by the required due dates. Reports required at the close of a project are generally due within 90 days of the project end date to include:

- A final technical report to include the following:
  - Project manager's name, Project Title, Performance period
  - Significant results of the project
  - Examples of progress
  - Discussion of objectives and deliverables
- A final financial report to include the following:
  - Final Invoices
  - Cash flow close-out
- Resource Performance Evaluation(s) to include the following:
  - Technical difficulties and solutions
  - Activity verification
  - Feedback reports

- Equipment/Document specification updates and associated reports
- Capture Lessons-Learned
- Administrative closure

Additionally, the project manager ensures that the accounting and planning departments are provided with accurate equipment data, costs, and plant-in-service dates.

Details

The process is designed to finalize all activities across all of the process groups to formally close the project (Figure 6).

Essential Closure Items

Close Project **Initial**_____ **Date**_____
Close Contracts **Initial**_____ **Date**_____
Equipment Updates **Initial**_____ **Date**_____
Drawing/Modeling Updates **Initial**_____ **Date**_____
Final Reports **Initial**_____ **Date**_____
Historical Documentation **Initial**_____ **Date**_____

Critical Factors and Questions in the Project Closing Phase

- What was done right?
- What was done wrong?
- What future recommendations can be made?
- How, when, and to whom should the information be disseminated?
- Were the schedules realistic?
- Was the level of detail correct?
- Was it easy to evaluate performance from the schedule?
- Was tracking easily accomplished?

Figure 6. Project closure process

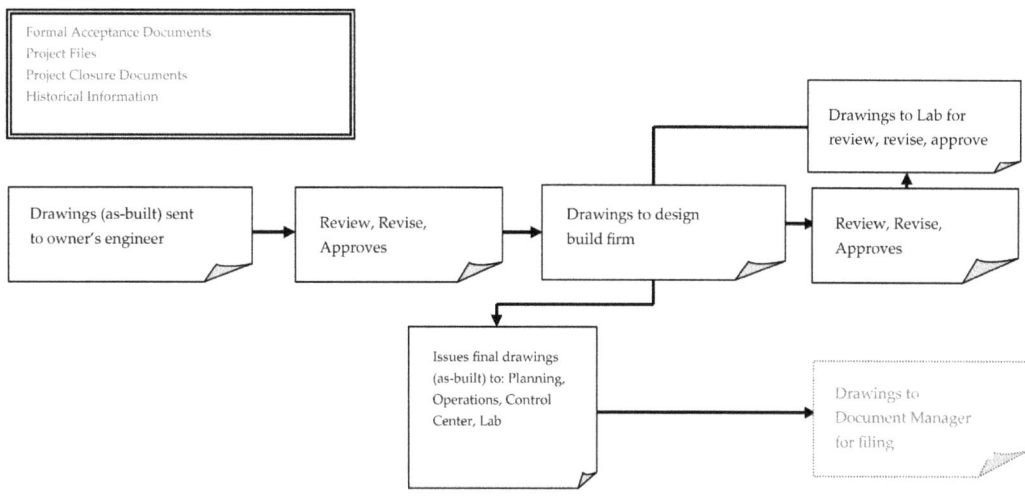

Formal Acceptance Documents
Project Files
Project Closure Documents
Historical Information

Drawings (as-built) sent to owner's engineer → Review, Revise, Approves → Drawings to design build firm → Review, Revise, Approves → Drawings to Lab for review, revise, approve

Issues final drawings (as-built) to: Planning, Operations, Control Center, Lab → Drawings to Document Manager for filing

Responsibility Codes: CL1~6

Construction Complete → System Control Energizes equipment → Notification of completed work and energized equipment to project manager → Project Manager closes file. → Accounting unitization process

- Lists Assets
- Associated Lump Sum
- Maintenance Cost

-Audits work orders
-Transfers CWIP to plant-in-service via accounting

New assets tracked against monthly maintenance costs.

New equipment/facilities recorded in Equipment Management System

Contractor sends final invoice

Accounts payable ← Project manager/contractor develop final negotiations of change orders/costs ← Accounting reviews all project documentation (Close work orders/POs, tax information, AIA documentation) → Accounting unitization process

Compilation of References

Aakre, D., & Haugen, R. (2009). *Wind turbine lease considerations for landowners*. North Dakota State University Extension Service Report supported by the U.S. Department of Energy Wind Powering America Program. Award Number DE-FG36-07GO47010.

Aft, L. S. (1998). *Fundamentals of industrial quality control*. Boca Raton, FL: CRC Press.

Agor, W. (1984). *Intuitive management: Integrating left and right brain management skills*. Englewood Cliffs, NJ: Prentice-Hall.

Ahlstrom, M., Grant, W., Jones, L., & Zavadil, L. (2003, November/December). The future of forecasting and utility operations, planning for improved system operations. *IEEE Power and Energy Magazine*, *3*(6), 57–64. doi:10.1109/MPAE.2005.1524621

Anaya-Lara, O. (2009). *Wind energy generation: Modeling and control*. Chichester, UK: John Wiley & Sons, Ltd.

Anderson, D. (1999). *An introduction to management science: Quantitative approaches to decision making* (9th ed.). South-Western College Publication.

Anderson, P. M., & Fouand, A. A. (2003). *Power system control and stability* (2nd ed.). Piscataway, NJ: IEEE Press.

ANSI/ISO/ASQ Q9001-2000. (2001). *Quality management systems – Requirements*. Milwaukee, WI: Quality Press, American National Standard, American Society for Quality (ASQ)..

Atkinson, G. (2000). Measuring corporate sustainability. *Journal of Environmental Planning*, *43*(2), 235–252.

Awerbuch, S. (1993). The surprising role of risk in utility integrated resource planning. *The Electricity Journal*, *6*, 20–33. doi:10.1016/1040-6190(93)90048-P

Axelrod, R., & Cohen, M. (2000). *Harnessing complexity: Organizational implications of a scientific frontier*. New York, NY: Basic Books Publishers.

Barker, P. P., & Short, T. A. (1996). Findings of recent experiments involving natural and triggered lightning. IEEE/Power Engineering Society, Transmission and Distribution Conference, Los Angeles, CA.

Barrios, L., & Rodriguez, A. (2004). Behavioural and environmental correlates mortality at on-shore wind turbines. *Journal of Applied Ecology*, *41*, 72–81. doi:10.1111/j.1365-2664.2004.00876.x

Bently, J. (1999). *Introduction to reliability and quality engineering* (2nd ed.). Upper Saddle River, NJ: Prentice-Hall.

Bergen, A., & Vittal, A. (2000). *Power system analysis* (2nd ed.). Upper Saddle River, NJ: Prentice Hall.

Besterfield, D. H. (1999). *Total quality management* (2nd ed.). Upper Saddle River, NJ: Prentice Hall.

Bhattacharjya, D., & Schachter, R. (2012). Formulating asymmetric decision problems as decision circuits. *Decision Analysis*, *9*(2), 138–145. doi:10.1287/deca.1110.0226

Blair, N., & Hand, M. (2008). *Power system modeling of 20% wind-generated electricity by 2030*. National Renewable Energy Laboratory conference paper presented at the Power Engineering Society, Pittsburg, PA.

Bohan, B., Ferrar, B., & Luigs, M. (September 2011). Addressing the big data concern in the utilities sector. *Utility Automation Engineering T&D Magazine*. Retrieved 15 May, 2012, from http://www.elp.com/elp/en-us/index

Breckenridge, T. (February 13, 2010). Lake Erie wind turbine backers ready to study wildlife impact, court manufacturer. *The Plain Dealer*. Retrieved 19 May, 2010, from http://www.plaindealer.com

Breyfogle, F. W. III (2000). *Managing Six Sigma*. New York, NY: John Wiley & Sons.

Brown, G. M., Szabados, B., Hoolbloom, G. J., & Poloujadoff, M. E. (1992). High-power cycloconverter drive for doubly-fed induction motors. *IEEE Transactions on Power Electronics, 39*(3), 230–240.

Budget of the U.S. Government. (FY 2013). *Performance and management section*. Analytical Perspectives Volume. Retrieved 29 May, 2012, from http://www.whitehouse.gov/omb/budget/analytical-perspectives

Busch, J. F., & Eto, J. (1996). Estimates of avoided costs for electric utility demand-side planning. *Energy Source, 18*.

Cao, H., Hu, C., & Jiang, C. (2011). Integrated stochastic resource planning of human capital supply chains. *Interfaces, 41*(5), 414–435. doi:10.1287/inte.1110.0596

Carlson, B. (2012). MISO unlocks billions in savings through the application of operations research for energy and ancillary service markets. *Interfaces, 42*(1), 58–73. doi:10.1287/inte.1110.0601

Chen, F. H. (2000). *Soil engineering: Testing, design, and remediation*. Boca Raton, FL: CRC Press.

Chesbrough, H. (2011). *Open services innovation: Rethinking your business to grow and compete in a new era*. San Francisco, CA: Jossey-Bass. doi:10.1007/978-88-470-1980-5

Chinchilla, M., Arnaltes, S., & Burgos, J. C. (2006). Control of permanent magnet generators applied to variable-speed wind energy systems connected to the Grid. *IEEE Transactions on Energy Converters, 21*(1), 130–135. doi:10.1109/TEC.2005.853735

Christopher, R. (May 2012). Survive and thrive: Adapting to the multigenerational workplace. *Quality Progress*, pp. 48-49.

Clapp, A. L. (2006). *NESC handbook: A discussion of the national electric safety code* (6th ed.). New York, NY: The Institute of Electrical and Electronic Engineers, Inc.

Clymer, J. R. (2009). *Simulation-based engineering of complex systems*. New York, NY: John Wiley.

Cokins, G. (May-June 2012). Why do large companies fail? *Analytics Magazine*, pp. 16-19.

COMAP. (2009). *For all practical purposes: Mathematical literacy in today's world* (8th ed.). New York, N.Y: W.H. Freeman & Company.

Conklin, J. D. (April 2012). Next in line: Always look ahead to the next project for maximum quality gains. *Quality Progress*, pp. 44-47.

Cooper, R. B. (1981). *Introduction to queuing theory* (2nd ed.). New York, NY: John Wiley.

Cooper, R. G. (1993). *Winning at new product development: Accelerating the process from idea to launch*. Reading, MA: Addison-Wesley.

Cowen, D., & Jensen, J. (1999). *Remote sensing of urban/suburban infrastructure and socio-economic attributes* (pp. 611–622). Columbia, SC: American Society for Photogrammetry and Remote Sensing.

Damborg, S. (2003). *Public attitudes towards wind power*. Danish Wind Energy Association. Retrieved 4 May, 2010, from http://www.windpower.org/en/articles

Davenport, T., Harris, J., & Morison, R. (2010). *Analytics at work: Smarter decision, better results*. Boston, MA: Harvard Business Press.

Decision Process. (2005). *McGraw-hill concise encyclopedia of science and technology* (5th ed., pp. 697–698). New York, NY: McGraw-Hill.

DOE/GO-102008-2567. (2008). *20% wind energy by 2030: Increasing wind energy's contribution to U.S. electricity supply*. Washington, DC: U.S. Department of Energy, Office of Energy Efficiency and Renewable Energy.

Duffuaa, S., Raouf, A., & Campbell, J. (1999). *Planning and control maintenance systems: Modeling and analysis*. New York, NY: John Wiley and Sons, Inc.

Edwards, F. W. (1993). *Substation grounding system evaluation*. Conducted by Virginia Power. Presented at the Southeastern Electric Exchange. New Orleans, LA.

Emil, S., & Scanlan, R. (1986). *Wind effects on structures: An introduction to wind engineering* (2nd ed.). New York, NY: John C. Wiley & Sons.

Energy Storage Systems. (2007). *Encyclopedia of science and technology* (10th ed.). New York, NY: McGraw-Hill.

ENIRDGnet. (May 31, 2003). *The driving European forces and trends: Concepts and Opportunities of DG.* Retrieved 4 May, 2010, from http://www.wp1.leader@dgnet.org

Erickson, W. P., Johnson, G. D., Strickland, D. P., Young, K. J. Jr, Sernka, K. J., & Good, R. E. (2001). *Collisions with wind turbines: A summary of existing studies and comparisons to other sources for avian collision mortality in the United States.* Washington, D.C.: Western Eco-systems Technology, Inc. National Wind Coordinating Committee Research Document.

Evans, W., Howe, R. W., & Wolf, A. (2002). *Effects of wind turbines on birds and bats in Northeastern Wisconsin.* University of Wisconsin-Green Bay submitted to the Wisconsin Public Service Commission and Madison Gas & Electric Company, Green Bay, WI.

Evans, J. R., & Lindsay, W. M. (2005). *The management and control of quality* (6th ed.). Florence, KY: South-Western College Publishers.

Everett, G., Philpott, A., Vatn, K., & Gjessing, R. (2010). Norske Skog improves global profitability using operations research. *Interfaces, 40*(1), 58–70. doi:10.1287/inte.1090.0471

Farret, F. A., & Simoes, M. G. (2006). *Integration of alternative sources of energy.* Chichester, UK: John Wiley & Sons, Ltd.

Feigenbaum, A. V. (1991). *Total quality control*, 3rd ed. New York, NY: McGraw-Hill Green Energy News. (April 19, 2010). *Pentagon attempts to stop wind farm development.* Retrieved 28 April, 2010, from http://www.renewable-energy-news.info

Figge, F., Hahn, T., Schaltegger, S., & Wagner, M. (2002). The sustainability balanced scorecard – Linking sustainability management to business strategy. *Business Strategy and the Environment, 11*(5), 269–284. doi:10.1002/bse.339

Fink, D. G., & Beaty, W. H. (2001). *Standard handbook for electrical engineers* (14th ed.). New York, NY: McGraw-Hill Inc.

FIST. Facility Instructions, Standards, and Techniques. (2005). *Maintenance scheduling in electrical equipment,* Volume 4-1B. U.S. Department of the Interior, Bureau of Reclamation, Denver, CO.

French, S. (2012). Expert judgment, meta-analysis, and participating risk analysis. *Decision Analysis, 9*(2), 119–127. doi:10.1287/deca.1120.0234

Fukahori, M. (1995). Optimization of an electrical encapsulate. *Journal of Quality Engineering Forum, 3*(2).

Fuller, F. M. (1983). *Engineering of pile installations.* New York, NY: McGraw-Hill.

Gattiker, T. (March 2012). Rethinking design. *Quality Progress*, pp. 33-37.

George, M. L. (2003). *Lean Six Sigma for service.* New York, NY: McGraw-Hill.

Ghorai, M., & Reddy, N. (2008). *Meeting the IESO interconnection requirements for Ontarian wind farms.* Presented at the European Wind Energy Conference, Brussels, Belgium. Retrieved 23 August, 2010, from http://www.ewec2008.info

Glover, F., Kelly, J. P., & Laguna, M. (1996). New advances and applications of combining simulation and optimization. In J. M. Charnes (Ed.), *Proceedings of the 1996 Winter Simulation Conference*, (pp. 144-152).

Glover, J. D., & Sarma, M. S. (2001). *Power system analysis and design* (3rd ed.). Toronto, Canada: Thompson Learning.

Godlewski, E., Lee, G., & Cooper, K. (2012). Systems dynamics transforms fluor project and change management. *Interfaces, 42*(1), 17–32. doi:10.1287/inte.1110.0595

Gojanovic, T. (April 2012). Theory of evolution, *Quality Progress*, pp. 12-13.

Good, J. (2006). The aesthetics of wind energy. *Human Ecology Review, 13*(1), 76–80.

Grady, J. O. (1994). *System integration.* Boca Raton, FL: CRC Press.

Grauers, A. (1996). Efficiency of a three wind energy generator system. *IEEE Transactions on Energy Conversion, 11*(3), 650–657. doi:10.1109/60.537038

Gross, D., & Harris, C. M. (1997). *Fundamentals of queuing theory* (3rd ed.). New York, NY: John Wiley.

Grubaum, R., & Pernot, J. (1999). *Thyristor-controlled series compensation: A state of the art approach for optimization of transmission over power links.* Retrieved 5 July, 2010, from http://www.abb.com/FACTS

Grumbine, R. E., & Xu, J. (2011). Mekong hydropower development. *Science, 332,* 178–179. doi:10.1126/science.1200990

Gunther, E. (2007). Field and device technologies: Customer portals, home area networks and connected devices. *GRID INTEROP 2007 Forum Proceedings* (Gunther-1), Richland, WA.

Hall, E. (2007). *Beyond culture.* New York, NY: Anchor Publisher.

Hall, J., Ellner, P. M., & Mosleh, A. (2010). Reliability growth management methods and statistical methods for discrete-use system. *Technometrics, 52*(4), 379–389. doi:10.1198/TECH.2010.08068

Hammack, L. (April 12, 2010). Construction of state's first wind farm to resume this month in Highland. *The Roanoke Times.* Retrieved 10 May, 2010, from http://www.roanoke.com

Hankey, R. S., Liggett, W. D., & McNerney, R. A. (2000). *The changing structure of the electric power industry 2000: An update. Energy Information Administration, Office of Coal, Nuclear, Electric and Alternative Fuels.* U.S. Department of Energy.

Harness, R. (2000). *Raptor electrocutions and distribution pole types.* Fort Collins, CO: North American Wood Pole Coalition Technical Bulletin.

Harry, M., & Schroeder, R. (2000). *Six Sigma, the breakthrough management strategy.* New York, NY: Currency/ Doubleday.

Hawkins, D., & Rothleder, M. (2006). *Evolving role of wind forecasting in market operation at the CAISO. Power System Conference and Exposition* (pp. 234–238). Piscataway, NJ: The Institute of Electrical and Electronic Engineers.

Hofmann, W., & Okafor, F. (2001). Optimal control of doubly-fed full controlled induction generator with high frequency. *Annual Conference of the IEEE Industrial Electronics Society,* Vol. 2, (pp. 1213-1218).

IEEE Standard 1159. (1995). *Recommended practices for monitoring electric power quality.* Piscataway, NJ: The Institute of Electrical and Electronic Engineers.

IEEE Standard 519. (1992). *Recommended practices and requirements for harmonic control in electrical power systems.* Piscataway, NJ: The Institute of Electrical and Electronic Engineers.

IEEE Standard 902. (1998). *IEEE guide for maintenance, operation, and safety of industrial and commercial power systems.* Piscataway, NJ: The Institute of Electrical and Electronic Engineers, Inc.

IEEE Standard C37.118. (2005). *Enclosed field discharge circuit breakers for rotating electric machinery.* Piscataway, NJ: The Institute of Electrical and Electronic Engineers.

ISO Standard 10005. (2005). *Quality management – Guidelines for quality plans.* New York, NY: American National Standards Institute.

ISO Standard 10006. (2007). *Quality management – Guidelines to quality in project management.* New York, NY: American National Standards Institute.

ISO/TR Standard 10017. (2003). *Guidance on statistical techniques for ISO 9001:2000.* New York, NY: American National Standards Institute.

Iward, J., & Mwasha, A. (2012). The effects of ISO certification on organization workmanship performance. *Quality Management Journal, 19*(1), 53–67.

Jacobsen, J. (April 2010). *Wind power company gets to the root of an icy issue.* Making the Case for Quality: The Knowledge Center. Retrieved 1 April, 2011, from http://www.asq.org/knowledge-center/index.html

Johnson, B. W. (2005). *Utility restoration cost recovery.* Washington, DC: Edison Electric Institute.

Jondahl, D. W., Rockfield, L. H., & Cupp, G. M. (1991). *Connector performance of new vs. service aged conductor.* IEEE/Power Engineering Society Transmission and Distribution Conference, Los Angeles, CA.

Jones, L., & Cheung, K. (March 2012). You're going to need a smarter crystal ball. *Electric Light and Power Magazine.* Retrieved 1 June, 2012, from http://www.elp.com/elp/en-us/index

Jones, D., & Womack, J. (1996). *Lean thinking: Banish waste and create wealth in your corporation.* New York, NY: Simon & Schuster.

Jones, R. (2007). *Project management survival: A practical guide for leading, managing, and delivering challenging projects.* Philadelphia, PA: Kogan Page.

Jordan, T. E., & Constanti, D. J. (June 1995). The use of non-invasive electromagnetic techniques for focusing environmental investigations. *The Professional Geologist,* pp 4-9.

Juran, J. M. (1999). *Juran's quality handbook* (5th ed.). New York, NY: McGraw-Hill.

Juran, J. M. (1999). *Quality control handbook* (5th ed.). New York, NY: McGraw-Hill, Inc.

Justus, D. (2005). *Case study 5: Wind power integration into electricity systems,* International Energy Technology Collaboration and Climate Change Mitigation. Organization for Economic Co-operation and Development International Energy Agency. Retrieved 2 May, 2010, from http://www.oecd.org/env/cc/

Kahn, E. (1991). *Electric utility planning and regulation,* 2nd ed. Washington, DC: American Council for an Energy-Efficient Economy.

Kaplan, R. S., & Norton, D. P. (1996). Linking the balanced scorecard to strategy. *Management Review, 39*(1), 53–79.

Kendall, K. E., & Kendall, J. E. (1994). *Systems analysis and design* (3rd ed.). Upper Saddle River, NJ: Prentice Hall.

Kerzner, H. (1992). *Project management: A systems approach to planning, scheduling, and controlling* (4th ed.). New York, NY: Van Reinhold.

Kretschmer, R. K., & Hundrieser, K. E. (1996). Reliability: What level and what price? *Public Utility Fortnightly, 139*(20).

Kubiak, T. M. (June 2012). The significance of simulation. *Quality Progress,* pp. 44-45.

Lamprecht, J. L. (2005). *Applied data analysis for process improvement.* Milwaukee, WI: ASQ Quality Press.

Ledel, D., & Baker, C. (May 2012). Modern transformer maintenance, life extension strategies. *Utility Automation Engineering T&D Magazine.* Retrieved 15 May, 2012, from http://www.elp.com/elp/en-us/index

Leonhardi, D. (November/December 2011). Soft skills: The 'killer app' for analytics. *Analytics Magazine,* pp. 34-37

Levasseur, R. E. (2010). People skills: Ensuring project success- A change management perspective. *Interfaces, 40*(2), 159–162. doi:10.1287/inte.1090.0473

Levasseur, R. E. (2011). People skills: Optimizing team development and performance. *Interfaces, 41*(2), 204–208. doi:10.1287/inte.1100.0519

Levasseur, R. E. (2012). People skills: Leading virtual teams – A change management perspective. *Interfaces, 42*(2), 213–216. doi:10.1287/inte.1120.0634

Levine, E. S. (March-April 2012). Challenges for public sector analytics. *Analytics Magazine,* pp. 27-29.

Liberatore, M., & Luo, W. (2012). The analytics movement: Implications for operations research. *Interfaces, 40*(4), 313–324. doi:10.1287/inte.1100.0502

Liberatore, W., & Luo, W. (2012). The analytics movement: Implications for operations research. *Interfaces, 40*(4), 313–324. doi:10.1287/inte.1100.0502

Lippman, S. A., & McCardle, K. F. (2012). Risk aversion and impatience. *Decision Analysis, 9*(1), 31–40. doi:10.1287/deca.1110.0224

Logan, D. M. (1990). Decision analysis in engineering-economic modeling. *Energy, 15*(7/8), 677–696. doi:10.1016/0360-5442(90)90014-S

Lyncole Company. (2005). *Soil resistivity testing four point Wenner method.* Application Note LEP-1001. Retrieved 10 October, 2009, from http://www.lyncole.com

Malcolm, S. A., & Zenios, S. A. (1994). Robust optimization for power system capacity expansion under uncertainty. *Journal of Operations Research, 45*(9), 1040–1049.

Mann, D., & Domb, E. (1997). *40 inventive (business) principles with examples.* Retrieved 8 November, 2008, from http://www.triz-journal.com

Martin, J. N. (1997). *System engineering guidebook: A process for developing systems and products.* Boca Raton, FL: CRC Press.

Matas, J., Castilla, M., Gueirrero, J. M., Garcia de Vicuna, L., & Miret, J. (2008). Feedback linearization of direct-drive synchronous wind turbines via a sliding mode approach. *IEEE Transactions on Power Electronics, 23*(3), 1093–1103. doi:10.1109/TPEL.2008.921192

Mattson, C. A., Mullur, A. A., & Messac, A. (2004). Smart pareto filter: Obtaining a minimal representation of multi-objective design space. *Engineering Optimization, 36*, 721–740. doi:10.1080/0305215042000274942

McCalley, J., Vooris, T. V., Jiang, Y., & Meliopoulis, A. P. (2003). *Risk-based maintenance allocation and scheduling for bulk transmission system equipment.* Retrieved 3 January, 2006 from http://www.pserc.org

McCrary, S. W. (2007). Validation of project time decision-support tools and processes. *Journal of Information Technology, 23*(2).

McLay, L., Rothschild, C., & Guikema, S. (2012). Robust adversarial risk analysis: A level-k approach. *Decision Analysis, 9*(1), 41–54. doi:10.1287/deca.1110.0221

Meehan, B. (July 2011). As season utility staff retire, will they take wisdom with them? *Utility Automation Engineering T&D Magazine.* Retrieved 15 May, 2012, from http://www.elp.com/elp/en-us/index

Megahed, F., Woodall, W. H., & Camelio, J. H. (2011). A review and perspective on control charting with image data. *Journal of Technology, 43*(2), 83–98.

Meibom, P., Ravn, H., Soder, L., & Weber, C. (2004). *Market integration of wind power.* RISØ National Laboratory conference paper presented at the 8th International Conference on Probabilistic Methods Applied to Power Systems, Ames, Iowa.

Melnyk, S. A., Stewart, D. M., & Swink, M. (2004). Metrics and performance measurements in operations management: Dealing with the metrics maze. *Journal of Operations Management, 22*(9), 209. doi:10.1016/j.jom.2004.01.004

Meyer, M., Van Deventer, L., Wykes, C., & Cawood, E. (2011). Innovation decision support in a petrochemical production environment. *Interfaces, 41*(1), 79–92. doi:10.1287/inte.1100.0528

Miettinen, K. (1999). *Nonlinear multi-objective optimization.* Kluwer Academic Press.

Molina, M. J., & Rowland, F. S. (1974). Stratospheric sink for chlorofluoromethanes: Chlorine atom-catalyzed destruction of ozone. *Nature, 249*(5460), 810–812. doi:10.1038/249810a0

Monroe, E., Pan, R., Anderson-Cook, C. M., Montgomery, D. C., & Borror, C. M. (2010). Sensitivity analysis of optimal designs for accelerated life testing. *Journal of Quality Technology, 42*(2), 121–135.

Morren, J., de Haan, S. W. H., Kling, W. L., & Ferreira, J. A. (2006). Wind turbines emulating inertia and supporting primary frequency control. *IEEE Transactions on Power Systems, 21*(1), 433–434. doi:10.1109/TPWRS.2005.861956

Moubray, J. (1997). *Reliability centered maintenance* (2nd ed.). New York, NY: Industrial Press, Inc.

Mueller, R. (2005). 10 steps to help reduce SF_6 emissions in T&D. *Utility Automation and Engineering/T&D.* Retrieved 4 June, 2009, from http://www.elp.com/index

Muller, S., Deicke, M., & De Doncker, R. W. (2002, May-June). Doubly-fed induction generator systems for wind turbine applications. *IEEE Industry Applications Magazine, 8*(3), 26–33. doi:10.1109/2943.999610

Nanda, V. (February 2010). Preempting problems. *Six Sigma Forum Magazine.* Retrieved 1 April, 2011, from http://asq.org/knowledge-center/index.html

Nejati, M., & Ghasemi, S. (May 2012). On the right course: DMAIC can help organizations navigate tough decisions on social responsibility projects. *Quality Progress*, pp. 29-31.

Nestler, S., & Leong, J. (January-February 2009). O.R. credentialing. *OR/MS Today Magazine*, pp. 32-36.

Nichols, D., & Von Hippel, D. (2002). *Best practices guide; Integrated resources planning for electricity.* Retrieved 24 September, 2010, from http://www.info.usaid.gov

Nour, N. E., & Bhevaraju, M. P. (1989). *A comparison between two decision analysis methods.* EPRI Report, RP2537-2

O'Brien, J., Havers, J., & Stubbs, F. (1996). *The standard handbook of heavy construction* (3rd ed.). New York, NY: McGraw-Hill.

Ogilvy, J. (2002). *Creating better futures: Scenario planning as a tool for a better tomorrow.* New York, NY: Oxford University Press.

Papoulis, A. (1991). *Probability, random variables, and stochastic processes* (3rd ed.). New York, NY: McGraw-Hill, Inc.

Park, B., & Meier, R. (2007). Reality-based construction project management: A constraint-based 4D simulation environment. *Journal of Information Technology, 23*(1).

Peace, G. S. (1993). *Taguchi methods: A hands-on approach to quality engineering,* (pp. 1-236; 292-312). Reading, MA: Addison-Wesley.

Pertti, J., Antti, M., Kimmo, K., et al. (2007). *Using advanced AMR system in low voltage distribution network management.* Presented at the 19th International Conference on Electrical Distribution. Vienna, Austria. Retrieved 23 February, 2010, from http://www.cired2007.be/pdf/ustopselectedreportss6.pdf

Phadke, M. (1989). *Quality engineering using robust design.* Englewood Cliffs, NJ: Prentice Hall.

Pietroforte, R. (1997). Communication and governance in the building process. *Construction Management and Economics, 23,* 71–82. doi:10.1080/014461997373123

Polinder, H., Vanderpijl, F. F. A., De Vilder, G. J., & Tavner, P. J. (2006). Comparison of direct-drive and gear generator concepts for wind turbines. *IEEE Transactions on Energy Conversion, 21*(3), 725–733. doi:10.1109/TEC.2006.875476

Ramshaw, R., & van Heeswijk, R. G. (1990). *Energy conversion: Electric motors and generators.* Philadelphia, PA: Saunders College Publishing.

Retseptor, G. (2003). *40 inventive principles in quality management.* Retrieved 24 March, 2009, from http://www.triz-journal.com

Rodriguez-Amenedo, J. L., Arnalte, S., & Burgos, J. C. (2002). Automatic generation control of a wind farm with variable speed turbines. *IEEE Transactions on Power Electronics, 17*(2), 279–284.

Rossi, R. H., & Freeman, H. E. (1993). *Evaluation: A systematic approach.* Beverley Hills, CA: Sage Publications.

RUS 1728-810. (1998). *Electrical transmission specifications and drawings 34.5-kV through 69-kV.* United States Department of Agriculture. Rural Utility Services.

RUS Bulletin 1724E-203. (1994). *Guide for upgrading rural utility service transmission lines.* United States Department of Agriculture Rural Utility Services.

Sage, A. P., & Rouse, W. B. (2009). *Handbook of system engineering* (2nd ed.). New York, NY: John Wiley.

Sanchez, L., & Pan, R. (2011). An enhanced parenting process: Predicting reliability in product's design phase. *Quality Engineering, 23,* 378–387. doi:10.1080/08982112.2011.603110

Savitz, A. W., Besly, M., & Booth, K. (2002). *2002 sustainability survey report* PriceWaterhouseCoopers LLP. Retrieved 15 November, 2010, from http://www.pwc.com/fas/pdf/sustainability%survey%20report.pdf

Schaar, T., & Wilson, R. D. (2010). *Quality basics simplify complex engineering document management challenge.* Making the Case for Quality: The Knowledge Center for the American Society for Quality. Retrieved 15 May, 2010, from http://asq.org/knowledge-center/index.html

Schermerhorn, J. R. Jr. (1993). *Management for productivity* (4th ed.). New York, NY: John Wiley & Sons.

Schlabbach, J., & Berka, T. (2001). *Reliability centered maintenance of M.V. circuit breakers. Power Tech Proceedings* (Vol. 4). IEEE Porto.

Scholtes, P. R., Joiner, B. L., & Streibel, B. J. (1996). *The team handbook* (2nd ed.). Madison, WI: Oriel, Inc.

Schroer, B. (2004). Simulation as a tool in understanding the concepts of lean manufacturing. *Simulation, 80*(3), 171–175. doi:10.1177/0037549704045049

Senge, P. M. (2005). *Presence: An explanation of profound change in people, organizations, and society*. New York, NY: Doubleday/Currency.

Senge, P. M. (2006). *The fifth discipline: The art and practice of the learning organization*. New York, NY: Doubleday Publication. doi:10.1002/pfi.4170300510

Senthilmaran, K. (August 2010). *Electric utility deploys powerful approach for continuous improvements*. Making the Case for Quality: American Society for Quality Knowledge Center. Retrieved 1 April, 2011, from http://www.asq.org/knowledge center/index.html

Shanbhag, U., Infanger, G., & Glynn, P. (2011). A complementary framework for forward contracting under uncertainty. *Operations Research, 59*(4), 810–834. doi:10.1287/opre.1110.0947

Short, T. A. (2006). *Electric power distribution and systems*. Boca Raton, FL: CRC Press Taylor & Francis Group.

Simpson, P., & Van Bossuyt, T. R. (1996). Tree caused electrical outages. *Journal of Arboriculture, 22*(3), 117.

Sirower, M. L. (1997). *The synergy trap: How companies lose the acquisition game*. New York, NY: Free Press.

Smedlund, A. (2012). Value co-creation in service platforms business models. *Service Science, 4*(1), 79–88. doi:10.1287/serv.1110.0001

Stewart, M. T., & Gay, M. C. (1986). Evaluation of transient electromagnetic soundings for deep detection of conductive fields. *Ground Water, 24*, 351–356. doi:10.1111/j.1745-6584.1986.tb01011.x

Subramanian, N., & Vasanthi, V. (1990, March). Design of tower foundations. *Indian Concrete Journal*, (March): 135–141.

Switzer, K. W. (1996). Eleven practical tips for grounding substations. *Electrical Construction and Maintenance Magazine*. Retrieved 18 December, 2010, from http://ecmweb.com/mag/ electric_eleven_practical_tips/ index.html

Taguchi, G. (2001). Robust design by simulation: Standard SN ratio. *Quality Engineering, 9*(2).

Taguchi, G., Chowdhury, S., & Wu, Y. (1999). *Robust engineering*. New York, NY: McGraw-Hill.

Taguchi, G., Chowdhury, S., & Wu, Y. (2005). *Taguchi's quality engineering handbook*. Hoboken, NJ: John Wiley and Sons, Inc.

Tegen, S., & Lantz, E. (2009). *Social acceptance of wind power in the United States: Evaluating stakeholder perspective*. National Renewable Energy Laboratory poster session at the Wind Powering America Conference, Chicago, IL.

Thomas, B. G., & Bollapragada, S. (2010). General Electric uses an integrated framework product costing, demand forecasting, and capacity planning of new photovoltaic technology products. *Interfaces, 40*(5), 353–367. doi:10.1287/inte.1100.0518

Thomas, K. (1975). *The handbook of industrial and organizational psychology* (Dunnette, M., Ed.). Chicago, IL: Rand McNally.

Thomas, K. (1975). *The handbook of industrial and organizational psychology* (Dunnette, M., Ed.). Chicago, IL: Rand McNally.

Tillman, R. F. (2001). Loading power transformers. In Grigsby, L. L. (Ed.), *The electric power engineering handbook*. Boca Raton, FL: CRC Press.

UK HM Treasury. (2006). *Stern review on the economics of climate change*. Retrieved 20 June, 2009, from http://hm-treasury.gov.uk/sternreview-index/html

Van Cleve, F. B., & Copping, A. E. (2010). *Offshore wind energy permitting: A survey of U.S. project developers*. Richland, WA: U.S. Department of Energy, Pacific Northwest National Laboratory (PNNL-20024).

Vanessa, I. (2007). Swiss precision. *Industrial Engineer, 39*.

Wack, P. (1985). Scenarios: Uncharted waters ahead. *Harvard Business Review*, 72.

Wang, L., & Lin, L. (2007). A methodological framework for the triple bottom line accounting and management of industrial enterprises. *International Journal of Production Research*, *45*(5), 1063–1088. doi:10.1080/00207540600635136

Wexler, M. (2010). Riding on the global warming express. *National Wildlife*, *48*(5).

Wiser, R., Barbose, G., & Holt, E. (2010). *Supporting solar power in renewables portfolio standards: Experience from the United States*. Lawrence Berkeley National Laboratory. Retrieved 9 January, 2011, from http://www.eetd.lbl.gov/ea/ems/reports/lbnl-3984e.pdf

Wisler, D. (2003). Engineering – What you don't necessarily learn in school. *Mechanical Engineering Magazine Online*. American Society of Mechanical Engineers. Retrieved 15 February, 2009, from http://www.memagazine.org/contents/current/webonly/

Wood, A. J., & Wollenberg, B. F. (1996). *Power generation, operations, and control* (2nd ed.). New York, NY: John Wiley.

Woods, R. A. (2009). Industry output and employment projections to 2018. [Washington, DC: Bureau of Labor Statistics, U.S. Department of Labor.]. *Monthly Labor Review*, *132*(11), 53–81.

Ye, Z., Tang, L., & Xie, M. (2011). A burn-in scheme based on the percentiles of the residual life. *Journal of Quality Technology*, *43*(4), 334–345.

Yoon, K., & Kim, G. (1989). Multiple attribute decision analysis with imprecise information. *IIE Transactions*, *21*(1), 21–25. doi:10.1080/07408178908966203

Young, W., Stamp, J., & Dillinger, J. (2003). *Communication vulnerabilities and mitigations in wind power SCADA systems*. Paper presented at the American Wind Energy Association WINDPOWER 2003 Conference, Austin, TX.

Zhang, L., & Watthanasarn, N. C. (1998). A matrix converter excited doubly-fed induction machine as a wind power generator. *7th International Conference on Power Electronics and Variable Speed Drives*, Vol. 456, (pp. 532-537).

Zou, C., Jiang, W., & Tsung, F. (2011). A LASSO-based diagnostic framework for multivariate statistical process control. *Technometrics*, *53*(3), 297–309. doi:10.1198/TECH.2011.10034

About the Authors

Reginald Wilson is an Industrial Operations Specialist with more than twenty years of electrical system assessments and methodical business process philosophy experience in Fortune 250 companies. He has held visiting Lecturer positions teaching courses in managerial business processes, circuit analysis, and electronic instrumentation. Currently, he is a Senior Management Systems Analyst for the Skagit Hydroelectric Project with *Seattle City Light*, and also the preceding president/owner of *Redawil Engineering Company* which has consulted and collaborated with several electrical utility organizations throughout North America by assisting with their process improvement efforts. Wilson has authored several articles and conference papers concerning quality measurement, equipment implementation processes, financial analysis, and strategic management. His unique research experience is in financial control systems and the engineering optimization field of electrical transmission and distribution development. A Certified Quality Process Analyst, and Certified Six Sigma Green Belt, Wilson holds a Bachelor's degree in Engineering and a Master's degree in Industrial Operations from Lawrence Technological University in Southfield, Michigan (USA). He is an active member of the Institute for Operations Research and the Management Sciences, the Institute of Industrial Engineers, and the American Society for Quality.

Hisham Younis holds a Master's degree in Industrial Engineering and a PhD in Industrial and Manufacturing Engineering from Wayne State University in Detroit, Michigan (USA). He is a Senior Lecturer in Optimization Systems and Operations Management at the Graduate School of Engineering at Wayne State University and Lawrence Technological University. He previously held senior industrial management positions at Ford Motor Company and Visteon as a global leader in process assurance, failure mode effects analysis, and lean engineering initiatives. Younis has consulted with numerous energy companies and technical manufacturing organizations and provided advanced algorithms to solve large scale optimization problems across several disciplines. His research interests includes linear optimization using block pivoting for fast convergence, global programming of quadratic functions, and scale parameters of the Weibull distribution using linear regression models. He is an active committee member of the Society of Automotive Engineers (U.S. Tag 13) and the Institute of Electrical and Electronic Engineers (IEC TC 56 and IEC Z1) with voting authority to release reliability and maintainability industry standards.

Index

A

act to minimize movement (Lean) 205
air pollutant emissions 77-78
All Aluminum Alloy Conductor (AAAC) 308
All Aluminum Conductor (AAC) 308
alternating current (AC) 307
alternative analysis process 30, 35
ALTW-WPS Project 41, 43
Aluminum Conductor Steel Reinforced (ACSR) 181, 308
area control error (ACE) 16, 333
ARGENTA 301, 310, 316, 318
attribute matrix 11, 13, 19
attribute ordinal ranking 19
average idle cost 341-345
average instantaneous utilization 196, 207, 343, 345-347

B

Below Grade (BG) 164
Brute Force 217-218, 220-221, 223, 226, 252, 295, 322, 339-341, 343-344, 346, 349, 357

C

Capacitor Voltage Transformers (CCVT) 310
chlorofluorocarbons (CFC) 133
circuit breaker 48-51, 66, 122, 125-127, 131, 133-136, 138, 247, 256, 259, 275, 295, 301-302, 304, 309-310, 320, 338
comparative process model 21
compression splices 177
cost-effectiveness 28, 52
creative tension 31, 38, 52
crucial generation-load balance 334
cycloconverter 35, 52

D

Day Ahead (DA) 180
decision analysis 44, 64, 84, 86, 117, 193, 206, 254, 295-299, 302, 310-311, 313, 316-317, 322-323
decision parameter 8, 161
decision theory 60, 295, 300, 313
decreasing-times processing algorithm (DTA) 214
department of natural resources (DNR) 202
distributed generator 2-3

E

electrical footprint 31-32, 41, 46, 90-91, 94-95, 97-99, 103, 107, 111, 131, 138-139, 148, 156, 180, 182, 280, 327-328, 334
electromagnetic method 150
elevations 84, 273
entities 4, 8, 10-11, 23, 29-30, 56, 60, 80, 90, 97, 99, 116, 126, 130, 160, 162, 166, 170, 172, 180, 189, 191, 193, 195, 201-203, 205-206, 217-218, 221, 223, 234, 243, 254, 277, 326, 330, 333, 338-340, 346, 348, 350, 353, 355-356, 359
environmental policies 28, 74
environmental studies 6, 9, 12-13, 19, 34, 146, 202
Equipotentiality 66, 156, 363

F

Failure Mode and Effects Analysis (FEMA) 126
failure range analysis 42
fall-ins 179
farthest-from-the-source (FFS) 101
Federal Electric Reliability Commission (FERC) 237
Federal Energy Regulatory Commission (FERC) 81
first-come-first-served (FIFO) 101
forecast modeling 94
fossil-fuel dependency 28, 52
foundations 17, 147-148, 150, 157, 165-166, 168, 174, 191, 273, 354, 356, 361

Lightning Source UK Ltd.
Milton Keynes UK
UKHW050830291119
354427UK00010B/354/P